20 Lectures on Eigenvectors, Eigenvalues, and Their Applications

ORANGE GROVE TEXT *PLUS*

UNIVERSITY PRESS OF FLORIDA

Florida A&M University, Tallahassee
Florida Atlantic University, Boca Raton
Florida Gulf Coast University, Ft. Myers
Florida International University, Miami
Florida State University, Tallahassee
New College of Florida, Sarasota
University of Central Florida, Orlando
University of Florida, Gainesville
University of North Florida, Jacksonville
University of South Florida, Tampa
University of West Florida, Pensacola

20 Lectures on Eigenvectors, Eigenvalues, and Their Applications

Problems in Chemical Engineering

L. E. Johns

UNIVERSITY PRESS OF FLORIDA

Gainesville · Tallahassee · Tampa · Boca Raton
Pensacola · Orlando · Miami · Jacksonville · Ft. Myers · Sarasota

Contents

Part I: Elementary Matrices   1

1   Getting Going   3

2   Independent and Dependent Sets of Vectors   29

Acknowledgment

In writing out these lectures in this way I was guided by what I imagined Charles Petty and Anthony DeGance, students in the 60's and 70's, and Ranganathan Narayanan, a colleague from the 80's to the present, might ask me if I were teaching this material to them.

Thanks are due Debbie Sandoval and Santiago A. Tavares for the best possible help.

What Sets this Book Apart?

First it has a detailed readers guide. Setting that aside, we answer the question:

There is no formal mathematics in this book, the proofs presented are mostly sketches. But plausibility is not slighted and the geometric interpretation of the results obtained is a theme that is maintained throughout the book.

We present the theory of finite dimensional spaces in Part I, first because there are many interesting problems that can be formulated and solved in finite dimensions and second because the main ideas in n dimensional spaces can be illustrated by drawing pictures in two dimensions.

Then, without stretching the readers imagination too much, in Part II we carry these pictures over to infinite dimensional spaces. Indeed, often what is of interest in an infinite dimensional space is a finite dimensional subspace.

In both finite and infinite dimensional spaces our search is always for a problem specific basis, hence eigenvectors and eigenfunctions become a second theme. We make the jump from Part I to Part II by introducing difference approximations to the solutions to the diffusion equation.

Now whether we are solving problems in Part I or in Part II, what we do is always the same. We solve an eigenvalue problem, then all the steps come down to summation in Part I or integration in Part II. No series, finite or infinite, is differentiated in order to solve a problem. For instance the method of separation of variables is used only to solve eigenvalue problems, never to solve initial value problems.

An explanation of domain perturbations is presented in order to extend the problems that can be solved to domains other than cubes, cylinders and spheres.

Chemical engineering students need to solve problems having physical origins. Thus problems of physical interest are presented in every lecture and the readers will meet many of these problems in their other courses, in their research or they will find these problems to be the first problems they would solve in learning a new subject. This is a theme.

We often obtain linear problems via perturbation methods and due to this, there is a strong emphasis on solvability conditions and hence on inner products. Solvability is thus a theme as is the use of the eigenfunctions themselves to reveal patterns in nonlinear problems.

A short list of the examples presented in the lectures, and some of what they illustrate, follows:

Boiling curves
{
Linear approximation
Matrix multiplication
}

Greatest number of reactions among M molecules made up A atoms
{
Linear independence
Rank
Determinant
}

Kremser equation
{
Spectral decomposition
}

Dynamic stripping cascade
{
Generalized eigenvectors
}

Chemostat
Stirred tank reactor
{
Eigenvalues
Branch points
Hopf bifurcations
}

Isomerization reactions
Draining tanks
{
Eigenvectors
}

Difference approximations
{
Gerschgorin's theorem
}

Chromatography $\Big\{$ Power moments

Electrical potential $\Big\{$ Multipole expansions

Activator-inhibitor kinetics $\Big\{$ Eigenvalues

Petri dish $\Big\{$ Solvability

Size of a confined autothermal heat source $\Big\{$ Eigenvalues

Saffman-Taylor problem
Rayleigh-Taylor problem
$\left\{\begin{array}{l} \text{Separation of variables} \\ \text{Integral constraints} \\ \text{Patterns derived from eigenfunctions} \end{array}\right.$

Energy of a quantum ball
Solute dispersion
Oscillations of an inviscid drop
$\left\{\begin{array}{l} \text{Separation of variables} \\ \text{Eigenvalues and eigenfunctions} \\ \text{Cartesian, cylindrical and spherical coordinates} \end{array}\right.$

Many home problems stem from these examples and many of the home problems are not questions but stories leading to the reader to derive some well known results.

Reader's Guide

Before we outline the main ideas presented in each lecture we present an overview stating how the lectures fit together.

Part I has to do with problems whose solutions lie in finite dimensional spaces. Part II has to do with problems whose solutions are functions.

Thus in Part I the subject is matrices, in Part II the subject is ∇^2 and the differential equations arising upon use of the method of separation of variables.

The applications are indicated in the opening statement: "What Sets this Book Apart?"

Diffusion is a theme of Part II, but only because ∇^2 would not be so interesting if there were no diffusion. Of course other themes could have been chosen from classical physics.

The first five lectures bring the reader to the point where they understand the basic facts about the eigenvectors and eigenvalues of matrices.

The sixth lecture then uses these ideas to write the solution to systems of constant coefficient ordinary differential equations.

The seventh and eighth lectures are applications of the sixth lecture to problems the reader might have learned something about as an undergraduate.

The ninth lecture makes the transition to Part II by solving difference approximation to the diffusion equation. The idea is that the expansion of the solution in the eigenvectors of a matrix carries over to the solution of the diffusion equation itself in Part II.

Lectures 11, 14, 16 and 17 present the basic facts about the eigenvalues and eigenvectors of ∇^2 in a bounded domain. Lectures 12 and 13 explain a little about problems in unbounded domains, mostly by the use of the method of moments to derive simple facts about concentration distributions.

In Lecture 11 formulas for ∇ and ∇^2 are presented in various coordinate systems.

In Lecture 14 the dependence of the eigenfunctions and eigenvalues of ∇^2 on the boundary conditions is illustrated in one space dimension.

Lecture 15 presents two applications, one to activator-inhibitor kinetics, the other to the construction of the solution to a nonlinear reaction-diffusion problem.

Lecture 16 repeats Lecture 14, but now we have three space dimensions and volume and surface sources are taken into account.

The method of separation of variables is presented in Lecture 17 and two well-known stability problems are solved in Lecture 18.

Lecture 19 is about the second order ordinary differential equations that present themselves upon the use of separation of variables. Their solutions by Frobenius' method appears in Lecture 20.

Lecture 20 presents applications of the eigenfunctions and eigenvalues of ∇^2 to problems I like in Cartesian, cylindrical and spherical coordinates. The problems are not about summing infinite series but instead illustrate the physical significance of the eigenvalues and eigenfunctions themselves.

Lecture 1: Getting Going

Our aim is to introduce linear problems and to suggest how they might arise.

A linear problem can be written

$$\mathcal{L}\vec{x} = \vec{f}$$

where $\vec{x}$ lies in the space of unknown vectors and $\vec{f}$ lies in the space of vectors that drive the problem. We seek $\vec{x}$ such that $\mathcal{L}$ carries $\vec{x}$ into $\vec{f}$, where $\mathcal{L}$ is a linear operator in the sense that

$$\mathcal{L}\left(\vec{x} + \vec{y}\right) = \mathcal{L}\vec{x} + \mathcal{L}\vec{y}$$

The simplest linear problems are those where $\vec{x} = \underline{x} = \begin{pmatrix} x_1 \\ x_2 \\ \vdots \\ x_n \end{pmatrix}$ and

$$\vec{f} = \underline{f} = \begin{pmatrix} f_1 \\ f_2 \\ \vdots \\ f_m \end{pmatrix}$$

Then $\mathcal{L} = A = \begin{pmatrix} a_{11} & a_{12} & \cdots & a_{1n} \\ & \text{etc.} & & \end{pmatrix}$, an $m \times n$ matrix, and we write our problem $A\underline{x} = \underline{f}$.

The most important point in Lecture 1 is that A should be viewed as a set of n columns

$$A = \begin{pmatrix} \underline{a}_1 & \underline{a}_2 & \cdots & \underline{a}_n \end{pmatrix}$$

each lying in the space where $\underline{f}$ lies.

What we are mostly interested in is learning whether or not there is a solution and if there is, how many.

So we start with $A\underline{x} = \underline{f}$ and write it

$$\underline{a}_1 x_1 + \underline{a}_2 x_2 + \cdots + \underline{a}_n x_n = \underline{f}$$

This causes us to ask our question in the following way: Can $\underline{f}$ be written as a linear combination of the set of vectors $\underline{a_1}, \underline{a_2}, \cdots, \underline{a_n}$? If it can then the coefficients in the expansion of $\underline{f}$ are the elements of $\underline{x}$, our answer.

Lecture 2: Independent and Dependent Sets of Vectors

We have a vector $\underline{f}$ lying in the space C_m of dimension m and we have a set of vectors $\underline{a}_1, \underline{a}_2, \cdots, \underline{a}_n$ also lying in C_m, and we want to know how much of C_m can be included in the linear combinations of $\underline{a}_1, \underline{a}_2, \cdots, \underline{a}_n$. Then we want to know if $\underline{f}$ lies in that part of C_m.

The main idea is introduced: linear independence. It is illustrated by the question: how many independent chemical reactions can be written among M molecules made up of A atoms?

The function $\det$ is introduced where $\det$ acting on a square matrix produces a scalar, called its determinant.

We define the rank of a matrix, denoted r, where the matrix may or may not be square. And we identify a set of r basis columns. The basis columns are independent, r is the largest number of independent columns that can be found and any set of r independent columns is a set of basis columns. Every set of $r + 1$ columns is dependent and each of the remaining $n - r$ columns must be a linear combination of the r basis columns.

Lecture 3: Vector Spaces

The idea of a vector space is presented, its dimension is defined and the idea of a basis is introduced.

Two vector spaces associated to an $m \times n$ matrix A, are introduced: $\operatorname{Im} A$ and $\operatorname{Ker} A$. The dimension of $\operatorname{Im} A$ is r, the dimension of $\operatorname{Ker} A$ is $n - r$. $\operatorname{Im} A$ is the collection of all vectors $A\underline{x}$, $\operatorname{Ker} A$ is the collection of all vectors $\underline{x}$ such that $A\underline{x} = \underline{0}$.

The solvability condition for $A\underline{x} = \underline{b}$ is then given. Thus if $\underline{b} \in \operatorname{Im} A$ then $A\underline{x} = \underline{b}$ is is solvable and the general solution is $\underline{x}_0$, a particular solution, plus an arbitrary vector lying in $\operatorname{Ker} A$.

Lecture 4: Inner Products

To be able to extend what we now know to problems beyond matrix problems and free ourselves of the determinant function which only applies to matrix problems, we introduce the idea of an inner product, paying most attention to the case where A is $n \times n$.

We are looking for a way to tell if a vector $\underline{b}$ lies in the subspace $\operatorname{Im} A$, where the dimension of $\operatorname{Im} A$ is r.

Now we can find $\text{Ker}\, A$, by solving

$$A\underline{x} = \underline{0}$$

It is of dimension $n - r$.

To find $\text{Im}\, A$ we introduce a new matrix A^*, called the adjoint of A, where A^* depends on the inner product in which we are working.

Having A^* we find $\text{Ker}\, A^*$ by solving $A^*\underline{x} = \underline{0}$, where $\text{Ker}\, A^*$ has dimension $n - r$. And $\text{Ker}\, A^*$ has an interesting geometric property: it is orthogonal to $\text{Im}\, A$. Thus $\underline{b} \in \text{Im}\, A$ if and only if $\underline{b}$ is orthogonal to all $\underline{y} \in \text{Ker}\, A^*$.

Hence to find out if the problem

$$A\underline{x} = \underline{b}$$

is solvable we must test $\underline{b}$ against $n - r$ independent solutions of $A^*\underline{y} = \underline{0}$ whence $A\underline{x} = \underline{b}$ is solvable if and only if $\underline{b}$ is orthogonal to each of them.

Lecture 5: Eigenvectors

Here we denote by A an $n \times n$ matrix and we ask if there are vectors $\underline{x}$ that are mapped by A without change of direction, i.e., we ask for $\underline{x}$'s such that

$$A\underline{x} = \lambda\underline{x}$$

This is the eigenvalue problem for A and solutions $(\underline{x} \neq \underline{0},\, \lambda)$ are called eigenvectors and eigenvalues. If $\underline{x}$ is a solution, likewise $c\underline{x}$ for any c.

Writing $A\underline{x} = \lambda\underline{x}$ as

$$(A - \lambda I)\,\underline{x} = \underline{0}$$

we see that in order that solutions $\underline{x} \neq \underline{0}$ can be found the λ's must be such that the rank of $\text{Ker} \, (A - \lambda I)$ is less than n, i. e., we must have

$$\det \, (A - \lambda I) = 0$$

This equation has n roots $\lambda_1, \lambda_2, \cdots, \lambda_d$ repeated $m_1, m_2, \cdots, m_d$ times, where $m_1 + m_2 + \cdots + m_d = n$ The m's are called the algebraic multiplicities of the λ's and to each λ there is at least one $\underline{x} \neq \underline{0}$.

The number of independent solutions corresponding to a root λ is the dimension of $\text{Ker} \, (A - \lambda I)$.

Denoting these dimensions $n_1, n_2, \cdots, n_d$, corresponding to $\lambda_1, \lambda_2, \cdots, \lambda_d$ we have n_1 independent eigenvectors corresponding to λ_1, etc. and $n_1 \leq m_1$, etc.

Now we pay most of our attention to the plain vanilla case where $d = n$ and $n_1 = 1 = m_1$, etc.

Then we have n distinct eigenvalues and the corresponding n eigenvectors are independent and form a basis.

Introducing an inner product, we can derive A^*, the adjoint of A, and write its eigenvalue problem

$$A^* \underline{y} = \mu \, \underline{y}$$

whereupon we find the μ's are the complex conjugates of the λ's and the set of eigenvectors $\{\underline{y}_1, \underline{y}_2, \cdots, \underline{y}_n\}$ is the set of vectors orthogonal to the set $\{\underline{x}_1, \underline{x}_2, \cdots, \underline{x}_n\}$, viz., $\langle \underline{y}_i, \underline{x}_j \rangle = 0$

Thus any vector $\underline{x}$ can be expanded in two ways

$$\underline{x} = \sum c_i \underline{x}_i$$

where $c_i = \langle \underline{y}_i, \underline{x} \rangle$ and

$$\underline{x} = \sum d_i \, \underline{y}_i$$

where $d_i = \langle \, \underline{x}_i, \, \underline{x} \, \rangle$.

We solve the linear stripping cascade problem and derive a symmetric form of the Kremser equation.

Lecture 6: The Solution of Differential and Difference Equations

We are going to learn how to solve the system of differential equations

$$\frac{d\underline{x}}{dt} = A \, \underline{x}, \quad t > 0$$

where $\underline{x} = \begin{pmatrix} x_1 \\ x_2 \\ \vdots \\ x_n \end{pmatrix}$ denotes the time dependent unknowns, $\underline{x}$ at $t = 0$ is specified, and A denotes an $n \times n$ matrix of constants.

The ordinary case is where A has a complete set of n independent eigenvectors, $\underline{x}_1, \underline{x}_2, \cdots, \underline{x}_n$ corresponding to eigenvalues $\lambda_1, \lambda_2, \cdots, \lambda_n$. To solve our problem we are going to write $\underline{x}(t)$ in terms of $\underline{x}_1, \underline{x}_2, \cdots, \underline{x}_n$, viz.,

$$\underline{x}(t) = c_1(t) \, \underline{x}_1 + c_2(t) \, \underline{x}_2 + \cdots + c_n(t) \, \underline{x}_n$$

and try to find $c_1(t), c_2(t) \cdots$

To do this we introduce an inner product, denoted $\langle \, , \, \rangle$, and in this inner product we derive the adjoint of A, viz., A^*.

Then the eigenvectors of A^*, viz., $\underline{y}_1, \underline{y}_2, \cdots, \underline{y}_n$, with the normalization $\langle \underline{y}_1, \underline{x}_1 \rangle = 1$, etc. form a set of vectors biorthogonal to the set of eigenvectors of A.

Thus we have

$$c_1 = \langle \underline{y}_1, \underline{x}(t) \rangle$$

$$c_2 = \langle \underline{y}_2, \underline{x}(t) \rangle$$

etc.

and to derive the equation for c_1 we take the steps

$$\langle \underline{y}_1, \frac{d\underline{x}}{dt} \rangle = \langle \underline{y}_1, A\underline{x} \rangle \implies \frac{d}{dt}\langle \underline{y}_1, \underline{x} \rangle = \langle A^*\underline{y}_1, \underline{x} \rangle \implies$$

$$\frac{d}{dt} c_1 = \langle \overline{\lambda}_1 \underline{y}_1, \underline{x} \rangle = \lambda_1 \langle \underline{y}_1, \underline{x} \rangle = \lambda_1 c_1$$

and hence we have

$$c_1(t) = c_1(t=0)\, e^{\lambda_1 t}$$

And our solution is

$$\underline{x}(t) = \langle \underline{y}_1, \underline{x}(t=0) \rangle\, e^{\lambda_1 t}\, \underline{x}_1 + \langle \underline{y}_2, \underline{x}(t=0) \rangle\, e^{\lambda_2 t}\, \underline{x}_2 + \text{etc.}$$

If instead of a differential equation we have a difference equation, viz.,

$$\underline{x}(n+1) = A\,\underline{x}(n) \qquad n = 0, 1, 2, \ldots$$

we find

$$\underline{x}(n) = \langle \underline{y}_1, \underline{x}(n=0) \rangle\, \lambda_1^n\, \underline{x}_1 + \langle \underline{y}_2, \underline{x}(n=0) \rangle\, \lambda_2^n\, \underline{x}_2 + \text{etc.}$$

Stability of solutions to our differential equation requires all the eigenvalues of A to lie in the left half of the complex plane, viz., $\operatorname{Re}\lambda < 0$, whereas stability of solutions to our difference equation requires all eigenvectors to lie inside the unit circle $|\lambda| < 1$.

We then explain what to do if eigenvectors are missing. For example if λ_1 is a root of algebraic multiplicity 2 but geometric multiplicity 1, i.e., $\dim \operatorname{Ker}(A - \lambda_1 I) = 1$ so that there is only one independent eigenvector corresponding to λ_1, we are going to be short one eigenvector and we will not have an eigenvector basis for our space. What we do to overcome this difficulty is to introduce generalized eigenvectors. Thus we write

$$A\,\underline{x}_1 = \lambda_1\,\underline{x}_1$$

and

$$A\,\underline{x}_2 = \underline{x}_1 + \lambda_1\,\underline{x}_2$$

And hence writing

$$\underline{x}(t) = c_1(t)\,\underline{x}_1 + c_2(t)\,\underline{x}_2$$

we find

$$\frac{dc_1}{dt} = \lambda_1\,c_1 + c_2$$

and

$$\frac{dc_2}{dt} = \lambda_1\,c_2$$

whereupon c_1 and c_2 can be found sequentially, and the factor $te^{\lambda_1 t}$ appears.

We use this to solve a dynamic linear stripping cascade problem.

Lecture 7: Simple Chemical Reactor Models

We put to use what we have been learning and we do this in the context of a chemostat and a very simple stirred tank reactor, but a reactor that retains the interesting physics of these reactors. Thus as the reaction proceeds it releases heat, it uses up reactants and it speeds up as the temperature increases.

The stirred tank reactor model is two dimensional and therefore 2×2 matrices turn up in the investigation of the stability of its steady states. The eigenvalues of a 2×2 matrix depend on its trace T and determinant D and in the $D - T$ plane stability obtains in the fourth quadrant, $D > 0, T < 0$.

The model may be a bit too simple, but then we only need to solve quadratic equations to see what is going on.

Lecture 8: The Inverse Problem

In this lecture we assume we have a model

$$\frac{d\underline{x}}{dt} = A\underline{x}$$

and that we can run experiments where we measure $\underline{x}(t)$ vs t.

The aim is to derive the elements of the matrix A from the measurements. The main idea is that there are straight line paths, $\underline{x}(t)$ vs t, and that if we can find the directions of these straight lines, we have the eigenvectors of A, and hence we can derive A from its spectral expansion

$$A = \sum \lambda_i \, \underline{x}_i \, \underline{y}_i^T$$

We illustrate this by solving the problem of measuring reaction rate coefficients in a system of isomers, the Wei and Prater problem.

Lecture 9: More Uses of Gerschgorin's Circle Theorem

This is a lecture on difference approximations to the solution of the diffusion equation, first, to present some ideas about diffusion, second, to illustrate the use of Gerschgorin's theorem in estimating the eigenvalues of a matrix and, third, to present the method of solution which will be carried over to the diffusion equation itself in Part II.

Lecture 10: A Word To The Reader Upon Leaving Finite Dimensional Vector Spaces

This lecture presents a warning that we are leaving behind solutions that are finite sums and moving ahead to solutions that are infinite sums, and that it is now important that we do not substitute our proposed solutions into our equations, i.e., integration is the rule not differentiation.

Lecture 11: The Differential Operator ∇^2

The main idea is to learn how to write ∇^2 in orthogonal coordinate systems.

We begin by defining the gradient operator ∇ in terms of the derivative of a function along a curve. Then we write ∇ in Cartesian coordinates and in any orthogonal coordinate system derived from Cartesian coordinates.

We introduce the surface gradient operator, ∇_S, so that we can differentiate a function defined only on a surface and we go on and obtain a formula for the mean curvature of a surface.

We present a formula for ∇^2 in an arbitrary orthogonal coordinate system and we work out the details in cylindrical and spherical coordinate systems. Our emphasis is on the variation of the base vectors from one point in space to a nearby point.

Then in order to solve problems on domains close to domains we like, we explain how domain perturbations are carried out.

Lecture 12: Diffusion in Unbounded Domains

We begin our study of diffusion, and ∇^2, by deriving formulas for the power moments of a concentration field in an unbounded domain. Viewing the concentration as the probability of finding a solute molecule at a certain point at a certain time, we introduce its mean, its variance, etc. We then present an example where the effect of convection on the variance can be derived.

We use power moments to explain how chromatographic separations work.

We introduce a random walk model.

And we present the point source solution to the diffusion equation and explain superposition.

Lecture 13: Multipole Expansions

We continue solving problems in an unbounded domain and derive solutions to the problem of steady diffusion from a source near the origin to a sink at infinity.

We introduce the monopole, dipole, quadrupole, etc. moments of the source and expand the solution in these moments.

By doing this we obtain solutions to Poisson's equation. We derive the electrical potential due to a set of charges and thus the electrostatic potential energy of two charge distributions.

Lecture 14: One Dimensional Diffusion in Bounded Domains

We solve a one dimensional diffusion problem, viz.,

$$\frac{\partial c}{\partial t} = \frac{\partial^2 c}{\partial x^2}$$

on a bounded domain, say $0 \leq x \leq 1$. The solute concentration is specified at $t = 0$, via

$$c = c_0 \left(x \right) \geq 0$$

This is the source of the solute. The sinks are at $x = 0$ and $x = 1$ where we specify a variety of homogeneous boundary conditions.

To solve our diffusion problem we introduce an eigenvalue problem

$$\frac{d^2\psi}{dx^2} + \lambda^2 \psi = 0, \quad 0 \le x \le 1$$

and try to decide what homogeneous boundary conditions we ought to require ψ to satisfy at $x = 0$ and $x = 1$ in order that we can use the eigenfunctions and eigenvalues in solving for c.

To help us do this we introduce two integration by parts formulas:

$$\int_0^1 \phi \frac{d\psi^2}{dx^2} \, dx = \phi \frac{d\psi}{dx}\bigg|_0^1 - \int_0^1 \frac{d\phi}{dx} \frac{d\psi}{dx} \, dx$$

and

$$\int_0^1 \phi \frac{d^2\psi}{dx^2} \, dx = \left(\phi \frac{d\psi}{dx} - \frac{d\phi}{dx} \psi \right)\bigg|_0^1 + \int_0^1 \psi \frac{d^2\phi}{dx^2} \, dx$$

Our expectation is that by solving our eigenvalue problem we will find an infinite set of orthogonal eigenfunctions in an inner product denoted $\langle \, , \, \rangle$ and by the theory of Fourier series we expect to be able to expand the solution to our diffusion problem as a linear combination of these functions.

Thus we write our solution

$$c(x,t) = \sum c_i(t) \, \psi_i(x)$$

where

$$c_i(t) = \langle \, \psi_i, \, c \, \rangle$$

and we derive the equation for c_i, via

$$\frac{\partial c}{\partial t} = \frac{\partial^2 c}{\partial x^2} \implies \overline{\psi_i} \frac{\partial c}{\partial t} = \overline{\psi_i} \frac{\partial^2 c}{\partial x^2} \implies \int_0^1 \overline{\psi_i} \frac{\partial c}{\partial t} \, dx = \int_0^1 \overline{\psi_i} \frac{\partial^2 c}{\partial x^2} \implies$$

$$\frac{dc_i}{dt} = \left\{ \overline{\psi}_i \frac{\partial c}{\partial x} - \frac{\partial \overline{\psi}_i}{\partial x} c_i \right\} \Bigg|_0^1 + \int_0^1 c \, \frac{\partial^2 \overline{\psi}_i}{\partial x^2} \, dx$$

It is at this point that we decide how to choose the boundary conditions that ψ_i must satisfy. Thus if c is specified at $x = 0$ and $x = 1$, we set $\psi_i = 0$ at $x = 0$ and $x = 1$ to eliminate the unknown $\dfrac{\partial c}{\partial x}$ at $x = 0, 1$. If $\dfrac{\partial c}{\partial x}$ is specified at $x = 0$ and c is specified at $x = 1$ we chose $\dfrac{\partial \psi}{\partial x} = 0$ at $x = 0$ and $\psi = 0$ at $x = 1$. The plan is now apparent. We choose ψ at the boundary to remove the indeterminacy in the equation for c_i, whereupon the λ^2's and ψ's depend on the boundary conditions satisfied by c.

We present several examples differing from one another only in the boundary conditions at $x = 0$ and $x = 1$.

Lecture 15: Two Examples of Diffusion in One Dimension

Two examples of diffusion in one dimension are presented. The first is an activator–inhibitor model which illustrates an instability caused by diffusion, viz., the inhibitor diffuses away before it can arrest the growth of a perturbation. The second is our Petri Dish problem, first introduced in Lecture 1, where we now explain how to find a solution branch which appears as some input to the problem advances beyond its critical value. We see that the amplitude of the branch depends on the input variable in different ways for different nonlinearities.

Lecture 16: Diffusion in Bounded, Three Dimensional Domains

In this lecture the use of the eigenfunctions of ∇^2 to solve inhomogeneous problems on bounded, three dimensional domains is explained.

Our first job is to use Green's two theorems to help us discover the boundary conditions that the eigenfunctions must satisfy and then to derive the important facts about the eigenvalues and the eigenfunctions.

We then indicate how diffusion eigenvalues can be used to estimate critical conditions in nonlinear problems, say, the critical size of a region confining an autothermal heat source.

Lecture 17: Separation of Variables

To solve the solute diffusion problem

$$\frac{\partial c}{\partial t} = \nabla^2 c + Q$$

in a bounded domain, we introduce the eigenvalue problem

$$\nabla^2 \psi + \lambda^2 \psi = 0$$

where ψ satisfies homogeneous conditions on the boundary of our domain. The specified function Q assigns sources and sinks of solute on the domain. Other sources and sinks may be assigned at the boundary of our domain.

The simplest domain shapes that we can deal with are those where we would introduce Cartesian, cylindrical or spherical coordinates.

Now separation of variables is the method ordinarily used to solve the eigenvalue problem. And our aim here is to explain how it works in simple cases.

In Cartesian, cylindrical and spherical coordinates we substitute

$$\psi = X\left(x\right) Y\left(x\right) Z\left(x\right)$$

$$\psi = R\left(r\right) \Theta\left(\theta\right) Z\left(z\right)$$

and

$$\psi = R\left(r\right) \Theta\left(\theta\right) \Phi\left(\phi\right)$$

and obtain

$$\frac{\partial^2 X}{\partial x^2} + \alpha^2 X = 0 \tag{1}$$

$$\frac{\partial^2 Y}{\partial y^2} + \beta^2 Y = 0 \tag{2}$$

and

$$\frac{\partial^2 Z}{\partial z^2} + \gamma^2 Z = 0, \tag{3}$$

$$\frac{\partial^2 Z}{\partial z^2} + \gamma^2 Z = 0 \tag{4}$$

$$\frac{\partial^2 \Theta}{\partial \theta^2} + m^2 \Theta = 0 \tag{5}$$

and

$$\left(\frac{\partial^2}{\partial r^2} + \frac{1}{r} \frac{\partial}{\partial r} \right) R + \left(\lambda^2 - \frac{m^2}{r^2} - \gamma^2 \right) R = 0 \tag{6}$$

and

$$\frac{d^2 \Phi}{d\phi^2} + m^2 \Phi = 0 \tag{7}$$

$$\frac{1}{\sin\theta} \frac{d}{d\theta} \left(\sin\theta \frac{\partial\Theta}{\partial\theta} \right) + \left(\ell(\ell+1) - \frac{m^2}{\sin^2\theta} \right) \Theta = 0 \tag{8}$$

and

$$\left(\frac{d^2}{dr^2} + \frac{2}{r} \right) R + \left(\lambda^2 - \frac{\ell(\ell+1)}{r^2} \right) R = 0 \tag{9}$$

Eqs. (4), (5) and (7) are just like Eqs. (1), (2) and (3). Eqs. (6), (8) and (9) are new.

The reader ought to observe that Eqs. (1), (2) and (3) are independent of one another but by the time we get to Eqs. (7), (8) and (9), Eq. (7) must be solved first so that the m's are available in Eq. (8) then Eq. (8) must be solved so that the ℓ's are available in Eq. (9) and λ^2 appears only in Eq. (9), and it is independent of m^2.

We work out two simple two dimensional problems in order to see what changes occur as we go from one coordinate system to another.

First our diffusion problem is set on a rectangle of sides a and b, c is specified on the perimeter and Q is specified on the domain. We find two sets of orthogonal functions, viz.,

$$X = \sin \frac{m\pi x}{a}, \qquad \alpha^2 = \frac{m^2\pi^2}{a^2} \qquad m = 1, 2, \ldots$$

and

$$Y = \sin \frac{n\pi y}{b}, \qquad \beta^2 = \frac{n^2\pi^2}{b^2} \qquad n = 1, 2, \ldots$$

Cartesian coordinates are special. These two sets of orthogonal functions are all that we need to solve our problem and to obtain $c(x, y, t)$ we take the same steps we took in the one dimensional case, in Lecture 14.

Second our diffusion problem is now set on a circle of radius R_0. And again we assume c is specified on the circumference, but now bounded at the origin and periodic in θ.

Thus we obtain a set of orthogonal angular functions

$$\Theta_m(\theta) = \frac{1}{\sqrt{2\pi}} e^{i\,m\theta}, \qquad m = \cdots, -2, -1, 0, 1, 2, \cdots$$

and these can be used to expand the θ variation of our solution.

The corresponding radial functions satisfy

$$\left(\frac{d^2}{dr^2} + \frac{1}{r}\frac{d}{dr} - \frac{m^2}{r} + \lambda^2 \right) R = 0$$

and we see something new: for each value of m^2 we will have a corresponding set of radial eigenfunctions and the set will differ as m^2 differs.

Lecture 18: Two Stability Problems

In this Lecture we see that eigenvalues and eigenfunctions of ∇^2 are themselves of great interest, whether or not a series solution to a diffusion problem is being sought.

To make this point we solve two stability problems, the Saffman-Taylor problem and the

Rayleigh-Taylor problem. In both cases we imagine the setting to be cylinder of circular cross section bounding a porous solid. Two immiscible fluids fill the pores and in one case a less viscous fluid is displacing a more viscous fluid, in the other case a heavy fluid lies above a light fluid.

The eigenvalues of ∇^2 tell us the critical value of the input variable of interest, the eigenfunctions tell us the pattern we ought to see at critical.

In the second problem the distinction between free and pinned edges bears on the possibility of separating variables.

Lecture 19: Ordinary Differential Equations

Separation of variables leads to second order, linear, ordinary differential equations. The simple facts about these equations are presented in this lecture, before the method of Frobenius for solving these equations is outlined in Lecture 20. Our problem is to find u where u satisfies

$$L u = f$$

$$B_0 u = a_0 \quad \text{at} \quad x = 0$$

and

$$B_1 u = a_1 \quad \text{at} \quad x = 1$$

where L is a linear, second order differential operator and B is a linear combination of u and u'.

Two independent solutions of $Lu = 0$ are introduced and the general solution to $Lu = f$ is presented. Then the homogeneous problem

$$L u = 0$$

$$B_0 u = 0 \quad \text{at} \quad x = 0$$

and

$$B_1 u = 0 \quad \text{at} \quad x = 1$$

is taken up and conditions that it has solutions other than zero are presented.

This leads to solvability conditions for the inhomogeneous problem.

The Green's function is introduced.

The simple facts about the eigenvalue problem

$$L\psi + \lambda^2 \psi = 0$$

$$B_0 \psi = 0 \quad \text{at} \quad x = 0$$

and

$$B_1 \psi = 0 \quad \text{at} \quad x = 1$$

are derived.

Lecture 20: Eigenvalues and Eigenfunctions of ∇^2 in Cartesian, Cylindrical and Spherical Coordinate Systems

Solutions to the eigenvalue problem for ∇^2 are presented in this lecture. The coordinate systems of interest are Cartesian, cylindrical and spherical.

The method of Frobenius is presented and first used to obtain a power series expansion for the Bessel function $I_0(x)$. The coefficients in the series define the nature of the functions so obtained. To emphasize this point we derive the zeros of $J_0(z)$ and $\cos z$ from the coefficients in their power series expansions.

Then the bounded solutions of the associated Legendre equation are worked out and the spherical harmonics are introduced.

Applications to the problem of solute dispersion due to a velocity gradient, to the problem of small amplitude oscillations of a nonviscous sphere and to the energies of a quantum ball in a gravitational field are presented.

Part I

Elementary Matrices

This is the title of an old book by Frazer, Duncan and Collar; while this book is by no means elementary the title does fit Part I of these lectures.

Lecture 1

Getting Going

In this lecture we introduce some ideas which point to the direction of our future work. The next to last section presents the facts about the solutions to linear algebraic equations.

The last section suggests that we can learn something about nonlinear problems by solving certain linear problems.

1.1 Boiling curves

To determine a boiling curve for a liquid solution, we heat the liquid and draw off an equilibrium vapor. We denote the species making up the liquid by $1, 2, \ldots, n$ in order of decreasing volatility or increasing boiling point. Ordinarily the vapor is enriched in the more volatile species and so, as the boiling goes on, the liquid composition shifts in favor of the less volatile species. To write a model of this we denote by x_i and y_i the mole fractions of species i in the liquid and in the equilibrium vapor and by N the number of moles of liquid being heated, then we write

$$\frac{dx_i}{ds} = -y_i + x_i \quad i = 1, \ldots, n$$

where

$$s = -\ell n \left\{ \frac{N(t)}{N(t=0)} \right\}$$

where s increases as t increases, and where, at constant pressure, $y_1, \ldots, y_n$ and T are determined by $x_1, \ldots, x_n$, assuming the phases remain in equilibrium as the liquid is boiled off. Writing the phase equilibrium equations at constant pressure as $y_i = f_i(x_1, x_2, \ldots, x_n)$ and assuming $\sum y_i = 1$ whenever $\sum x_i = 1$ and $y_i = 0$ whenever $x_i = 0$ we see that if $\sum x_i = 1$ and $x_i \geq 0$ at any point in the boiling process then $\sum x_i = 1$ and $x_i \geq 0$ at any subsequent point.

It is enough to let $n = 3$. Then we can represent the state of the liquid geometrically by $\vec{x} = x_1 \vec{i} + x_2 \vec{j} + x_3 \vec{k}$ or algebraically by $\underline{x} = \begin{pmatrix} x_1 \\ x_2 \\ x_3 \end{pmatrix}$ and observe that the state moves on the plane $\sum x_i = 1$ but remains in the positive octant, i.e., the state space is the equilateral triangle whose vertices lie at $\vec{i}, \vec{j}$ and $\vec{k}$, or at

$$\begin{pmatrix} 1 \\ 0 \\ 0 \end{pmatrix}, \quad \begin{pmatrix} 0 \\ 1 \\ 0 \end{pmatrix} \quad \text{and} \quad \begin{pmatrix} 0 \\ 0 \\ 1 \end{pmatrix}$$

The state space is not a vector space, indeed no sum of two vectors in the state space lies in the state space; it is a subset but not a subspace of R^3.

Should we wish to solve our system of equations, we first ought to develop some guidelines to what the solution looks like. The simplest guide posts are the points where the system comes to rest, the so called critical points, steady state points, equilibrium points, etc. These points are defined by $\dfrac{dx_i}{ds} = 0$, $i = 1, 2, \ldots, n$, and here this corresponds to the equations $-y_i + x_i = 0$, $i = 1, 2, \ldots, n$, i.e., to the homogeneous azeotropes. The simplest of these lie at the vertices of the triangle, (points $1, 2, 3$), next are the binary azeotropes lying on the edges (e.g., point 4) then the full ternary azeotropes on the face of the triangle (e.g., point 5), viz,

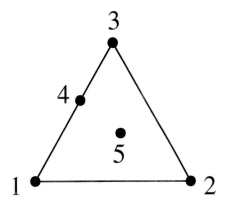

If we can decide how points move in the neighborhood of these rest points, that is whether they are attracted to or repelled by the rest points, we can begin to make a qualitative sketch of the family of solutions to our problem. This property of a critical point is referred to as its stability and we can get some information on stability by assuming the system is displaced a small amount from a rest point and then determining whether this small displacement is strengthened or weakened. To do this we construct a linear approximation to our model in the neighborhood of a rest point of interest and then investigate its solution.

Denote by x_i^0, $i = 1, 2, \ldots, n$ a solution to the equations

$$0 = -f_i\left(x_1^0, x_2^0, \cdots, x_n^0\right) + x_i^0, \quad i = 1, 2, \ldots, n$$

Then approximate the model when $\underline{x}$ is near $\underline{x}^0$ by writing $\underline{x} = \underline{x}^0 + \underline{\xi}$ and retain only terms linear in $\underline{\xi}$. By doing this we obtain

$$\frac{d\underline{\xi}}{ds} = A\underline{\xi}$$

where A is an $n \times n$ matrix whose elements are $a_{ij} = -\frac{\partial f_i}{\partial x_j}\left(x_1^0, x_2^0, \cdots, x_n^0\right) + \delta_{ij}$

This is a system of linear differential equations; what its solutions look like is determined by the matrix A. To understand stability problems is one of our main interests in studying matrices but it is far from our only interest as we will soon see. But for now, what is important is that this example introduces the multiplication $A\underline{\xi}$. The reader can carry out the calculation described above and learn the rule determining the product.

How to Think about the Multiplication of a Vector by a Matrix

A column of n complex numbers is called a column vector and is denoted $\underline{x} = \begin{pmatrix} x_1 \\ x_2 \\ \vdots \\ x_n \end{pmatrix}$.

It belongs to the vector space denoted C^n. It is also an $n \times 1$ matrix and its transpose $\underline{x}^T = \begin{pmatrix} x_1 & x_2 & \dots & x_n \end{pmatrix}$ is a $1 \times n$ matrix. In the foregoing we multiplied a column vector on the left by a matrix, the result of doing this being another column vector. The rule for doing this is:

the product of an $m \times n$ matrix A and an $n \times p$ matrix B is an $m \times p$ matrix C where $c_{ij} = \sum_{k=1}^{n} a_{ik} b_{kj}$ and where a_{ij} lies in the ith row and the jth column of the matrix A.

This formula defines matrix multiplication but to see what is going on when two matrices are multiplied it is useful to think about a matrix in terms of its columns or its rows instead of in terms of its elements. Indeed an $m \times n$ matrix A is made up of n columns, each a column vector lying in C^m. Denoting these $\underline{a}_1, \underline{a}_2, \dots, \underline{a}_n$ where

$$\underline{a}_1 = \begin{pmatrix} a_{11} \\ a_{21} \\ \vdots \\ a_{m1} \end{pmatrix}, \quad \underline{a}_2 = \begin{pmatrix} a_{12} \\ a_{22} \\ \vdots \\ a_{m2} \end{pmatrix}, \quad \dots, \quad \underline{a}_n = \begin{pmatrix} a_{1n} \\ a_{2n} \\ \vdots \\ a_{mn} \end{pmatrix}$$

we can write

$$A = \begin{pmatrix} \underline{a}_1 & \underline{a}_2 & \dots & \underline{a}_n \end{pmatrix}$$

whence the product $A\underline{x}$ is the column vector

$$x_1 \underline{a}_1 + x_2 \underline{a}_2 + \dots + x_n \underline{a}_n \in C^m$$

where

$$\underline{x} = \begin{pmatrix} x_1 \\ x_2 \\ \vdots \\ x_n \end{pmatrix} \in C^n$$

The product $A\underline{x}$ is then the linear combination of the columns of A determined by coefficients taken to be the elements of $\underline{x}$. Likewise in the product AB, each column of B belongs to C^n and determines the coefficients for the linear combination of the columns of A that adds up to the corresponding column of the product. So the jth column of a product AB is a linear combination of the columns of A, the coefficients being the elements of the jth column of B, indeed $AB = (A\underline{b}_1 \ A\underline{b}_2 \ \ldots)$. Again each row of BA is a linear combination of the rows of A, the coefficients for constructing the ith row of BA being the elements of the ith row of B. Ordinarily if AB is defined BA is not and vice versa, the exception is when A and B are square. Then AB and BA need not be equal. This way of looking at matrix multiplication will help us understand the solvability conditions for linear algebraic equations, viz., $A\underline{x} = \underline{b}$. In fact, if $m > n$, the picture

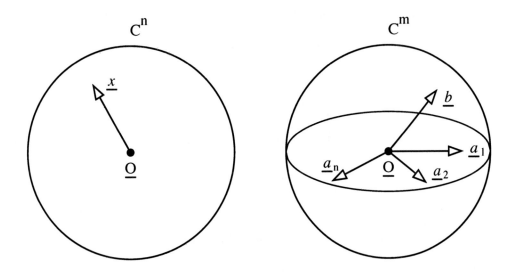

suggests the need for a solvability condition.

Just as

$$A\underline{x} = \begin{pmatrix} \underline{a}_1 \ \underline{a}_2 \ \cdots \ \underline{a}_n \end{pmatrix} \begin{pmatrix} x_1 \\ x_2 \\ \vdots \\ a_n \end{pmatrix} = \underline{a}_1 x_1 + \underline{a}_2 x_2 + \cdots + \underline{a}_n x_n$$

so also

$$AX = \begin{pmatrix} \underline{a}_1 \ \underline{a}_2 \ \cdots \ \underline{a}_n \end{pmatrix} \begin{pmatrix} \underline{y}_1^T \\ \underline{y}_2^T \\ \vdots \\ \underline{y}_n^T \end{pmatrix} = \underline{a}_1 \underline{y}_1^T + \underline{a}_2 \underline{y}_2^T + \cdots + \underline{a}_n \underline{y}_n^T$$

where $\underline{y}_1^T$ is the first row of X, $\underline{y}_2^T$ the second, etc. and where $\underline{a}_1 \underline{y}_1^T$ is a matrix each of whose columns is a multiple of $\underline{a}_1$. This way of looking at a matrix multiplication, instead of $AX = \begin{pmatrix} A\underline{x}_1 \ A\underline{x}_2 \ \cdots \end{pmatrix}$ as above, will be useful later on in turning the solutions to the eigenvalue problem for A into a spectral representation of A.

Columns and rows turn up in a symmetric way. For every result about a set of columns there is a corresponding result about a set of rows. We emphasize columns and write a system of algebraic equations as $A\underline{x} = \underline{b}$ but we can use the transpose operation, denoted T, where the columns of A become the rows of A^T, viz., $A^T = \begin{pmatrix} \underline{a}_1^T \\ \underline{a}_2^T \\ \vdots \\ \underline{a}_n^T \end{pmatrix}$, and write the problem $A\underline{x} = \underline{b}$ as $\underline{x}^T A^T = \underline{b}^T$.

Before leaving the boiling curve problem we can observe that the temperature of the liquid plays no role. Because we ordinarily expect the temperature to increase as the boiling proceeds we expect a family of solution curves for a plain vanilla liquid system to look as follows

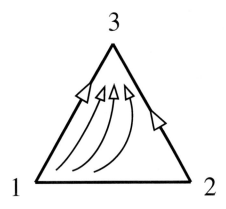

If there are binary and ternary azeotropes we first observe that binary azeotropes come in two kinds, maximum boiling which are stable and minimum boiling which are unstable. Knowing whether a binary azeotrope is maximum or minimum boiling allows us to determine what the system does on the binary edges. But it may not determine what happens in the interior even in the absence of ternary azeotropes. Thus if the $1, 2$ binary has a maximum boiling azeotrope we expect either

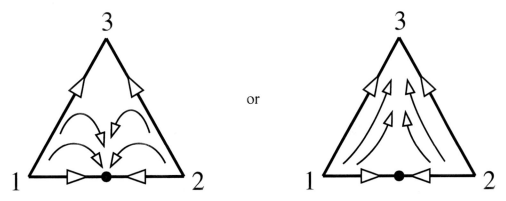

or

depending on whether the 1-2 azeotrope boils at a higher or lower temperature than does 3.

We might guess that if we calculate the temperature associated with each state and plot the isotherms on the state diagram we can sketch the solution curves of our problem by insisting only that the temperature not decrease as the boiling goes on. The temperature then is a sort of potential for this problem.

The arrows on the edges 1-3 and 2-3 show that the vertex 3 is stable and therefore has at least a small region of attraction. (Arrows pointing at vertex 3 will lie on the two edges that converge on vertex 3 as long as there is not a 1-3 or 2-3 maximum boiling azeotrope.) This makes the first of the two figures doubtful. Assuming the 1-2 azeotrope is stable whenever its boiling point is higher than the boiling point at vertex 3, and knowing that vertex 3 is stable, we can speculate that this boiling point ordering is sufficient that an unstable ternary azeotrope mediates the dynamics of the system, anchoring a boundary separating the regions of attraction of vertex 3 and the 1-2 azeotrope.

There is a theory, called index theory, an account of which can be found in Coddington and Levinson's book "*Theory of Ordinary Differential Equations*," which gives global information about questions of this sort. It establishes conditions that must be satisfied by the sum over the local stability at each of a set of multiple critical points. This theory requires ideas beyond what we intend to explain in these lectures and is therefore a direction for advanced study.

1.2 A Simple Evaporator

The solution of the differential equation

$$\frac{dx}{dt} = a(t)x + b(t)$$

where $x\,(t = t_0)$ and $b\,(t)$ are assigned is

$$x\,(t) = x\,(t = t_0)\,e^{\int_{t_0}^{t} a(\lambda)d\lambda} + \int_{t_0}^{t} e^{\int_{\tau}^{t} a(\lambda)d\lambda}\,b(\tau)d\tau$$

This formula tells us that the value of x at time t is determined by its value at time t_0 and the values of b on the interval (t_0, t). It shows that the contributions of the two sources to the solution are independent and additive. This is one way, but not the only way, of stating the principle of superposition. It exhibits the main way in which linear problems are special.

If a is constant the formula is

$$x\,(t) = x\,(t = t_0)\,e^{a(t - t_0)} + \int_{t_0}^{t} e^{a(t - \tau)}\,b(\tau)d\tau$$

and if b is also constant it is

$$x\left(t\right) = x\left(t = t_0\right) e^{a\left(t - t_0\right)} + \frac{b}{a}\left\{e^{a\left(t - t_0\right)} - 1\right\}$$

We can make use of this formula in studying the dynamics of a simple evaporator. The problem is to concentrate a nonvolatile solute in a feed stream by boiling off the volatile solvent. The stream to be concentrated, specified by its feed rate F [#/hr], concentration x_F [mass fraction] and temperature T_F [°F], is run into a tank containing a heat exchanger of area A [ft^2] and heat transfer coefficient U [Btu/hr ft^2 °F] supplied with steam condensing at temperature T_s. The pressure in the system is determined by the conditions under which the solvent being boiled off is condensed. This is done in a heat exchanger of area A_c and heat transfer coefficient U_c, supplied with cooling water at temperature T_c.

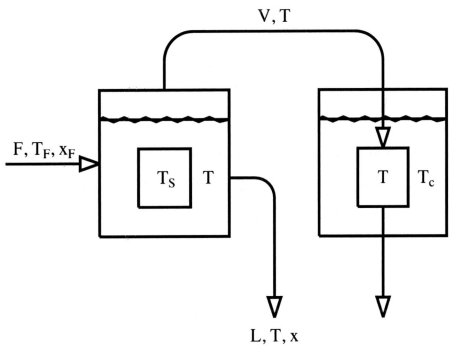

The simplest model corresponds to concentrating a dilute solution. In this case we assume that the physical properties of the solution are those of the solvent and add that its heat capacity $c_p \left[\dfrac{\text{Btu}}{\text{\# °F}}\right]$ and latent heat $\lambda \left[\dfrac{\text{Btu}}{\text{\#}}\right]$ are constant. Then under steady conditions we write

$$0 \; = \; F - L - V$$

$$0 \; = \; x_F F - x L$$

$$0 \; = \; h_F F + U A (T_s - T) - h L - H V$$

$$0 \; = \; H V - h V - U_c A_c (T - T_c)$$

where T is the boiling point of the solvent, x the concentration of the product and h and H $\left[\dfrac{\text{Btu}}{\#} \right]$ the enthalpies of the liquid and the vapor streams. The pressure is the vapor pressure of the solvent at temperature T. If F, x_F, T_F, T_s, $U A$, T_c and $U_c A_c$ (the operating variables) are set then the number of equations equals the number of unknowns and we can determine x, L, V and T (the performance variables). Indeed eliminating L and introducing c_p and λ we get

$$0 \; = \; x_F F - x (F - V)$$

$$0 \; = \; c_p (T_F - T) F + U A (T_s - T) - \lambda V$$

$$0 \; = \; \lambda V - U_c A_c (T - T_c)$$

and we see that as long as

$$U A T_s + c_p F T_F > (U A + c_p F) T_c$$

then $T > T_c$ and a pressure is established so that a boiling, i.e., $V > 0$, solution to these equations is obtained. We say, then, that the pressure is established so that the heat balance balances, i.e., so that the heat supplied at the evaporator equals the heat removed at the condenser:

$$c_p (T_F - T) F + U A (T_s - T) = U_c A_c (T - T_c)$$

Indeed taking $F = 1000, T_F = 100, U A = 2000, T_s = 300, U_c A_c = 2000, T_c = 50, c_p = 1$ and $\lambda = 1000$ we find $T = 160, V = 220$, whereas if the feed is colder and faster, viz., $F = 2000, T_F = 50$, we find $T = 133, V = 166$. Assuming P to be the vapor pressure of water we can understand the sensitivity of P to the operating conditions in an evaporator.

Now let the system be in a steady state corresponding to assigned values of the operating variables and suppose that certain of these are changed to new values at $t = 0$. Then, while we know how to determine the new steady state reached as $t \to \infty$ we need to answer the question: how does the system make the transition from the old to the new steady state?

Letting $M[\#]$ denote the amount of well mixed solution held in the evaporator and assuming that L is adjusted to hold M fixed, we replace the left hand sides of the original equations by $\dfrac{dM}{dt} = 0$, $M\dfrac{dx}{dt}$, $M\dfrac{dh}{dt} = c_p M\dfrac{dT}{dt}$ and 0, this last by assuming the condenser hold up to be small. Then eliminating L we find

$$ c_p M \frac{dT}{dt} = c_p F\left(T_F - T\right) + UA\left(T_s - T\right) - U_c A_c\left(T - T_c\right) $$

The value of T at $t = 0$ is the old steady state value while the operating variables take their new values at $t = 0$.

In this simple model T vs t can be found using the formula introduced at the beginning of this section. Then V vs t is determined by

$$ V = \frac{U_c A_c}{\lambda}\left(T - T_c\right) $$

and, using this, x vs t can be found by solving

$$ M\frac{dx}{dt} = x_p F - \left(F - V\right) x $$

again using our now favorite formula. Setting $M = 1000$ the reader can determine T vs t, V vs t and x vs t as the evaporator makes the transition from the steady state corresponding to $T = 160$ to that corresponding to $T = 133$.

It is helpful in dealing with problems like this to scale the variables so that only dimensionless variables appear. In doing this there may be a variety of time scales and each may suggest a useful approximation. Here the problem is so simple that the only important time scale is $\dfrac{c_p M}{c_p F + UA + U_c A_c}$ and this determines how fast the system responds to step changes.

1.3   A Less Simple Evaporator

If in the foregoing we make the value of $U_c A_c$ very large then the solvent condenses at the temperature T_c and the evaporator operates at constant pressure. The problem remains interesting when the pressure is fixed if we include in the model the possibility of a boiling point rise. As a dilute solution can exhibit a significant boiling point rise we retain all the simplifying approximations in the foregoing save one: we now assume the boiling point of the solution to be $T_c + \beta x$ where βx is the boiling point rise. Then as long as the solution is boiling, i.e., $V > 0$, we can write

$$M\frac{dx}{dt} = \left(x_F - x\right)F + xV$$

$$Mc_p\frac{dT}{dt} = UA\left(T_s - T\right) + c_p\left(T_F - T\right)F - \lambda V$$

and

$$T = T_c + \beta x$$

whereas if $V = 0$ we write instead

$$M\frac{dx}{dt} = \left(x_F - x\right)F$$

$$Mc_p\frac{dT}{dt} = UA\left(T_s - T\right) + c_p\left(T_F - T\right)F$$

and

$$T < T_c + \beta x$$

When the solution is boiling, the model contains the nonlinear term xV but only two of the three equations are differential equations. To determine V we use $T = T_c + \beta x$ and hence $Mc_p\frac{dT}{dt} = Mc_p\beta\frac{dx}{dt}$ to conclude that

$$UA\left(T_s - T\right) + c_p\left(T_F - T\right) - \lambda V = c_p\beta\left\{\left(x_F - x\right)F + xV\right\}$$

This formula determines V as a function of x and T and can be used to eliminate V from the differential equations. We will return to this problem and examine the stability of its steady solutions to small upsets.

1.4   The Hilbert Matrix

To approximate an assigned function $f(x)$ on the interval $a \leq x \leq b$ by a polynomial of degree n, viz., by $P_n(x) = a_0 + a_1 x + \cdots + a_n x^n$, the problem is to find the $n+1$ coefficients $a_0, a_1, \cdots, a_n$. The error is $f(x) - P_n(x)$ and if we determine $a_0, a_1, \cdots, a_n$ to make the integral square error, $\int_a^b \{f(x) - P_n(x)\}^2 \, dx$, as small as possible, we find, on setting the derivatives of this with respect to $a_0, a_1, \cdots, a_n$ to zero that

$$\sum_{j=0}^n \left\{ \int_a^b x^{j+i} dx \right\} a_j = \int_a^b f(x) x^i dx, \qquad i = 0, 1, \ldots, n$$

This is a system of $n + 1$ equations in $n + 1$ unknowns which, when $a = 0$ and $b = 1$, can be written

$$\begin{pmatrix} \frac{1}{1} & \frac{1}{2} & \cdots & \frac{1}{n+1} \\ \frac{1}{2} & \frac{1}{3} & \cdots & \frac{1}{n+2} \\ \vdots & \vdots & \vdots & \vdots \\ \frac{1}{n+1} & \frac{1}{n+2} & \cdots & \frac{1}{2n+1} \end{pmatrix} \begin{pmatrix} a_0 \\ a_1 \\ \vdots \\ a_n \end{pmatrix} = \begin{pmatrix} \int_0^1 f(x) dx \\ \int_0^1 f(x) x dx \\ \vdots \\ \int_0^1 f(x) x^n dx \end{pmatrix}$$

The matrix on the left hand side is called the *Hilbert matrix* and the corresponding equations are remarkable for how difficult they are to solve for values of n that are not large. Problems such as this require for their solution the use of correction methods designed to improve approximations obtained by elimination methods. The determinants of the 2×2 and 3×3 Hilbert matrices are $1/12$ and $1/2160$, where the numerators are $4 - 3 = 1$ and $81 - 80 = 1$. Now, if $1/3$ is replaced by $33/100$, where $\frac{1}{3} - \frac{33}{100} = \frac{1}{300}$, the determinant of the altered 3×3 Hilbert matrix is $63/10^6$, which is about 10% of $1/2160$.

If we approximate $f(x)$ by $a_0 + a_1 x$ and denote $\frac{1}{b-a} \int_a^b (\ ) \, dx$ by $(\ )_{\text{avg}}$, we get

$$a_0 = \frac{(x^2)_{\text{avg}} (f)_{\text{avg}} - (x)_{\text{avg}} (xf)_{\text{avg}}}{(x^2)_{\text{avg}} - (x)_{\text{avg}}^2}$$

and

$$a_1 = \frac{(xf)_{\text{avg}} - (x)_{\text{avg}} (f)_{\text{avg}}}{(x^2)_{\text{avg}} - (x)_{\text{avg}}^2}$$

Now let X and Y be random variables, the values x of X lying in $[a, b]$, the values y of Y lying in $[c, d]$. And let $f(X, Y)$ be the joint probability density: $f(x, y)dxdy$ being the probability that the point (X, Y) lies in the rectangle $(x, x + dx) \times (y, y + dy)$. The expected value of any function $G(X, Y)$ is

$$E\big(G(X, Y)\big) = \int_a^b \int_c^d G(x, y) f(x, y) dxdy$$

To approximate Y by $a_0 + a_1 X$ we seek to determine a_0 and a_1 so that $E\big((Y - (a_0 + a_1 X))^2\big)$ is least. Then as

$$E\left((Y - (a_0 + a_1 X))^2\right) = E\left(Y^2\right) - 2a_1 E\left(XY\right) -$$

$$2a_0 E\left(Y\right) + a_1^2 E\left(X^2\right) + 2a_0 a_1 E\left(X\right) + a_0^2$$

we find, on setting the derivatives of this with respect to a_0 and a_1 to zero and solving for a_0 and a_1, that

$$a_0 = \frac{E\left(X^2\right) E\left(Y\right) - E\left(X\right) E\left(XY\right)}{E\left(X^2\right) - E\left(X\right)^2}$$

and

$$a_1 = \frac{E\left(XY\right) - E\left(X\right) E\left(Y\right)}{E\left(X^2\right) - E\left(X\right)^2}$$

These formulas state in another way what we found just above. If X and Y are uncorrelated, we have $E(XY) = E(X)E(Y)$ whence $a_0 = E(Y), a_1 = 0$.

Defining the variance σ^2 and the correlation coefficient ρ as

$$\sigma_X^2 = E\left((X - E(X))^2\right) = E\left(X^2\right) - E(X)^2$$

$$\sigma_Y^2 = E\left(Y^2\right) - E(Y)^2$$

and

$$\rho_{XY}\sigma_X\sigma_Y = E\left[(X - E(X))(Y - E(Y))\right]$$

$$= E(XY) - E(X)E(Y)$$

we can write

$$a_0 = E(Y) - E(X)a_1$$

and

$$a_1 = \rho_{XY}\frac{\sigma_Y}{\sigma_X}$$

The least value of $E\left((Y - (a_0 + a_1 X))^2\right)$ is then $\sigma_Y^2\left(1 - \rho_{XY}^2\right)$. If X and Y are uncorrelated this is σ_Y^2, whereas if they are perfectly correlated it is 0.

We will use some simple ideas about random variables and probability densities when we deal with the diffusion of a solute in a solvent.

1.5 More Equations than Unknowns

Instead of having the values of a function everywhere on an interval, we may have its values only at a discrete set of points. Call these values $y_1, y_2, \ldots, y_n$ corresponding to $x = x_1, x_2, \ldots, x_n$. Then we can try to find a polynomial of degree $n - 1$ that fits this information. Thus, writing

$$P_{n-1}(x) = a_0 + a_1 x + \cdots + a_{n-1}x^{n-1}$$

we determine $a_0, a_1, \ldots, a_{n-1}$ via

$$y_i = P_{n-1}(x_i), \qquad i = 1, \ldots, n$$

which we write as

$$V \begin{pmatrix} a_0 \\ a_1 \\ \vdots \\ a_{n-1} \end{pmatrix} = \begin{pmatrix} y_1 \\ y_2 \\ \vdots \\ y_n \end{pmatrix}$$

where

$$V = \begin{pmatrix} 1 & x_1 & x_1^2 & \cdots & x_1^{n-1} \\ 1 & x_2 & x_2^2 & \cdots & x_2^{n-1} \\ \vdots & \vdots & \vdots & \vdots & \vdots \\ 1 & x_n & x_n^2 & \cdots & x_n^{n-1} \end{pmatrix}$$

and where V is called the *Vandermonde matrix*. This is a system of n equations in n unknowns. Ordinarily it has one and only one solution, but the solution may be sensitive to small changes in the y's.

If we measure our function at a set of m points, where $m > n$, our problem is

$$\begin{pmatrix} 1 & x_1 & x_1^2 & \cdots & x_1^{n-1} \\ \vdots & \vdots & \vdots & \vdots & \vdots \\ 1 & x_n & x_n^2 & \cdots & x_n^{n-1} \\ \vdots & \vdots & \vdots & \vdots & \vdots \\ 1 & x_m & x_m^2 & \cdots & x_m^{n-1} \end{pmatrix} \begin{pmatrix} a_0 \\ a_1 \\ a_2 \\ \vdots \\ a_{n-1} \end{pmatrix} = \begin{pmatrix} y_1 \\ \vdots \\ y_n \\ \vdots \\ y_m \end{pmatrix}$$

and this is a system of m equations in n unknowns. It is overdetermined and we suspect that it does not have a solution, either because our function is not really a polynomial of degree $n-1$ or, if it is, that errors in the data preclude us from seeing this. In fact, what we will find is that a system of m equations in n unknowns may (i) not have a solution or (ii) may have exactly one solution or (iii)

may have many solutions. This also summarizes the possibilities if $m = n$, whereas if $m < n$, an underdetermined system, the second possibility must be excluded.

We write this

$$V\underline{a} = \underline{y}$$

where V is the $m \times n$ *Vandermonde matrix*, $\underline{a}$ is the $n \times 1$ column of unknown coefficients and $\underline{y}$ is the $m \times 1$ column of measured values of our function. If V and $\underline{y}$ are the results of measurements and $m > n$ it is unlikely that there is a solution and we can look for an approximation, that is a value of $\underline{a}$ such that $V\underline{a}$ is as close as possible to $\underline{y}$. The error in the ith equation is

$$y_i - \sum_{j=0}^{n-1} a_j x_i^j$$

and the sum of the squares of the errors is

$$\sum_{i=1}^{m} \left(y_i - \sum_{j=0}^{n-1} x_i^j a_j \right) = \sum_{i=1}^{m} \left(\underline{y} - V\underline{a} \right)_i^2$$

$$= \left(\underline{y} - V\underline{a} \right)^T \left(\underline{y} - V\underline{a} \right)$$

$$= \underline{y}^T \underline{y} - \underline{y}^T V\underline{a} - (V\underline{a})^T \underline{y} + (V\underline{a})^T \underline{a}$$

$$= \underline{y}^T \underline{y} - 2\underline{a}^T V^T \underline{y} + \underline{a}^T V^T V\underline{a}$$

To find $a_0, a_1, \ldots, a_{n-1}$ so that the sum of the squares of the errors takes its least value, we set the derivative of this expression with respect to each $a_k, k = 0, \ldots, n-1$, to zero getting n equations:

$$\sum_{j=0}^{n-1} \left\{ \sum_{i=1}^{m} x_i^k x_i^j \right\} a_i = \sum_{i=1}^{m} x_i^k y_i$$

which can be written

$$V^T V \underline{a} = V^T \underline{y}$$

This is a system of n equations in n unknowns where the elements of the $n \times n$ coefficient matrix $V^T V$ are

$$\left(V^T V\right)_{ij} = \sum_{k=1}^{m} x_k^{i+j}, \qquad i, j = 0, \ldots, n-1$$

and this is just what we would expect to turn up in this, the discrete problem, knowing that the elements of the corresponding matrix in the continuous problem are $\int_a^b x^{i+j} dx$

It is not easy to get an accurate solution to the problem $V^T V \underline{a} = V^T \underline{y}$, as $V^T V$, like the Hilbert matrix, does not work well when elimination methods are used. To see what is going on suppose that $x_1, x_2, \ldots, x_m$ is an increasing sequence of positive numbers. Then the columns of $V^T V$, viz.,

$$\begin{pmatrix} \sum x_i^0 \\ \sum x_i^1 \\ \sum x_i^2 \\ \vdots \end{pmatrix}, \quad \begin{pmatrix} \sum x_i^1 \\ \sum x_i^2 \\ \sum x_i^3 \\ \vdots \end{pmatrix} \quad \cdots, \quad \begin{pmatrix} \sum x_i^{n-1} \\ \sum x_i^n \\ \sum x_i^{n+1} \\ \vdots \end{pmatrix}$$

lie in the positive cone of R^n and their directions converge to a limiting direction as n grows large. Linear independence is retained for all n, but just barely as n grows large. The reader can see this simply by letting $m = 4, n = 3$ and $x_1 = 1, x_2 = 2, x_3 = 3$ and $x_4 = 4$.

To see why nearly dependent columns lead to uncertainties in numerical work, observe that the solution to

$$\begin{pmatrix} 1 & 1 \\ 0 & \varepsilon \end{pmatrix} \begin{pmatrix} x \\ y \end{pmatrix} = \begin{pmatrix} 1 \\ 0 \end{pmatrix}$$

is $x = 1, y = 0$ whereas it is $x = 0, y = 1$ if $\begin{pmatrix} 1 \\ 0 \end{pmatrix}$ on the RHS is replaced by $\begin{pmatrix} 1 \\ \varepsilon \end{pmatrix}$.

For later work we record the observation that $V^T V \underline{a} = V^T \underline{y}$ can be written in terms of the error, $\underline{y} - V\underline{a}$, as

$$\left(\underline{y} - V\underline{a}\right)^T V = \underline{0}^T$$

1.6 Getting Ready for Lectures 2 and 3

It is important to build up a picture of the facts about the solutions to the problem $A\underline{x} = \underline{b}$. We begin to do this in the hope that the readers will add to it as they go along.

Write $A\underline{x} = \underline{b}$ as

$$x_1\underline{a}_1 + x_2\underline{a}_2 + \cdots + x_n\underline{a}_n = \underline{b}$$

where $\underline{x} \in C^n$, where $\underline{a}_1, \underline{a}_2, \ldots, \underline{a}_n, \underline{b} \in C^m$ and where r denotes the largest number of independent vectors in $\{\underline{a}_1, \underline{a}_2, \ldots, \underline{a}_n\}$.

The simplest case is $n = 1, r = 1, m = 2$. Then the picture corresponding to the problem $x\underline{a}_1 = \underline{b}$ is

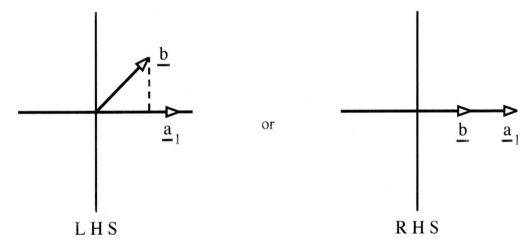

LHS   RHS

On the LHS there is no value of x satisfying $x\underline{a}_1 = \underline{b}$ while on the RHS there is exactly one value of x. On the LHS we can determine the value of x so that $x\underline{a}_1$ is as close as possible to $\underline{b}$ but $x\underline{a}_1$ cannot equal $\underline{b}$.

If $n = 2, r = 2, m = 3$ the problem is $x\underline{a}_1 + y\underline{a}_2 = \underline{b}$, the picture is

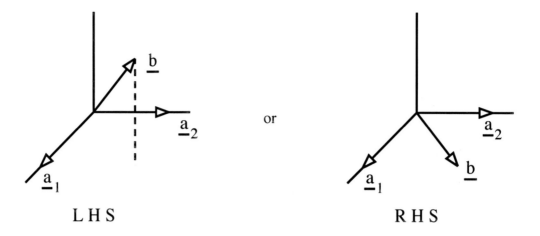

LHS   or   RHS

and the conclusions are as before: on the LHS there are no values of x and y such that $x\underline{a}_1 + y\underline{a}_2 = \underline{b}$, while on the RHS there is exactly one value of x and one value of y. But if $r = 1$ the picture is

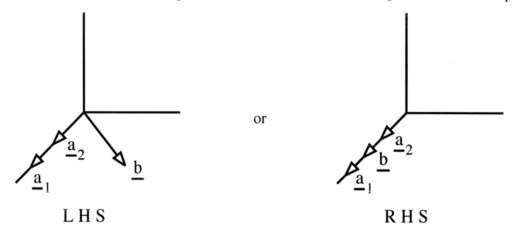

LHS   or   RHS

and again on the LHS there are no values of x and y such that $x\underline{a}_1 + y\underline{a}_2 = \underline{b}$. The RHS is new: as before x and y can be determined, but now this is possible in many ways.

These pictures lead us to certain conclusions about the solutions in terms of the numerical values of n, r and m. If $r < m$ the problem has solutions for some values of $\underline{b}$ but not for others. If the problem has a solution and if $r = n$ then it is the only solution, but if $r < n$ there are many solutions.

Certainly r cannot exceed n and, as $\underline{a}_1, \ldots, \underline{a}_n \in C^m$, r cannot exceed m either. Using this the reader can draw more conclusions about the solutions when $n < m, n = m$ and $n > m$.

1.7 A Source of Linear Problems: Boiling an Azeotrope

Assuming that most of the interesting problems a student will face are nonlinear, we ought to indicate at least one source of linear problems.

Suppose that upon writing a model to explain or predict an experimental observation, viz., the output variable, we have to solve a nonlinear problem. Included in the specification of the problem will be the values of the input variables.

Often a simple steady solution to our problem can be found where, possibly, the nonlinear terms vanish and we would then like to know if we can see this solution if we run the experiment.

To answer this question we add to the simple base solution we have a small correction and substitute the sum into our nonlinear equation in order to obtain an equation for the correction. Upon discarding squares, etc., of small quantities, this will be a linear equation and our aim will be to discover if the small displacement grows or dies out in time.

If the displacement grows, our base solution will be called unstable and we will not see it in an experiment.

If all displacements die out our base solution will be stable to small displacements and we may be able to see it in an experiment.

Ordinarily there will be ranges of inputs where stability obtains and ranges of inputs where it does not. The critical values of the inputs divide these ranges. Hence we may decide to run a sequence of experiments where we increase an input to its critical value and ask what we expect to see if the input is advanced just beyond its critical value.

By asking this question, we are led to derive a sequence of linear problems which are inhomogeneous versions of the homogeneous stability problem and which introduce solvability questions.

Boiling an Azeotrope

As a simple example, recall that in our boiling problem we have

$$\frac{d\vec{x}}{ds} = -\vec{y} + \vec{x}$$

where $\vec{x}$ is specified at $s = 0$ and where, at constant pressure, $\vec{y}$ is known as a function of $\vec{x}$. Hence, if we are boiling an azeotrope, i.e., if $\vec{x}$ at $s = 0$ is an azeotropic composition so that at $s = 0$ we have $\vec{y} = \vec{x}$, then, for all s we have the solution

$$\vec{x} = \vec{x}(t = 0)$$

and we can ask: *is this what we see in an experiment?*

The easy case is $n = 2$, where we have

$$\frac{dx}{ds} = -y + x$$

and y as function of x looks one of two ways

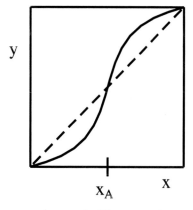

maximum boiling azeotrope.

or

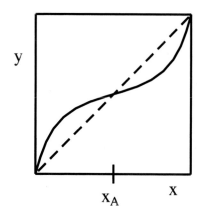

minimum boiling azeotrope.

And starting a boiling experiment at $x = x_A$ we predict $x = x_A$ for all s.

To see if this solution is stable, we substitute $x = x_A + \xi$, ξ small, and obtain

$$\frac{d\xi}{ds} = -f'(x_A)\xi + \xi$$

Now we observe that $t = 0$ corresponds to $s = 0$ and that s is a time like variable. Hence we have stability if $f'(x_A) > 1$, instability if $f'(x_A) < 1$ and we find that a maximum boiling azeotrope can be boiled off at constant composition. But a minimum boiling azeotrope cannot be sustained. Here it is not that we expect a composition fluctuation to occur during boiling, instead the problem lies in the preparation of the initial state.

Boiling Point Rise Model

Going back to our boiling point rise model where the inputs, viz., M, x_F, T_F, F, c_p, λ, UA, T_s, T_c and β, are held fixed and where the outputs are x, T and V, denote by $x_0, T_0, V_0 > 0$, a boiling steady state. Is it stable? To see, substitute

$$V = V_0 + \varepsilon V_1,$$

$$x = x_0 + \varepsilon x_1,$$

and

$$T = T_0 + \varepsilon T_1$$

into the model presented earlier, where ε is small, and obtain equations for x_1, T_1 and V_1. Eliminate V_1 from the first two equations, then eliminate T_1 via $T_1 = \beta x_1$ and draw the conclusion that for any small initial displacement of the steady state, x_1 goes to zero as t increases.

Petri Dish Problem

To introduce a solvability question we present the following model, where c denotes the concentration of a solute in a one dimensional domain and $F(c)$ denotes its rate of formation. All variables are scaled and at first we say only that $F(0) = 0$ and $F'(0) > 0$. Thus an excursion away from $c = 0$ reinforces itself. Our model is

$$\frac{\partial c}{\partial t} = \frac{\partial^2 c}{\partial x^2} + \lambda F(c)$$

where $c = 0$ at $x = 0, 1$, i.e., there is a solute sink at the ends of our domain, and where λ denotes the strength of the source.

Our aim is to find the value of λ at which diffusion to the solute sinks at $x = 0, 1$ can no longer control the solute source on the domain.

We have a solution $c = 0$ for all λ and we might wish to know for what values of λ we can observe this solution. There are two simple things we can do, both leading to the same conclusion. First we can introduce a small perturbation, viz., $c = 0 + c_1$ where c_1 is small and find that c_1 satisfies

$$\frac{\partial c_1}{\partial t} = \frac{\partial^2 c_1}{\partial x^2} + \lambda F'(0) c_1$$

where $c_1 = 0$ at $x = 0, 1$.

This is a linear problem and we can seek solutions of the form

$$c_1 = \psi(x) e^{\sigma t}$$

where σ is the growth rate of a perturbation whose spatial dependence is $\psi(x)$. Then σ and ψ solve the homogeneous problem

$$\frac{d^2 \psi}{dx^2} + \lambda F'(0) \psi - \sigma \psi = 0$$

where $\psi = 0$ at $x = 0, 1$, and this problem has solutions other than $\psi = 0$ only for special values of σ.

The solutions are $\psi = \sin \pi x, \sin 2\pi x, \ldots$ corresponding to

$$\sigma = \lambda F'(0) - \pi^2, \ \lambda F'(0) - 4\pi^2, \ldots$$

and these σ's are the growth rates of an independent set of perturbations spanning all allowable perturbations. Because the strength of diffusion increases as the spatial variation increases, we see that $\sin \pi x$ is the most dangerous perturbation.

At $\lambda = 0$, the greatest value of σ is found to be $-\pi^2$. And for all $\lambda > 0$ the greatest σ is $\lambda F'(0) - \pi^2$, whereupon the greatest σ becomes zero at $\lambda = \pi^2/F'(0)$. This then is the critical value of λ beyond which the solution $c = 0$ is not stable.

The second thing we can do is to look only at steady solutions, viz., solutions to

$$\frac{d^2 c}{dx^2} + \lambda F(c) = 0$$

where $c = 0$ at $x = 0, 1$, and observe that one such solution is $c = 0$ for all λ. Then we can ask: if we have the solution $c = 0$ at some λ can we advance it if we advance λ, i.e., can we find $\dfrac{dc}{d\lambda}$?

Denoting $\dfrac{dc}{d\lambda}$ by $\dot{c}$ we have

$$\frac{d^2\dot{c}}{dx^2} + \lambda F'(0)\dot{c} = 0$$

where $\dot{c} = 0$ at $x = 0, 1$.

This is the forgoing, viz., ψ, problem at $\sigma = 0$ and it has only the solution $\dot{c} = 0$ for all $\lambda < \lambda_{crit}$, hence we keep finding only the solution $c = 0$ as λ increases from zero until we reach $\lambda = \lambda_{crit}$, whereupon a solution $\dot{c} \neq 0$ appears, signaling something new may be happening.

Now we can go back to the problem for σ and differentiate with respect to λ, obtaining

$$\frac{d^2\dot{\psi}}{dx^2} + \lambda F'(0)\dot{\psi} - \sigma\dot{\psi} = \dot{\sigma}\psi - F'(0)\psi$$

where $\dot{\psi} = 0$ at $x = 0, 1$.

The corresponding homogeneous equation is the equation for ψ and, at λ_{crit}, it has a solution other that $\psi = 0$. Hence $\dot{\psi}$, not zero, must exist and, therefore, the equation for $\dot{\psi}$ must be solvable. The solvability condition then determines $\dot{\sigma}$. The reader can multiply the $\dot{\psi}$ equation by ψ, the ψ equation by $\dot{\psi}$, subtract and integrate the difference over $0 \leq x \leq 1$, learning that $\dot{\sigma}$ is positive at $\lambda = \lambda_{crit}$. Thus we have $\sigma = 0$ and $\dfrac{d\sigma}{d\lambda} > 0$ at $\lambda = \lambda_{crit}$. We therefore anticipate seeing a nonzero solution branch for $\lambda > \lambda_{crit}$.

We will return to the Petri Dish problem later on and try to decide what the nonzero solution looks like for λ just beyond λ_{crit}. And we can do this by solving only linear equations.

1.8 Home Problems

1. A graph is a set of points connected pairwise by directed line segments. If there are n points and a line segment runs from point i to point j then the ij element in an $n \times n$ matrix is

set to 1 otherwise it is 0. The resulting matrix is called a connection matrix. What do the powers of a connection matrix tell us about the graph? The powers of a connection matrix are easy to determine because its columns are made up of zeros and ones. Each column of the product of a matrix A multiplied on the right by a connection matrix is simply the sum of certain columns of A.

2. Suppose the elements of the square matrix A themselves are square matrices. Then A is called a block matrix. Write A as LU where L is blockwise lower triangular and U is blockwise upper triangular, its diagonal blocks being I.

3. Determine T vs. t, V vs. t and x vs. t in the simple evaporator problem presented in Lecture 1.

4. In the nonlinear evaporator model, set $UA = 2000$, $T_S = 300$, $T_c = 40$, $\beta = 1000$, $c_p = 1$ and $\lambda = 1000$. Then if $F = 1000$, $x_F = \dfrac{1}{10}$ and $T_F = 100$ the steady solution corresponds to boiling: $T = 165$, $V = 200$, $x = \dfrac{1}{8}$. But if $F = 4000$ and $T_F = 50$ the steady solution corresponds simply to heating: $T = 133$, $V = 0$, $x = \dfrac{1}{10}$. Set $M = 1000$ and determine how the system makes the boiling-to-nonboiling transition if it is in the first steady state when $t < 0$ and then at $t = 0$ the above step changes in F and T_F are made. A simple Euler approximation will do the job. It is instructive to see how $T = T_c + \beta x$ is used to determine V at each step. Once V is zero the calculation can be continued without approximation.

Lecture 2

Independent and Dependent Sets of Vectors

The $m \times n$ matrix A is assigned. In this and the next two lectures we show how to determine whether or not a vector $\underline{x} \in C^n$ can be found so that the vector $A\underline{x} \in C^m$ equals an assigned vector $\underline{b} \in C^m$. Assuming the problem $A\underline{x} = \underline{b}$ has a solution, we then show how to write its general solution.

2.1 Linear Independence

The main idea is linear independence. This, or its opposite, linear dependence, is a property of a set of vectors. Indeed, starting with a set of vectors, $\{\underline{v}_1, \underline{v}_2, ..., \underline{v}_n\}$, we can create additional vectors by making linear combinations of the assigned vectors using arbitrary complex numbers, viz., $c_1\underline{v}_1 + c_2\underline{v}_2 + ... + c_n\underline{v}_n$. The set $\underline{v}_1, \underline{v}_2, ..., \underline{v}_n$ is said to be independent if the only way we can create the vector $\underline{0}$ is by setting $c_1, c_2, ..., c_n$ to zero. In other words the set of vectors $\underline{v}_1, \underline{v}_2, ..., \underline{v}_n$ is said to be independent iff

$c_1, c_2, ..., c_n$ all zero

is the only solution to the equation

$$c_1\underline{v}_1 + c_2\underline{v}_2 + ... + c_n\underline{v}_n = \underline{0}$$

otherwise it is said to be dependent.

The idea of linear dependence can be stated in terms of linear combinations in the following way: at least one vector of the set $\underline{v}_1, \underline{v}_2, ..., \underline{v}_n$ is a linear combination of the others if and only if the set is dependent, i.e., if and only if the equation $c_1\underline{v}_1 + c_2\underline{v}_2 + ... + c_n\underline{v}_n = \underline{0}$ is satisfied for $c_1, c_2, ..., c_n$ other than $c_1, c_2, ..., c_n$ all zero.

The idea of linear independence is not special to sets of column vectors and is defined as above for vectors in general, using the corresponding zero vector; indeed it pertains to sets of row vectors as well as to sets of column vectors.

2.2   Independent Chemical Reactions

To illustrate this idea suppose we have M molecules, $m = 1, 2, ..., M$, participating in R chemical reactions, $r = 1, 2, ..., R$. Let ν_{rm} denote the stoichiometric coefficient of molecule m in reaction r. Then we can identify the molecule m with the column $\underline{\nu}_m = \begin{pmatrix} \nu_{1m} \\ \nu_{2m} \\ \vdots \\ \nu_{Rm} \end{pmatrix}$ and the reaction r with the row $(\nu_{r1}\ \nu_{r2}\ ...\ \nu_{rM})$. Independent reactions then correspond to independent rows. A question of interest in chemical equilibrium calculations is this: for an assigned set of molecules how many independent reactions can we write? Writing our columns (or rows) as the columns (or rows) of a matrix ν, where $\nu = (\underline{\nu}_1\ \underline{\nu}_2\ ...\ \underline{\nu}_M)$, called the stoichiometric or reaction-molecule matrix, we must determine the greatest number of independent rows in this $R \times M$ matrix. Indeed we must determine whether the greatest number of independent rows increases indefinitely or is bounded as the number of rows increases indefinitely.

It turns out that matrices have a surprising property: the greatest number of independent rows cannot exceed the greatest number of independent columns (and vice versa); and this, willy nilly, cannot exceed the total number of columns. The number of molecules then is a bound on the greatest number of independent reactions that can be written.

If we take into account that each molecule is made up of atoms from the set $a = 1, 2, ..., A$ and denote by α_{ma} the number of atoms a in molecule m, then as each atom is conserved in each reaction we have for reaction r and atom a

$$\sum_{m=1}^{M} \nu_{rm}\alpha_{ma} = 0$$

for all r and all a. We can identify the atom a with the column $\underline{\alpha}_a = \begin{pmatrix} \alpha_{1a} \\ \alpha_{2a} \\ \vdots \\ \alpha_{Ma} \end{pmatrix}$ whence the

conservation conditions are $\nu\underline{\alpha}_a = \underline{0}$, $a = 1, 2, ..., A$ and these can be written

$$\underline{\nu}_1\alpha_{1a} + \underline{\nu}_2\alpha_{2a} + ... + \underline{\nu}_M\alpha_{Ma} = \underline{0}, \ a = 1, 2, ..., A$$

Each independent condition of this kind reduces by one the greatest number of independent columns in the set $\underline{\nu}_1, \underline{\nu}_2, ..., \underline{\nu}_M$. Assuming the atoms to be distributed independently over the molecules, i.e., assuming the set $\underline{\alpha}_1, \underline{\alpha}_2, ..., \underline{\alpha}_A$ to be independent , the greatest number of independent reactions that can be written using M molecules made up of A atoms is $M - A$, hence the requirement that balanced reactions be written reduces the greatest number of independent reactions from M to $M - A$. But even the bound M corresponding to arbitrary reactions is interesting and it has nothing to do with the requirement that the stoichiometric coefficients be integers. It holds assuming the ν_{rm} to be arbitrary complex numbers and cannot be lowered by the restriction to integers.

To every set of n column vectors in C^m there corresponds a set of m row vectors in C^n generated by writing the column vectors as the columns of an $m \times n$ matrix A. The row vectors are then the rows of A. If we rephrase the question as to the greatest number of independent vectors in the set $\underline{a}_1, \underline{a}_2, ..., \underline{a}_n$ to a corresponding question about the greatest number of independent columns in the matrix A we will get information not only about the columns of A but also about the rows of A as well. But the rows of A are the columns of A^T so answers to questions about n column vectors in C^m are also answers to questions about a corresponding set of m column vectors in C^n.

2.3 Looking at the Problem $A\underline{x}=\underline{b}$ from the Point of View of Linearly Independent Sets of Vectors

To see what linear independence has to do with our main problem of determining whether or not solutions of $A\underline{x} = \underline{b}$ exist and, if they do, writing them, we let $\underline{a}_1, \underline{a}_2, ..., \underline{a}_n$ denote the columns of A and write the problem $A\underline{x} = \underline{b}$ as

$$x_1\underline{a}_1 + x_2\underline{a}_2 + ... + x_n\underline{a}_n = \underline{b}$$

which requires $\underline{b}$ to be a linear combination of $\underline{a}_1, \underline{a}_2, ..., \underline{a}_n$. The corresponding homogeneous problem, $A\underline{x} = \underline{0}$, can be written

$$x_1\underline{a}_1 + x_2\underline{a}_2 + ... + x_n\underline{a}_n = \underline{0}$$

We see therefore that $A\underline{x} = \underline{0}$ has solutions other than $x_1, x_2, ..., x_n$ all zero iff the set of vectors $\underline{a}_1, \underline{a}_2, ..., \underline{a}_n$ is dependent, i.e., not independent, and that $A\underline{x} = \underline{b}$ has solutions iff on joining $\underline{b}$ to the set $\underline{a}_1, \underline{a}_2, ..., \underline{a}_n$ we do not increase the greatest number of independent columns.

Our first problem, therefore, is to determine the greatest number of independent vectors in a set of n vectors, $\underline{a}_1, \underline{a}_2, ..., \underline{a}_n$ in C^m. To do this we introduce the determinant of a square matrix. The determinant is a function defined on square matrices mapping each square matrix into a complex number.

Let $A = (a_{ij}) = (\underline{a}_1\ \underline{a}_2\ ...\ \underline{a}_n)$ be an $n \times n$ matrix, then the determinant of A, denoted det A, is defined by

$$\det A = \Sigma \pm a_{\alpha_1 1}a_{\alpha_2 2}...a_{\alpha_n n}$$

where we have chosen to write the column indices in their natural order and where the sum is over all sets of integers $\alpha_1, \alpha_2, ..., \alpha_n$ that are permutations of $1, 2, ..., n$, the + sign to be used if the permutation is even, the $-$ sign if it is odd. This then is a sum of $n!$ terms, each term being a product of n factors, where each row and each column is represented once in each term. This definition leads to the following four properties from which our conclusions can be drawn:

(i) $\det A = \det A^T$

(ii) $\det(\ldots \underline{a}_i \ldots \underline{a}_j \ldots) = -\det(\ldots \underline{a}_j \ldots \underline{a}_i \ldots)$

(iii) $\det(\ldots \sum c_k \underline{b}_k \ldots) = \sum c_k \det(\ldots \underline{b}_k \ldots)$

(iv) $\det\left(\ldots \underline{a}_i + \sum\limits_{k \neq i} c_k \underline{a}_k \ldots\right) = \det(\ldots \underline{a}_i \ldots)$

In (ii) two columns are interchanged; in (iii) a fixed column is written as a linear combination of arbitrary column vectors; in (iv) a linear combination of other columns is added to the ith column. All else is held fixed. The proofs of these properties and their corollaries, such as $\det A = 0$ if $\underline{a}_i = \underline{a}_j$ for any $i \neq j$, $\det A = 0$ if $\underline{a}_i = 0$ for any i, etc., come easily out of the definition and either can be supplied by the reader or can be found in Shilov's book "*Linear Algebra*."

Property (i) is important in turning column theorems into row theorems and vice-versa. Properties (ii), (iii) and (iv) are properties of columns and therefore properties of the columns of A^T. As the rows of A are the columns of A^T they are also properties of the rows of A.

If $\underline{a}_1, \underline{a}_2, ..., \underline{a}_n$ is a dependent set of vectors in C^n then $\det A = 0$. This is so as we can write one of a dependent set of vectors as a linear combination of the others and then use this combination in (iii) to show that $\det A$ is zero. We can restate this as: if $\det A \neq 0$ then $\{\underline{a}_1, \underline{a}_2, ..., \underline{a}_n\}$ is independent.

Each term in the expansion of $\det A$ contains one factor from the jth column. Of the $n!$ terms, $(n-1)!$ contain the common factor a_{1j}, $(n-1)!$ contain the common factor a_{2j}, etc. Writing the first of these n sets of $(n-1)!$ terms as $a_{1j}A_{1j}$, the second as $a_{2j}A_{2j}$, etc., we find that

$$\det A = a_{1j}A_{1j} + a_{2j}A_{2j}\ldots = \sum_{i=1}^{n} a_{ij}A_{ij}$$

and we notice that

$$\frac{\partial \det A}{\partial a_{ij}} = A_{ij}$$

This is the expansion of det A via its jth column and it can be written for each $j = 1, 2, ..., n$. Each factor $A_{ij}, i = 1, 2, ..., n$ is called the cofactor of the corresponding element $a_{ij}, i = 1, 2, ..., n$ and by definition their values do not depend on the values of the elements in the jth column. When this construction is carried out for each column, $j = 1, 2, ..., n$, it generates n^2 elements A_{ij}. Then letting

$$\underline{A}_1 = \begin{pmatrix} A_{11} \\ A_{21} \\ \vdots \\ A_{n1} \end{pmatrix}, \quad \underline{A}_2 = \begin{pmatrix} A_{12} \\ A_{22} \\ \vdots \\ A_{n2} \end{pmatrix}, \text{etc.}$$

where $\underline{A}_1$ is the column of cofactors of $\underline{a}_1$, the first column of A, etc, we can write

$$\det A = \underline{A}_j^T \underline{a}_j, \; j = 1, 2, \dots, n$$

The matrix whose columns are $\underline{A}_1, \underline{A}_2, ..., \underline{A}_n$ is called the matrix of the cofactors of A and its transpose, denoted adj A, where adj $A = (\underline{A}_1 \underline{A}_2 ... \underline{A}_n)^T$, is called the adjugate of A.

It turns out that $A_{ij} = (-1)^{i+j} M_{ij}$ where M_{ij}, a minor of A, is the determinant of the $(n-1) \times (n-1)$ submatrix of A obtained by deleting its ith row and j^{th} column. It is worth stating that, unless $n = 2$ or 3, it is not practical to evaluate determinants directly from the definition, requiring the evaluation of $n!$ terms, nor by the expansion in cofactors, requiring the evaluation of $(n-1)!$ terms n times, etc.

We now have two sets of columns $\{\underline{a}_1, \underline{a}_2, ..., \underline{a}_n\}$ and $\{\underline{A}_1, \underline{A}_2, ..., \underline{A}_n\}$, which satisfy

$$\underline{A}_j^T \underline{a}_k = \det A, \; k = j$$

$$\underline{A}_j^T \underline{a}_k = 0, \; k \neq j$$

The second formula is the expansion, via the jth column, of the determinant obtained by writing

$\underline{a}_k$ in place of $\underline{a}_j$ as the jth column of A and hence is zero. The multiplication on the left hand side is a column multiplied on the left by a row. The product is a scalar, indeed $\underline{A}_j^T \underline{a}_k = \underline{a}_k^T \underline{A}_j$.

2.4 Biorthogonal Sets of Vectors

Two sets of n vectors belonging to C^n satisfying the conditions

$$\underline{A}_i^T \underline{a}_j \neq 0, \ i = j$$

$$\underline{A}_i^T \underline{a}_j = 0, \ i \neq j$$

are called biorthogonal sets. This is a useful idea. It's usefulness stems from the observation that if we expand a vector in one of the sets, the coefficients in the expansion can be determined simply by operating on the expansion using vectors of the other set. Indeed to solve the problem $A\underline{x} = \underline{b}$ where $\underline{x}$ and $\underline{b}$ belong to C^n and $\det A \neq 0$, we rewrite the equation as

$$x_1 \underline{a}_1 + x_2 \underline{a}_2 + ... + x_n \underline{a}_n = \underline{b}$$

and multiply both sides by $\underline{A}_j^T$ to obtain

$$x_j \underline{A}_j^T \underline{a}_j = \underline{A}_j^T \underline{b}$$

whereupon

$$x_j = \frac{\underline{A}_j^T \underline{b}}{\det A}, \ j = 1, ..., n.$$

This is Cramer's rule and it can be written

$$x = \frac{(\underline{A}_1 \underline{A}_2 \cdots \underline{A}_n)^T}{\det A} \underline{b}$$

$$= \frac{\text{adj } A}{\det A} \underline{b} \equiv A^{-1} \underline{b}$$

Indeed the vectors $\underline{A}_1, \underline{A}_2, ..., \underline{A}_n$ enable us to select from expressions such as $x_1 \underline{a}_1 + x_2 \underline{a}_2 + ... + x_n \underline{a}_n$ any coefficient we wish to look at. This is familiar from analytic geometry where if $\vec{r} = x\vec{i} + y\vec{j} + z\vec{k}$ then $x = \vec{i} \cdot \vec{r}$ as $\left\{ \vec{i}, \vec{j}, \vec{k} \right\}$ is its own biorthogonal set. Cramer's rule produces the unique solution to $A\underline{x} = \underline{b}$ when A is square and $\det A \neq 0$.

Before going on we write our results in another way: the column formulas for the expansion of a determinant, viz.,

$$\underline{A}_j^T \underline{a}_j = \sum_{i=1}^{n} a_{ij} A_{ij} = \det A, \quad j = 1, ..., n$$

and

$$\underline{A}_j^T \underline{a}_k = \sum_{i=1}^{n} a_{ik} A_{ij} = 0, \ j = 1, ..., n \ , \quad k = 1, ..., n, \ k \neq j$$

can be written

$$(\text{adj } A)A = (\det A)I$$

where $I = (\delta_{ij})$.

Now there are row formulas for the expansion of a determinant which can be obtained either by going back to the definition and factoring an element in the ith row out of each term or by writing the column formulas using A^T in place of A. The row formulas

$$\det A = \sum_{j=1}^{n} a_{ij} A_{ij}, \ i = 1, ..., n$$

and their corollaries

$$0 = \sum_{j=1}^{n} a_{kj} A_{ij}, i = 1, ..., n, \ k = 1, ..., n, \ k \neq i$$

obtained by expanding, via the ith row, the determinant obtained by replacing the ith row of A by its kth row, can be written

$$A\,(\,\mathrm{adj}\,A) = (\det A)\,I$$

The readers may wish to satisfy themselves that a_{ij} is multiplied by one and the same coefficient, denoted A_{ij} , whether it appears in a row or column expansion. If $\det A = 0$, we see that $A(\,\mathrm{adj}\,A) = 0$ and hence that each column of $\mathrm{adj}\,A$ is a solution of $A\underline{x} = \underline{0}$.

2.5   The Number of Linearly Independent Vectors in a Set of Vectors and the Rank of a Matrix

The determinant, defined only on square matrices, can be used to determine the greatest number of independent columns in an $m \times n$ matrix $A = (\underline{a}_1\underline{a}_2...\underline{a}_n)$, where $\underline{a}_j \in C^m, j = 1, ..., n$, and where m need not be equal to n. To do this we introduce square submatrices of A of order k by deleting all but k rows and all but k columns and then we calculate the determinants, called minors of order k, of all these submatrices. Using this information we define the rank of A to be the order of the largest non-vanishing minor of A and denote it by r; we call the set of r columns of A running through that minor a set of basis columns. The rank is unique but, as more than one minor of order r may be non-vanishing, a set of basis columns need not be unique, however each such set is made up of r columns, and each such set is independent. To see this let $\underline{a}_1, \underline{a}_2, ..., \underline{a}_r$ be a set of r basis columns. (In the problem $A\underline{x} = \underline{b}$ interchanging columns of A and interchanging corresponding elements of $\underline{x}$ leave the problem unchanged.) Then to see if the set is independent we investigate the solutions of

$$c_1\underline{a}_1 + c_2\underline{a}_2 + ... + c_r\underline{a}_r = \underline{0}$$

Looking at the r equations corresponding to the basis minor we see by Cramer's rule that their only solution is $c_1, c_2, ..., c_r$ all zero. This then is the only solution to the full set of m equations.

What we have established is this: in an $m \times n$ matrix there is at least one set of r columns, the basis columns, that is independent, and r, the rank of the matrix, cannot exceed the smaller of m or n. To go on we require a result telling us how the columns not in a set of basis columns depend on the basis columns. Indeed what we need is the basis minor theorem. As stated and proved in Shilov's book *"Linear Algebra,"* on p. 25, this theorem tells us that any column of a matrix can be written as a linear combination of any set of basis columns. This most important result in linear algebra is surprisingly easy to prove. As a set of basis columns is independent each such expansion must be unique. The basis minor theorem tells us directly that any set of $r + 1$ or more columns of which r are basis columns is dependent. Indeed any set of $r + 1$ or more columns in a matrix of rank r must be dependent. The argument for this can be found early in the next lecture. Hence the greatest number of independent columns in an $m \times n$ matrix of rank r is r. Of course any subset of an independent set of vectors is also independent.

Because of property (i) we see that the ranks of A and A^T coincide, every square submatrix of A being the transpose of a square submatrix of A^T. The foregoing argument in terms of A^T then shows that r is also the greatest number of independent rows in A. Indeed if $m > n$ the greatest number of independent rows cannot exceed n whatever values are assigned to the a_{ij}.

We also see that if A is square, i.e., $n \times n$, and $\det A = 0$ then its rank is at most $n - 1$ whence the columns of A must be dependent. This is the converse of the earlier result that if $\det A \neq 0$ its columns are independent.

To wind this lecture down we develop some useful results having to do with the determinant.

2.6 Derivatives of Determinants

Let the elements a_{ij} of a matrix A be functions of t. Then $\det A$ is a function of t and its derivative is

$$\frac{d}{dt} \det A = \sum_i \sum_j \frac{\partial \det A}{\partial a_{ij}} \frac{da_{ij}}{dt}$$

Using

$$\frac{\partial \det A}{\partial a_{ij}} = A_{ij}$$

we get the important formula

$$\frac{d}{dt} \det A = \sum_i \sum_j \frac{da_{ij}}{dt} A_{ij}$$

This can be written as the sum of n determinants in two ways, using column or row expansions.

Now let $\underline{x}_1(t), \underline{x}_2(t), ..., \underline{x}_n(t)$ be a set of n solutions of

$$\frac{d\underline{x}}{dt} = A(t)\underline{x}$$

and let $W = \det\left(\underline{x}_1(t)\ \underline{x}_2(t) \ldots \underline{x}_n(t)\right) = \det\left(x_{ij}(t)\right)$. Then, as above,

$$\frac{dW}{dt} = \sum_i \sum_j \frac{dx_{ij}}{dt} X_{ij}$$

But this is

$$\frac{dW}{dt} = \sum_i \sum_j \sum_k a_{ik} x_{kj} X_{ij}$$

and hence, using $\sum_j x_{kj} X_{ij} = W\delta_{ki}$, we find

$$\frac{dW}{dt} = \operatorname{tr} A(t) W$$

where $\operatorname{tr} A$, called the trace of A, is $a_{11} + a_{22} + \cdots + a_{nn}$. As

$$W(t) = W(t_0)\, e^{\int_{t_0}^t \operatorname{tr} A(t)\, dt}$$

we conclude that $W(t)$ is either always zero or never zero. This is an important result in the theory of differential equations. It is required in Lecture 19. The determinant $W(t)$ is called the Wronskian of the solutions.

Derivatives of Determinants: The Trace Formula

If $A = (\underline{a}_1 \underline{a}_2 ... \underline{a}_n)$ and $B = (\underline{b}_1 \underline{b}_2 ... \underline{b}_n)$ then the ij element of $A^T B$ is $\underline{a}_i^T \underline{b}_j$ whence $\mathrm{tr}\,(A^T B) = \sum_{i=1}^n \underline{a}_i^T \underline{b}_i$. Using this, the formula for the derivative of a determinant, viz.,

$$\frac{d}{dt} \det A = \sum_j \sum_i \frac{da_{ij}}{dt} A_{ij} = \sum_j \underline{A}_j^T \frac{d\underline{a}_j}{dt}$$

can be written

$$\frac{d}{dt} \det A = \mathrm{tr}\left((\mathrm{adj}\, A) \frac{dA}{dt} \right)$$

2.7   Work for the Reader

We give here some simple results which the readers can verify and some not so simple results.

A square matrix is called diagonal if its elements off the main diagonal vanish; it is called upper or lower triangular if its elements below or above the main diagonal vanish. The determinant of each such matrix is the product of its diagonal elements.

The determinant of a product of square matrices is the product of the determinants of the factors. This is a particular instance of a result by which a minor of a product can be expressed as a sum of products of minors of the factors, see p. 91 of Shilov's "*Linear Algebra.*"

The reader can discover that $(AB)^T = B^T A^T$ and then that $\mathrm{rank}\, AB \leq \mathrm{rank}\, A$ as each column of AB is a linear combination of the columns of A. As $\mathrm{rank}\, AB = \mathrm{rank}\,(AB)^T \leq \mathrm{rank}\, B^T = \mathrm{rank}\, B$ we also see that $\mathrm{rank}\, AB \leq \mathrm{rank}\, B$.

If $AA^{-1} = I$ and $BB^{-1} = I$ then $ABB^{-1}A^{-1} = I$ and so $(AB)^{-1} = B^{-1}A^{-1}$. This result assumes A and B are square. If AB is square but A and B are not, more work is required.

Ordinarily we can write a square matrix A as a product LU where L is lower triangular and U is upper triangular having 1's on its main diagonal. This is easy to do column by column: on writing

$$(\underline{a}_1\ \underline{a}_2\ \cdots\ \underline{a}_n) = (\underline{\ell}_1\ \underline{\ell}_2\ \cdots\ \underline{\ell}_n) \begin{pmatrix} 1 & u_{12} & \cdots \\ 0 & 1 & \cdots \\ \vdots & \vdots & \cdots \\ 0 & 0 & \cdots \end{pmatrix}$$

we can derive the columns of L and U recursively via

$$\underline{a}_1 = \underline{\ell}_1$$

$$\underline{a}_2 = u_{12}\underline{\ell}_1 + \underline{\ell}_2$$

$$\underline{a}_3 = u_{13}\underline{\ell}_1 + u_{23}\underline{\ell}_2 + \underline{\ell}_3$$

$$etc.$$

Thus, because $\ell_{12} = 0, u_{12}$ can be determined to be a_{12}/a_{11}, etc. What appears on the diagonal of L is $a_{11}, \dfrac{a_{11}a_{22} - a_{12}a_{21}}{a_{11}}, \ldots$, i.e., the ratios of the determinants of the upper left hand submatrices of A. If one of these is zero the calculation, as indicated above, cannot go on. This may happen whether or not $\det A = 0$. For instance $\begin{pmatrix} 0 & 1 \\ 1 & 1 \end{pmatrix}$ cannot be so expanded but the problem disappears if the columns are interchanged.

The readers can show that if A is tridiagonal then L and U are bidiagonal. The readers can also satisfy themselves that the recipe for the determination of L and U can be improved by calculating the columns of L and the rows of U in the following sequence: first column of L, first row of U, second column of L, second column of U, etc. Indeed if A is partitioned into blocks, the decomposition can be carried out blockwise. In doing this the blocks of A must satisfy certain minimum conditions, e.g., the diagonal blocks must be square. The equation $LU\underline{x} = \underline{b}$ is easy to solve.

2.8 Looking Ahead

Again we denote by $\underline{a}_1, \underline{a}_2, \ldots, \underline{a}_n$ the n columns of an $m \times n$ matrix A and we say that the set $\{\underline{a}_1, \underline{a}_2, \ldots, \underline{a}_n\}$ is independent iff $c_1\underline{a}_1 + c_2\underline{a}_2 + \cdots + c_n\underline{a}_n = \underline{0}$ has only the solutions: all c's $= 0$.

The reader should then believe that if $\{\underline{a}_1, \underline{a}_2, \ldots, \underline{a}_n\}$ is independent, the equation $A\underline{x} = \underline{0}$ has only the solution $\underline{x} = \underline{0}$ and if $\{\underline{a}_1, \underline{a}_2, \ldots, \underline{a}_n, \underline{b}\}$ is independent, the equation $A\underline{x} = \underline{b}$ has no solution.

Suppose the $R \times M$ matrix ν, having rank r, is our reaction-molecule matrix and the $M \times A$ matrix $\alpha = (\underline{\alpha}_1, \ldots, \underline{\alpha}_A)$ is our molecule-atom matrix. We will see in Lecture 3 that the equation $\nu\underline{x} = \underline{0}$ has $M - r$ independent solutions. Thus if

$$\nu\underline{a}_\alpha = \underline{0}, \quad \alpha = 1, \ldots, A$$

accounts for A of these, we must have $M - r \geq A$ and thus

$$r \leq M - A$$

where r is the greatest number of independent reactions.

2.9 Home Problems

1. Denote by $D(x_1, x_2, \ldots, x_n)$ the determinant of the Vandermonde matrix

$$\begin{pmatrix} 1 & x_1 & x_1^2 & \cdots & x_1^{n-1} \\ 1 & x_2 & x_2^2 & \cdots & x_2^{n-1} \\ \vdots & \vdots & \vdots & & \vdots \\ 1 & x_n & x_n^2 & \cdots & x_n^{n-1} \end{pmatrix}$$

Then derive the formula

$$D(x_1, x_2, \ldots, x_n) = \pi(x_i - x_j) \quad 1 \leq j < i \leq n$$

This is a problem in Shilov's book.

To do this expand D by its last row and observe that it is a polynomial of degree $n - 1$ in x_n whose coefficients depend on $x_1, x_2, \ldots, x_{n-1}$. The $n - 1$ zeros of this polynomial are $x_1, x_2, \ldots, x_{n-1}$, and it can be written

$$D(x_1, x_2, \ldots, x_n) = c \prod_{i=1}^{n-1} (x_n - x_i)$$

where c is the coefficient of x_n^{n-1} and is therefore $D(x_1, x_2, \ldots, x_{n-1})$.

2. A determinant of order k can be written as a linear combination of its first minors, determinants of order $k - 1$. Let A be an $m \times n$ matrix. If all its minors of order k vanish then so too all its minors of order $k + 1, k + 2$, etc. Prove this.

3. Suppose the elements of an $n \times n$ matrix A are polynomials in a scalar λ. Then $\det A$ is also a polynomial in λ. Denote by the term first minors all the minors of order $n - 1$, by the term second minors all the minors of order $n - 2$, etc. Then second minors are first minors of first minors, etc.

Using the formula for the derivative of a determinant and the formula for the expansion of a determinant in terms of its first minors show that: $\dfrac{d}{d\lambda} \det A$ is a linear homogeneous function of the first minors of A, $\dfrac{d^2}{d\lambda^2} \det A$ is a linear homogeneous function of the second minors of A, etc.

4. Because you see 6 molecules and 3 atoms in the set of 5 reactions listed below, you believe that at most 3 of the reactions are independent. Calculate the rank of the reaction-molecule matrix.

$$C + H_2O \longrightarrow CO + H_2$$

$$C + 2H_2O \longrightarrow CO_2 + 2H_2$$

$$C + 2H_2 \longrightarrow CH_4$$

$$C + CO_2 \longrightarrow 2CO$$

$$CO + H_2O \longrightarrow CO_2 + H_2$$

5. Suppose $m = 3 = n$ and expand $J = \det(I + \varepsilon A)$ in powers of ε

Answer:

$$J = 1 + \varepsilon \operatorname{tr} A + \varepsilon^2 \begin{pmatrix} +a_{11}a_{33} - a_{13}a_{31} \\ +a_{11}a_{22} - a_{12}a_{21} \\ +a_{22}a_{33} - a_{23}a_{32} \end{pmatrix} + \varepsilon^3 \det A$$

6. Given R reactions among M molecules composed of A atoms, multiply the $R \times M$ reaction-molecule matrix by the corresponding $M \times A$ molecule-atom matrix obtaining an $R \times A$ matrix all of whose elements are?

An example is:

$$C + O_2 \longrightarrow CO_2 \qquad R = 3, M = 4, A = 2$$

$$C + \frac{1}{2} O_2 \longrightarrow CO$$

$$CO + \frac{1}{2} O_2 \longrightarrow CO_2$$

7. A scalar valued function of an $n \times n$ matrix A, say $f(A)$, is said to be a scalar invariant of A iff

$$f\left(QAQ^T\right) = f(A)$$

for all Q's such that $Q^{-1} = Q^T$.

Expanding $\det(\lambda I + A)$ as

$$\lambda^n + \lambda^{n-1}I_1(A) + \cdots + \lambda I_{n-1}(A) + I_n(A)$$

show that $I_1(A), \ldots, I_n(A)$ are scalar invariants of A. They are called the principal invariants of A.

Define the gradient of a scalar function of A, denoted $f_A(A)$, having components $\dfrac{\partial f}{\partial A_{ij}}$, by

$$\frac{d}{ds}f(A+sC)\bigg|_{s=0} = \frac{\partial f}{\partial A_{ij}}C_{ij} = \operatorname{tr}\left(f_A(A)C^T\right)$$

for any C.

Derive

$$\det{}_A(A) = \det A \left(A^{-1}\right)^T$$

and

$$\operatorname{tr}{}_A(A) = I$$

Lecture 3

Vector Spaces

3.1 Vector Spaces

A set of vectors on which rules for addition and scalar multiplication are defined is called a vector space if the sum of any two vectors in the set is in the set and the product of any scalar and any vector in the set is in the set. The set of columns of m complex numbers, denoted C^m, is a vector space.

Let the vectors $\underline{v}_1, \ldots, \underline{v}_n$ belong to C^m. Then the set of all linear combinations of $\underline{v}_1, \ldots, \underline{v}_n$, i.e., the set of all vectors $c_1\underline{v}_1 + c_2\underline{v}_2 + \ldots + c_n\underline{v}_n$ corresponding to all ways of choosing $c_1, \ldots, c_n$ is a vector space. It is a subspace of C^m; it is called the manifold spanned by $\underline{v}_1, \ldots, \underline{v}_n$ and it is denoted $[\underline{v}_1, \underline{v}_2, \ldots, \underline{v}_n]$. A set of independent vectors in a vector space that spans the space is called a basis for the space. Each vector in the space has a unique expansion in a set of basis vectors and the coefficients in the expansion are called the components of the vector in the basis.

If there are n vectors in a basis for a space then every set of $n + 1$ vectors in the space is dependent. Indeed if $\underline{v}_1, \ldots, \underline{v}_n$ is a basis then $\underline{u}_1, \ldots, \underline{u}_{n+1}$ must be dependent. To see this let

$$\underline{u}_j = \sum_{i=1}^{n} \xi_{ij}\underline{v}_i, \qquad j = 1, \ldots, n+1$$

47

and observe that the equation

$$c_1 \underline{u}_1 + c_2 \underline{u}_2 + \cdots + c_{n+1} \underline{u}_{n+1} = \underline{0}$$

can be written

$$\sum_{i=1}^{n} \underline{v}_i \sum_{j=1}^{n+1} \xi_{ij} \, c_j = \underline{0}$$

whence we have

$$\sum_{j=1}^{n+1} \xi_{ij} \, c_j = 0, \qquad i = 1, \ldots, n$$

and denoting $\begin{pmatrix} \xi_{1j} \\ \xi_{2j} \\ \vdots \\ \xi_{nj} \end{pmatrix}$ by $\underline{\xi}_j \in C^n, j = 1, 2, \ldots, n+1$, this is

$$c_1 \underline{\xi}_1 + c_2 \underline{\xi}_2 + \cdots + c_{n+1} \underline{\xi}_{n+1} = \underline{0}$$

As the rank of the $n \times (n+1)$ matrix $\left(\underline{\xi}_1 \underline{\xi}_2 \cdots \underline{\xi}_{n+1} \right)$ cannot exceed n, the set of columns $\underline{\xi}_1, \underline{\xi}_2, \cdots, \underline{\xi}_{n+1}$ must be dependent and so too therefore the set of vectors $\underline{u}_1, \underline{u}_2, \cdots, \underline{u}_{n+1}$. This tells us: if there are n vectors in some basis for a space then every basis for the space is made up of n vectors. We go on and define the dimension of the space to be n. It follows directly that any set of n independent vectors in a space of dimension n is a basis for the space.

3.2   The Image and the Kernel of a Matrix: The Geometric Meaning of Its Rank

There are two subspaces associated to an $m \times n$ matrix A that are important to us, one a subspace of C^m the other of C^n. Both depend for their identification on the basis columns of A. We denote by r the rank of A, and assume the columns $\underline{a}_1, \underline{a}_2, \cdots, \underline{a}_r$ to be a set of basis columns. Then we denote the set of vectors $A\underline{x} \in C^m$, for all $\underline{x} \in C^n$, by Im A and call it the image of A. Because $c_1 A\underline{x}_1 + c_2 A\underline{x}_2 = A(c_1\underline{x}_1 + c_2\underline{x}_2)$, Im A is a vector space and hence a subspace of C^m. Because $A\underline{x} = x_1\underline{a}_1 + x_2\underline{a}_2 + \cdots + x_n\underline{a}_n$, we can write Im $A = [\underline{a}_1, \underline{a}_2, \ldots, \underline{a}_n]$ and hence by the basis minor theorem Im $A = [\underline{a}_1, \underline{a}_2, \ldots, \underline{a}_r]$. As $\{\underline{a}_1, \underline{a}_2, \ldots, \underline{a}_r\}$ is independent, it is a basis for Im A. Hence the dimension of Im A is r and the rank of a matrix, an algebraic quantity, turns out to have a geometric interpretation as the dimension of its image.

This leads directly to results such as rank $AB \leq$ rank A inasmuch as Im AB cannot lie outside Im A.

The geometric interpretation of the rank of A leads to a practical way of determining its value. Indeed we do not change the rank of A by carrying out operations on the columns of A that do not change the dimension of Im A. The following operations satisfy this requirement and are therefore rank preserving:

1. interchange two columns

2. multiply a column by a non-zero number

3. add a multiple of a column to another column

These rank preserving column operations can be used to produce from A a matrix whose rank, i.e., number of independent columns, can be established by inspection. The idea is to create zeros in the first row in columns $2, \ldots, n$, then in the second row in columns $3, \ldots, n$, then etc.

If the problem $A\underline{x} = \underline{b}$ has a solution then $\underline{b} = x_1\underline{a}_1 + x_2\underline{a}_2 + \cdots + x_n\underline{a}_n$ for certain values of $x_1, x_2, \ldots, x_n$ and $\underline{b} \in$ Im A, if it does not, then $\underline{b} \neq x_1\underline{a}_1 + x_2\underline{a}_2 + \cdots + x_n\underline{a}_n$ for any values of $x_1, x_2, \ldots, x_n$. and $\underline{b} \notin$ Im A. The solvability condition then is this: $A\underline{x} = \underline{b}$ is solvable iff $\underline{b} \in$ Im A; in terms of ranks this is: $A\underline{x} = \underline{b}$ is solvable iff rank $(\underline{a}_1\underline{a}_2 \ldots \underline{a}_n\underline{b}) =$ rank $(\underline{a}_1\underline{a}_2 \ldots \underline{a}_n)$. Indeed if rank $(\underline{a}_1\underline{a}_2 \ldots \underline{a}_n\underline{b}) =$ rank A then $\{\underline{a}_1, \underline{a}_2, \ldots, \underline{a}_r\}$ is a set of basis columns for $(\underline{a}_1 \underline{a}_2 \ldots \underline{a}_n\underline{b})$ and by the basis minor theorem $A\underline{x} = \underline{b}$ is solvable. Conversely if $A\underline{x} = \underline{b}$ is solvable then $\underline{b}$ is a linear combination of $\underline{a}_1, \underline{a}_2, \ldots, \underline{a}_n$ and hence the manifolds $[\underline{a}_1, \underline{a}_2, \ldots, \underline{a}_n, \underline{b}]$ and $[\underline{a}_1, \underline{a}_2, \ldots, \underline{a}_n]$ are identical and so therefore are their dimensions. The practical evaluation of rank makes the rank test for solvability practical as well.

The corresponding homogeneous problem $A\underline{x} = \underline{0}$ is always solvable; for there to be solutions other than $\underline{x} = \underline{0}$, the rank of A must be less than the number of its columns for then by the basis minor theorem each of $n-r$ columns can be written as a linear combination of the r basis columns.

We denote the set of vectors $\underline{x}$ in C^n satisfying $A\underline{x} = \underline{0}$ by $\operatorname{Ker} A$, called the kernel of A. As $A\left(c_1\underline{x}_1 + c_2\underline{x}_2\right) = \underline{0}$ if $A\underline{x}_1 = \underline{0} = A\underline{x}_2$, $\operatorname{Ker} A$ is a vector space and hence a subspace of C^n. It is the solution space for $A\underline{x} = \underline{0}$. To identify it and determine its dimension we need to find a basis, i.e., a set of independent solutions of $A\underline{x} = \underline{0}$ which span all solutions of $A\underline{x} = \underline{0}$. To do this we write $A\underline{x} = x_1\underline{a}_1 + x_2\underline{a}_2 + \cdots + x_n\underline{a}_n = \underline{0}$ and observe that $\underline{a}_{r+1}, \underline{a}_{r+2}, \cdots, \underline{a}_n$ belong to $\operatorname{Im} A$ for which $\{\underline{a}_1, \underline{a}_2, \cdots, \underline{a}_r\}$ is a basis. Hence each vector $\underline{a}_{r+1}, \underline{a}_{r+2}, \cdots, \underline{a}_n$ has a unique expansion in terms of $\underline{a}_1, \underline{a}_2, \cdots, \underline{a}_r$ and writing these expansions

$$c_1\underline{a}_1 + c_2\underline{a}_2 + \cdots + c_r\underline{a}_r + \underline{a}_{r+1} = \underline{0}$$

$$d_1\underline{a}_1 + d_2\underline{a}_2 + \cdots + d_r\underline{a}_r + \underline{a}_{r+2} = \underline{0}$$

etc.

we read off $n - r$ independent solutions of $A\underline{x} = \underline{0}$ as

$$
\begin{pmatrix} c_1 \\ \vdots \\ c_r \\ 1 \\ 0 \\ \vdots \\ 0 \end{pmatrix}, \quad
\begin{pmatrix} d_1 \\ \vdots \\ d_r \\ 0 \\ 1 \\ \vdots \\ 0 \end{pmatrix}, \quad \text{etc.}
$$

We denote these fundamental solutions of $A\underline{x} = \underline{0}$ by $\underline{x}_1, \underline{x}_2, \ldots, \underline{x}_{n-r}$. The set of vectors $\underline{x}_1, \underline{x}_2, \ldots, \underline{x}_{n-r}$ is independent by inspection. And if $\underline{x} = \begin{pmatrix} \xi_1 \\ \xi_2 \\ \vdots \\ \xi_n \end{pmatrix}$ is any solution of

$A\underline{x} = \underline{0}$, then $\xi_1\underline{a}_1 + \xi_2\underline{a}_2 + \cdots + \xi_n\underline{a}_n = \underline{0}$ and this implies, using the expansions of

$\underline{a}_{r+1}, \underline{a}_{r+2}, \ldots$ in terms of $\underline{a}_1, \underline{a}_2, \ldots, \underline{a}_r$ and the fact that $\{\underline{a}_1, \underline{a}_2, \ldots, \underline{a}_r\}$ is independent, that

$$\begin{pmatrix} \xi_1 \\ \xi_2 \\ \vdots \\ \xi_r \end{pmatrix} - \xi_{r+1} \begin{pmatrix} c_1 \\ c_2 \\ \vdots \\ c_r \end{pmatrix} - \xi_{r+2} \begin{pmatrix} d_1 \\ d_2 \\ \vdots \\ d_r \end{pmatrix} - \cdots = \begin{pmatrix} 0 \\ 0 \\ \vdots \\ 0 \end{pmatrix}$$

and hence that

$$\underline{x} - \xi_{r+1} \, \underline{x}_1 - \xi_{r+2} \, \underline{x}_2 - \cdots = \underline{0}$$

This tells us that $\{\underline{x}_1, \underline{x}_2, \ldots, \underline{x}_{n-r}\}$ is a basis for $\operatorname{Ker} A$, which we can now write as $[\underline{x}_1, \underline{x}_2, \ldots, \underline{x}_{n-r}]$, and that the dimension of $\operatorname{Ker} A$, the solution space of $A\underline{x} = \underline{0}$, is $n - r$, the number of columns of A less its rank.

As simple examples the readers may satisfy themselves that the homogeneous equation

$$\underline{f}^T \underline{x} = \underline{0}$$

has $n - 1$ independent solutions if $\underline{f} \neq \underline{0}$, whereas the set of two homogeneous equations

$$\underline{f}_1^T \underline{x} = 0$$

and

$$\underline{f}_2^T \underline{x} = 0,$$

rewritten as

$$\begin{pmatrix} \underline{f}_1^T \\ \underline{f}_2^T \end{pmatrix} \underline{x} = \begin{pmatrix} 0 \\ 0 \end{pmatrix},$$

has $n - 2$ independent solutions if $\underline{f}_1$ and $\underline{f}_2$ are independent, etc. We will see that the solutions to the first equation may be interpreted as the set of vectors perpendicular to $\underline{f}$, the solutions to the second as the set of vectors perpendicular to $\underline{f}_1$ and $\underline{f}_2$, etc.

3.3   The Facts about the Solutions to the Problem $A\underline{x}=\underline{b}$

We can sum up the facts about the problem $A\underline{x} = \underline{b}$. Even though all the foregoing is phrased in terms of matrix operators and column vectors the conclusions hold as well for all other linear operator problems so we will state them generally:

The problem $A\vec{x} = \vec{b}$ has a solution iff $\vec{b} \in$ Im A. If $\vec{b} \in$ Im A and $\vec{x}_0$ satisfies $A\vec{x} = \vec{b}$ then so also does $\vec{x}_0 + \vec{y}$ for any $\vec{y} \in$ Ker A. And all solutions can be so written, for if $A\vec{x}_1 = \vec{b}$ then $\vec{y}_1 = \vec{x}_1 - \vec{x}_0 \in$ Ker A and $\vec{x}_1 = \vec{x}_0 + \vec{y}_1$. If $\vec{x} = \vec{0}$ is the only solution of $A\vec{x} = \vec{0}$, then if $\vec{b} \in$ Im A, the solution to $A\vec{x} = \vec{b}$ is unique.

For a matrix of n columns and rank r the general solution of $A\underline{x} = \underline{b}$ depends on $n-r$ constants and is $\underline{x}_0 + c_1\underline{x}_1 + c_2\underline{x}_2 + \cdots + c_{n-r}\underline{x}_{n-r}$ where $\underline{x}_0$ is any particular solution and $\underline{x}_1, \underline{x}_2, \cdots, \underline{x}_{n-r}$ is a fundamental system of solutions of $A\underline{x} = \underline{0}$. In this the set of fundamental solutions, viz., $\{\underline{x}_1, \underline{x}_2, \cdots, \underline{x}_{n-r}\}$, may be replaced by any basis for Ker A.

Working out the following problem (this problem is on page 71 in Shilov's "Linear Algebra") will help the reader get all this straightened out. The problem is to determine the solution to a system of four equations in five unknowns:

$$x_1 + x_2 + x_3 + x_4 + x_5 = 7$$

$$3x_1 + 2x_2 + x_3 + x_4 - 3x_5 = -2$$

$$x_2 + 2x_3 + 2x_4 + 6x_5 = 23$$

$$5x_1 + 4x_2 + 3x_3 + 3x_4 - x_5 = 12$$

The rank of the 4×5 coefficient matrix A is 2. So Im A is a 2 dimensional subspace of R^4 while Ker A is a 3 dimensional subspace of R^5. Because the solvability condition is satisfied there are two independent equations. It helps therefore to drop two dependent equations, for they must be satisfied if the remaining two independent equations are satisfied. Then taking the basis minor to be in the upper left hand corner and transposing dependent columns to the right hand side, we write

$$\begin{pmatrix} 1 \\ 3 \end{pmatrix} x_1 + \begin{pmatrix} 1 \\ 2 \end{pmatrix} x_2 = \begin{pmatrix} 7 \\ -2 \end{pmatrix} - x_3 \begin{pmatrix} 1 \\ 1 \end{pmatrix} - x_4 \begin{pmatrix} 1 \\ 1 \end{pmatrix} + x_5 \begin{pmatrix} 1 \\ -3 \end{pmatrix}$$

From here a particular solution and the fundamental system of solutions to the homogeneous equations can be obtained easily. A particular solution is obtained by setting x_3, x_4 and x_5 to zero and then the fundamental system is obtained by dropping $\begin{pmatrix} 7 \\ -2 \end{pmatrix}$ and first setting $x_3 = 1$, $x_4 = 0, x_5 = 0$, then setting $x_3 = 0, x_4 = 1, x_5 = 0$, etc.

3.4   Systems of Differential Equations

To illustrate that what we have been doing has more significance than what its face value would suggest define p_{ij}, a polynomial differential operator, by

$$p_{ij} = a_{ij} + b_{ij} \frac{d}{dt} + c_{ij} \frac{d^2}{dt^2} + \cdots$$

and suppose that we have n differential equations which determine n functions $u_1(t), u_2(t), \ldots, u_n(t)$, via

$$\begin{pmatrix} p_{11} & p_{12} & \cdots & p_{1n} \\ p_{21} & p_{22} & \cdots & p_{2n} \\ \vdots & \vdots & & \vdots \\ p_{n1} & p_{n2} & \cdots & p_{nn} \end{pmatrix} \begin{pmatrix} u_1 \\ u_2 \\ \vdots \\ u_n \end{pmatrix} = \begin{pmatrix} b_1(t) \\ b_2(t) \\ \vdots \\ b_n(t) \end{pmatrix}$$

or $P\underline{u} = \underline{b}$. Then $D = \det P$ is a polynomial differential operator, as is P_{ij}, the cofactor of p_{ij} in the matrix P.

Now the differential operators p_{ij} can be added and multiplied as if they were complex numbers and so we can write

$$\sum_i p_{ij} P_{ij} = D, \quad j = 1, 2, \ldots, n$$

and

$$\sum_i p_{ik} P_{ij} = 0, \quad k, j = 1, 2, \ldots, n, \ k \neq j$$

or in better notation

$$\underline{P}_j^T \underline{p}_k = D\delta_{jk}$$

Then writing $P\underline{u} = \underline{b}$ as

$$\underline{p}_1 u_1 + \underline{p}_2 u_2 + \cdots + \underline{p}_n u_n = \underline{b}$$

we can use the selection properties of $\underline{P}_1, \underline{P}_2, \ldots$ to write

$$Du_1 = \underline{P}_1^T \underline{b}$$

$$Du_2 = \underline{P}_2^T \underline{b}$$

etc.

The result is that we have turned n differential equations in n unknowns, of order equal to the highest order among the p_{ij}, into one differential equation in one unknown of order equal to the order of D. Only the right hand sides of the equations determining $u_1, u_2, \ldots$ differ.

3.5 Example

Let y satisfy

$$\frac{d^2y}{dx^2} + \lambda y = f(x)$$

then, of course, there is a solution for any f.

Add to this

$$y = 0 \quad \text{at} \quad x = 0, 1$$

and ask again: is there a solution?

If λ is not $\pi^2, 4\pi^2, 9\pi^2$, etc., the homogeneous problem

$$\frac{d^2y}{dx^2} + \lambda y = 0, \quad y = 0 \quad \text{at} \quad x = 0, 1$$

has only the solution $y = 0$, whereupon our problem has a solution for all f's and it is unique.

If λ is one of the values $\pi^2, 4\pi^2, 9\pi^2$, etc., then the homogeneous problem has the solution

$$y = \sin \sqrt{\lambda} x$$

and our problem may or may not have a solution.

If we introduce a difference approximation to $\dfrac{d^2y}{dx^2}$, we can use the results of this lecture. Otherwise, we need a new idea.

3.6 Home Problems

1. We have S atomic or molecular species, $s = 1, 2, \ldots, S$ participating in E elementary reactions, $e = 1, 2, \ldots, E$ and we denote the stoichiometric coefficients in each elementary reaction by

$$\underline{\nu}_e = \begin{pmatrix} \nu_{1e} \\ \nu_{2e} \\ \vdots \\ \nu_{Se} \end{pmatrix}$$

and its rate by $\dot{\xi}_e$. Here $\underline{\nu}_e$ specifies a reaction not a species and ν_{se} is a net stoichiometric coefficient as a species may be written more than one time in an elementary reaction.

If the rank of the $S \times E$ matrix $\nu = (\underline{\nu}_1 \, \underline{\nu}_2 \, \ldots \, \underline{\nu}_E)$ is R then an independent set of R columns can be selected from $\left[\underline{\nu}_1, \underline{\nu}_2, \ldots, \underline{\nu}_E\right]$, say $\{\underline{\mu}_1, \underline{\mu}_2, \ldots, \underline{\mu}_R\}$, and as each of the $\underline{\mu}_i$'s may differ from all of the $\underline{\nu}_j$'s the corresponding set of reactions is called a set of apparent reactions. Denote their rates $\dot{\eta}_r$, $r = 1, 2, \ldots, R$.

Then as unique coefficients C_{re} can be found such that

$$\underline{\nu}_e = \sum_{r=1}^{R} C_{re} \underline{\mu}_r$$

the rate of production of the species, $\sum_{e=1}^{E} \underline{\nu}_e \dot{\xi}_e$, can be written in terms of the rates of the apparent reactions as

$$\sum_{e=1}^{E} \dot{\xi}_e \sum_{r=1}^{R} C_{re} \underline{\mu}_r = \sum_{r=1}^{R} \underline{\mu}_r \dot{\eta}_r$$

where

$$\dot{\eta}_r = \sum_{e=1}^{E} C_{re} \dot{\xi}_e$$

This last equation is the rule for writing rate laws for apparent reactions in terms of rate laws for elementary reactions. As an example take the following set of elementary reactions

$$O_3 + O_3 \rightleftarrows O_1 + O_2 + O_3$$

$$O_3 + O_1 \rightleftarrows O_1 + O_1 + O_2$$

$$O_3 + O_2 \rightleftarrows O_1 + O_2 + O_2$$

$$O_3 + O_1 \rightleftarrows O_2 + O_2$$

$$O_2 + O_1 \rightleftarrows O_1 + O_1 + O_1$$

$$O_2 + O_3 \rightleftarrows O_1 + O_1 + O_3$$

and show that it is equivalent to the apparent reactions

$$O_3 \rightleftarrows 3O_1$$

$$2O_3 \rightleftarrows 3O_2$$

by writing $\dot{\eta}_1$ and $\dot{\eta}_2$ in terms of $\dot{\xi}_1, \dot{\xi}_2, \ldots, \dot{\xi}_6$.

2. Given a real number a and an estimate of a^{-1}, the iteration formula

$$x_{i+1} = x_i \left(2 - ax_i\right)$$

produces a sequence of approximations converging to a^{-1} if the initial estimate of a^{-1} is close enough. This is Newton's iteration as it is used to find the roots of

$$1 - \frac{1}{ax} = 0$$

Because division is not required to determine the sequence of approximations, we can try to use this iteration formula in matrix inversion. To see why it might work suppose that B is an estimate of A^{-1}, differing from it by a small amount Δ, so that AB and BA are close to I. Then we can write

$$A(B + \Delta) = I$$

which leads to

$$A\Delta = I - AB$$

and then to

$$BA\Delta = B - BAB$$

and we thereby discover that Δ is approximately

$$B - BAB$$

Using this, a new estimate of A^{-1} can be determined from B as

$$B + \Delta = 2B - BAB$$

and this is the iteration formula

$$B_{i+1} = 2B_i - B_iAB_i$$

Let A be the $n \times n$ Hilbert matrix where $n = 5$. Find an approximate inverse and then try to improve it using the above iteration formula.

Here is another way to think about this. Suppose we can find approximations to the solution of $A\underline{x} = \underline{b}$, possibly by using an elimination procedure corrupted by round-off errors.

Let X_0 be an approximation to X, the solution to $AX = I$. Then the error, $E_0 = X - X_0$, satisfies $AE_0 = R_0$ where the residual R_0 is $I - AX_0$.

Now let Δ_0 be an approximation to E_0 and define an improved approximation via $X_1 = X_0 + \Delta_0$. The new error $E_1 = X - X_1$ satisfies $AE_1 = R_1$ where the residual R_1 is $I - AX_1 = R_0 - A\Delta_0$.

Let Δ_1 be an approximation to E_1 and define $X_2 = X_1 + \Delta_1$. The error is $E_2 = X - X_2$, the residual is $R_2 = I - AX_2 = R_1 - A\Delta_1$ and $AE_2 = R_2$.

Etc.

Because $AE_0 = R_0$, we have $E_0 = XR_0$ and we can estimate E_0 to be X_0R_0. Taking this to be Δ_0 and using

$$R_1 = R_0 - A\Delta_0$$

we can estimate R_1 as

$$R_0 - AX_0R_0 = R_0 - (I - R_0)R_0 = R_0^2.$$

So if R_0 is small, R_1 is smaller yet and

$$X_1 = X_0 + \Delta_0$$

is approximately

$$X_0 + X_0R_0 = X_0 + X_0(I - AX_0) = 2X_0 - X_0AX_0$$

This is the same iteration formula as before.

$$B_{i+1} = 2B_i - B_i A B_i$$

3. To find the inverse of $\begin{pmatrix} A & \underline{b} \\ \underline{c}^T & d \end{pmatrix}$ in terms of A^{-1}, find $\begin{pmatrix} W & \underline{x} \\ \underline{y}^T & z \end{pmatrix}$ so that

$$\begin{pmatrix} A & \underline{b} \\ \underline{c}^T & d \end{pmatrix} \begin{pmatrix} W & \underline{x} \\ \underline{y}^T & z \end{pmatrix} = \begin{pmatrix} I & \underline{0} \\ \underline{0}^T & 1 \end{pmatrix}$$

To do this show that W and $\underline{y}^T$ satisfy

$$AW + \underline{b}\,\underline{y}^T = I$$

and

$$\underline{c}^T W + d\,\underline{y}^T = \underline{0}^T$$

Then write the first of these as

$$W + A^{-1}\underline{b}\,\underline{y}^T = A^{-1}$$

multiply this by $\underline{c}^T$ and use the second to determine $\underline{y}^T$ via

$$\underline{y}^T = \frac{\underline{c}^T A^{-1}}{-d + \underline{c}^T A^{-1}\underline{b}}$$

and then W via

$$W = A^{-1} - \frac{A^{-1}\underline{b}\,\underline{c}^T A^{-1}}{-d + \underline{c}^T A^{-1}\underline{b}}$$

Indeed $\begin{pmatrix} A & \underline{b} \\ \underline{c}^T & d \end{pmatrix}$ has an inverse iff $-d + \underline{c}^T A^{-1} \underline{b} \neq 0$. In the same way find $\underline{x}$ and z.

This is the bordering algorithm used to invert $(n+1) \times (n+1)$ matrices in terms of the inverses of $n \times n$ matrices. Use it to invert

$$\begin{pmatrix} 1 & 0 & 0 & 1 \\ 2 & 1 & 0 & 1 \\ 3 & 2 & 1 & 1 \\ 1 & 1 & 1 & 1 \end{pmatrix}$$

4. To see what happens in solving $A\underline{x} = \underline{b}$ when A is nearly singular write the LU decomposition of A as

$$A = LU = \begin{pmatrix} L_1 & \underline{0} \\ \underline{\ell}^T & 1 \end{pmatrix} \begin{pmatrix} U_1 & \underline{u} \\ \underline{0}^T & \varepsilon \end{pmatrix}$$

where here 1's lie on the diagonal of L instead of on the diagonal of U and where $\det A = \varepsilon \det U_1$. Requiring 1's on the diagonal of L instead of on the diagonal of U introduces no new idea; but requiring the small quantity ε to be in the lower right hand corner of U may require interchanging some columns and/or rows of A.

Then write $A\underline{x} = \underline{b}$ as

$$\begin{pmatrix} L_1 U_1 & L_1 \underline{u} \\ \underline{\ell}^T U_1 & \underline{\ell}^T \underline{u} + \varepsilon \end{pmatrix} \begin{pmatrix} \underline{x}_1 \\ x \end{pmatrix} = \begin{pmatrix} \underline{b}_1 \\ b \end{pmatrix}$$

and multiply this out to get

$$L_1 U_1 \underline{x}_1 + L_1 \underline{u} \, x = \underline{b}_1$$

and

$$\underline{\ell}^T U_1 \, \underline{x}_1 + \left(\underline{\ell}^T \underline{u} + \varepsilon \right) x = b$$

Write the first of these

$$\underline{x}_1 + U_1^{-1} \, \underline{u} \, x = \left(L_1 U_1 \right)^{-1} \underline{b}_1$$

multiply this by $\underline{\ell}^T U_1$ and use the second to find x via

$$x = \frac{b - \underline{\ell}^T L_1^{-1} \underline{b}_1}{\varepsilon}$$

whence

$$\underline{x} = \begin{pmatrix} \underline{x}_1 \\ x \end{pmatrix} = \begin{pmatrix} \left(L_1 U_1 \right)^{-1} \underline{b}_1 \\ 0 \end{pmatrix} + \begin{pmatrix} -U_1^{-1} \underline{u} \\ 1 \end{pmatrix} \frac{b - \underline{\ell}^T L_1^{-1} \underline{b}_1}{\varepsilon}$$

This formula is the main result of this problem. It tells us that the closer ε is to zero the better job we must do in the determination of $b - \underline{\ell}^T L_1^{-1} \underline{b}_1$. Write this formula as

$$\underline{x} = \begin{pmatrix} \left(L_1 U_1 \right)^{-1} \underline{b}_1 \\ 0 \end{pmatrix} + \frac{1}{\varepsilon} \, \underline{\psi}^T \underline{b} \, \underline{\phi}$$

where $\underline{\phi} = \begin{pmatrix} U_1^{-1} \underline{u} \\ -1 \end{pmatrix}$ and $\underline{\psi} = \begin{pmatrix} \left(L_1^{-1} \right)^T \underline{\ell} \\ -1 \end{pmatrix}$. Observe that if $\underline{\psi}^T \underline{b} = 0$ then $\underline{x}$ is independent of ε. Show that this obtains when $\underline{b} \in \text{Im } A \, (\varepsilon = 0)$ by observing that $A \, (\varepsilon = 0) \, \underline{\phi} = \underline{0}$ and $A^T \, (\varepsilon = 0) \, \underline{\psi} = \underline{0}$.

5. The 5×5 Hilbert matrix is

$$
\begin{pmatrix}
1 & 1/2 & 1/3 & 1/4 & 1/5 \\
1/2 & 1/3 & 1/4 & 1/5 & 1/6 \\
1/3 & 1/4 & 1/5 & 1/6 & 1/7 \\
1/4 & 1/5 & 1/6 & 1/7 & 1/8 \\
1/5 & 1/6 & 1/7 & 1/8 & 1/9
\end{pmatrix}
$$

Retaining the fractions, find the inverse. Round to four decimal places, write the decimals as fractions and find the inverse.

6. Let $v_1, v_2, \ldots, v_n$ be a set of n variables and let X be a dimensionless variable made up using products of powers of these variables via

$$
X = v_1^{p_1} \, v_2^{p_2} \, \cdots \, v_n^{p_n}
$$

Denote the three fundamental dimensions by M, L and T and write

$$
[v_1] = \mathrm{M}^{a_1} \mathrm{L}^{b_1} \mathrm{T}^{c_1}
$$

$$
[v_2] = \mathrm{M}^{a_2} \mathrm{L}^{b_2} \mathrm{T}^{c_2}
$$

etc.

then

$$
[X] = \left\{ \mathrm{M}^{a_1} \mathrm{L}^{b_1} \mathrm{T}^{c_1} \right\}^{p_1} \left\{ \mathrm{M}^{a_2} \mathrm{L}^{b_2} \mathrm{T}^{c_2} \right\}^{p_2} \cdots \left\{ \mathrm{M}^{a_n} \mathrm{L}^{b_n} \mathrm{T}^{c_n} \right\}^{p_n}
$$

where $p_1, p_2, \ldots, p_n$ must satisfy

$$
\begin{pmatrix} a_1 & a_2 & \ldots & a_n \\ b_1 & b_2 & \ldots & b_n \\ c_1 & c_2 & \ldots & c_n \end{pmatrix}
\begin{pmatrix} p_1 \\ p_2 \\ \vdots \\ p_n \end{pmatrix}
=
\begin{pmatrix} 0 \\ 0 \\ 0 \end{pmatrix}
$$

to make X dimensionless. This is a system of three homogeneous equations in n unknowns.

Each solution to these equations leads to a dimensionless variable. Independent solutions produce independent dimensionless variables. The number of independent dimensionless variables is the number of independent solutions. And because the sum of independent solutions is a solution so also the product of dimensionless variables is a dimensionless variable.

As an example, find an independent set of dimensionless variables in terms of which data on the flow of a liquid in a pipe can be correlated. The variables of interest are: the pressure drop across the pipe, the velocity of the flow, the length and diameter of the pipe, the viscosity, density and surface tension of the liquid and the acceleration due to gravity. Take the dimensions of these variables to be

$$[\Delta p] = M^1 L^{-1} T^{-2}$$

$$[v] \ = M^0 L^1 T^{-1}$$

$$[\ell] \ = M^0 L^1 T^0$$

$$[d] \ = M^0 L^1 T^0$$

$$[\mu] \ = M^1 L^{-1} T^{-1}$$

$$[\rho] \ = M^1 L^{-3} T^0$$

$$[\sigma] \ = M^1 L^0 T^{-2}$$

$$[g] \ = M^0 L^1 T^{-2}$$

7. A set of variables $P_1, P_2, \ldots$ is said to be dimensionally independent if the dimension of none of the variables can be written as a product of powers of the dimensions of the others.

Suppose

$$[P_1] = M^{\alpha_1} L^{\beta_1} T^{\gamma_1}$$

etc.

and write the definition of dimensional independence in terms of the rank of the matrix

$$\begin{pmatrix} \alpha_1 & \alpha_2 & \cdots & \alpha_n \\ \beta_1 & \beta_2 & \cdots & \beta_n \\ \gamma_1 & \gamma_2 & \cdots & \gamma_n \end{pmatrix}$$

8. The dimensions of P are independent of the dimensions of P_1 and P_2 if there are no solutions, p and q, to

$$\begin{pmatrix} \alpha \\ \beta \\ \gamma \end{pmatrix} = \begin{pmatrix} \alpha_1 & \alpha_2 \\ \beta_1 & \beta_2 \\ \gamma_1 & \gamma_2 \end{pmatrix} \begin{pmatrix} p \\ q \end{pmatrix}$$

where

$$[P] = M^{\alpha} L^{\beta} T^{\gamma}$$

etc.

How can this be stated in terms of the rank of a matrix?

Are the dimensions of pressure independent of the dimensions of viscosity and velocity?

9. The differential equations determining the growth rate of a small disturbance superimposed on the base state of a rotating fluid layer under an adverse temperature gradient are

$$
\begin{pmatrix}
D^2 - a^2 - \Pr\sigma & 0 & \dfrac{\beta}{\kappa}d^2 \\[2ex]
0 & D^2 - a^2 - \sigma & \dfrac{2\Omega}{\nu}dD \\[2ex]
-\dfrac{g\alpha}{\nu}d^2a^2 & -\dfrac{2\Omega}{\nu}d^3D & (D^2 - a^2)(D^2 - a^2 - \sigma)
\end{pmatrix}
\begin{pmatrix}
\Theta \\[2ex] Z \\[2ex] W
\end{pmatrix}
=
\begin{pmatrix}
0 \\[2ex] 0 \\[2ex] 0
\end{pmatrix}
$$

where D denotes $\dfrac{d}{dz}$.

Write this as a single differential equation in Θ or Z or W.

Lecture 4

Inner Products

4.1   Inner Products

An inner product on C^n is a function that assigns a complex number to every pair of vectors in C^n. The complex number assigned to $\underline{x}$ and $\underline{y}$ is denoted $\langle \underline{x}, \underline{y} \rangle$ and is required to satisfy the following conditions

$$\langle \underline{x}, \underline{y} \rangle = \overline{\langle \underline{y}, \underline{x} \rangle}$$

$$\langle \underline{x}, \ c_1\underline{y}_1 + c_2\underline{y}_2 \rangle = c_1 \langle \underline{x}, \underline{y}_1 \rangle + c_2 \langle \underline{x}, \underline{y}_2 \rangle$$

and

$$\langle \underline{x}, \underline{x} \rangle > 0 \quad \text{unless} \quad \underline{x} = \underline{0}$$

where an overbar denotes a complex conjugate. This definition is not special to C^n.

All inner products on C^n take the form

$$\langle \underline{x}, \underline{y} \rangle = \overline{\underline{x}}^T G \underline{y}$$

where G is an $n \times n$ matrix satisfying $\overline{G}^T = G$ (Hermitian) and $\overline{\underline{x}}^T G \underline{x} > 0$ (positive definite) for all $\underline{x} \neq \underline{0}$. Then to each Hermitian positive definite matrix G there corresponds an inner product on C^n. The simplest inner product, which we call the plain vanilla inner product, corresponds to $G = I$ and therein $\langle \underline{x}, \underline{y} \rangle = \overline{\underline{x}}^T \underline{y} = \sum_{i=1}^{n} \overline{x}_i y_i$

In a specific inner product two vectors $\underline{x}$ and $\underline{y}$ are said to be perpendicular if $\langle \underline{x}, \underline{y} \rangle = 0$; when this is so $\underline{x}$ and $G\underline{y}$ are perpendicular in the plain vanilla inner product.

Defining an inner product makes it easy to use the idea of biorthogonal sets of vectors and we do this at every opportunity as it greatly simplifies our work. We can also formulate a solvability condition that is not specific to matrix problems, unlike the rank condition presented earlier. To do this in C^n let A be an $n \times n$ matrix. Then A maps vectors $\underline{x} \in C^n$ into vectors $A\underline{x} \in \text{Im } A$, a subspace of C^n of dimension r.

4.2 Adjoints

What we seek is a new test to tell us when an arbitrary vector belongs to $\text{Im } A$. At first all we have to work with is $\text{Ker } A$, a subspace of C^n having the interesting dimension $n - r$ but otherwise bearing no special relation to $\text{Im } A$. Indeed $\text{Ker } A$ may be wholly inside $\text{Im } A$ or wholly outside $\text{Im } A$ or .... What we do then is this: we define a matrix whose kernel helps us identify $\text{Im } A$. To do this we fix an inner product on C^n. Then in this inner product A acquires a companion, denoted A^*, and called the adjoint of A, by the requirement that

$$\langle A^* \underline{x}, \underline{y} \rangle = \langle \underline{x}, A\underline{y} \rangle$$

for all $\underline{x}$ and $\underline{y}$ in C^n. Writing this out using $\langle \underline{x}, \underline{y} \rangle = \overline{\underline{x}}^T G \underline{y}$ we can determine a formula for A^*. It is

$$A^* = G^{-1} \overline{A}^T G$$

and this shows how A^* depends on the inner product in which we are working. If $G = I$ then $A^* = \overline{A}^T$

Now the rank of A^* is equal to the rank of A. Indeed it is easy to see that the rank of $\overline{A}^T$

is equal to the rank of A; it is less easy to see, but no less true, that the rank of $G^{-1}\overline{A}^T G$ is also equal to the rank of A. Bezout's theorem in Gantmacher's book "Theory of Matrices," can be used to produce an algebraic proof of this. The reader can produce a geometric proof by showing that the dimensions of the subspaces $\operatorname{Im} G^{-1}\overline{A}^T G$ and $\operatorname{Im} \overline{A}^T$ are equal. In doing this it is useful to observe that, because G is not singular, if the set of vectors $\{\underline{x}_1, \underline{x}_2, \dots\}$ is independent then so also the set of vectors $\{G^{-1}\underline{x}_1, G^{-1}\underline{x}_2, \dots\}$.

Fixing an inner product on C^n, we can identify an $n \times n$ matrix A^*, the adjoint of A. And as A^* maps C^n into itself it defines two subspaces of C^n: $\operatorname{Im} A^*$, of dimension r, and $\operatorname{Ker} A^*$, of dimension $n - r$.

4.3   Solvability Conditions

Now we need to see what these subspaces tell us about the subspaces $\operatorname{Im} A$ and $\operatorname{Ker} A$. To do this we first let S be an arbitrary subspace of C^n. Then if we fix an inner product on C^n, S acquires a companion, denoted $S^\perp$ and called the *orthogonal complement of S,* by the requirement that $S^\perp$ be the set of all vectors in C^n perpendicular to each vector in S. It is a subspace of C^n and it depends on the inner product being used. We write the definition of $S^\perp$ as

$$S^\perp = \left\{ \underline{y} : \langle \underline{y}, \underline{x} \rangle = 0 \quad \forall \underline{x} \in S \right\}$$

The readers can verify that $S^{\perp\perp} = S$, they can also verify that if $\dim S = r$ then $\dim S^\perp = n - r$ by using the fact that an $r \times n$ system of homogeneous equations of rank r has $n - r$ independent solutions. Then $\underline{x} \in S$ iff $\underline{x} \perp S^\perp$.

In terms of A^* and $\left(\operatorname{Ker} A^* \right)^\perp$ we can formulate our main result; it is

$$\operatorname{Im} A = \left(\operatorname{Ker} A^* \right)^\perp$$

Schematically this is:

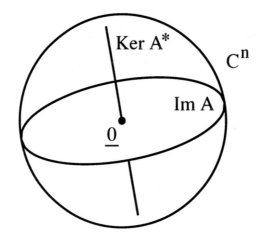

The solvability condition for the problem $A\underline{x} = \underline{b}$ is the requirement that $\underline{b} \in$ Im A. Our main result tells us that this is also the requirement that $\underline{b} \in \left(\text{Ker } A^* \right)^{\perp}$. To use this we fix an inner product on C^n and obtain A^*. We then determine Ker A^* by finding $n - r$ independent solutions to $A^* y = \underline{0}$ and, using these, decide whether or not $\underline{b} \perp$ Ker A^*. Hence the solvability condition for the problem $A\underline{x} = \underline{b}$ is this: Either $\langle \underline{b}, y \rangle = 0$ for all y such that $A^* y = \underline{0}$, whence $A\underline{x} = \underline{b}$ is solvable or $\langle \underline{b}, y \rangle \neq 0$ for some y such that $A^* y = \underline{0}$, whence $A\underline{x} = \underline{b}$ is not solvable.

The proof that Ker $A^* = (\text{Im } A)^{\perp}$ is simple. It is just this: If $y \in$ Ker A^* and $\underline{x} \in$ Im A then $\underline{x} = A\underline{z}$ and $\langle y, \underline{x} \rangle = \langle y, A\underline{z} \rangle = \langle A^* y, \underline{z} \rangle = 0$ whence Ker $A^* \subset (\text{Im } A)^{\perp}$. If $y \in (\text{Im } A)^{\perp}$ and $\underline{z} \in C^n$ then $A\underline{z} \in$ Im A and $\langle y, A\underline{z} \rangle = 0 = \langle A^* y, \underline{z} \rangle$; hence setting $\underline{z} = A^* y$ we have $A^* y = \underline{0}$ whence $(\text{Im } A)^{\perp} \subset$ Ker A^*.

4.4   Invariant Subspaces

A subspace S is said to be A invariant if $A\underline{x} \in S$ whenever $\underline{x} \in S$. If S is A invariant then $S^{\perp}$ is A^* invariant. To see this let $y \in S^{\perp}$ then if $\underline{x} \in S$, $A\underline{x} \in S$ and $\langle A^* y, \underline{x} \rangle = \langle y, A\underline{x} \rangle = 0$ hence $A^* y \in S^{\perp}$. If we fix an inner product, obtain A^* and find that $A^* = A$, then A is called *self-adjoint* (in that inner product). Self-adjoint matrices (and self-adjoint operators in general) satisfy special and useful conditions. Many of these come from the fact that if S is A invariant and $A^* = A$ then $S^{\perp}$ is also A invariant. This lies at the heart of the fact that a self-adjoint matrix has a complete set of eigenvectors. Eigenvectors are introduced in Lecture 5 (on page

85)and are important ingredients in solving linear operator problems. Given a matrix A, then, we may want to determine if there is an inner product in which it is self-adjoint. This amounts to determining a positive definite, Hermitian matrix G such that $A^* = A$. The condition on G is that $A = G^{-1}\overline{A}^T G$ or $GA = (\overline{GA})^T$. Matrices A for which there is a solution to this equation are called *symmetrizable*. They are self-adjoint in the corresponding inner product.

To this point our operators have been $n \times n$ matrices mapping C^n into itself. We conclude this lecture by producing the solvability condition for the problem $A\underline{x} = \underline{b}, \underline{x} \in C^n, \underline{b} \in C^m$. Then we introduce projection operators. Projections are basic to what are called generalized inverses which can be used to construct $\underline{x}$ such that $A\underline{x}$ is as close as possible to $\underline{b}$ when $\underline{b} \notin$ Im A.

Let $\langle \ , \ \rangle_m$ and $\langle \ , \ \rangle_n$ denote inner products on C^m and C^n then A, an $m \times n$ matrix mapping C^n into C^m, has an adjoint A^*, an $n \times m$ matrix mapping C^m into C^n. It is defined by requiring

$$\langle A^*\underline{y}, \underline{x} \rangle_n = \langle \underline{y}, A\underline{x} \rangle_m$$

for all $\underline{y} \in C^m, \underline{x} \in C^n$ and therefore it can be obtained via the formula

$$A^* = G_n^{-1}\overline{A}^T G_m$$

By arguments not unlike the above we discover that

1. Im A and Ker A^* are subspaces of C^m of dimensions r and $m - r$ whereas Im A^* and Ker A are subspaces of C^n of dimensions r and $n - r$, and

2. Im $A = \left(\text{Ker } A^* \right)^\perp$ and Im $A^* = (\text{Ker } A)^\perp$.

The solvability condition for $A\underline{x} = \underline{b}$ is as before: $A\underline{x} = \underline{b}$ is solvable iff $\langle \underline{b}, \underline{y} \rangle_m = 0$ for all $\underline{y}$ such that $A^*\underline{y} = \underline{0}$.

As an example of the use of this to determine solvability, let $A = \begin{pmatrix} 1 & 1 & 1 & 1 & 1 \\ 3 & 2 & 1 & 1 & -3 \\ 0 & 1 & 2 & 2 & 6 \\ 5 & 4 & 3 & 3 & -1 \end{pmatrix}$ and

$$b = \begin{pmatrix} 7 \\ -2 \\ 23 \\ 12 \end{pmatrix}. \text{ Then taking } G_m = I_m \text{ and } G_n = I_n \text{ we have } A^* = \overline{A}^T = \begin{pmatrix} 1 & 3 & 0 & 5 \\ 1 & 2 & 1 & 4 \\ 1 & 1 & 2 & 3 \\ 1 & 1 & 2 & 3 \\ 1 & -3 & 6 & -1 \end{pmatrix}$$

$$\text{whence } \operatorname{Ker} A^* = \left[y_1 = \begin{pmatrix} -3 \\ 1 \\ 1 \\ 0 \end{pmatrix}, \quad y_2 = \begin{pmatrix} -2 \\ -1 \\ 0 \\ 1 \end{pmatrix} \right].$$

As b passes the test $\langle b, y_1 \rangle_m = 0 = \langle b, y_2 \rangle_m$, $Ax = b$ is solvable, confirming our earlier conclusion.

4.5   Example: Functions Defined on $0 \leq x \leq 1$

Denote by f and g two real valued functions, vanishing at $x = 0$ and $x = 1$ and belonging to the set of smooth functions defined on the interval $[0, 1]$. Then

$$\int_0^1 f(x)g(x)\, dx = \langle f, g \rangle$$

defines an inner product on these functions and L where

$$L = \frac{d^2}{dx^2}$$

is a linear differential operator acting on this set of functions.

Now L is self-adjoint and so too $L + \pi^2$ and hence to decide if the problem

$$\frac{d^2 u}{dx^2} + \pi^2 u = f(x)$$

where $u = 0$ at $x = 0, 1$ has a solution, we observe that $\operatorname{Ker}(L + \pi^2)$ is spanned by $\sin \pi x$. Hence our problem is solvable for all right hand sides such that

$$\int_0^1 \sin \pi x \, f(x) \, dx = 0$$

i.e., it is solvable for every function odd about $\frac{1}{2}$, but for no others.

4.6   Projections

Let S be any assigned subspace of C^m and let $\underline{y}$ be any fixed vector of C^m. Then we can ask: what vector in S is closest to $\underline{y}$? The answer is given by the projection theorem.

To introduce the projection of C^m onto S we construct $S^\perp$, the orthogonal complement of S, using the inner product at hand, and then define P, the projection of C^m onto S by

$$P = I \text{ on } S$$

and

$$P = 0 \text{ on } S^\perp$$

Then as $I - P$ satisfies

$$I - P = 0 \text{ on } S$$

and

$$I - P = I \text{ on } S^\perp$$

we see that $I - P$ is the projection of C^m onto $S^\perp$. Indeed we have $P^2 = P$ and $P^* = P$; to see that $P^* = P$ we observe that

$$0 = \left\langle (I - P)\underline{y}, \, P\underline{z} \right\rangle = \left\langle \underline{y}, \, \left(I - P^*\right) P\underline{z} \right\rangle \qquad \forall \underline{y}, \underline{z} \in C^m$$

Thus we have $P^*P = P$ and therefore $P^* = I$ on S. But $\operatorname{Ker} P^* = \{\operatorname{Im} P\}^\perp = S^\perp$; hence $P^* = 0$ on $S^\perp$ and we have $P^* = P$.

The projection theorem tells us that Py is the vector in S that is closest to y. This is established later. The difference $y - Py = (I - P)y$ lies in $S^\perp$ and is, therefore, perpendicular to every vector in S. This is the error in approximating y by Py.

If $\dim S = r$ then $\dim S^\perp = m - r$ and so if $\underline{b}$ is assigned and $\underline{a}$ is required to be its best approximation in S then $\underline{a}$ is defined by the conditions

$$\underline{a} \in S$$

and

$$\underline{b} - \underline{a} \in S^\perp$$

This requires $\underline{a}$ to satisfy $m - r + r = m$ conditions and $\underline{a}$ is uniquely defined as a function of $\underline{b}$.

To construct P we take any set of r independent vectors in S, where $\dim S = r$, say $\underline{f}_1, \underline{f}_2, \ldots, \underline{f}_r$ and let F denote the $m \times r$ matrix

$$F = \left(\underline{f}_1\, \underline{f}_2\, \cdots\, \underline{f}_r\right)$$

where F is of rank r. Then for any $y \in C^m$ we have

$$Py = a_1\underline{f}_1 + a_2\underline{f}_2 + \cdots + a_r\underline{f}_r$$
$$= F \begin{pmatrix} a_1 \\ a_2 \\ \vdots \\ a_r \end{pmatrix} = F\underline{a}$$

where the coefficients $a_1, a_2, \ldots, a_r$ are uniquely determined. Now as $(I - P)y \in S^\perp$ we have

$$\left\langle \underline{f}_i, (I - P)y \right\rangle = \underline{0} \quad i = 1, 2, \ldots, r$$

This is

$$\underline{\overline{f}}_i^T G_m \left(\underline{y} - F\underline{a} \right) = \underline{0} \quad i = 1, 2, \dots, r$$

and denoting

$$\begin{pmatrix} \underline{\overline{f}}_1^T \\ \underline{\overline{f}}_2^T \\ \vdots \\ \underline{\overline{f}}_r^T \end{pmatrix} G_m = \overline{F}^T G_m$$

by F^* we have

$$F^* \underline{y} - F^* F \underline{a} = \underline{0}$$

and hence we get

$$P\underline{y} = F\underline{a} = F \left(F^* F \right)^{-1} F^* \underline{y}$$

Because this must be true for all $\underline{y} \in C^m$ we have for P the formula

$$P = F \left(F^* F \right)^{-1} F^*$$

where, as F^* is an $r \times m$ matrix of rank r, $F^* F$ is an $r \times r$ matrix of rank r.

The use of the notation F^* for $\overline{F}^T G_m$ is not too far fetched. If A maps C^n into C^m and inner products $\langle \ , \ \rangle_m$ and $\langle \ , \ \rangle_n$ are defined on C^m and C^n via positive definite, Hermitian matrices G_m and G_n, then A^* mapping C^m into C^n is given by

$$A^* = G_n^{-1} \overline{A}^T G_m$$

This is the way in which F^* is to be understood. Its leading factor is I_r.

4.7 The Projection Theorem: Least Squares Approximations

Let S be a subspace of C^m and let $\langle \ , \ \rangle$ denote an inner product defined on C^m in terms of a positive definite, Hermitian matrix G_m. Let $S^\perp$ denote the orthogonal complement of S and P and $I - P$ denote the projections of C^m onto S and $S^\perp$.

We can write any vector $\underline{y} \in C^m$ as the sum of a vector in S and another vector in $S^\perp$. The expansion is unique, it is

$$\underline{y} = P\underline{y} + (I - P)\underline{y}$$

and it leads to the Pythagorean Theorem:

$$||\underline{y}||^2 = \langle y, y \rangle = \left\langle P\underline{y} + (I - P)\underline{y}, \ P\underline{y} + (I - P)\underline{y} \right\rangle$$

$$= \left\langle P\underline{y}, P\underline{y} \right\rangle + \left\langle (I - P)\underline{y}, \ (I - P)\underline{y} \right\rangle$$

$$= ||P\underline{y}||^2 + ||(I - P)\underline{y}||^2$$

where $||\underline{y}|| = \langle y, y \rangle^{1/2}$ is the length of the vector $\underline{y}$.

We can also write any vector $\underline{y} - \underline{s} \in C^m$, where $\underline{y} \in C^m$ and $\underline{s} \in S$, as the sum of a vector in S and another in $S^\perp$. The expansion is

$$\underline{y} - \underline{s} = \{-\underline{s} + P\underline{y}\} + \{\underline{y} - P\underline{y}\}$$

and using the Pythagorean theorem we can conclude

$$||\underline{y} - \underline{s}||^2 = ||\underline{y} - P\underline{y}||^2 + ||\underline{s} - P\underline{y}||^2$$

This tells us that for any vector $\underline{s} \in S$

$$||\underline{y} - \underline{s}||^2 \geq ||\underline{y} - P\underline{y}||^2$$

and hence that $P\underline{y}$ lies at least as close to $\underline{y}$ as does any other vector in S. The vector $P\underline{y}$ is then the best approximation to $\underline{y}$ in S. The error in this approximation, $\underline{y} - P\underline{y}$, lies in $S^{\perp}$ and is perpendicular to all the vectors in S. This establishes the projection theorem, save for the question of uniqueness.

We can investigate this best approximation problem in another way which shows how it is the same as the least squares problem.

Again let $\underline{y}$ be any vector in C^m and S be a subspace of C^m. Then if $\left(\underline{f}_1, \underline{f}_2, \ldots, \underline{f}_r\right)$ is a set of r independent vectors spanning S we can write any vector $\underline{s} \in S$ as

$$\underline{s} = a_1\underline{f}_1 + a_2\underline{f}_2 + \cdots + a_r\underline{f}_r$$

$$= F\underline{a}$$

where

$$F = \left(\underline{f}_1 \ \underline{f}_2 \ \cdots \ \underline{f}_r\right)$$

and

$$\underline{a} = \begin{pmatrix} a_1 \\ a_2 \\ \vdots \\ a_r \end{pmatrix}$$

The problem is to determine $a_1, a_2, \ldots, a_r$ so that $\underline{s} \in S$ is the best approximation to $\underline{y}$.

To make the square of the length of the error vector, viz.,

$$\left\langle \underline{y} - \sum a_i\underline{f}_i, \ \underline{y} - \sum a_j\underline{f}_j \right\rangle,$$

least, we look for its stationary points by setting its derivatives with respect to $a_k, k = 1, \ldots, r$, to zero. To do this we write out the error vector as

$$\left\langle \underline{y} - \sum a_i \underline{f}_i, \ \underline{y} - \sum a_j \underline{f}_j \right\rangle = \left\langle \underline{y} - F\underline{a}, \ \underline{y} - F\underline{a} \right\rangle = \left(\overline{\underline{y}}^T - \overline{\underline{a}}^T \overline{F}^T \right) G \left(\underline{y} - F\underline{a} \right)$$

use

$$\underline{a} = \operatorname{Re} \underline{a} + i \operatorname{Im} \underline{a}$$

and

$$\overline{\underline{a}}^T = \operatorname{Re} \underline{a}^T - i \operatorname{Im} \underline{a}^T$$

and then differentiate this with respect to both $\operatorname{Re} a_i$ and $\operatorname{Im} a_i$. Using

$$\frac{\partial \operatorname{Re} \underline{a}}{\partial \operatorname{Re} a_i} = \underline{e}_i, \qquad \frac{\partial \operatorname{Re} \underline{a}^T}{\partial \operatorname{Re} a_i} = \underline{e}_i^T, \qquad \frac{\partial \operatorname{Im} \underline{a}}{\partial \operatorname{Re} a_i} = \underline{0}, \quad \text{etc.}$$

we find, on setting the derivative with respect to $\operatorname{Re} a_i$ to zero, that

$$\left(\overline{\underline{y}}^T - \overline{\underline{a}}^T \overline{F}^T \right) G \left(-F\underline{e}_i \right) + \left(-\underline{e}_i^T \overline{F}^T G \left(\underline{y} - F\underline{a} \right) \right) = 0, \quad i = 1, 2, \dots, r$$

and hence

$$\operatorname{Re} \left\{ \overline{F}^T G \left(\underline{y} - F\underline{a} \right) \right\} = \underline{0}$$

Likewise on setting the derivative with respect to $\operatorname{Im} a_i$ to zero, we find

$$\left(\overline{\underline{y}}^T - \overline{\underline{a}}^T \overline{F}^T \right) G \left(-iF\underline{e}_i \right) + i \, \underline{e}_i^T \overline{F}^T G \left(\underline{y} - F\underline{a} \right) = 0, \quad i = 1, 2, \dots, r$$

and hence

$$\operatorname{Im} \left\{ \overline{F}^T G \left(\underline{y} - F\underline{a} \right) \right\} = \underline{0}$$

Our result then is

$$\overline{F}^T G \underline{y} = \overline{F}^T G F \underline{a}$$

but $\overline{F}^T G = F^*$ and therefore, we get

$$\underline{a} = \left(F^* F\right)^{-1} F^* \underline{y}$$

This formula solves the least squares problem. The solution to the best approximation problem is $F\underline{a}$ where

$$F\underline{a} = F\left(F^* F\right)^{-1} F^* \underline{y} = P\underline{y}$$

and this is our earlier result.

4.8   Generalized Inverses

Supose A is an $m \times n$ matrix mapping C^n into C^m. The problem

$$A\underline{x} = \underline{y}$$

has a solution iff $\underline{y} \in \text{Im } A$.

Suppose dim Im A is less than m, then there are vectors in C^m not in Im A. If $\underline{y} \notin \text{Im } A$, there is no solution to our problem but we can ask for the best approximation, i.e., a vector $\underline{x} \in C^n$ such that $A\underline{x} \in \text{Im } A$ is as close as possible to $\underline{y}$.

To do this we introduce an inner product $\langle \ , \ \rangle$ on C^m, i.e., we select a positive definite, Hermitian matrix G_m. Then we denote the length of a vector $\underline{y} \in C^m$ by $||\underline{y}||$ where

$$||\underline{y}||^2 = \langle \underline{y}, \underline{y} \rangle$$

To find $\underline{x}$ such that $A\underline{x}$ is the best approximation to $\underline{y}$, we first determine $A\underline{x}$ via

$$A\underline{x} = P\underline{y}$$

where P, the projection of C^m onto Im A, can be constructed using the basis columns of A. The error $\underline{y} - A\underline{x} = \underline{y} - P\underline{y} = (I - P)\,\underline{y}$ is then perpendicular to Im A.

The vector $P\underline{y}$, the best approximation to $\underline{y}$ in Im A, is uniquely determined, hence so too $A\underline{x}$. But $\underline{x}$ need not be unique. Because $P\underline{y} \in$ Im A, the solvability condition is satisfied and we can begin to determine $\underline{x}$ by letting $\underline{x}_0$ be a particular solution to $A\underline{x} = P\underline{y}$. Then the general solution is $\underline{x}_0 + \underline{\xi}$ where $\underline{\xi} \in$ Ker $A \subset C^n$. Now $\underline{\xi}$ can be any vector in Ker A, hence $\underline{x}_0 + \underline{\xi}$ can be any vector in a plane parallel to Ker A. This plane is simply Ker A translated by $\underline{x}_0$ and it is independent of the particular solution $\underline{x}_0$ we happen to be using.

To make the solution to the best approximation problem unique we make $\underline{x}_0 + \underline{\xi}$ lie as close to $\underline{0}$ as possible. To do this we let $I - Q$ be the projection of C^n onto Ker A, Q being the projection of C^n onto (Ker $A)^{\perp}$. Then the vector $(I - Q)\,(\underline{x}_0 + \underline{\xi})$ is the closest vector in Ker A to $\underline{x}_0 + \underline{\xi}$ and their difference, $Q\,(\underline{x}_0 + \underline{\xi}) = Q\underline{x}_0$ is independent of $\underline{\xi}$. Hence, every solution is the same distance from Ker A as every other, this distance being the length of $Q\underline{x}_0$. But $Q\underline{x}_0 = \underline{x}_0 - (I - Q)\,\underline{x}_0 \in ($ Ker $A)^{\perp}$ is also a solution, due to $(I - Q)\,\underline{x}_0 \in$ Ker A, and, because its projection on Ker A is $\underline{0}$, it must be the solution closest to $\underline{0}$. So by requiring $A\underline{x}$ to be the best approximation to $\underline{y}$ and $\underline{x}$ to be as short as possible we get a unique solution to our best approximation problem, viz., if $\underline{x}_0$ is any vector satisfying

$$A\underline{x}_0 = P\underline{y}$$

then the unique solution $\underline{x}$ is

$$\underline{x} = Q\underline{x}_0$$

Indeed because $Q\underline{x}_0$ is independent of $\underline{x}_0$, i.e., $Q\,(\underline{x}_0 + \underline{\xi}) = Q\underline{x}_0 \; \forall \underline{\xi} \in$ Ker A, it is the unique solution to our problem.

It remains only to express $\underline{x}$ in terms of A and $\underline{y}$ and thereby define a generalized inverse, denoted A^I, such that A^I, mapping C^m into C^n, satisfies the requirement that to any $\underline{y} \in C^m$, $\underline{x} = A^I\underline{y}$ is the shortest vector in C^n such that $A\underline{x}$ is the best approximation in Im A to $\underline{y}$.

To do this let $\underline{f}_1, \underline{f}_2, \ldots, \underline{f}_r$ be as in the construction of P, i.e., a set of r independent vectors spanning Im A, then we can write A as

$$A = FR$$

where R is an $r \times n$ matrix whose columns are the coefficients in the expansion of the columns of A in the basis $\underline{f}_1, \underline{f}_2, \ldots, \underline{f}_r$ for Im A. Hence R is unique and of rank r. Because F^* is defined to be $\overline{F}^T G_m$, we take R^* to be the $n \times r$ matrix $G_n^{-1}\overline{R}^T$ and find that

$$A^* = G_n^{-1}\overline{A}^T G_m = G_n^{-1}\overline{R}^T\overline{F}^T G_m$$

$$= R^* F^*$$

where the r independent columns of R^* span Im A^*. Then P, where $P = F\left(F^*F\right)^{-1}F^*$, is the projection of C^m onto Im A and Q, where $Q = R^*\left(R^*R^*\right)^{-1}R^*$, is the projection of C^n onto Im A^*. And observing that $R^{**} = R$, viz.,

$$R^{**} = \left(\overline{R}^*\right)^T G_n = \left(\overline{G_n^{-1}\overline{R}^T}\right)^T G_n = R$$

we can write Q as

$$Q = R^*\left(RR^*\right)^{-1}R$$

These formulas for A, A^*, P and Q establish the important facts that

$$PA = A, \qquad AQ = A$$

and

$$QA^* = A^*, \qquad A^* P = A^*$$

Then defining A^I via

$$A^I = R^* \left(RR^* \right)^{-1} \left(F^* F \right)^{-1} F^*$$

we get

$$QA^I = A^I, \qquad A^I P = A^I$$

$$AA^I = P$$

$$A^I A = Q$$

$$AA^I A = A$$

and

$$A^I AA^I = A^I$$

These formulas tell us that for any $\underline{y} \in C^m$, we have $AA^I \underline{y} = P\underline{y}$ and so $A^I \underline{y}$ is a solution of

$$A\underline{x} = P\underline{y}$$

Indeed, as $QA^I \underline{y} = A^I \underline{y}$, $A^I \underline{y}$ is the shortest such solution. Hence A^I is the generalized inverse of A, i.e., $A^I \underline{y}$ is the shortest vector such that $AA^I \underline{y}$ is the best approximation to $\underline{y}$.

4.9 Home Problems

1. Let $\{\underline{x}_1, \underline{x}_2, \ldots, \underline{x}_n\}$ be a basis for C^n. Then introduce an inner product and let $\{\underline{y}_1, \underline{y}_2, \ldots, \underline{y}_n\}$ be its biorthogonal set. Show that the elements of a matrix A in the basis $\{\underline{x}_1, \underline{x}_2, \ldots, \underline{x}_n\}$ are

$$A_{ij} = \langle\, \underline{y}_i, \; A\underline{x}_j \,\rangle$$

where the elements of a matrix A in a basis $\{\underline{x}_1, \underline{x}_2, \ldots, \underline{x}_n\}$ are defined by

$$A\underline{x}_j = \sum A_{ij}\, \underline{x}_i$$

Using

$$A\underline{x} = \sum \underline{x}_i \langle\, \underline{y}_i, \; A\underline{x} \,\rangle$$

show that the elements of the product AB are

$$(AB)_{ij} = \sum \langle\, \underline{y}_i, \; A\underline{x}_k \,\rangle \langle\, \underline{y}_k, \; B\underline{x}_j \,\rangle = \sum A_{ik}\, B_{kj}$$

2. Let $A : C^n \to C^m$ and let G_n and G_m be the weighting factors in the corresponding inner products. Then $A^*A : C^n \to C^n$ and $AA^* : C^m \to C^m$. Show that $\left(A^*A\right)^* = AA^*$ and $\left(AA^*\right)^* = A^*A$. Be careful as there are three different *'s here.

3. Let A be the matrix in the numerical example in Lectures 3 and 4 where $m = 4$, $n = 5$ and $r = 2$. Then A has many generalized inverses each one corresponding to definite inner products on C^4 and C^5. Determine the generalized inverse of A when $G_4 = I_4$ and $G_5 = I_5$. Use this to solve the example problem in Lecture 3.

4. The problem $A\underline{x} = \underline{y}$ where

$$A = \begin{pmatrix} 1 & 1 & 1 & 1 \\ 1 & -1 & 1 & 1 \\ 1 & 1 & -1 & 1 \\ 1 & 1 & 1 & -1 \end{pmatrix}$$

and

$$\underline{y} = \begin{pmatrix} 1 \\ 1 \\ 1 \\ 1 \end{pmatrix}$$

has a unique solution. It is $\underline{x} = \begin{pmatrix} 1 \\ 0 \\ 0 \\ 0 \end{pmatrix}$. Determine c_1 and c_2 so that $A\{c_1\,\underline{x}_1 + c_2\,\underline{x}_2\}$ is the best approximation to $\underline{y}$ where

$$\underline{x}_1 = \begin{pmatrix} 1 \\ 1 \\ 0 \\ 0 \end{pmatrix} \quad \text{and} \quad \underline{x}_2 = \begin{pmatrix} 1 \\ 0 \\ 1 \\ 0 \end{pmatrix}$$

Is $c_1\,\underline{x}_1 + c_2\,\underline{x}_2$ the best approximation to $\underline{x}$ in $[\underline{x}_1,\ \underline{x}_2]$?

Lecture 5

Eigenvectors

5.1 Eigenvectors and Eigenvalues

Eigenvectors are defined for operators that map vector spaces into themselves. Operators that do this can act repetitively, and hence their squares, cubes, etc., are defined. Their action can be understood in terms of their invariant subspaces and these may be built up out of their eigenvectors.

Let A be an $n \times n$ matrix mapping C^n into itself. Then a vector $\underline{x} \neq \underline{0}$ which satisfies

$$A\underline{x} = \lambda\underline{x}$$

for some complex number λ is called an eigenvector of A and λ is called the corresponding eigenvalue. Each eigenvector of A spans a one dimensional A-invariant subspace of C^n and each vector in this span is also an eigenvector of A corresponding to λ. Eigenvectors are not unique, their lengths being arbitrary.

To determine the eigenvectors of A we write $A\underline{x} = \lambda\underline{x}$ as

$$(A - \lambda I)\underline{x} = \underline{0}$$

and observe that solutions other than $\underline{x} = \underline{0}$ can be found only for certain values of λ. To find the eigenvectors we must first find the eigenvalues, viz., the values of λ such that the solution space of our homogeneous problem contains vectors other than $\underline{0}$, i.e., such that dim Ker $(A - \lambda I) > 0$ and therefore that dim Im $(A - \lambda I) < n$. This is satisfied iff the rank of $A - \lambda I$ is less than n which in turn is satisfied iff

$$\det (A - \lambda I) = 0$$

Each value of λ satisfying this equation is an eigenvalue of A and the corresponding eigenvectors make up the solution space Ker $(A - \lambda I)$, called the eigenspace corresponding to the eigenvalue λ. The number of independent eigenvectors corresponding to λ is dim Ker $(A - \lambda I)$ and this is n less the rank of $(A - \lambda I)$. It is called the geometric multiplicity of the eigenvalue λ.

To determine the eigenvalues of A we let $\Delta(\lambda) = \det(A - \lambda I)$, write $\lambda I - A = (\lambda \underline{e}_1 - \underline{a}_1 \ \lambda \underline{e}_2 - \underline{a}_2 \ \cdots \ \lambda \underline{e}_n - \underline{a}_n)$ and expand $\Delta(\lambda)$ using property (iii) of determinants, page 33, to get

$$\Delta(\lambda) = \det (\lambda \underline{e}_1 \ \lambda \underline{e}_2 - \underline{a}_2 \ \cdots) + \det (-\underline{a}_1 \ \lambda \underline{e}_2 - \underline{a}_2 \ \cdots)$$

$$= \det (\lambda \underline{e}_1 \ \lambda \underline{e}_2 \ \lambda \underline{e}_3 - \underline{a}_3 \ \cdots) + \det (\lambda \underline{e}_1 - \underline{a}_2 \ \lambda \underline{e}_3 - \underline{a}_3 \ \cdots) +$$

$$\det (-\underline{a}_1 \ \lambda \underline{e}_2 \ \lambda \underline{e}_3 - \underline{a}_3 \ \cdots) + \det (-\underline{a}_1 \ - \underline{a}_2 \ \lambda \underline{e}_3 - \underline{a}_3 \ \cdots)]$$

$$= \text{etc.}$$

and conclude that $\Delta(\lambda)$ is a monic polynomial of degree n in λ. We write it

$$\Delta(\lambda) = \lambda^n - \Delta_1 \lambda^{n-1} + \Delta_2 \lambda^{n-2} - \cdots + (-1)^n \Delta_n$$

where the coefficient Δ_i is the sum of the $i \times i$ principal minors of A and where a minor is a principle minor if the elements on its diagonal are also on the diagonal of A. The coefficients in $\Delta(\lambda)$ can be written in terms of the eigenvalues. The coefficient Δ_i is the sum of the products of the eigenvalues taken i at a time, e.g.,

$$\Delta_1 = \operatorname{tr} A = \lambda_1 + \lambda_2 + \cdots + \lambda_n$$

and

$$\Delta_n = \det A = \lambda_1 \lambda_2 \cdots \lambda_n$$

where tr and $\det$ denote trace and determinant.

The polynomial $\Delta(\lambda)$ has n roots in C, counting each root according to its multiplicity. To be definite we call its distinct roots the eigenvalues of A and we denote them

$$\lambda_1, \ \lambda_2, \ \cdots, \ \lambda_d$$

denoting their algebraic multiplicities $m_1, \ m_2, \ \ldots, \ m_d$ where $m_1 + m_2 + \cdots + m_d = n$.

The geometric multiplicity of each eigenvalue, i.e., the greatest number of independent eigenvectors corresponding to that eigenvalue, cannot exceed its algebraic multiplicity. Indeed if we let $n_1 = \dim \operatorname{Ker}(A - \lambda_1 I)$ and determine $\Delta(\lambda)$ in a basis for C^n whose first n_1 vectors span $\operatorname{Ker}(A - \lambda_1 I)$, we see that $\Delta(\lambda)$ contains the factor $(\lambda - \lambda_1)^{n_1}$ whence $n_1 \leq m_1$. This may cause the reader to look at section 5.9 on page 109.

We introduce the eigenvectors of A in the hope of constructing a basis for C^n which will simplify certain calculations that we plan to make. But we will not always be able to find an eigenvector basis for C^n. To make the distinction we need to make, we call an eigenvalue problem plain vanilla if it leads to n algebraically simple eigenvalues. Then the geometric multiplicity of each eigenvalue is the same as its algebraic multiplicity and this value is one. In fact we will go on and call an eigenvalue problem plain vanilla whenever the geometric multiplicity of each eigenvalue is also its algebraic multiplicity.

There are eigenvalue problems that are not plain vanilla. But this is the exception, not the rule. To see why this is so and why it is important, we look first at the eigenvalue problem in the simplest case, $n = 2$. To do this we write out the characteristic polynomial of a 2×2 matrix A as

$$\lambda^2 - (\operatorname{tr} A) \lambda + \det A$$

and observe that the characteristic equation has either two distinct roots or one double root. The matrix A then has either two eigenvalues each of algebraic multiplicity one or one eigenvalue of algebraic multiplicity two. In the first instance we write

$$A \underline{x}_1 = \lambda_1 \underline{x}_1$$

and

$$A \underline{x}_2 = \lambda_2 \underline{x}_2$$

and derive the important fact that $\underline{x}_1$ and $\underline{x}_2$ are independent. Indeed $c_1 = 0 = c_2$ is the only solution to $c_1 \underline{x}_1 + c_2 \underline{x}_2 = \underline{0}$ for if

$$c_1 \underline{x}_1 + c_2 \underline{x}_2 = \underline{0}$$

then

$$c_1 \lambda_1 \underline{x}_1 + c_2 \lambda_2 \underline{x}_2 = \underline{0}$$

and so

$$c_2 (\lambda_2 - \lambda_1) \underline{x}_2 = \underline{0}$$

As $\lambda_1 \neq \lambda_2$ and $\underline{x}_2 \neq \underline{0}$ we find only $c_2 = 0$ and likewise only $c_1 = 0$. And so eigenvectors corresponding to distinct eigenvalues are independent. By an easy extension of this we see that if $n > 2$ and $\lambda_1, \lambda_2, \cdots, \lambda_d$ are distinct eigenvalues then any set of d eigenvectors, one corresponding to each of $\lambda_1, \lambda_2, \cdots, \lambda_d$, is independent. This is worked out in section 5.8 on page 107.

Because $\{\underline{x}_1, \underline{x}_2\}$ is an independent set of vectors it is a basis for C^2. Introducing an inner product on C^2 we can construct its biorthogonal set, denoted $\left\{\underline{y}_1,\ \underline{y}_2\right\}$, via

$$\langle \underline{y}_i,\ \underline{x}_j \rangle = \delta_{ij}$$

The set of vectors $\left\{\underline{y}_1,\ \underline{y}_2\right\}$ is independent and likewise a basis for C^2; indeed the matrix

$$\begin{pmatrix} \underline{\bar{y}}_1^T G \\ \underline{\bar{y}}_2^T G \end{pmatrix}$$

is the inverse of the matrix $(\underline{x}_1\ \underline{x}_2)$.

We remind the reader that the idea of biorthogonal sets is a powerful idea. If any vector $\underline{x}$ is expanded as

$$\underline{x} = c_1 \underline{x}_1 + c_2 \underline{x}_2$$

or as

$$\underline{x} = d_1 \underline{y}_1 + d_2 \underline{y}_2$$

then the coefficients are simply

$$c_1 = \langle \underline{y}_1,\ \underline{x} \rangle, \qquad c_2 = \langle \underline{y}_2,\ \underline{x} \rangle$$

and

$$d_1 = \langle \underline{x}_1,\ \underline{x} \rangle, \qquad d_2 = \langle \underline{x}_2,\ \underline{x} \rangle$$

But there is even more. If we introduce A^*, the adjoint of A, in the same inner product used to construct $\underline{y}_1$ and $\underline{y}_2$, then $\underline{y}_1$ and $\underline{y}_2$ turn out to be eigenvectors of A^*, viz.,

$$A^*\underline{y}_i = \langle \underline{x}_1, A^*\underline{y}_i \rangle \underline{y}_1 + \langle \underline{x}_2, A^*\underline{y}_i \rangle \underline{y}_2$$

$$= \langle A\underline{x}_1, \underline{y}_i \rangle \underline{y}_1 + \langle A\underline{x}_2, \underline{y}_i \rangle \underline{y}_2$$

$$= \overline{\lambda}_1 \langle \underline{x}_1, \underline{y}_i \rangle \underline{y}_1 + \overline{\lambda}_2 \langle \underline{x}_2, \underline{y}_i \rangle \underline{y}_2$$

and hence

$$A^*\underline{y}_1 = \overline{\lambda}_1 \underline{y}_1$$

and

$$A^*\underline{y}_2 = \overline{\lambda}_2 \underline{y}_2$$

This tells us that the eigenvalues of A and A^* are complex conjugates {i.e., if $\det (\lambda I - A) = 0$ then $\det \left(\overline{\lambda} I - A^* \right) = 0$ where $A^* = G^{-1}\overline{A}^T G$ } and that their eigenvectors form biorthogonal sets.

The first case is complete: when two eigenvalues of algebraic multiplicity one turn up, the corresponding eigenvectors determine a basis for C^2. If one eigenvalue of multiplicity two is obtained this continues to be true if $\dim \text{Ker} (A - \lambda_1 I) = 2$ as then we can find two independent eigenvectors, viz., any two independent vectors in C^2, and write

$$A\underline{x}_1 = \lambda_1 \underline{x}_1$$

and

$$A\underline{x}_2 = \lambda_1 \underline{x}_2$$

and go on as before. But if $\dim \text{Ker} (A - \lambda_1 I) = 1$ a complication arises: we cannot find two independent eigenvectors.

This corresponds to the rank of $A - \lambda_1 I$ having the value one instead of zero whence both Im $(A - \lambda_1 I)$ and Ker $(A - \lambda_1 I)$ are one dimensional subspaces of C^2 and there is at most one independent eigenvector. Denoting this $\underline{x}_1$ we have Ker $(A - \lambda_1 I) = [\underline{x}_1]$. And we observe that Ker $(A - \lambda_1 I) =$ Im $(A - \lambda_1 I)$ otherwise λ_1 cannot be a double root. Indeed, as Im $(A - \lambda_1 I)$ is one dimensional and A invariant, any vector in Im $(A - \lambda_1 I)$ is an eigenvector of A corresponding to an eigenvalue other than λ_1 unless Ker $(A - \lambda_1 I) =$ Im $(A - \lambda_1 I)$. This is established again in section 5.2.

5.2 Generalized Eigenvectors

So, being short an eigenvector and observing that $\underline{x}_1 \in$ Im $(A - \lambda_1 I)$ we seek a vector $\underline{x}_2$ satisfying

$$(A - \lambda_1 I)\,\underline{x}_2 = \underline{x}_1$$

Now, to find $\underline{x}_2$, a solvability condition must be satisfied because $(A - \lambda_1 I)\,\underline{x} = \underline{0}$ has a solution other than $\underline{0}$, viz., $\underline{x}_1$. However $\underline{x}_1 \in$ Im $(A - \lambda_1 I)$ and hence the solvability condition is satisfied. And therefore a solution, $\underline{x}_2$, can be found. It is called a generalized eigenvector. It satisfies $(A - \lambda_1 I)^2\,\underline{x}_2 = \underline{0}$ but not $(A - \lambda_1 I)\,\underline{x}_2 = \underline{0}$. And it is independent of $\underline{x}_1$: if

$$c_1\underline{x}_1 + c_2\underline{x}_2 = \underline{0}$$

then

$$c_1\lambda_1\underline{x}_1 + c_2\left\{\underline{x}_1 + \lambda_1\underline{x}_2\right\} = \underline{0}$$

or

$$c_2\underline{x}_1 = \underline{0}$$

and we find only $c_1 = 0 = c_2$. Of course $\underline{x}_2$ is not uniquely determined; a multiple of $\underline{x}_1$ can be added to a particular $\underline{x}_2$ to produce another possible $\underline{x}_2$.

This illustrates the main idea: when we cannot find enough eigenvectors to make up a basis for our space we generalize the eigenvector problem in such a way that to each eigenvalue of algebraic multiplicity m there corresponds m eigenvectors and generalized eigenvectors. The only new idea required when n is greater than two is that an eigenvector may generate a chain of more than one generalized eigenvector and there may be more than one chain corresponding to each eigenvalue.

In C^2 the solutions to the generalized eigenvalue problem for A satisfy

$$A\underline{x}_1 = \lambda_1 \underline{x}_1$$

and

$$A\underline{x}_2 = \underline{x}_1 + \lambda_1 \underline{x}_2$$

where $\{\underline{x}_1,\ \underline{x}_2\}$ is independent and a basis for C^2. The corresponding biorthogonal set, $\left\{\underline{y}_1,\ \underline{y}_2\right\}$, is also independent and a basis for C^2. Making the calculation $A^* \underline{y}_i$ we find

$$A^* \underline{y}_1 = \overline{\lambda}_1 \underline{y}_1 + \underline{y}_2$$

and

$$A^* \underline{y}_2 = \overline{\lambda}_1 \underline{y}_2$$

The readers need to carry out this calculation by expanding $A^* \underline{y}_i$ in $\left\{\underline{y}_1,\ \underline{y}_2\right\}$ to satisfy themselves that the eigenvalue problem for A^* is required to generalize in just this way. The check on this is that the solvability condition for determining $\underline{x}_2$ is:

$$\underline{x}_1 \perp \mathrm{Ker}\ (A - \lambda_1 I)^* = \mathrm{Ker}\left(A^* - \overline{\lambda}_1 I\right) = \left[\underline{y}_2\right]$$

The problem of generalized eigenvectors leads to more possibilities than we found in C^2. To explain what can happen, let λ_1 be an eigenvalue of A of algebraic multiplicity m_1 and let the dimension of $\mathrm{Ker}\ (A - \lambda_1 I)$ be n_1 so that we have n_1 independent eigenvectors. If $n_1 < m_1$ then $\mathrm{Ker}\ (A - \lambda_1 I)$ and $\mathrm{Im}\ (A - \lambda_1 I)$ intersect in at least one vector $\underline{x}_1$ and we take $\underline{x}_1$ to be one of our eigenvectors. Using this vector we can determine a vector $\underline{y}_1$ such that $(A - \lambda_1 I)\underline{y}_1 = \underline{x}_1$ and $\underline{x}_1$ and $\underline{y}_1$ are the first two vectors in a chain. If $\underline{y}_1$ is not in $\mathrm{Im}\ (A - \lambda_1 I)$ the chain terminates, otherwise we can determine a vector $\underline{z}_1$ such that $(A - \lambda_1 I)\underline{z}_1 = \underline{y}_1$, etc. The vectors $\underline{x}_1,\ \underline{y}_1,\ \underline{z}_1,\ \dots$ satisfy the equations

$$A\underline{x}_1 = \lambda_1 \underline{x}_1$$

$$A\underline{y}_1 = \underline{x}_1 + \lambda_1 \underline{y}_1$$

$$A\underline{z}_1 = \underline{y}_1 + \lambda_1 \underline{z}_1$$

$$\vdots$$

which now generalize the eigenvalue problem. As there may be more than one chain corresponding to the eigenvalue λ_1, it is important in selecting a basis for $\mathrm{Ker}\ (A - \lambda_1 I)$ to first span $\mathrm{Ker}\ (A - \lambda_1 I) \cap \mathrm{Im}\ (A - \lambda_1 I)$. The condition that a vector lie in $\mathrm{Im}\ (A - \lambda_1 I)$ is that it be orthogonal to $\mathrm{Ker}\ (A - \lambda_1 I)^*$.

We can illustrate the main idea in the case $n = 2$. Suppose λ_1 is an eigenvalue of algebraic multiplicity 2 and geometric multiplicity 1. Then we have

$$\mathrm{Ker}\ (A - \lambda_1 I) = [\underline{x}_1]$$

and

$$\dim \mathrm{Im}\ (A - \lambda_1 I) = 1$$

whereupon

$$\mathrm{Im}\ (A - \lambda_1 I) = [\underline{x}]$$

and we see that

$$(A - \lambda_1 I)\,\underline{x} = c\,\underline{x}$$

hence

$$A\underline{x} = (\lambda_1 + c)\,\underline{x}$$

whence c must be zero and $\underline{x}$ must be a multiple of $\underline{x}_1$. And we conclude that

$$\mathrm{Im}\,(A - \lambda_1 I) = \mathrm{Ker}\,(A - \lambda_1 I)$$

5.3 The Generalized Eigenvector Corresponding to a Double Eigenvalue

Let λ_1 be a double root of $\det\left(A - \lambda I\right) = 0$. { If A is real and λ_1 is not real then $\overline{\lambda}_1$ is also a double root. } Let $\underline{x}_1$ be an eigenvector corresponding to λ_1 and suppose there is no other solution of $\left(A - \lambda_1 I\right)\underline{x} = \underline{0}$ independent of $\underline{x}_1$, i.e., $\mathrm{Ker}\left(A - \lambda_1 I\right) = \left[\underline{x}_1\right]$. Then $\dim\,\mathrm{Im}\left(A - \lambda_1 I\right) = n - 1$ and we can see that $\underline{x}_1 \in \mathrm{Im}\left(A - \lambda_1 I\right)$. Indeed if $\underline{x}_1 \notin \mathrm{Im}\left(A - \lambda_1 I\right)$, then, in a basis made up of $\underline{x}_1$ and $n - 1$ independent vectors in $\mathrm{Im}\left(A - \lambda_1 I\right)$, A has the representation

$$\begin{pmatrix} \lambda_1 & \underline{0}_{n-1}^T \\ \underline{0}_{n-1} & A_{n-1\,n-1} \end{pmatrix}$$

due to the fact that $\underline{x}_1 \notin \mathrm{Im}\left(A - \lambda_1 I\right)$ and $\mathrm{Im}\left(A - \lambda_1 I\right)$ is A invariant. But as λ_1 is a double root it must be a root of $\det\left(A_{n-1\,n-1} - \lambda I_{n-1\,n-1}\right) = 0$ whence A must have a second eigenvector corresponding to λ_1, it must lie in $\mathrm{Im}\left(A - \lambda_1 I\right)$ and it must be independent of $\underline{x}_1$. But this is not so and we conclude that $\underline{x}_1 \in \mathrm{Im}\left(A - \lambda_1 I\right)$. This is important because it is the solvability condition for the problem

$$\left(A - \lambda_1 I\right)\underline{x} = \underline{x}_1$$

and hence we can find $\underline{x}_2$ such that

$$\left(A - \lambda_1 I\right)\underline{x}_2 = \underline{x}_1$$

Indeed to any particular solution $\underline{x}_2$ may be added any multiple of $\underline{x}_1$ but all such $\underline{x}_2$'s are independent of $\underline{x}_1$. That is, if

$$c_1\underline{x}_1 + c_2\underline{x}_2 = \underline{0}$$

then

$$\left(A - \lambda_1 I\right)\left\{c_1\underline{x}_1 + c_2\underline{x}_2\right\} = c_2\underline{x}_1 = \underline{0}$$

whence $c_2 = 0$ and so too $c_1 = 0$.

So, if λ_1 is a double root of $\det\left(A - \lambda I\right) = 0$ and $\dim\mathrm{Ker}\left(A - \lambda_1 I\right) = 1$ then $\mathrm{Ker}\left(A - \lambda_1 I\right) \subset \mathrm{Im}\left(A - \lambda_1 I\right)$ and we can write

$$A\underline{x}_1 = \lambda_1\underline{x}_1$$

and

$$A\underline{x}_2 = \underline{x}_1 + \lambda_1\underline{x}_2$$

And $\underline{x}_2 \notin \mathrm{Im}\left(A - \lambda_1 I\right)$ otherwise there would be a vector $\underline{x}_3$ such that $\left(A - \lambda_1 I\right)\underline{x}_3 = \underline{x}_2$ and λ_1 would be a triple root of $\det\left(A - \lambda I\right) = 0$. As $\underline{x}_1 \in \mathrm{Im}\left(A - \lambda_1 I\right)$ but $\underline{x}_2 \notin \mathrm{Im}\left(A - \lambda_1 I\right)$, $\underline{x}_1$, but not $\underline{x}_2$, is perpendicular to $\mathrm{Ker}\left(A^* - \overline{\lambda}_1 I\right)$.

If we denote by $\underline{y}_2$ a nonzero solution to $A^*\underline{y} = \overline{\lambda}_1 I$, then

$$\left[\underline{y}_2\right] = \mathrm{Ker}\left(A^* - \overline{\lambda}_1 I\right) = \left\{\mathrm{Im}\left(A - \lambda_1 I\right)\right\}^{\perp}$$

and there is no other solution independent of $\underline{y}_2$. Because $\left\langle \underline{x}_1, \underline{y}_2 \right\rangle = 0$ and $\left\langle \underline{x}_2, \underline{y}_2 \right\rangle \neq 0$ we can require $\underline{y}_2$ to satisfy $\left\langle \underline{x}_2, \underline{y}_2 \right\rangle = 1$. Now as $\underline{x}_1 \in \mathrm{Im}\left(A - \lambda_1 I\right)$ we conclude:

$\underline{y}_2 \perp [\underline{x}_1] = \mathrm{Ker}\,(A - \lambda_1 I)$ and hence $\underline{y}_2 \in \mathrm{Im}\,(A^* - \bar{\lambda}_1 I)$. Using this we can let $\underline{y}_1$ denote a solution to $(A^* - \bar{\lambda}_1 I)\underline{y}_1 = \underline{y}_2$. Then $\underline{y}_1$ is independent of $\underline{y}_2$, $\underline{y}_1 \notin \mathrm{Im}\,(A^* - \bar{\lambda}_1 I)$ and we can write

$$A^* \underline{y}_1 = \bar{\lambda}_1 \underline{y}_1 + \underline{y}_2$$

and

$$A^* \underline{y}_2 = \bar{\lambda}_1 \underline{y}_2$$

By calculating $\lambda_1 \langle \underline{y}_1, \underline{x}_2 \rangle$ we can see that $\langle \underline{x}_1, \underline{y}_1 \rangle = 1$ while $\underline{y}_1$ is determined only up to an additive multiple of $\underline{y}_2$. It remains only to select the constant c so that $\langle \underline{x}_2, \underline{y}_1 + c\underline{y}_2 \rangle = 0$ and rename $\underline{y}_1 + c\underline{y}_2$ as $\underline{y}_1$, then $\{\underline{x}_1, \underline{x}_2\}$ and $\{\underline{y}_1, \underline{y}_2\}$ are biorthogonal sets in C^n.

It is ordinarily not true that $[\underline{x}_1, \underline{x}_2]$ and $[\underline{y}_1, \underline{y}_2]$ coincide, this obtains only if $[\underline{x}_1, \underline{x}_2]$ is A^* invariant and then $\{\underline{y}_1, \underline{y}_2\}$ can be determined directly as the biorthogonal set to $\{\underline{x}_1, \underline{x}_2\}$ in $[\underline{x}_1, \underline{x}_2]$.

A picture may help:

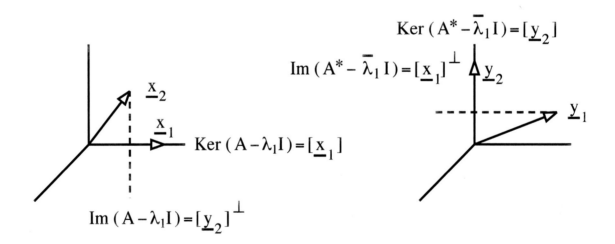

5.4 Complete Sets of Eigenvectors

Our expectation when we solve an eigenvalue problem in C^n must be: either we will determine a set of n independent eigenvectors and hence a basis for C^n or we will not. A set of n independent eigenvectors is called a complete set. There are sufficient conditions for this; one is that the eigenvalues of A turn out to be simple roots of $\Delta(\lambda) = 0$. This requires A to have n distinct eigenvalues. Another is that $A^* = A$ in some inner product. If this is so we can determine an eigenvector $\underline{x}_1$ in the usual way and then observe that as $[\underline{x}_1]$ is A invariant so also is $[\underline{x}_1]^\perp$. Restricting A to this $n - 1$ dimensional subspace we can then start over and determine an eigenvector, $\underline{x}_2$, of the restriction of A to $[\underline{x}_1]^\perp$ in the usual way. This will be the second eigenvector of A and it will satisfy $\langle \underline{x}_1, \underline{x}_2 \rangle = 0$. If $n > 2$ we can continue this to determine a set of n mutually orthogonal eigenvectors, orthogonal in the inner product in which $A^* = A$.

To decide the likelihood of turning up n independent eigenvectors we go back to the case $n = 2$ where $\Delta(\lambda) = \det(\lambda I - A) = 0$ is the quadratic equation

$$\lambda^2 - (\operatorname{tr} A)\lambda + \det A = 0$$

and where $\operatorname{tr} A = a_{11} + a_{22}$ and $\det A = a_{11}a_{22} - a_{21}a_{12}$. This equation has a double root iff $(\operatorname{tr} A)^2 - 4 \det A = 0$; otherwise it has two simple roots. If a_{11}, a_{12}, a_{21} and a_{22} are real numbers then the double root is real and it corresponds to the one dimensional locus $(\operatorname{tr} A)^2 - 4 \det A = 0$ in the $\det A$, $\operatorname{tr} A$ plane separating the region corresponding to two simple real roots and the region corresponding to two simple complex roots (which are complex conjugates). Two simple roots is generic, being realized almost everywhere in the $\det A$, $\operatorname{tr} A$ plane; the alternative, a double root, turns up only on a set of measure zero. This continues to be true for $n > 2$. Our emphasis then is on the ordinary and simplest possibility, we take up exceptions by example. What we require at the outset is that A determine a basis for C^n made up of independent eigenvectors; we refer to this as a complete set of eigenvectors, and n simple eigenvalues is sufficient but not necessary for this.

Before going on we introduce a simple way to find all the eigenvectors lying in one dimensional eigenspaces. Let A be an $n \times n$ matrix. The corresponding eigenvalues are the roots of $\Delta(\lambda) =$

$\det (\lambda I - A)$ where we write

$$\Delta (\lambda) = \lambda^n - \Delta_1 \lambda^{n-1} + \Delta_2 \lambda^{n-2} - \cdots + (-1)^n \Delta_n$$

Letting $B (\lambda) = \mathrm{adj} (\lambda I - A)$, we see that the elements of $B (\lambda)$ are polynomials of degree $n - 1$ in λ and so we can write

$$B (\lambda) = \lambda^{n-1} I - B_1 \lambda^{n-2} + \cdots + (-1)^{n-1} B_{n-1}$$

And, as

$$(\lambda I - A) \, \mathrm{adj} \, (\lambda I - A) = \det (\lambda I - A) \, I$$

we have

$$(\lambda I - A) B (\lambda) = \Delta (\lambda) I,$$

and we see that corresponding to any eigenvalue, say λ_1, where $\Delta (\lambda_1) = 0$, the non-vanishing columns of $B (\lambda_1)$ are eigenvectors of A. Now $B (\lambda_1)$ is a matrix whose rank is either one or zero depending on whether the rank of $(\lambda_1 I - A)$ is either $n - 1$ or less than $n - 1$. If the rank of $(\lambda_1 I - A)$ is $n-1$ we have $\dim \mathrm{Ker} (\lambda_1 I - A) = 1$ and then there is one independent eigenvector and a candidate can be found among the columns of $B (\lambda_1)$. This is all that is required if λ_1 is a simple eigenvalue. To determine $B_1, B_2, \ldots, B_{n-1}$ and hence $B (\lambda)$, we can equate the coefficients of the powers of λ on the two sides of

$$(\lambda I - A) \left(\lambda^{n-1} I - B_1 \lambda^{n-2} + \cdots + (-1)^{n-1} B_{n-1} \right) = \lambda^n I - \Delta_1 \lambda^{n-1} + \Delta_2 \lambda^{n-2} \cdots$$

Doing this we get

$$B_1 = \Delta_1 I - A$$

$$B_2 = \Delta_2 I - AB_1$$

etc.

In Gantmacher's book, "Theory of Matrices," a method is explained for determining the sequences $\Delta_1, \Delta_2, \ldots$ and $B_1, B_2, \ldots$ simultaneously.

5.5 The Spectral Representation of a Matrix and a Derivation of the Kremser Equation

Henceforth we let n be arbitrary and assume, unless an exception is made, that we have a complete set of eigenvectors. Then the algebraic multiplicity of the eigenvalues is not important and we can first denote the eigenvectors $\underline{x}_1, \underline{x}_1, \ldots, \underline{x}_n$ and then denote the corresponding eigenvalues $\lambda_1, \lambda_2, \ldots, \lambda_n$ where $\lambda_1, \lambda_2, \ldots, \lambda_n$ may not be distinct complex numbers. Upon solving the eigenvalue problem $A\underline{x} = \lambda\underline{x}$ we obtain set of independent eigenvectors. Then we introduce an inner product in C^n and construct its biorthogonal set. Denoting this $\{\underline{y}_1, \underline{y}_2, \ldots, \underline{y}_n\}$ we require

$$\langle\, \underline{y}_i, \underline{x}_j \,\rangle = \delta_{ij}, \qquad i, j = 1, 2, \ldots, n$$

Each of the sets of vectors $\{\underline{x}_1, \underline{x}_2, \ldots, \underline{x}_n\}$ and $\{\underline{y}_1, \underline{y}_2, \ldots, \underline{y}_n\}$ is a basis for C^n. Now the set of n^2 equations $\langle\, \underline{y}_i, \underline{x}_j \,\rangle = \overline{\underline{y}}_i^T G \underline{x}_j = \delta_{ij}$ can be written

$$\begin{pmatrix} \overline{\underline{y}}_1^T G \\ \overline{\underline{y}}_2^T G \\ \vdots \\ \overline{\underline{y}}_n^T G \end{pmatrix} \begin{pmatrix} \underline{x}_1 & \underline{x}_2 & \cdots & \underline{x}_n \end{pmatrix} = I$$

and we see that the matrix

$$\begin{pmatrix} \underline{\bar{y}}_1^T G \\ \underline{\bar{y}}_2^T G \\ \vdots \\ \underline{\bar{y}}_n^T G \end{pmatrix}$$

is the inverse of the matrix $\begin{pmatrix} \underline{x}_1 & \underline{x}_2 & \cdots & \underline{x}_n \end{pmatrix}$. Indeed the vectors $\underline{\bar{y}}_1^T G$, $\underline{\bar{y}}_2^T G$, ... are independent of G.

Writing the n vector equations

$$A\underline{x}_i = \lambda_i \underline{x}_i, \qquad i = 1, 2, \ldots, n$$

as the matrix equation

$$A\begin{pmatrix} \underline{x}_1 & \underline{x}_2 \cdots & \underline{x}_n \end{pmatrix} = \begin{pmatrix} \lambda_1 \underline{x}_1 & \lambda_2 \underline{x}_2 & \cdots & \lambda_n \underline{x}_n \end{pmatrix}$$

we get a formula for A, viz.,

$$A = \begin{pmatrix} \lambda_1 \underline{x}_1 & \lambda_2 \underline{x}_2 & \cdots & \lambda_n \underline{x}_n \end{pmatrix} \begin{pmatrix} \underline{\bar{y}}_1^T G \\ \underline{\bar{y}}_2^T G \\ \vdots \\ \underline{\bar{y}}_n^T G \end{pmatrix}$$

And when this is written out, we have

$$A = \lambda_1 \underline{x}_1 \underline{\bar{y}}_1^T G + \lambda_2 \underline{x}_2 \underline{\bar{y}}_2^T G + \cdots + \lambda_n \underline{x}_n \underline{\bar{y}}_n^T G$$

This is called the spectral representation of A. Letting $P_1 = \underline{x}_1 \underline{y}_1^T G$, $P_2 = \underline{x}_2 \underline{y}_2^T G$, etc., and using $\underline{\bar{y}}_i^T G \underline{x}_j = \delta_{ij}$, we find

$$P_i P_i = P_i$$

and

$$P_i P_j = 0, \qquad i \neq j$$

and we write

$$A = \lambda_1 P_1 + \lambda_2 P_2 + \cdots + \lambda_n P_n$$

We say P_i selects $\underline{x}_i$ because $P_i \underline{x}_i = \underline{x}_i$ and $P_i \underline{x}_j = \underline{0}$, $i \neq j$, viz., $P_i \sum c_j \underline{x}_j = c_i \underline{x}_i$.

This formula simplifies certain calculations via the multiplication rules for the P_i; indeed it can be used to derive powers of A via

$$A^p = \lambda_1^p P_1 + \lambda_2^p P_2 + \cdots + \lambda_n^p P_n, \quad p = 0, 1, 2, \ldots$$

where $I = P_1 + P_2 + \cdots + P_n$. It can also be used to define polynomials and power series in A in terms of polynomials and power series in λ, i.e., if $f(\lambda)$ is a polynomial or power series in λ then $f(A) = f(\lambda_1) P_1 + f(\lambda_2) P_2 + \cdots$. The formula holds as well for $p = -1, -2, \ldots$ if $\lambda_i \neq 0$, $i = 1, 2, \ldots, n$.

As an example of its use, the balance and equilibrium equations corresponding to the linear, n stage, counter current separating cascade sketched below

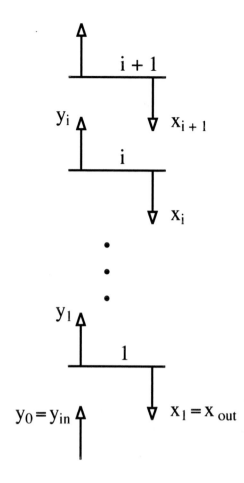

are:

$$Lx_{i+1} + Vy_{i-1} = Lx_i + Vy_i$$

$$y_i^* = mx_i$$

and

$$y_i - y_{i-1} = E\left(y_i^* - y_{i-1}\right)$$

where y_0 and x_{n+1} are the compositions of the V and L phase feed streams and y_n and x_1 are the compositions of the V and L phase product streams. We write these equations

$$\begin{pmatrix} x_{i+1} \\ y_i \end{pmatrix} = A \begin{pmatrix} x_i \\ y_{i-1} \end{pmatrix}$$

Then by stepping through the cascade we determine its input-output formula to be

$$\begin{pmatrix} x_{\text{in}} \\ y_{\text{out}} \end{pmatrix} = A^n \begin{pmatrix} x_{\text{out}} \\ y_{\text{in}} \end{pmatrix}$$

where A is the 2×2 matrix

$$A = \begin{pmatrix} 1 + \frac{mV}{L}E & -\frac{V}{L}E \\ mE & 1 - E \end{pmatrix}.$$

The eigenvalues of A, viz., $\lambda_1 = 1$ and $\lambda_2 = 1 + E\left(\frac{mV}{L} - 1\right)$, are simple unless $\frac{mV}{L} - 1 = 0$. The corresponding eigenvectors are

$$\underline{x}_1 = \begin{pmatrix} 1 \\ m \end{pmatrix}, \qquad \underline{x}_2 = \begin{pmatrix} V/L \\ 1 \end{pmatrix}$$

and in the plain vanilla inner product, viz., $G = I$, we find

$$\underline{y}_1 = \frac{1}{1 - \frac{mV}{L}} \begin{pmatrix} 1 \\ -\frac{V}{L} \end{pmatrix}, \qquad \underline{y}_2 = \frac{1}{1 - \frac{mV}{L}} \begin{pmatrix} -m \\ 1 \end{pmatrix}$$

As a result we have

$$\begin{pmatrix} x_{\text{in}} \\ y_{\text{out}} \end{pmatrix} = \left\{ \frac{1}{1 - \frac{mV}{L}} \begin{pmatrix} 1 & -\frac{V}{L} \\ m & -\frac{mV}{L} \end{pmatrix} + \frac{\left(1 + E\left(\frac{mV}{L} - 1\right)\right)^n}{1 - \frac{mV}{L}} \begin{pmatrix} -\frac{mV}{L} & -\frac{V}{L} \\ -m & 1 \end{pmatrix} \right\} \begin{pmatrix} x_{\text{out}} \\ y_{\text{in}} \end{pmatrix}$$

This is equivalent to the Kremser equation and the overall material balance, but in a symmetric form. What is important is that we have constructed a useful representation of A^n and we did not need a concrete value of n to do this.

If $\dfrac{mV}{L} - 1$ is zero, then $\lambda_1 = 1$ is a double root and to it there corresponds only one independent eigenvector, $\underline{x}_1 = \begin{pmatrix} 1 \\ m \end{pmatrix}$. The spectral representation of A must take this into account. And to do this for a 2×2 matrix A we write

$$A\underline{x}_1 = \lambda_1 \underline{x}_1$$

and

$$A\underline{x}_2 = \underline{x}_1 + \lambda_1 \underline{x}_2$$

where $\underline{x}_2$ is a generalized eigenvector, here $\underline{x}_2 = \begin{pmatrix} 1 \\ 0 \end{pmatrix}$. Then we write this

$$A\begin{pmatrix} \underline{x}_1 & \underline{x}_2 \end{pmatrix} = \begin{pmatrix} \lambda_1 \underline{x}_1 & \underline{x}_1 + \lambda_1 \underline{x}_2 \end{pmatrix}$$

and using $\begin{pmatrix} \underline{x}_1 & \underline{x}_2 \end{pmatrix}^{-1} = \begin{pmatrix} \underline{\bar{y}}_1^T G \\ \underline{\bar{y}}_2^T G \end{pmatrix}$, where $\{ \underline{x}_1 \ \underline{x}_2 \}$ and $\{ \underline{y}_1 \ \underline{y}_2 \}$ are biorthogonal sets, we have

$$A = \lambda_1 \underline{x}_1 \underline{\bar{y}}_1^T G + \underline{x}_1 \underline{\bar{y}}_2^T G + \lambda_1 \underline{x}_2 \underline{\bar{y}}_2^T G$$

$$= \lambda_1 P_1 + P_{12} + \lambda_1 P_2$$

$$= \lambda_1 I + P_{12}$$

where the multiplication rules are now $P_1 P_{12} = P_{12}$, $P_{12} P_1 = 0$, $P_2 P_{12} = 0$, $P_{12} P_2 = P_{12}$ and $P_{12} P_{12} = 0$. Hence we have $A^n = \lambda_1^n I + n\lambda_1^{n-1} P_{12}$ and this can be used to derive the Kremser equation when the equilibrium and operating lines are parallel.

5.6 The Adjoint Eigenvalue Problem

This restates what we already know in the case $n = 2$. Thus if A has a complete set of eigenvectors, then, in the same inner product in which we determine $\{\underline{y}_1, \underline{y}_2, \cdots, \underline{y}_n\}$, we can obtain A^* and calculate $A^*\underline{y}_i$. Expanding $A^*\underline{y}_i$ in $\{\underline{y}_1, \underline{y}_2, \cdots, \underline{y}_n\}$ we get

$$A^*\underline{y}_i = \sum_{j=1}^{n} \langle \underline{x}_j, A^*\underline{y}_i \rangle \underline{y}_j$$

$$= \sum_{j=1}^{n} \langle A\underline{x}_j, \underline{y}_i \rangle \underline{y}_j$$

$$= \overline{\lambda}_i \underline{y}_i$$

This tells us that the vectors $\underline{y}_i$ satisfy the eigenvalue problem for A^*, viz.,

$$A^*\underline{y}_i = \overline{\lambda}_i \underline{y}_i, \quad i = 1, 2, \cdots, n$$

The sets of eigenvectors of A and A^* are biorthogonal, while the sets of eigenvalues are complex conjugates. Indeed $\det(A - \lambda I) = 0$ implies $\det\left(A^* - \overline{\lambda}I\right) = \det\left(G^{-1}\overline{A}^T G - \overline{\lambda}I\right) = 0$.

5.7 Eigenvector Expansions and the Solution to the Problem $A\underline{x}=\underline{b}$

We plan to use what we have learned in this lecture in the next lecture to write the solution to differential and difference equations. Before we do this, and to illustrate the useful fact that we can solve problems by expanding their solutions in a convenient basis, we return to the problem $A\underline{x} = \underline{b}$ and assume it has a solution. Then expanding $\underline{x}$ in the set of eigenvectors of A, assumed to be complete, we write

$$\underline{x} = \sum c_i \underline{x}_i$$

and our job is to determine the coefficients c_i, where $c_i = \langle \underline{y}_i, \underline{x} \rangle$. We can find c_i by calculating the inner product of $\underline{y}_i$ and both sides of $A\underline{x} = \underline{b}$. Indeed we have

$$\langle \underline{y}_i, A\underline{x} \rangle = \langle \underline{y}_i, \underline{b} \rangle$$

$$\langle A^* \underline{y}_i, \underline{x} \rangle = \langle \underline{y}_i, \underline{b} \rangle$$

$$\langle \overline{\lambda}_i \underline{y}_i, \underline{x} \rangle = \langle \underline{y}_i, \underline{b} \rangle$$

$$\lambda_i \langle \underline{y}_i, \underline{x} \rangle = \langle \underline{y}_i, \underline{b} \rangle$$

whence, assuming $\lambda_i \neq 0$, we get

$$\langle \underline{y}_i, \underline{x} \rangle = \frac{\langle \underline{y}_i, \underline{b} \rangle}{\lambda_i}$$

and so we conclude that

$$\underline{x} = \sum \frac{\langle \underline{y}_i, \underline{b} \rangle}{\lambda_i} \underline{x}_i$$

is the solution of $A\underline{x} = \underline{b}$. We see that each coefficient c_i is determined independent of the other coefficients.

The subspaces of C^n important to the problem $A\underline{x} = \underline{b}$, viz., Im A and Ker A, can be thought about in terms of the eigenvectors of A and A^*: Im A is the span of the eigenvectors of A corresponding to eigenvalues that are not zero, Ker A^* is the span of the eigenvectors of A^* corresponding to eigenvalues that are zero and Ker A is the span of the eigenvectors of A corresponding to eigenvalues that are zero. For instance if $\lambda_1 = 0$ in the foregoing, the solvability condition is $\langle \underline{y}_1, \underline{b} \rangle = 0$ and if that is satisfied $\frac{\langle \underline{y}_1, \underline{b} \rangle}{\lambda_1}$ is indeterminate and can be replaced by an arbitrary constant c_1. The solution then contains an arbitrary multiple of $\underline{x}_1$, a basis vector for Ker A.

If corresponding to λ_1 ($m_1 = 2$, $n_1 = 1$) we have an eigenvector $\underline{x}_1$ and a generalized eigenvector $\underline{x}_2$, we write

$$\underline{x} = \langle \underline{y}_1, \underline{x} \rangle \underline{x}_1 + \langle \underline{y}_2, \underline{x} \rangle \underline{x}_2 + \cdots$$

and observe that

$$\langle\, \underline{y}_i,\ A\,\underline{x}\,\rangle = \langle\, \underline{y}_i,\ \underline{b}\,\rangle, \qquad i = 1, 2$$

imply

$$\langle\, \underline{y}_2,\ \underline{x}\,\rangle + \lambda_1 \langle\, \underline{y}_1,\ \underline{x}\,\rangle = \langle\, \underline{y}_1,\ \underline{b}\,\rangle$$

and

$$\lambda_1 \langle\, \underline{y}_2,\ \underline{x}\,\rangle = \langle\, \underline{y}_2,\ \underline{b}\,\rangle$$

whereupon, if λ_1 is not zero, we obtain $\langle\, \underline{y}_1,\ \underline{x}\,\rangle$ and $\langle\, \underline{y}_2,\ \underline{x}\,\rangle$ and write the solution to $A\underline{x} = \underline{b}$ accordingly.

5.8 Solvability Conditions and the Solution to Perturbation Problems

Solvability conditions are important in perturbation calculations. To see why this is so, suppose a matrix of interest, A, is close to a matrix A^0 whose eigenvalue problem has been solved resulting in a complete set of eigenvectors and simple eigenvalues:

$$A = A^0 + \varepsilon A^1 + \varepsilon^2 A^2 + \cdots$$

where ε is small and

$$A^0 \underline{x}_i^0 = \lambda_i^0 \underline{x}_i^0, \quad i = 1, 2, \ldots, n$$
$$\left(A^0\right)^* \underline{y}_i^0 = \overline{\lambda}_i^0 \underline{y}_i^0, \quad i = 1, 2, \ldots, n$$

Then writing

$$\underline{x}_i = \underline{x}_i^0 + \varepsilon \underline{x}_i^1 + \varepsilon^2 \underline{x}_i^2 + \cdots$$

and

$$\lambda_i = \lambda_i^0 + \varepsilon \lambda_i^1 + \varepsilon^2 \lambda_i^2 + \cdots$$

substituting into $A\underline{x}_i = \lambda_i \underline{x}_i$ and equating the coefficients of 1, ε, ε^2, $\ldots$ to zero we find

$$\left(A^0 - \lambda_i^0 I\right) \underline{x}_i^0 = \underline{0}$$

$$\left(A^0 - \lambda_i^0 I\right) \underline{x}_i^1 = \left(\lambda_i^1 I - A^1\right) \underline{x}_i^0$$

$$\left(A^0 - \lambda_i^0 I\right) \underline{x}_i^2 = \left(\lambda_i^2 I - A^2\right) \underline{x}_i^0 + \left(\lambda_i^1 I - A^1\right) \underline{x}_i^1$$

etc.

The first problem determines $\underline{x}_i^0$ and λ_i^0. And at every succeeding order the matrix $A^0 - \lambda_i^0 I$ appears and the homogeneous problem $\left(A^0 - \lambda_i^0 I\right) \underline{x} = \underline{0}$ has a non zero solution, viz., $\underline{x}_i^0$. To determine the first corrections, $\underline{x}_i^1$ and λ_i^1, we turn to the second problem. To get $\underline{x}_i^1$ requires that a solvability condition be satisfied. This is the requirement that $\left(\lambda_i^1 I - A^1\right) \underline{x}_i^0$ belong to $\text{Im}\left(A^0 - \lambda_i^0 I\right)$ and hence be perpendicular to $\text{Ker}\left(A^0 - \lambda_i^0 I\right)^*$. But this is $\left[\underline{y}_i^0\right]$ and hence the solvability condition

$$\left\langle \underline{y}_i^0, \left(\lambda_i^1 I - A^1\right) \underline{x}_i^0 \right\rangle = 0$$

determines λ_i^1 as

$$\lambda_i^1 = \left\langle \underline{y}_i^0, A^1 \underline{x}_i^0 \right\rangle$$

and so to first order

$$\lambda_i = \lambda_i^0 + \left\langle \underline{y}_i^0, A^1 \underline{x}_i^0 \right\rangle \varepsilon$$

Continuing the calculation requires no new ideas but a lot of tedious work. Indeed to determine $\underline{x}_i^1$ we use the solution of $A\underline{x} = \underline{b}$ written above putting $A^0 - \lambda_i^0 I$ in place of A and $\left(\lambda_i^1 I - A^1 \right)\underline{x}_i^0$ in place of $\underline{b}$. Because the eigenvalues and eigenvectors of $A^0 - \lambda_i^0 I$ are $\lambda_j^0 - \lambda_i^0$ and $\underline{x}_j^0$, $\quad j = 1, 2, \ldots, n$, we get

$$\underline{x}_i^1 = \sum_{j \neq i} \frac{\left\langle \underline{y}_j^0, \left(\lambda_i^1 I - A^1 \right) \underline{x}_i^0 \right\rangle}{\lambda_j^0 - \lambda_i^0} \underline{x}_j^0 + c_i^1 \underline{x}_i^0$$

This is required to determine λ_i^2; the readers may wish to satisfy themselves that λ_i^2, where $\lambda_i^2 = \left\langle \underline{y}_i^0, A^2 \underline{x}_i^0 \right\rangle - \left\langle \underline{y}_i^0, \left(\lambda_i^1 I - A^1 \right) \underline{x}_i^1 \right\rangle$, is independent of the value assigned to c_i^1.

The calculation becomes more interesting as complexities arise. Suppose λ_1^0 turns out to be a double root and $\mathrm{Ker}\left(A^0 - \lambda_1^0 I \right)$ is two dimensional so that A^0 retains a complete set of eigenvectors. On perturbation, λ_1^0 is likely to split into two simple roots λ_1 and λ_2 to which correspond independent eigenvectors $\underline{x}_1$ and $\underline{x}_2$. Now $\underline{x}_1$ and $\underline{x}_2$ approach definite vectors $\underline{x}_1^0$ and $\underline{x}_2^0$ in $\mathrm{Ker}\left(A^0 - \lambda_1^0 I \right)$ as ε goes to zero but we cannot know in advance what these limits are and hence we cannot select $\underline{x}_1^0$ and $\underline{x}_2^0$ in advance out of all of the possibilities in $\mathrm{Ker}\left(A^0 - \lambda_1^0 I \right)$. This means that at the outset when we write $\underline{x}_1 = \underline{x}_1^0 + \varepsilon \underline{x}_1^1 + \cdots$ and $\underline{x}_2 = \underline{x}_2^0 + \varepsilon \underline{x}_2^1 + \cdots$ we do not know $\underline{x}_1^0$ and $\underline{x}_2^0$ and we must determine their values as we go along. We do not explore this and other complications. That would deflect us from our elementary goals.

5.9   More Information

Here is more information on the material in this lecture.

1. Early in the lecture we said that if there is an inner product in which $A^* = A$ then A has a complete set of eigenvectors. Indeed in that inner product A has n mutually perpendicular eigenvectors. This establishes their linear independence and linear independence if not orthogonality is retained as we introduce other inner products. The reader may go on and show that the corresponding eigenvalues are real by calculating $\lambda_i \left\langle \underline{x}_i, \underline{x}_i \right\rangle$. The condition that $A^* = A$ is $GA = \left(\overline{GA} \right)^T$.

The converse of this is true. If A has a complete set of eigenvectors and real eigenvalues then $A^* = A$ in some inner product. Denoting the eigenvectors $\underline{x}_1,\ \underline{x}_2,\ \ldots,\ \underline{x}_n$ and the corresponding eigenvalues $\lambda_1,\ \lambda_2,\ \ldots,\ \lambda_n$ we can write

$$A\left(\underline{x}_1\,\underline{x}_2\ \ldots\ \underline{x}_n\right) = \left(\underline{x}_1\,\underline{x}_2\ \ldots\ \underline{x}_n\right)\ \text{diag}\left(\lambda_1\,\lambda_2\ \ldots\ \lambda_n\right)$$

hence letting $X = \left(\underline{x}_1\ \underline{x}_2\ \ldots\ \underline{x}_n\right)$ we have

$$AX\overline{X}^T = X\ \text{diag}\left(\lambda_1\,\lambda_2\ \ldots\ \lambda_n\right)\overline{X}^T$$

so if $G = X\overline{X}^T$ then $G = \overline{G}^T$, $\underline{x}^T G\underline{x} > 0$ for all $\underline{x} \neq \underline{0}$ and

$$AG = \left(\overline{AG}\right)^T$$

This result tells us that the requirement $A^* = A$ in some inner product is necessary and sufficient that A have a complete set of eigenvectors and real eigenvalues. The readers may ask: why is it that the λ's must be real?

2. If $\lambda_1,\ \lambda_2,\ \ldots,\ \lambda_d$ are distinct eigenvalues then any set of d eigenvectors, each eigenvector corresponding to a different eigenvalue, is independent. This is true whatever the multiplicities of $\lambda_1\,\lambda_2\ \ldots\ \lambda_d$. If the set of eigenvectors is $\left\{\underline{x}_1,\ \underline{x}_2,\ \ldots,\ \underline{x}_d\right\}$ then to determine whether it is independent or dependent we must solve the equation

$$c_1\underline{x}_1 + c_2\underline{x}_2 + \cdots + c_d\underline{x}_d = \underline{0}$$

To do this we multiply by $\left(A - \lambda_1 I\right)$ getting

$$c_2\left(\lambda_2 - \lambda_1\right)\underline{x}_2 + \cdots + c_d\left(\lambda_d - \lambda_1\right)\underline{x}_d = \underline{0}$$

and then by $\left(A - \lambda_2 I\right),\ \ldots$ ultimately getting

$$c_d\left(\lambda_d - \lambda_1\right)\left(\lambda_d - \lambda_2\right)\cdots\left(\lambda_d - \lambda_{d-1}\right)\underline{x}_d = \underline{0}$$

or $c_d = 0$. Likewise $c_1 = 0, c_2 = 0$, etc. More is true: If n_1 independent eigenvectors correspond to λ_1, n_2 to $\lambda_2, \ldots$ then the set of $n_1 + n_2 + \cdots$ eigenvectors is independent.

Even more is true. If corresponding to each distinct eigenvalue we determine a set of independent eigenvectors and then corresponding to each of these a chain of generalized eigenvectors, then all of these eigenvectors and generalized eigenvectors are independent. The idea is that by multiplying a linear combination of these vectors by $(A - \lambda_1 I)$ sufficiently many times we can remove from it all vectors corresponding to λ_1. To see how this works take the simple example where $\underline{x}_1$ and $\underline{x}_2$ correspond to λ_1 and $\underline{x}_3$ to λ_3. Then write

$$A\underline{x}_1 = \lambda_1 \underline{x}_1$$

$$A\underline{x}_2 = \underline{x}_1 + \lambda_1 \underline{x}_2$$

and

$$A\underline{x}_3 = \lambda_3 \underline{x}_3$$

To determine c_1, c_2 and c_3 so that

$$c_1 \underline{x}_1 + c_2 \underline{x}_2 + c_3 \underline{x}_3 = \underline{0}$$

multiply this by $(A - \lambda_1 I)$ to get

$$c_2 \underline{x}_1 + c_3(\lambda_3 - \lambda_1)\underline{x}_3 = \underline{0}$$

and then again by $(A - \lambda_1 I)$ to get

$$c_3(\lambda_3 - \lambda_1)^2 \underline{x}_3 = \underline{0}.$$

By doing this we discover that $c_3 = 0$ and this implies $c_2 = 0$ and so $c_1 = 0$.

3. Any set of independent vectors in C^n determines a unique biorthogonal set in its span. For example if $\underline{x}_1$ and $\underline{x}_2$ are independent in C^n then we can determine $\underline{y}_1$ and $\underline{y}_2$ in $[\underline{x}_1, \underline{x}_2]$

so that $\langle\, \underline{y}_1,\ \underline{x}_1\,\rangle = 1, \langle\, \underline{y}_1,\ \underline{x}_2\,\rangle = 0, \langle\, \underline{y}_2,\ \underline{x}_1\,\rangle = 0,$ and $\langle\, \underline{y}_2,\ \underline{x}_2\,\rangle = 1.$ Indeed writing $\underline{y}_1 = a\underline{x}_1 + b\underline{x}_2,\ \underline{y}_2 = c\underline{x}_1 + d\underline{x}_2$ the four biorthogonality conditions determine a, b, c and d uniquely. If $\underline{x}_1, \underline{x}_2,$ and $\underline{x}_3,$ are independent we can determine $\underline{y}_1, \underline{y}_2,$ and $\underline{y}_3$ in $\left[\underline{x}_1,\ \underline{x}_2,\ \underline{x}_3\right]$ so that $\langle\, \underline{y}_i,\ \underline{x}_j\,\rangle = \delta_{ij}$ but now $\underline{y}_1$ and $\underline{y}_2$ need not lie in $\left[\underline{x}_1,\ \underline{x}_2\right].$

5.10   Similarity or Basis Transformations

We introduce the idea that a matrix represents a linear operator in a specified basis. Let $\vec{x}$ be a vector (possibly a column vector) and $\left\{\vec{e}_1,\ \vec{e}_2,\ \ldots,\ \vec{e}_n\right\}$ be a basis for the n dimensional vector space in which $\vec{x}$ resides. Then we can write

$$\vec{x} = x_1\vec{e}_1 + x_2\vec{e}_2 + \cdots + x_n\vec{e}_n$$

and denote by $\underline{x} = \begin{pmatrix} x_1 \\ x_2 \\ \vdots \\ x_n \end{pmatrix}$ the column vector representing $\vec{x}$ in the basis $\left\{\vec{e}_1,\ \vec{e}_2,\ \ldots,\ \vec{e}_n\right\}.$ If $\left\{\vec{f}_1,\ \vec{f}_2,\ \ldots,\ \vec{f}_n\right\}$ is a second basis written in terms of the first by

$$\vec{f}_1 = p_{11}\,\vec{e}_1 + p_{21}\,\vec{e}_2 + \cdots + p_{n1}\,\vec{e}_n$$

$$\vec{f}_2 = p_{12}\,\vec{e}_1 + p_{22}\,\vec{e}_2 + \cdots + p_{n2}\,\vec{e}_n$$

etc.

then the column vectors representing the vectors $\vec{f}_1,\ \vec{f}_2,\ \ldots,\ \vec{f}_n$ in the basis $\left\{\vec{e}_1,\ \vec{e}_2,\ \ldots,\ \vec{e}_n\right\}$ are the columns of a matrix denoted P. And because $\left\{\vec{f}_1,\ \vec{f}_2,\ \ldots,\ \vec{f}_n\right\}$ is independent so also the set of columns of P and hence $\det P \neq 0.$ If we now write

$$\vec{x} = y_1\vec{f}_1 + y_2\vec{f}_2 + \cdots + y_n\vec{f}_n$$

then $\vec{x}$ is represented in the basis $\left\{ \vec{f_1},\ \vec{f_2},\ \ldots,\ \vec{f_n} \right\}$ by the column vector $\underline{y} = \begin{pmatrix} y_1 \\ y_2 \\ \vdots \\ y_n \end{pmatrix}$ and a

simple calculation shows that

$$\underline{x} = P\underline{y}$$

The formula $\underline{x} = P\underline{y}$ determines the column vector $\underline{x}$ representing a vector $\vec{x}$ in the basis $\left\{ \vec{e_1},\ \vec{e_2},\ \ldots,\ \vec{e_n} \right\}$ in terms of the column vector $\underline{y}$ representing the vector $\vec{x}$ in another basis $\left\{ \vec{f_1},\ \vec{f_2},\ \ldots,\ \vec{f_n} \right\}$. The columns of the transformation matrix P are the column vectors representing the second basis vectors in the first basis. Each vector $\vec{x}$ is represented by many column vectors corresponding to many bases and each column vector represents many vectors again corresponding to many bases but the representation is one-to-one in a fixed basis.

If L is a linear operator (possibly an $n \times n$ matrix) acting in this vector space we can write

$$L\vec{e_1} = a_{11}\,\vec{e_1} + a_{21}\,\vec{e_2} + \cdots + a_{n1}\,\vec{e_n}$$

$$L\vec{e_2} = a_{12}\,\vec{e_1} + a_{22}\,\vec{e_2} + \cdots + a_{n2}\,\vec{e_n}$$

etc.

and denote by A the matrix whose columns are the column vectors representing $L\vec{e_1},\ L\vec{e_2},\ \ldots$ in the basis $\left\{ \vec{e_1},\ \vec{e_2},\ \ldots,\ \vec{e_n} \right\}$. We call this the matrix representing L in the basis $\left\{ \vec{e_1},\ \vec{e_2},\ \ldots,\ \vec{e_n} \right\}$. Likewise denoting by B the matrix representing L in the basis $\left\{ \vec{f_1},\ \vec{f_2},\ \ldots,\ \vec{f_n} \right\}$ we find:

$$A = PBP^{-1}$$

The formula $A = PBP^{-1}$ determines the matrix A representing L in the basis $\left\{ \vec{e_1},\ \vec{e_2},\ \ldots,\ \vec{e_n} \right\}$ in terms of the matrix B representing L in another basis $\left\{ \vec{f_1},\ \vec{f_2},\ \ldots,\ \vec{f_n} \right\}$. Each linear operator L is represented by many matrices corresponding to many bases and all display the same information but this information is easier to obtain in some bases than it is in others. Indeed if $A, \underline{x}$ and $\underline{b}$

represent L, $\vec{x}$ and $\vec{b}$ in the basis $\{\vec{e}_1, \vec{e}_2, \ldots, \vec{e}_n\}$ while in $\{\vec{f}_1, \vec{f}_2, \ldots, \vec{f}_n\}$ the representation is $B, \underline{y}$ and $\underline{c}$ then the equation $L\vec{x} = \vec{b}$ is represented in C^n by both $A\underline{x} = \underline{b}$ and $B\underline{y} = \underline{c}$. And one of these may be easier to solve than the other.

Using $A = PBP^{-1}$ and the theorem that the determinant of a product is the product of the determinants of its factors we see that

$$\det\left(\lambda I - A\right) = \det\left(\lambda I - B\right)$$

and hence we define the characteristic polynomial of L to be the characteristic polynomial of any matrix that represents it. We then define the eigenvalues of L to be the eigenvalues of any matrix that represents it. The eigenvalues of any two matrices A and B, where $A = PBP^{-1}$ for any nonsingular matrix P, are the same. The eigenvalues of L are independent of the basis used for their determination.

If L has an invariant subspace of dimension k, then, using a basis whose first k vectors lie in this subspace, the matrix representing L in this basis reflects this structure by exhibiting an $(n-k) \times k$ block of zeros in its lower left hand corner. Its determinant then factors as the product of the determinants of its upper left hand $k \times k$ block and its lower right hand $(n-k) \times (n-k)$ block. This establishes the result that the geometric multiplicity of an eigenvalue cannot exceed its algebraic multiplicity. To see this let $\dim \operatorname{Ker}\left(A - \lambda_1 I\right) = n_1$ and let the first n_1 vectors in a basis be a basis for $\operatorname{Ker}\left(A - \lambda_1 I\right)$, then $\det\left(\lambda I - A\right)$ contains the factor $\left(\lambda - \lambda_1\right)^{m_1}$ where m_1 cannot be less than n_1.

Let the linear operator L be the $n \times n$ matrix A. Then it is represented by itself in the natural

basis, $\left\{\begin{pmatrix} 1 \\ 0 \\ \vdots \\ 0 \end{pmatrix}, \cdots, \begin{pmatrix} 0 \\ 0 \\ \vdots \\ 1 \end{pmatrix}\right\}$. If A has a complete set of eigenvectors $\underline{x}_1, \underline{x}_2, \ldots, \underline{x}_n$

its matrix in this basis is the diagonal matrix of the corresponding eigenvalues. Such a basis is called a diagonalizing basis. If to the eigenvalue λ_1 repeated twice, there corresponds only the eigenvector $\underline{x}_1$, we construct a generalized eigenvector $\underline{x}_2$ and in the basis $\{\underline{x}_1, \underline{x}_2, \ldots\}$ the matrix of A has the block $\begin{pmatrix} \lambda_1 & 1 \\ 0 & \lambda_1 \end{pmatrix}$ in the upper left hand corner. Using a basis of eigenvectors and generalized eigenvectors, we find that the matrix representing A is block diagonal. To each eigenvalue λ_i of multiplicity $m_i, i = 1, 2, \ldots, d$, there corresponds an $m_i \times m_i$ block. Outside of these d blocks all elements vanish. Inside the ith block λ_i appears on the diagonal, 1 or 0 on the superdiagonal, and 0 elsewhere. The structure of the superdiagonal is determined by the chains of generalized eigenvectors. For instance if λ_1 is a threefold root to which corresponds only the eigenvector $\underline{x}_1$ then $\underline{x}_1$ generates the chain $\underline{x}_1 \to \underline{x}_2 \to \underline{x}_3$ and the corresponding block is

$\begin{pmatrix} \lambda_1 & 1 & 0 \\ 0 & \lambda_1 & 1 \\ 0 & 0 & \lambda_1 \end{pmatrix}$; but if there are two eigenvectors $\underline{x}_1$ and $\underline{x}_2$ and $\underline{x}_2$ generates the chain $\underline{x}_2 \to \underline{x}_3$

the block is $\begin{pmatrix} \lambda_1 & 0 & 0 \\ 0 & \lambda_1 & 1 \\ 0 & 0 & \lambda_1 \end{pmatrix}$.

The forms we have been talking about, including the purely diagonal form, are called Jordan forms. Such forms are either diagonal or as close to diagonal as we can get using basis transformations. In $A = PJP^{-1}$ the columns of the transformation matrix are the column vectors representing the eigenvectors and generalized eigenvectors of A in the natural basis.

Shilov's book "*Linear Algebra*" gives an algebraic account of this via polynomial algebras and their ideals. Gantmacher's book "*Theory of Matrices*" gives both an algebraic and a geometric explanation.

5.11   Home Problems

1. Derive the formula for the eigenvalues and eigenvectors of the block triangular matrix

$\begin{pmatrix} A & 0 \\ B & C \end{pmatrix}$ where the blocks on the diagonal are square and have simple eigenvalues. To do this, write the eigenvalue problem as

$$\begin{pmatrix} A & 0 \\ B & C \end{pmatrix} \begin{pmatrix} \underline{x} \\ \underline{y} \end{pmatrix} = \lambda \begin{pmatrix} \underline{x} \\ \underline{y} \end{pmatrix}$$

and solve it in terms of the solutions to the eigenvalue problems for A and C.

2. A linear separating cascade is run steadily in cocurrent flow. Show that

$$\begin{pmatrix} x_{i+1} \\ y_{i+1} \end{pmatrix} = \frac{1}{1 + \dfrac{mV}{L}E} \begin{pmatrix} 1 & \dfrac{V}{L}E \\ mE & 1 + \left(\dfrac{mV}{L} - 1\right)E \end{pmatrix} \begin{pmatrix} x_i \\ y_i \end{pmatrix}$$

and hence determine $\begin{pmatrix} x_{\text{out}} \\ y_{\text{out}} \end{pmatrix}$ in terms of $\begin{pmatrix} x_{\text{in}} \\ y_{\text{in}} \end{pmatrix}$ for an n stage cascade. The performance of such a cascade cannot exceed that of one equilibrium stage. Indeed if $E = 1$ the stepping matrix is singular and $\begin{pmatrix} x_i \\ y_i \end{pmatrix}$ cannot be determined from $\begin{pmatrix} x_{i+1} \\ y_{i+1} \end{pmatrix}$. And for good reason.

For a countercurrent cascade the Kremser equation is the corresponding formula for $\begin{pmatrix} x_{\text{in}} \\ y_{\text{out}} \end{pmatrix}$ in terms of $\begin{pmatrix} x_{\text{out}} \\ y_{\text{in}} \end{pmatrix}$. For fixed $x_{\text{in}}, y_{\text{in}}$, investigate x_{out} and y_{out} as n grows large. The results will depend on whether $\dfrac{mV}{L}$ is greater or less than 1 and on whether y_{in} is greater or less than mx_{in}.

Let $\dfrac{mV}{L} = 1$ and rederive the Kremser equation. In this instance the eigenvalue 1 is repeated and corresponds to only one independent eigenvector. Is the result the limit of the ordinary Kremser equation for $\dfrac{mV}{L} \neq 1$ as $\dfrac{mV}{L} \to 1$?

3. Determine the eigenvectors and the eigenvalues of the matrix

$$I + \underline{a}\,\underline{b}^T$$

Show that its trace is $n + \underline{b}^T \underline{a}$ and that its determinant is $1 + \underline{b}^T \underline{a}$.

4. Show that the eigenvalues of diagonal and triangular matrices are their diagonal elements.

5. If the simple Drude model is used to derive the potential energy of three molecules lying on a straight line, the matrix

$$\begin{pmatrix} a & -b & -c \\ -b & a & -d \\ -c & -d & a \end{pmatrix}$$

turns up. The numbers a, b, c and d are all positive and $a >> b, c, d$. The numbers b, c and d denote the dipole-dipole interactions.

The principal invariants of this matrix are

$$\Delta_1 = 3a$$

$$\Delta_2 = 3a^2 - \left\{ b^2 + c^2 + d^2 \right\}$$

$$\Delta_3 = a^3 - a\left\{ b^2 + c^2 + d^2 \right\} - 2bcd$$

and so its eigenvalues remain unchanged if b, c and d are interchanged.

Show that if $c = d = 0$ the eigenvalues are

$$\lambda_1 = a - b$$

$$\lambda_2 = a$$

$$\lambda_3 = a + b$$

while if only $d = 0$ the eigenvalues are

$$\lambda_1 = a - \sqrt{b^2 + c^2}$$

$$\lambda_2 = a$$

$$\lambda_3 = a + \sqrt{b^2 + c^2}$$

Indeed if $d \neq 0$ we might guess a fair approximation to be

$$\lambda_1 = a - \sqrt{b^2 + c^2 + d^2}$$

$$\lambda_2 = a$$

$$\lambda_3 = a + \sqrt{b^2 + c^2 + d^2}$$

and using this guess we get

$$\lambda_1 + \lambda_2 + \lambda_3 = \Delta_1$$

$$\lambda_1 \lambda_2 + \lambda_2 \lambda_3 + \lambda_3 \lambda_1 = \Delta_2$$

$$\lambda_1 \lambda_2 \lambda_3 = \Delta_3 + 2bcd$$

Write

$$\begin{pmatrix} a & -b & -c \\ -b & a & -d \\ -c & -d & a \end{pmatrix} = \begin{pmatrix} a & -b & -c \\ -b & a & 0 \\ -c & 0 & a \end{pmatrix} - d \begin{pmatrix} 0 & 0 & 0 \\ 0 & 0 & 1 \\ 0 & 1 & 0 \end{pmatrix}$$

and estimate the eigenvalues by a perturbation calculation. Carry this out to first order in d and then calculate $\lambda_1 + \lambda_2 + \lambda_3$, $\lambda_1 \lambda_2 + \lambda_2 \lambda_3 + \lambda_3 \lambda_1$ and $\lambda_1 \lambda_2 \lambda_3$. Are your estimates improved if d^2 is added to $b^2 + c^2$ wherever $b^2 + c^2$ appears?

6. Here is an $n \times n$ matrix:

$$
\begin{pmatrix}
1 & 1 & \cdots & 1 \\
1 & 1 & \cdots & 1 \\
\vdots & \vdots & & \vdots \\
1 & 1 & \cdots & 1
\end{pmatrix}
$$

Show that its eigenvalues are n and 0 where 0 is repeated $n-1$ times. Show that the corresponding eigenvectors are

$$
\begin{pmatrix}
1 \\
1 \\
\vdots \\
1
\end{pmatrix}
$$

and

$$
\begin{pmatrix}
1 \\
-1 \\
0 \\
0 \\
\vdots \\
0
\end{pmatrix},
\begin{pmatrix}
1 \\
0 \\
-1 \\
0 \\
\vdots \\
0
\end{pmatrix},
\cdots
\begin{pmatrix}
1 \\
0 \\
0 \\
0 \\
\vdots \\
-1
\end{pmatrix}
$$

Construct an orthogonal set of eigenvectors.

7. Let A, P and Q be $n \times n$ matrices where A and Q are known and P is to be determined. The equation for doing this is

$$
A^T P + P A = -Q
$$

This is a system of linear equations in the unknown elements of the matrix P.

As such it can be written

$$a\underline{p} = -\underline{q}$$

where a is an $n^2 \times n^2$ matrix and $\underline{p}$ and $\underline{q}$ are $n^2 \times 1$ vectors. To see what this equation looks like and to decide when it can be solved, suppose that the eigenvalue problems for A and A^T lead to the biorthogonal sets of eigenvectors $\{\underline{x}_1, \underline{x}_2, \ldots, \underline{x}_n\}$ and $\{\underline{y}_1, \underline{y}_2, \ldots, \underline{y}_n\}$ and the corresponding eigenvalues $\{\lambda_1, \lambda_2, \ldots, \lambda_n\}$. Then write

$$A^T P + PA = -Q$$

in the basis $\{\underline{y}_i\, \underline{y}_j^T\}$ and show that the linear operator

$$A^T(\quad) + (\quad)A$$

is non singular iff $\lambda_i + \lambda_j \neq 0 \ \ \forall\, i, j$.

8. Determine the eigenvalues and the eigenvectors of the matrix

$$I + \underline{a}\,\underline{b}^T + \underline{b}\,\underline{a}^T$$

taking $\underline{a}$ and $\underline{b}$ to be independent.

To do this let $\underline{x}_1, \underline{x}_2, \ldots, \underline{x}_{n-2}$ be independent and lie in $[\underline{a},\ \underline{b}]^{\perp}$. Then $\underline{x}_1, \underline{x}_2, \ldots, \underline{x}_{n-2}$ are eigenvectors corresponding to the eigenvalue 1.

The remaining two eigenvectors lie in $[\underline{a},\ \underline{b}]^{\perp\perp} = [\underline{a},\ \underline{b}]$. To find these and the corresponding eigenvalues put

$$\underline{x} = \alpha\,\underline{a} + \beta\,\underline{b}$$

and observe that

$$\{I + \underline{a}\,\underline{b}^T + \underline{b}\,\underline{a}^T\}\{\alpha\,\underline{a} + \beta\,\underline{b}\} = \lambda\{\alpha\,\underline{a} + \beta\,\underline{b}\}$$

requires

$$\alpha + \alpha\,\underline{b}^T\underline{a} + \beta\,\underline{b}^T\underline{b} = \lambda\,\alpha$$

and

$$\beta + \alpha\,\underline{a}^T\underline{a} + \beta\,\underline{a}^T\underline{b} = \lambda\,\beta$$

or

$$\begin{pmatrix} 1 + \underline{b}^T\underline{a} & \underline{b}^T\underline{b} \\ \underline{a}^T\underline{a} & 1 + \underline{a}^T\underline{b} \end{pmatrix} \begin{pmatrix} \alpha \\ \beta \end{pmatrix} = \lambda \begin{pmatrix} \alpha \\ \beta \end{pmatrix}$$

The determinant and the trace of the matrix on the left hand side are

$$\det = \left(1 + \underline{a}^T\underline{b}\right)^2 - \underline{a}^T\underline{a}\,\underline{b}^T\underline{b}$$

and

$$\mathrm{tr} = 2\left(1 + \underline{a}^T\underline{b}\right)$$

whence

$$\mathrm{tr}^2 - 4\det > 0$$

and this tells us that the remaining two eigenvalues are real and not equal.

As a simple example, let

$$\underline{a}^T\underline{b} = 0, \quad \underline{a}^T\underline{a} = 1, \quad \underline{b}^T\underline{b} = 1$$

then the eigenvalues of the matrix

$$\begin{pmatrix} 1 & 1 \\ 1 & 1 \end{pmatrix}$$

are 0 and 2 corresponding to $\alpha = 1, \beta = -1$ and $\alpha = 1, \beta = 1$.

Determine the eigenvectors and the eigenvalues of

$$I + \underline{a}\,\underline{b}^T + \underline{c}\,\underline{d}^T$$

taking $\underline{a}$ and $\underline{c}$ and $\underline{b}$ and $\underline{d}$ to be independent.

9. Let A and B be $m \times n$ matrices of ranks r and s. Then solutions, $\underline{x} \in R^n$, $\underline{x} \neq \underline{0}$, to the problem

$$A\underline{x} = \lambda B\underline{x}$$

can sometimes be of interest. Begin the study of this problem by finding a bound on the largest number of independent solutions corresponding to any $\lambda \neq 0$.

How can $\operatorname{Im} A \cap \operatorname{Im} B$ be determined?

10. Suppose A commutes with its adjoint A^*, viz.,

$$AA^* = A^*A$$

whereupon A is called normal.

Denote by $\underline{x}, \lambda$ an eigenvector and the corresponding eigenvalue of A. Assume the eigenspaces of A are one dimensional and prove that $\underline{x}$ is an eigenvector of A^*. Hence conclude that the eigenvectors of A form an orthogonal set of vectors.

Lecture 6

The Solution of Differential and Difference Equations

6.1 A Formula for the Solution to $d\underline{x}/dt = A\underline{x}$

Suppose the departure of $x_1(t), x_2(t), \ldots, x_n(t)$ from assigned initial values is determined by the set of n ordinary differential equations

$$\frac{dx_i}{dt} = \sum_{j=1}^{n} a_{ij} x_j, \quad i = 1, 2, \ldots, n.$$

Then introducing $\underline{x}(t)$ via

$$\underline{x}(t) = \begin{pmatrix} x_1(t) \\ \vdots \\ x_n(t) \end{pmatrix}$$

we can write this as

$$\frac{d\underline{x}}{dt} = A\underline{x}$$

where $\underline{x}(t=0)$ is assigned and where the constants a_{ij} are the elements of the $n \times n$ matrix A. We suppose, first, that A has a complete set of eigenvectors and construct the solution to this problem in terms of these eigenvectors.

To introduce the notation let the eigenvectors and eigenvalues of A and A^* satisfy

$$A\underline{x}_i = \lambda_i \underline{x}_i, \quad i = 1, \ldots, n$$

and

$$A^* \underline{y}_i = \overline{\lambda}_i \underline{y}_i, \quad i = 1 \ldots, n$$

where $\left\langle \underline{y}_i, \underline{x}_j \right\rangle = \delta_{ij}$.

Then to solve our problem we expand its solution in the eigenvectors of A as

$$\underline{x}(t) = \sum_{i=1}^{n} c_i(t) \underline{x}_i$$

and seek the coefficients $c_i(t)$ in this expansion, where $c_i(t) = \left\langle \underline{y}_i, \underline{x}(t) \right\rangle$.

Thus, what we do to solve linear problems requires three steps to be carried out: determine an eigenvector basis, expand the solution in this basis and determine the coefficients in this expansion. Its simplicity rests on the idea of biorthogonal sets of vectors.

To establish an equation satisfied by $c_i(t)$ we multiply both sides of $d\underline{x}/dt = A\underline{x}$ by $\overline{\underline{y}}_i^T G$ obtaining

$$\left\langle \underline{y}_i, \frac{d\underline{x}}{dt} \right\rangle = \left\langle \underline{y}_i, A\underline{x} \right\rangle$$

This leads to

$$\frac{d}{dt}\left\langle \underline{y}_i, \underline{x}\right\rangle = \left\langle A^*\underline{y}_i, \underline{x}\right\rangle$$

$$= \lambda_i \left\langle \underline{y}_i, \underline{x}\right\rangle$$

whence each coefficient $c_i(t)$ satisfies

$$\frac{dc_i}{dt} = \lambda_i c_i.$$

As a result we have

$$c_i(t) = c_i(t=0)e^{\lambda_i t}$$

and the solution to our problem is

$$\underline{x}(t) = \sum_{i=1}^{n}\left\langle \underline{y}_i, \underline{x}(t=0)\right\rangle e^{\lambda_i t}\underline{x}_i$$

Indeed if

$$\frac{d\underline{x}}{dt} = A\underline{x} + \underline{b}(t)$$

we need to add

$$\sum_{i=1}^{n}\int_{0}^{t} e^{\lambda_i(t-\tau)}\left\langle \underline{y}_i, \underline{b}(\tau)\right\rangle d\tau\, \underline{x}_i$$

to the foregoing to obtain the solution. And this may be discovered using the same steps that led to the solution of the problem where $\underline{b}(t) = \underline{0}$.

The problem as originally written requires that we determine the unknown functions $x_1(t), \ldots, x_n(t)$ simultaneously. These functions are the components of $\underline{x}(t)$ in the natural basis for the problem and each component ordinarily appears in each equation. So we look for a way to break this coupling. When A has a complete set of eigenvectors we can do this by expanding the solution $\underline{x}(t)$ in the eigenvector basis. Then the determination of the expansion coefficients $c_1(t), ..., c_n(t)$, unlike the determination of the natural components $x_1(t), ..., x_n(t)$, is a completely uncoupled problem.

We work out two examples:

Example (i)

Let $A = \begin{pmatrix} -1 & -1 \\ 1 & -1 \end{pmatrix}$ then $\operatorname{tr} A = -2$ and $\det A = 2$. The eigenvalues are $\lambda_1 = -1 + i$ and $\lambda_2 = -1 - i$ and the corresponding eigenvectors are $\underline{x}_1 = \begin{pmatrix} 1 \\ -i \end{pmatrix}$ and $\underline{x}_2 = \begin{pmatrix} 1 \\ i \end{pmatrix}$. Because A is real and λ_1 and $\underline{x}_1$ satisfy the eigenvalue problem so also do $\overline{\lambda}_1$ and $\overline{\underline{x}}_1$. And although we find, in the plain vanilla inner product, that $\langle \underline{x}_1, \underline{x}_2 \rangle = 0$, A^* is not equal to A. What in fact is true is that $A\overline{A}^T = \overline{A}^T A$, i.e., that A is normal. In the plain vanilla inner product the biorthogonal set is

$$\left\{ \underline{y}_1 = \frac{1}{2} \begin{pmatrix} 1 \\ -i \end{pmatrix}, \underline{y}_2 = \frac{1}{2} \begin{pmatrix} 1 \\ i \end{pmatrix} \right\}.$$

Hence the solution to $d\underline{x}/dt = A\underline{x}$ is

$$\underline{x}(t) = c_1 e^{(-1+i)t} \begin{pmatrix} 1 \\ -i \end{pmatrix} + c_2 e^{(-1-i)t} \begin{pmatrix} 1 \\ i \end{pmatrix}$$

where here c_1 denotes $c_1(t = 0)$, etc., and

$$c_1 = \left\langle \frac{1}{2} \begin{pmatrix} 1 \\ -i \end{pmatrix}, \underline{x}(t = 0) \right\rangle$$

and

$$c_2 = \left\langle \frac{1}{2} \begin{pmatrix} 1 \\ i \end{pmatrix}, \underline{x}(t = 0) \right\rangle.$$

When A is real and $\underline{x}(t = 0)$ is real then $\underline{x}(t)$ must be real for all values of t. If $\lambda_2 = \bar{\lambda}_1$ and we require $\underline{x}_2 = \bar{\underline{x}}_1$ then $\underline{y}_2 = \bar{\underline{y}}_1$ and hence $c_2 = \bar{c}_1$ and so the two terms adding to $\underline{x}(t)$, $c_1 e^{\lambda_1 t}\underline{x}_1$ and $c_2 e^{\lambda_2 t}\underline{x}_2$, are complex conjugates. As a result $\underline{x}(t)$ can be written as $2\mathrm{Re}\left\{c_1 e^{\lambda_1 t}\underline{x}_1\right\}$. In this example this is

$$2\mathrm{Re}\left\{c_1 e^{(-1+i)t}\begin{pmatrix} 1 \\ -i \end{pmatrix}\right\}$$

and, on writing

$$\lambda_1 = \mathrm{Re}\,\lambda_1 + i\,\mathrm{Im}\lambda_1 = -1 + i$$

$$\underline{x}_1 = \mathrm{Re}\,\underline{x}_1 + i\,\mathrm{Im}\underline{x}_1 = \begin{pmatrix} 1 \\ 0 \end{pmatrix} + i\begin{pmatrix} 0 \\ -1 \end{pmatrix}$$

and

$$c_1 = \rho e^{i\phi},$$

we have

$$2\rho e^{\mathrm{Re}\lambda_1 t}\left\{\cos\left(\mathrm{Im}\lambda_1 t + \phi\right)\mathrm{Re}\underline{x}_1 - \sin\left(\mathrm{Im}\lambda_1 t + \phi\right)\mathrm{Im}\underline{x}_1\right\}$$

which corresponds to a converging spiral in the x_1, x_2 plane.

Example (ii)

Let $A = \begin{pmatrix} 1 & -2 \\ 3 & -4 \end{pmatrix}$ then $\mathrm{tr}A = -3$, $\det A = 2$ and

$$\lambda_1 = -1,\ \underline{x}_1 = \begin{pmatrix} 1 \\ 1 \end{pmatrix},\ \lambda_2 = -2 \text{ and } \underline{x}_2 = \begin{pmatrix} 1 \\ \dfrac{3}{2} \end{pmatrix}.$$

The biorthogonal set in the plain vanilla inner product is

$$\left\{ \underline{y}_1 = 2 \begin{pmatrix} \dfrac{3}{2} \\ -1 \end{pmatrix}, \; \underline{y}_2 = 2 \begin{pmatrix} -1 \\ 1 \end{pmatrix} \right\}.$$

and the solution to $d\underline{x}/dt = A\underline{x}$ is

$$\underline{x}(t) = c_1 e^{-t} \begin{pmatrix} 1 \\ 1 \end{pmatrix} + c_2 e^{-2t} \begin{pmatrix} 1 \\ \dfrac{3}{2} \end{pmatrix}$$

where again c_1 denotes $c_1(t = 0)$, etc., and

$$c_1 = \left\langle 2 \begin{pmatrix} \dfrac{3}{2} \\ -1 \end{pmatrix}, \; \underline{x}(t = 0) \right\rangle$$

and

$$c_2 = \left\langle 2 \begin{pmatrix} -1 \\ 1 \end{pmatrix}, \; \underline{x}(t = 0) \right\rangle.$$

We see that as t increases the second term dies out exponentially fast compared to the first and for large enough values of t

$$\underline{x}(t) \sim c_1 e^{-t} \begin{pmatrix} 1 \\ 1 \end{pmatrix}$$

This tells us that $\underline{x}(t)$ approaches $\underline{0}$ from the direction $\begin{pmatrix} 1 \\ 1 \end{pmatrix}$. We will use this fact in Lecture 8 to help us turn experimental data into estimates of the elements of A.

There are special solutions to $d\underline{x}/dt = A\underline{x}$ called equilibrium solutions. These satisfy $A\underline{x} = \underline{0}$ as then $d\underline{x}/dt = \underline{0}$. An equilibrium solution is constant in time and can only be obtained by starting there. If $\det A \neq 0$ then $\underline{0}$ is the only solution to $A\underline{x} = \underline{0}$ and hence is the only equilibrium solution. All other solutions are always on the move and according to where they start are given by our formula

$$\underline{x}(t) = \sum_{i=1}^{n} \left\langle \underline{y}_i, \underline{x}(t=0) \right\rangle e^{\lambda_i t} \underline{x}_i.$$

Now these may or may not converge to the equilibrium solution as time grows large. If all do, we call the equilibrium solution asymptotically stable and a necessary and sufficient condition for this is that $\text{Re}\lambda_i < 0$, $i = 1, 2, \ldots, n$, i.e., all eigenvalues lie in the left half of the complex plane. If this is so $\underline{x}(t)$ goes to $\underline{0}$ exponentially fast as t grows large at a rate determined by the largest of the $\text{Re}\lambda_i$, $i = 1, 2, \ldots, n$.

6.2   Gerschgorin's Circle Theorem

If n is 2, the eigenvalues of A are roots of a quadratic equation and it is easy to see that asymptotic stability obtains iff $\text{tr}A < 0$ and $\det A > 0$. But as n increases beyond 2 eigenvalues are increasingly difficult to determine and what we need is a simple estimate of where the eigenvalues of A lie. The best of these is Gerschgorin's circle theorem and it is surprisingly easy to prove; it tells us that each eigenvalue of A lies on or inside at least one of n circles in the complex plane. In fact, there are two sets of n circles: they are

$$|\lambda - a_{ii}| \leq \sum_{j \neq i} |a_{ij}| \ , \ i = 1, ..., n$$

and

$$|\lambda - a_{ii}| \leq \sum_{j \neq i} |a_{ji}| \ , \ i = 1, ..., n$$

The first set of circles corresponds to the rows of A. The second set to the rows of A^T and hence to the columns of A. For instance, in the second example, the best estimate via Gerschgorin's theorem is that the eigenvalues cannot lie in the region outside the two circles of radius 2 centered on -4 and 1. And both sets of circles are required to determine this.

If A is diagonal, Gerschgorin's theorem predicts all its eigenvalues; if A is triangular the theorem predicts the eigenvalues a_{11} and a_{nn}; the more diagonally dominant the matrix, the better the estimates made by the theorem.

6.3 A Formula for the Solution to $\underline{x}(k + 1) = A\underline{x}(k)$

As a variation on the foregoing problem, where the evolution of $\underline{x}(t)$ is continuous in time, we also look at the problem where $\underline{x}(k)$ evolves discretely in time. Suppose $\underline{x}(k)$, $k = 1, 2, \ldots$ satisfies

$$\underline{x}(k + 1) = A\underline{x}(k)$$

where $\underline{x}(k = 0)$ is assigned. This is a system of linear constant coefficient difference equations and we can write its solution $\underline{x}(n) = A^n \underline{x}(0)$ and use the spectral representation of A to discover the nature of the solution. But we can also expand the solution in the eigenvectors of A and to do this we write

$$\underline{x}(k) = \sum_{i=1}^{n} c_i(k) \underline{x}_i$$

where $c_i(k) = \left\langle \underline{y}_i, \underline{x}(k) \right\rangle$. To find an equation satisfied by $c_i(k)$ we multiply both sides of $\underline{x}(k + 1) = A\underline{x}(k)$ by $\overline{\underline{y}}_i^T G$ to obtain

$$\left\langle \underline{y}_i, \underline{x}(k + 1) \right\rangle = \left\langle \underline{y}_i, A\underline{x}(k) \right\rangle = \left\langle A^* \underline{y}_i, \underline{x}(k) \right\rangle = \lambda_i \left\langle \underline{y}_i, \underline{x}(k) \right\rangle.$$

Hence we discover that

$$c_i(k + 1) = \lambda_i c_i(k)$$

and so

$$c_i(k) = \lambda_i^k c_i(0)$$

and therefore

$$\underline{x}(k) = \sum_{i=1}^{n} \left\langle \underline{y}_i, \underline{x}(0) \right\rangle \lambda_i^k \underline{x}_i$$

Using this formula we see that the evolution of $\underline{x}(k)$ as k increases depends on where the eigenvalues of A lie.

The equilibrium solutions of $\underline{x}(k+1) = A\underline{x}(k)$ satisfy $\underline{x} = A\underline{x}$ or $\{A - I\}\underline{x} = \underline{0}$ because then $\underline{x}(k+1) = \underline{x}(k)$ and $\underline{x}(k)$ remains constant. An equilibrium solution can only be obtained by starting there. If $\det\{A - I\} \neq 0$ then $\underline{0}$ is the only equilibrium solution. All other solutions are always on the move and are given by our formula. They may or may not converge to the equilibrium solution as k grows large. If all do we call the equilibrium solution asymptotically stable. A necessary and sufficient condition for this is that $|\lambda_i| < 1$, $i = 1, 2, \ldots, n$, where $|\lambda_i|^2 = \lambda_i \overline{\lambda_i} = (\mathrm{Re}\lambda_i)^2 + (\mathrm{Im}\lambda_i)^2$. The stability requirement is that the eigenvalues lie inside the unit circle in the complex plane.

6.4   The Stiffness Problem

What we have found then is this: stability for the problem $d\underline{x}/dt = A\underline{x}$ requires the eigenvalues of A to lie in the left half of the complex plane; stability for the problem $\underline{x}(k+1) = A\underline{x}(k)$ requires the eigenvalues of A to lie inside the unit circle. To see what these conditions have to do with one another, let $\underline{x}(t)$ denote the solution to $d\underline{x}/dt = A\underline{x}$. Then a simple Euler approximation to $\underline{x}(t)$ satisfies

$$\underline{x}(k+1) = (I + \Delta t A)\underline{x}(k)$$

whence $\underline{x}(t)$ and its approximation are given by

$$\sum_{i=1}^{n} \left\langle \underline{y}_i, \underline{x}(t=0) \right\rangle e^{\lambda_i t} \underline{x}_i$$

and

$$\sum_{i=1}^{n} \left\langle \underline{y}_i, \underline{x}(k=0) \right\rangle (1 + \Delta t \lambda_i)^k \underline{x}_i$$

This is so as $\underline{x}_i$ is an eigenvector of both A and $I + \Delta t A$ corresponding to the eigenvalues λ_i and $1 + \Delta t \lambda_i$.

Now the first thing to observe is that the approximation converges to $\underline{x}(t)$ as $\Delta t \to 0$ where the limit is taken holding $k\Delta t = t$ fixed. The second thing to observe is that, even assuming the stability of the differential equation, the difference equation is not stable for all values of Δt. Indeed we see that the difference equation is stable iff, $\forall \lambda_i, i = 1, 2, \ldots, n$,

$$|1 + \Delta t \lambda_i|^2 < 1$$

or

$$(1 + \Delta t \text{Re}\lambda_i)^2 + (\Delta t \text{Im}\lambda_i)^2 < 1.$$

That it is possible to satisfy this by making Δt sufficiently small, when $\text{Re}\lambda_i < 0$, is due to the fact that $\text{Im}\lambda_i$ is multiplied by $(\Delta t)^2$ whereas $\text{Re}\lambda_i$ is multiplied by Δt.

It is ordinarily true that the eigenvalue whose real part is most negative sets a bound on how large Δt may be. Indeed stability of the difference equation, in the case where the eigenvalues are real and negative, requires that $\forall \lambda_i$

$$-1 < 1 + \Delta t \lambda_i < 1$$

or

$$\Delta t < \frac{2}{|\lambda_i|}$$

So, the most negative eigenvalue, the eigenvalue associated with the term in $\underline{x}(t)$ that dies out most rapidly, controls the size of Δt in the difference approximation. In numerical work this is referred to as the stiffness problem. Problems where the real parts of the eigenvalues are widely separated so that insignificant parts of their solutions, at least for $t > \epsilon$, ϵ small, control approximations to their solutions are called stiff problems. In doing a calculation you can never get rid of the most negative eigenvalue due to the fact that numerical errors act like new initial conditions.

The results obtained in this lecture will be used in subsequent lectures to investigate problems where we can establish the fact of a complete set of eigenvectors. Before turning to this we take up a problem where the set of eigenvectors is not complete and where the use of generalized eigenvectors is required.

6.5   The Use of Generalized Eigenvectors

To show how the solution to such a problem can be found, we set $n = 2$ and suppose that λ_1 is a double root of $\Delta(\lambda) = 0$. Then if $\dim \operatorname{Ker}(A - \lambda_1 I) = 1$, we write

$$A\underline{x}_1 = \lambda_1 \underline{x}_1$$

$$A\underline{x}_2 = \underline{x}_1 + \lambda_1 \underline{x}_2$$

and

$$A^*\underline{y}_1 = \overline{\lambda}_1 \underline{y}_1 + \underline{y}_2$$

$$A^*\underline{y}_2 = \overline{\lambda}_1 \underline{y}_2$$

where $\left\langle \underline{y}_i, \underline{x}_j \right\rangle = \delta_{ij}$. To determine $\underline{x}(t)$ where $d\underline{x}/dt = A\underline{x}$ we write $\underline{x}(t) = c_1(t)\,\underline{x}_1 + c_2(t)\,\underline{x}_2$ where $c_i(t) = \left\langle \underline{y}_i, \underline{x}(t) \right\rangle$ and discover in the usual way that $c_1(t)$ and $c_2(t)$ satisfy:

$$\left\langle \underline{y}_1, \frac{d\underline{x}}{dt} \right\rangle = \frac{dc_1}{dt} = \left\langle \underline{y}_1, A\underline{x} \right\rangle = \left\langle \overline{\lambda}_1\underline{y}_1 + \underline{y}_2, \underline{x} \right\rangle = \lambda_1 c_1 + c_2$$

and

$$\left\langle \underline{y}_2, \frac{d\underline{x}}{dt} \right\rangle = \frac{dc_2}{dt} = \left\langle \underline{y}_2, A\underline{x} \right\rangle = \left\langle \overline{\lambda}_1\underline{y}_2, \underline{x} \right\rangle = \lambda_1 c_2$$

Solving these we get

$$c_1 = e^{\lambda_1 t} c_1(t=0) + \int_0^t e^{\lambda_1(t-\tau)} c_2(\tau)\,d\tau$$

and

$$c_2 = e^{\lambda_1 t} c_2(t=0)$$

whence

$$c_1 = e^{\lambda_1 t} c_1(t=0) + t e^{\lambda_1 t} c_2(t=0)$$

and as a result we find

$$\underline{x}(t) = \left\{ \left\langle \underline{y}_1, \underline{x}(t=0) \right\rangle e^{\lambda_1 t} + \left\langle \underline{y}_2, \underline{x}(t=0) \right\rangle t e^{\lambda_1 t} \right\} \underline{x}_1 + \left\langle \underline{y}_2, \underline{x}(t=0) \right\rangle e^{\lambda_1 t} \underline{x}_2$$

What we see then is this: when a pair of eigenvectors is replaced by an eigenvector and a generalized eigenvector the purely exponential time dependence $e^{\lambda_1 t}$ and $e^{\lambda_2 t}$ is replaced by $e^{\lambda_1 t}$ and $t e^{\lambda_1 t}$. If λ_1 were repeated three times, assuming $n > 2$, the number of possibilities increases. We may have three eigenvectors, two eigenvectors and a generalized eigenvector or an eigenvector and two generalized eigenvectors. The first corresponds to a complete set of eigenvectors, viz.,

$\dim \mathrm{Ker}\,(A - \lambda_1 I) = 3$; the second is like the above, $\dim \mathrm{Ker}\,(A - \lambda_1 I) = 2$ and one of the eigenvectors, but not the other, must lie in $\mathrm{Im}\,(A - \lambda_1 I)$ in order that it lead to a generalized eigenvector. The third possibility is new, $\dim \mathrm{Ker}\,(A - \lambda_1 I) = 1$ and $\mathrm{Ker}\,(A - \lambda_1 I)$ lies inside $\mathrm{Im}\,(A - \lambda_1 I)$; the readers can satisfy themselves that the time dependence is now given by $e^{\lambda_1 t}$, $te^{\lambda_1 t}$ and $\frac{1}{2}t^2 e^{\lambda_1 t}$.

6.6   Improving the Performance of a Linear Stripping Cascade

A problem where generalized eigenvectors are required turns up in the study of a simple stripping cascade operated as follows: Let M denote the heavy phase holdup in each stage of the cascade and suppose every T units of time we transfer the heavy phase contents of stage i to stage $i - 1$, taking M units of product from stage 1, adding M units of feed to stage n. By doing this we achieve a heavy phase throughput $L = M/T$. The light phase is run as before and strips the n stages of the cascade for a period of time T. Indeed this is the way sugar is stripped out of sugar beets using water.

If $y_{in} = 0$ and $E = 1$ we can determine, using the Kremser equation, that the ordinary performance of such a stripping cascade is predicted by

$$\frac{x_{out}}{x_{in}} = \frac{1}{1 + S + S^2 + \cdots + S^n}$$

where $S = mV/L$ is the stripping factor. We propose to show that the ordinary operation can be greatly improved upon.

In the newly proposed method of running the separation cascade, our equations are

$$M\frac{dx_i}{dt} = Vy_{i-1} - Vy_i, \quad i = 1, \ldots, n$$

which can be written

$$\frac{dx_i}{dt} = \frac{S}{T}\left(x_{i-1} - x_i\right), \quad i = 1, \ldots, n$$

where $x_n(t=0) = x_{in}$ and $x_1(t=T) = x_{out}$. Setting $\underline{x} = \begin{pmatrix} x_1 \\ x_2 \\ \vdots \\ x_n \end{pmatrix}$ we have

$$\frac{d\underline{x}}{dt} = \frac{S}{T} A \underline{x}$$

where

$$A = \begin{pmatrix} -1 & 0 & 0 & \cdots \\ 1 & -1 & 0 & \cdots \\ 0 & 1 & -1 & \cdots \\ & \cdots & & \end{pmatrix}$$

This matrix has the eigenvalue $\lambda_1 = -1$ repeated n times and, as $\dim \mathrm{Ker}\,(A - \lambda_1 I) = 1$, to it there corresponds only one independent eigenvector which we can take to be $\underline{x}_1 = \begin{pmatrix} 0 \\ 0 \\ \vdots \\ 1 \end{pmatrix}$. This eigenvector initiates a chain of generalized eigenvectors $\underline{x}_2, \ldots, \underline{x}_n$ via

$$A\underline{x}_2 = \underline{x}_1 + \lambda_1 \underline{x}_2$$

$$A\underline{x}_3 = \underline{x}_2 + \lambda_1 \underline{x}_3$$

etc.

or $(A - \lambda_1 I)^n \underline{x}_n = \underline{0}$ but $(A - \lambda_1 I)^{n-1} \underline{x}_n \neq \underline{0}$, $(A - \lambda_1 I)^{n-1} \underline{x}_{n-1} = 0$ but

$(A - \lambda_1 I)^{n-2} \underline{x}_{n-1} \neq \underline{0}$, etc. Indeed in this simple example we find

$$\underline{x}_2 = \begin{pmatrix} 0 \\ 0 \\ \vdots \\ 0 \\ 1 \\ 0 \end{pmatrix}, \quad \underline{x}_3 = \begin{pmatrix} 0 \\ 0 \\ \vdots \\ 1 \\ 0 \\ 0 \end{pmatrix}, \quad \cdots$$

For $n = 2$ we have $\underline{x}_1 = \begin{pmatrix} 0 \\ 1 \end{pmatrix}$, $\underline{x}_2 = \begin{pmatrix} 1 \\ 0 \end{pmatrix}$ and, in the plain vanilla inner product,

$\underline{y}_1 = \begin{pmatrix} 0 \\ 1 \end{pmatrix}, \underline{y}_2 = \begin{pmatrix} 1 \\ 0 \end{pmatrix}$. Putting this in our earlier formula, we get

$$x_1(t) = x_1(t = 0) e^{-\frac{S}{T}t}$$

and

$$x_2(t) = x_2(t = 0) e^{-\frac{S}{T}t} + x_1(t = 0) \frac{S}{T} t e^{-\frac{S}{T}t}$$

These formulas will take us through the startup period where for the first cycle we have $x_1(t = 0) = x_{in}$. Thereafter $x_1(t = 0)$ will be $x_2(t = T)$ for the preceding cycle. After some number of cycles we assume, and the reader can demonstrate, that the stripping cascade achieves a repetitive operation wherein $x_2(t = T)$, and indeed $x_1(t)$ and $x_2(t), 0 < t < T$, is repeated cycle after cycle. Then using $x_1(t = 0) = x_2(t = T)$ in the foregoing we get

$$\frac{x_{out}}{x_{in}} = \frac{1}{e^{2S} - Se^S}$$

and this is better than the Kremser equation prediction

$$\frac{1}{1 + S + S^2}$$

The two results for $n = 1$ are

$$\frac{1}{e^S}$$

and

$$\frac{1}{1 + S}$$

The readers can determine the general result.

6.7   Another Way to Solve $d\underline{x}/dt = A\underline{x}$

We present an alternative, solving $d\underline{x}/dt = A\underline{x}$ by using the Laplace transformation, assuming the eigenvalues of A to be distinct. To do this we write

$$(\lambda I - A) \,\mathrm{adj}\,(\lambda I - A) = \det(\lambda I - A)\, I$$

and

$$\left(\overline{\lambda} I - \overline{A}^T\right) \mathrm{adj}\left(\overline{\lambda} I - \overline{A}^T\right) = \det\left(\overline{\lambda} I - \overline{A}^T\right) I$$

and observe that if λ_i is a simple root of $\det(\lambda I - A)$ then the columns of $\mathrm{adj}\,(\lambda_i I - A)$ are all proportional to $\underline{x}_i$ while the columns of $\mathrm{adj}\left(\overline{\lambda}_i I - \overline{A}^T\right)$ are all proportional to $\underline{y}_i$. Because $\mathrm{adj}\left(\overline{\lambda}_i I - \overline{A}^T\right) = \left\{\overline{\mathrm{adj}\,(\lambda I - A)}\right\}^T$ we can write

$$\mathrm{adj}\,(\lambda_i I - A) = c_i \underline{x}_i \underline{\bar{y}}_i^T$$

for some non-zero constant c_i.

We also need to evaluate $\dfrac{d}{d\lambda} \det(\lambda I - A)\Big|_{\lambda = \lambda_i}$. We can do this by using our formula for differentiating a determinant, viz.,

$$\frac{d}{d\lambda}\det\left(\lambda I - A\right) = \text{tr}\left\{\text{adj}\left(\lambda I - A\right)\frac{d}{d\lambda}\left(\lambda I - A\right)\right\} = \text{tr}\left\{\text{adj}\left(\lambda I - A\right)\right\}$$

whence

$$\left.\frac{d}{d\lambda}\det\left(\lambda I - A\right)\right|_{\lambda=\lambda_i} = \text{tr}\left(c_i\underline{x}_i\underline{\bar{y}}_i^{\,T}\right) = c_i\underline{\bar{y}}_i^{\,T}\underline{x}_i$$

To determine the solution to $d\underline{x}/dt = A\underline{x}$ we take the Laplace transformation of both sides obtaining

$$s\mathcal{L}\left(\underline{x}\right) - \underline{x}\left(t=0\right) = A\mathcal{L}\left(\underline{x}\right)$$

and hence

$$\left(sI - A\right)\mathcal{L}\left(\underline{x}\right) = \underline{x}\left(t=0\right)$$

whence

$$\mathcal{L}\left(\underline{x}\right) = \frac{\text{adj}\left(sI - A\right)}{\det\left(sI - A\right)}\underline{x}\left(t=0\right)$$

The right hand side has simple poles at $s = \lambda_1, \lambda_2, \ldots, \lambda_n$. Thus $\mathcal{L}(\underline{x})$ can be written

$$\mathcal{L}\left(\underline{x}\right) = \sum_{i=1}^{n}\frac{\text{adj}\left(sI - A\right)}{\left.\dfrac{d}{ds}\det\left(sI - A\right)\;\right|_{\lambda=\lambda_i}}\frac{1}{s - \lambda_i}\underline{x}\left(t=0\right)$$

Then, using our formulas for the adjugate and for the derivative of the determinant, we have

$$\mathcal{L}\left(\underline{x}\right) = \sum_{i=1}^{n}\frac{\underline{x}_i\underline{\bar{y}}_i^{\,T}\underline{x}\left(t=0\right)}{\underline{\bar{y}}_i^{\,T}\underline{x}_i}\frac{1}{s - \lambda_i} = \sum_{i=1}^{n}\underline{x}_i\left\langle\underline{y}_i, \underline{x}\left(t=0\right)\right\rangle\frac{1}{s - \lambda_i}$$

on requiring $\bar{y}_i^T x_i = \left\langle \underline{y}_i, \underline{x}_i \right\rangle = 1$ and hence

$$\underline{x}(t) = \sum_{i=1}^{n} \underline{x}_i \left\langle \underline{y}_i, \underline{x}(t=0) \right\rangle e^{\lambda_i t}$$

6.8 The Solution to Higher Order Equations

Frazer, Duncan and Collar in their book "*Elementary Matrices*" propose a very nice way to construct solutions to quite general systems of linear differential equations. We outline the essential idea here in the hope that this provokes some readers to go and look at this great old book. Let $f_{ij}(\lambda)$, $i, j = 1, \ldots, n$, be n^2 polynomials in λ. Then $f_{ij}\left(\dfrac{d}{dt}\right)$ is a polynomial differential operator and our problem is to determine solutions to the system of differential equations

$$f\left(\frac{d}{dt}\right) x(t) = \underline{0}$$

where $f(\lambda) = \lambda^n A_0 + \lambda^{n-1} A_1 + \cdots$ is called a lambda matrix. We let $F(\lambda)$ denote the adjugate of $f(\lambda)$, i.e., $F(\lambda) = \operatorname{adj} f(\lambda)$, and write

$$f(\lambda) F(\lambda) = \Delta(\lambda) I$$

where $\Delta(\lambda) = \det f(\lambda)$. Then if λ_1 is a root of $\Delta(\lambda) = 0$ of algebraic multiplicity m_1 we have

$$\Delta(\lambda_1) = 0$$

$$\overset{(1)}{\Delta}(\lambda_1) = 0$$

$$\cdots$$

$$\overset{(m_1-1)}{\Delta}(\lambda_1) = 0$$

$$\overset{(m_1)}{\Delta}(\lambda_1) \neq 0$$

where $\overset{(1)}{\Delta}(\lambda) = d\Delta(\lambda)/d\lambda$, etc. We can then find a set of solutions corresponding to λ_1 by observing first that

$$f\left(\frac{d}{dt}\right)e^{\lambda t}F(\lambda) = e^{\lambda t}f(\lambda)F(\lambda) = e^{\lambda t}\Delta(\lambda)I$$

and then that

$$f\left(\frac{d}{dt}\right)\frac{d}{d\lambda}\left\{e^{\lambda t}F(\lambda)\right\} = \frac{d}{d\lambda}f\left(\frac{d}{dt}\right)e^{\lambda t}F(\lambda) = \left\{te^{\lambda t}\Delta(\lambda) + e^{\lambda t}\overset{(1)}{\Delta}(\lambda)\right\}I$$

$$etc$$

Whence all of the columns of

$$e^{\lambda_1 t}F(\lambda_1),$$

$$\frac{d}{d\lambda_1}\left\{e^{\lambda_1 t}F(\lambda_1)\right\},$$

$$\cdots$$

$$\frac{d^{m_1-1}}{d\lambda_1^{m_1-1}}\left\{e^{\lambda_1 t}F(\lambda_1)\right\}$$

satisfy

$$f\left(\frac{d}{dt}\right)\underline{x}(t) = 0$$

and so also for the remaining roots of $\Delta(\lambda)$.

Frazer, Duncan and Collar present the properties of the lambda matrix $f(\lambda)$ and its adjugate $F(\lambda)$ that are required to make this a workable method for writing the general solution to $f\left(\frac{d}{dt}\right)\underline{x}(t) = \underline{0}$.

6.9   Roots of Polynomials

The stability of an equilibrium point to a small displacement requires the real parts of the eigenvalues of some matrix to be negative. To decide stability then, we must determine the eigenvalues of this matrix and to do this we must first determine its characteristic polynomial and then the roots of this polynomial. If the problem depends on parameters and the dimension is large this can be a difficult calculation.

We would like to be able to determine the signs of the real parts of the eigenvalues, short of determining the eigenvalues themselves, either by looking at the elements of the matrix or, if that fails, by looking at the coefficients of its characteristic polynomial. Gerschgorin's circle theorem is a step in this direction but often it does not resolve the question; yet it always provides estimates of the eigenvalues that can be refined and hence it is always helpful. In what follows we let $n = 2$ and 3 and state necessary and sufficient conditions in terms of the coefficients of the characteristic polynomial that its roots, the eigenvalues, have negative real parts. The matrix is assumed to be real.

A useful reference is Porter's book: *Stability Criteria for Linear Dynamical Systems*. Besides providing a nice way of looking at this problem, this beautiful little book has a simple derivation of the Routh criteria.

In the case $n = 2$ the characteristic equation is

$$\lambda^2 - \Delta_1 \lambda + \Delta_2 = 0$$

where $\Delta_1 = T$ and $\Delta_2 = D$, T and D denoting trace and determinant. The necessary and sufficient condition that $\text{Re}\,\lambda_1$ and $\text{Re}\,\lambda_2$ be negative is: $T < 0,\ D > 0$.

For n = 3 the characteristic equation is

$$\lambda^3 - \Delta_1 \lambda^2 + \Delta_2 \lambda - \Delta_3 = 0$$

or

$$\lambda^3 - T\lambda^2 + S\lambda - D = 0$$

where we assume T, S and D to be real. Then λ_1, λ_2 and λ_3 satisfy

$$\lambda_1 + \lambda_2 + \lambda_3 = T = \lambda_1 + 2x$$

$$\lambda_1\lambda_2 + \lambda_2\lambda_3 + \lambda_3\lambda_1 = S = \lambda_1(2x) + x^2 + y^2$$

$$\lambda_1\lambda_2\lambda_3 = D = \lambda_1(x^2 + y^2)$$

where on the right hand side we assume $\lambda_2 = \overline{\lambda}_3 = x + iy$. It is easy to see that if λ_1, λ_2 and λ_3 are negative or have negative real parts then $T < 0$, $S > 0$ and $D < 0$.

So if T is not negative, or S is not positive or D is not negative we can conclude that not all of $\operatorname{Re}\lambda_1$, $\operatorname{Re}\lambda_2$ and $\operatorname{Re}\lambda_3$ can be negative.

If $T < 0$, $S > 0$ and $D < 0$ then

$$\lambda^3 - T\lambda^2 + S\lambda - D = 0$$

cannot have real positive roots. On substituting $\lambda_2 = \overline{\lambda}_3 = x + iy$, $y \neq 0$, we get

$$x^3 - 3xy^2 - T(x^2 - y^2) + Sx - D = 0$$

and

$$3x^2y - y^3 - T(2xy) + Sy = 0$$

and so, dividing the second by y and using the result to eliminate y^2 in the first, we get

$$-8x^3 + 8Tx^2 - 2(S + T^2)x + TS - D = 0$$

and this tells us: $T < 0, S > 0, D < 0$ and $TS - D < 0$ is sufficient that x not be positive. But

$$
\begin{aligned}
TS - D &= (\lambda_1 + \lambda_2 + \lambda_3)(\lambda_1 \lambda_2 + \lambda_2 \lambda_3 + \lambda_3 \lambda_1) - \lambda_1 \lambda_2 \lambda_3 \\
&= (\lambda_1 + 2x)(\lambda_1(2x) + x^2 + y^2) - \lambda_1(x^2 + y^2) \\
&= (\lambda_1^2 + x^2 + y^2)(2x) + (2x)^2 \lambda_1
\end{aligned}
$$

and so $\lambda_1 < 0$ and $x < 0$ is sufficient that $TS - D < 0$. As $TS - D < 0$ if $\lambda_1 < 0, \lambda_2 < 0$ and $\lambda_3 < 0$ we discover: the necessary and sufficient condition that $\text{Re}\lambda_1, \text{Re}\lambda_2$ and $\text{Re}\lambda_3$ all be negative is $T < 0, S > 0, D < 0$ and $TS - D < 0$.

Three real roots corresponds to

$$
4S^3 - S^2 T^2 + 27D^2 + 4DT^3 - 18TSD < 0;
$$

otherwise, i.e., $4S^3 - S^2 T^2 + 27D^2 + 4DT^3 - 18TSD > 0$, there is one real root and a complex conjugate pair.

If our cubic equation depends on a parameter and the parameter changes, then: if it has three real roots, one changes sign on crossing the plane $D = 0, T < 0, S > 0$; if it has one real root and a complex conjugate pair, the real root changes sign on crossing the plane $D = 0, T < 0, S > 0$ but now $4S^3 - S^2 T^2 + 27D^2 + 4DT^3 - 18TSD > 0$; if it has one real root and a complex conjugate pair, the real part of the complex conjugate pair changes sign on crossing the surface $TS - D = 0$, $T < 0, S > 0, D < 0$.

6.10 The Matrix e^A

The solution to

$$\frac{d\underline{x}}{dt} = A\underline{x}$$

can be written as

$$\underline{x}(t) = e^{tA}\underline{x}(t=0)$$

where

$$e^{tA} = I + tA + \frac{1}{2}t^2A^2 + \frac{1}{6}t^3A^3 + \cdots$$

and where the infinite sum converges to e^{tA} no matter any complications pertaining to the eigenvectors of A. However, there are questions regarding the rate of convergence of the series, yet if t is small we might get a fair estimate of $\underline{x}(t)$ by assuming $e^{tA} = I + tA$.

Now we may need to solve

$$\frac{d\underline{x}}{dt} = (A+B)\underline{x}$$

where A and B are large matrices and $AB \neq BA$.

The solution is

$$\underline{x}(t) = e^{t(A+B)}\underline{x}(t=0)$$

but $e^{t(A+B)}$ is not $e^{tA}e^{tB}$. In obtaining an estimate of $e^{t(A+B)}$ it is sometimes useful to observe that $e^{t(A+B)}$ and $e^{\frac{1}{2}tA}e^{tB}e^{\frac{1}{2}tA}$ agree through their first three terms in powers of t.

6.11 $A = A(t)$

Suppose our problem is to solve

$$\frac{d\underline{x}}{dt} = A(t)\,\underline{x}$$

where, as indicated, A depends on t.

To learn what is known about this problem, denote by

$$\{\underline{x}_1(t), \underline{x}_2(t), \cdots, \underline{x}_n(t)\}$$

a set of independent solutions and introduce $M(t)$, a fundamental solution matrix, via

$$M(t) = (\underline{x}_1(t)\ \underline{x}_2(t)\ \cdots\ \underline{x}_n(t))$$

Then every solution can be written

$$c_1\underline{x}_1(t) + c_2\underline{x}_2(t) + \cdots + c_n\underline{x}_n(t)$$

and hence every fundamental solution matrix can be written $M(t)\,C$ where $\det C \neq 0$.

A case of interest is that in which A is periodic, viz.,

$$A(t+T) = A(t)$$

whereupon $M(t+T)$ is a fundamental solution matrix and hence we have

$$M(t+T) = M(t)\,C.$$

If we define a matrix R such that

$$C = e^{TR}$$

and define $P(t)$ via

$$P(t) = M(t) e^{tR}$$

whereupon

$$M(t) = P(t) e^{-tR}$$

then, we find that $P(t)$ is periodic, viz.,

$$
\begin{aligned}
P(t+T) &= M(t+T) e^{-(t+T)R} \\
&= M(t) C e^{-(t+T)R} \\
&= M(t) e^{-tR} \\
&= P(t)
\end{aligned}
$$

Hence if we determine $M(t)$ for $0 \le t \le T$, then $C = M^{-1}(0) M(T) = e^{TR}$ implies R and $P(t) = M(t) e^{-tR}$ implies $P(t), 0 \le t \le T$. Thus we have $P(t)$ for all t and therefore also $M(t) = P(t) e^{tR}$ for all t, *i.e.*, $M(t), 0 \le t \le T$, implies $M(t)$ for all t.

Now $M(t)$ is stable if the eigenvalues of R have negative real parts. The simplest case is that in which the eigenvalues of C are distinct, for then the eigenvectors of C and R coincide and the eigenvalues of R, denoted μ, and C, denoted λ, satisfy

$$\lambda = e^{T\mu}.$$

Thus we have

$$\mathrm{Re}\,\mu < 0 \quad \text{iff } |\lambda| < 1$$

and we can decide the stability of M by looking at the eigenvalues of $C = M(0)^{-1} M(T)$.

As an example, the Mathieu equation, viz.,

$$\frac{d^2\psi}{dt^2} + \cos t\, \psi = 0$$

can be written

$$\frac{d}{dt}\begin{pmatrix} \psi \\ \dfrac{d}{dt}\psi \end{pmatrix} = \begin{pmatrix} 0 & 1 \\ -\cos t & 0 \end{pmatrix}\begin{pmatrix} \psi \\ \dfrac{d}{dt}\psi \end{pmatrix}$$

whereupon our matrix $A(t)$ is

$$\begin{pmatrix} 0 & 1 \\ -\cos t & 0 \end{pmatrix}$$

6.12   Home Problems

1. Let $A = \begin{pmatrix} -1 & 1 \\ -1 & -2 \end{pmatrix}$ and show that its eigenvalues lie inside its Gerschgorins circles.

   Then let $\underline{x}(t=0) = \begin{pmatrix} 1 \\ 1 \end{pmatrix}$ and sketch the solution to

   $$\frac{d\underline{x}}{dt} = A\underline{x}$$

   in the x_1, x_2 plane.

   Let $A = \begin{pmatrix} -3/2 & 1/4 \\ 1 & -3/2 \end{pmatrix}$ and $A = \begin{pmatrix} -3 & 4 \\ 1 & 1 \end{pmatrix}$ and repeat the above calculations.
   In the last problem $\lambda_1 = -1$ is a double root to which corresponds only one independent
   eigenvector $\underline{x}_1 = \begin{pmatrix} -2 \\ 1 \end{pmatrix}$. A generalized eigenvector $\underline{x}_2 = \begin{pmatrix} 1 \\ 0 \end{pmatrix}$ can be found so that
   $\{\underline{x}_1,\ \underline{x}_2\}$ is a basis.

2. Determine $\dfrac{x_{\text{out}}}{x_{\text{in}}}$ once a dynamic stripping cascade reaches a repetitive state if $\dfrac{1}{2}$ of the heavy phase on each stage is transferred to the stage below every $\dfrac{1}{2}T$ seconds. Is this better than transferring all of the heavy phase every T seconds? What is the result in the limit of transferring $\dfrac{1}{n}$ th of the heavy phase every $\dfrac{1}{n}T$ seconds as $n \to \infty$? Do this assuming a two stage cascade.

3. Determine $\dfrac{x_{\text{out}}}{x_{\text{in}}}$ for a three stage dynamic stripping cascade once a repetitive operation is established. The model is then

$$\frac{dx}{dt} = \frac{S}{T}A\underline{x}, \quad 0 \le t \le T$$

where $S = \dfrac{mV}{L}$, $T = \dfrac{M}{L}$, $A = \begin{pmatrix} -1 & 0 & 0 \\ 1 & -1 & 0 \\ 0 & 1 & -1 \end{pmatrix}$

and

$$x_1\,(t = 0) = x_2\,(t = T)$$

$$x_2\,(t = 0) = x_3\,(t = T)$$

$$x_3\,(t = 0) = x_{\text{in}}$$

4. The difference equation

$$x_{i+1} = c\,x_i\,(1 - x_i), \quad c > 0$$

is well known in the theory of deterministic chaos. Show that its constant solutions are $x_i = 0$ and $x_i = 1 - \dfrac{1}{c}$.

This is a simple model of population variation, x_i being the population of a species in year i scaled so that $0 < x_i < 1$. The interesting range of c is then $0 < c < 4$ as

$0 < x\,(1-x) < \dfrac{1}{4}$ when $0 < x < 1$.

Show that the solution $x_i = 0$ is stable to small upsets if and only if $0 < c < 1$.

Show that the solution $x_i = 1 - \dfrac{1}{c}$ is stable to small upsets if and only if $1 < c < 3$.

When $c = 3$ show that the eigenvalue of the linear approximation is -1. Because $(-1)\,(-1) = 1$ this leads to a stable period 2 solution, $x_{i+2} = x_i$, which takes the place of the unstable constant or period 1 solution, $x_{i+1} = x_i$. Determine the period 2 solution.

The range $3 < c < 4$ is interesting. As c increases beyond 3 there is a region of period 2 solutions, then a region of period 4 solutions, then a region of period 8 solutions, etc. The width of successive regions decreases geometrically until what is called deterministic chaos sets in. This is the period doubling route to chaos and it can be observed on a hand calculator.

5. The simple equilibrium stage model sketched below illustrates how a separation by chromatography works:

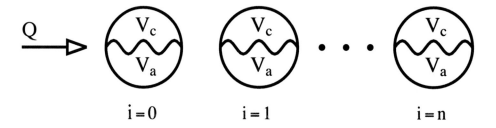

Denote by c and a the compositions of a dilute solute in the carrier and in the adsorbent phases, where V_c and V_a denote the volumes of the phases. Assume phase equilibrium holds in each stage, viz,. $c = Ka$, where strong binding corresponds to small values of K. The subscript i denotes the stage.

Then, scaling time by $\dfrac{V_c + \dfrac{V_a}{K}}{Q}$, i. e.,

$$t = \dfrac{V_c + \dfrac{V_a}{K}}{Q}\,\theta$$

and denoting by $\underline{c}$

$$\begin{pmatrix} c_0 \\ c_1 \\ c_2 \\ \vdots \end{pmatrix}$$

you have

$$\frac{d\underline{c}}{d\theta} = A\,\underline{c}$$

where

$$A = \begin{pmatrix} -1 & 0 & 0 & \cdots \\ 1 & -1 & 0 & \cdots \\ 0 & 1 & -1 & \cdots \\ \vdots & \vdots & \vdots & \end{pmatrix}$$

and where

$$\underline{c}\,(\theta = 0) = \begin{pmatrix} c_0\,(\theta = 0) \\ 0 \\ 0 \\ \vdots \end{pmatrix}$$

i.e., initially N moles of solute equilibrate in stage zero, the other stages and the inlet carrier being solute free, then the carrier is turned on at the volumetric flow rate Q.

The solute is swept through the cascade of stages and your job is to show that

$$\frac{c_i\,(\theta)}{c_0\,(\theta = 0)} = \frac{\theta^i e^{-\theta}}{i!}, \quad i = 0, 1, 2, \ldots$$

where

$$c_0\left(\theta = 0\right) = \frac{N}{V_c + \dfrac{V_a}{K}}$$

The distribution of solute over the stages at a given time θ is of most interest, viz., $c_i\left(\theta\right)$ vs i. Now $\dfrac{c_i\left(\theta\right)}{c_0\left(\theta = 0\right)}$ is the probability of finding a solute molecule in stage i at time θ, given it was in stage zero at time zero. Show that the average and variance of i, denoted $\bar{i}$ and σ^2, are given by

$$\bar{i} = \theta = \sigma^2$$

Then show that the speed of solute through the cascade, in stages per time, is

$$\frac{d\bar{i}}{dt} = \frac{Q}{V_c + \dfrac{V_a}{K}}$$

where $\dfrac{Q}{V_c}$ is the carrier speed.

Thus the stronger the binding the slower the speed. Hence different solutes having different K's move at different speeds.

6. The Stefan-Maxwell equations, viz.,

$$\nabla y_i = \sum_{\substack{j=1 \\ \neq i}}^{n} \frac{y_i \vec{N}_j - y_j \vec{N}_i}{c D_{ij}}$$

present us with a model for diffusion in an ideal gas at constant temperature and pressure, where c is the mole density of the gas, y_i and $\vec{N}_i$ are the mole fraction and mole flow rate per unit area of species i and $D_{ij} = D_{ji} > 0$ is the diffusion coefficient in an ideal gas made up of species i and j.

If diffusion takes place steadily in one spatial direction the vector notation is not necessary and we can write

$$\frac{dy_i}{dz} = \sum_{\substack{j=1 \\ \neq i}}^{n} \frac{y_i N_j - y_j N_i}{c D_{ij}}$$

where the N_i are constants independent of z. This equation can be used to find the N_i from measurements of the y_i at the two ends of a diffusion path in a two bulb diffusion experiment where a long, small diameter tube provides a diffusion path between two large, well mixed bulbs of gas at different compositions.

We can study this experiment under the assumption that the compositions in the bulbs remain constant in time if the bulb volumes are large and the tube cross sectional area is small. Then for a ternary ideal gas, denoting $\begin{pmatrix} y_1 \\ y_2 \\ y_3 \end{pmatrix}$ by $\underline{y}$, we can write the Stefan-Maxwell equations as

$$\frac{d}{dz}\underline{y} = \frac{1}{c}B\underline{y}$$

where B, viz.,

$$B = \begin{pmatrix} \dfrac{N_2}{D_{12}} + \dfrac{N_3}{D_{31}} & -\dfrac{N_1}{D_{12}} & -\dfrac{N_1}{D_{31}} \\[2ex] -\dfrac{N_2}{D_{12}} & \dfrac{N_3}{D_{23}} + \dfrac{N_1}{D_{12}} & -\dfrac{N_2}{D_{23}} \\[2ex] -\dfrac{N_3}{D_{31}} & -\dfrac{N_3}{D_{23}} & \dfrac{N_1}{D_{31}} + \dfrac{N_2}{D_{23}} \end{pmatrix}$$

is constant on a diffusion path.

Let

$$\sigma_1 = \frac{1}{D_{12}} + \frac{1}{D_{31}}, \qquad \delta_1 = \frac{1}{D_{12}} - \frac{1}{D_{31}}$$

$$\sigma_2 = \frac{1}{D_{23}} + \frac{1}{D_{12}}, \qquad \delta_2 = \frac{1}{D_{23}} - \frac{1}{D_{12}}$$

$$\sigma_3 = \frac{1}{D_{31}} + \frac{1}{D_{23}}, \qquad \delta_3 = \frac{1}{D_{31}} - \frac{1}{D_{23}}$$

and order the species so that $D_{31} > D_{12} > D_{23}$ then $\sigma_2 > \sigma_3 > \sigma_1 > 0$ and $\delta_2 > \delta_1 > 0 > \delta_3$.

Write the characteristic polynomial of B as

$$\lambda^3 - I\lambda^2 + II\lambda - III$$

and show that

$$I = \operatorname{tr} B = \underline{\sigma}^T \underline{N},$$

$$I^2 - 4II = \underline{N}^T \Delta \underline{N}$$

and

$$III = \det B = 0$$

where

$$\underline{\sigma} = \begin{pmatrix} \sigma_1 \\ \sigma_2 \\ \sigma_3 \end{pmatrix}$$

$$\underline{N} = \begin{pmatrix} N_1 \\ N_2 \\ N_3 \end{pmatrix}$$

and

$$\Delta = \begin{pmatrix} \delta_1^2 & -\delta_1\delta_2 & -\delta_3\delta_1 \\ -\delta_1\delta_2 & \delta_2^2 & -\delta_2\delta_3 \\ -\delta_3\delta_1 & -\delta_2\delta_3 & \delta_3^2 \end{pmatrix}$$

Then the eigenvalues of B are 0, $\dfrac{I \pm \sqrt{I^2 - 4II}}{2}$ and the eigenvectors of B and B^T corre-

sponding to the eigenvalue zero are $\begin{pmatrix} N_1 \\ N_2 \\ N_3 \end{pmatrix}$ and $\begin{pmatrix} 1 \\ 1 \\ 1 \end{pmatrix}$. Indeed it is worth observing that

II is proportional to $N_1 + N_2 + N_3$.

In the two bulb diffusion experiment the condition of constant pressure requires that $N_1 + N_2 + N_3 = 0$. Under this condition show that

$$I = \delta_1 N_2 - \delta_2 N_1$$

and

$$I^2 - 4II = I^2$$

Then the eigenvalues of B are $0, 0, I$. Show that the 2×2 minors of B are products of N_1 or N_2 or N_3 and $\dfrac{\delta_1 N_2}{D_{23}} - \dfrac{\delta_2 N_1}{D_{31}}$ and assuming that this is not zero show that the rank of B is two. Then the eigenvalue zero is repeated but to it there corresponds but one eigenvector.

Show that to the eigenvalue zero there corresponds the eigenvector $\begin{pmatrix} N_1 \\ N_2 \\ N_3 \end{pmatrix}$ and the

generalized eigenvector

$$\dfrac{1}{\dfrac{\delta_1 N_2}{D_{23}} - \dfrac{\delta_2 N_1}{D_{31}}} \begin{pmatrix} \dfrac{N_1}{D_{31}} \\[2mm] \dfrac{N_2}{D_{23}} \\[2mm] \delta_1 N_2 - \delta_2 N_1 - \dfrac{N_1}{D_{31}} - \dfrac{N_2}{D_{23}} \end{pmatrix}$$

and that to the eigenvalue I there corresponds the eigenvector $\begin{pmatrix} \delta_1 \\ \delta_2 \\ \delta_3 \end{pmatrix}$.

Using this information show how to predict the composition in one bulb of a two bulb diffusion experiment in terms of the composition in the other bulb and the values of N_1, N_2, N_3, D_{12}, D_{23} and D_{31}.

A full account of equimole counter diffusion in an ideal gas can be found in H.L. Toor's 1957 paper "*Diffusion in Three Component Gas Mixtures,*" A.I.Ch.E. J. $\underline{3}$ 198. He finds conditions where $N_1 = 0$ but $\dfrac{dy_1}{dz}$ is not zero, where $\dfrac{dy_1}{dz} = 0$ but N_1 is not zero and where $\operatorname{sgn} N_1 = \operatorname{sgn} \dfrac{dy_1}{dz}$.

7. If in Problem 6 the reactions $1 \rightleftarrows 2$ (with 3) take place at one end of the diffusion path then stoichiometry requires $N_1 + N_2 + N_3 = 0$ and the diffusion process is again represented by a matrix that does not have a complete set of eigenvectors. If, instead of this, a single reaction

$$\nu_1 A_1 + \nu_2 A_2 + \nu_3 A_3 = 0$$

takes place where $\nu_1 + \nu_2 + \nu_3 \neq 0$ then $N_1 + N_2 + N_3 \neq 0$, the eigenvalue zero is simple and the matrix B ordinarily has a complete set of eigenvectors.

Suppose that the stoichiometric coefficients are such that the eigenvalues of B are 0 and a complex conjugate pair. Then $\underline{y}$ vs. z will be a spiral in composition space.

Show that if a reservoir where the reaction $v_1 A_1 + v_2 A_2 + v_3 A_3 = 0$ takes place is fed by diffusion from a reservoir where $\underline{y}$ is fixed, the center of the spiral cannot lie in the physical part of composition space where $y_1 \geq 0$, $y_2 \geq 0$, $y_3 \geq 0$, $y_1 + y_2 + y_3 = 1$

8. The boiling curve for an ideal solution is obtained by solving

$$\frac{dx_i}{ds} = -y_i + x_i, \quad i = 1, 2, \ldots, n$$

where

$$y_i = \frac{P_i(T)}{P} x_i, \quad i = 1, 2, \ldots, n$$

and

$$\sum_{i=1}^{n} \frac{P_i(T)}{P} x_i = 1$$

where

$$P_1(T) > P_2(T) > \cdots > 0$$

and

$$\frac{dP_1}{dT} = P_1' > 0, \quad \frac{dP_2}{dT} = P_2' > 0, \text{ etc.}$$

The only rest states are the n pure component states. Determine the stability of each such state to a small perturbation and show that only the state $x_n = 1 = y_n$ is stable.

For example, one rest state is $x_1 = 1 = y_1$, $P_1(T) = P$, $x_i = 0 = y_i$, $P_i(T) < P, i = 2, \ldots, n$. The perturbation of x_1 must be negative, the perturbations of x_i, $i = 2, \ldots, n$ must be positive, and the perturbation of T must be positive.

9. The equations for the small transverse motions of a set of n particles, each of mass m, equally spaced on a string of fixed tension, viz.,

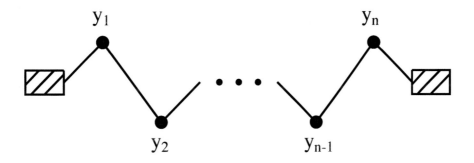

can be obtained from

$$L = \frac{1}{2}m\left\{\dot{y}_1^2 + \dot{y}_2^2 + \cdots + \dot{y}_n^2\right\} - \frac{1}{2}m\omega_0^2\left\{y_1^2 + (y_2 - y_1)^2 + \cdots + (y_n - y_{n-1})^2 + y_n^2\right\}$$

where L is the Lagrangian.

Show that L can be written

$$L = \frac{1}{2}m\,\underline{\dot{y}}^T I\underline{\dot{y}} - \frac{1}{2}m\omega_0^2\,\underline{y}^T A\underline{y}$$

where

$$\underline{y} = \begin{pmatrix} y_1 \\ y_2 \\ \vdots \\ y_n \end{pmatrix}$$

and

$$A = \begin{pmatrix} 2 & -1 & 0 & 0 & 0 & \ldots & 0 \\ -1 & 2 & -1 & 0 & 0 & \ldots & 0 \\ 0 & -1 & 2 & -1 & 0 & \ldots & 0 \\ \text{etc.} & & & & & & \end{pmatrix}$$

Show that Lagrange's equations of motion, i.e.,

$$\frac{d}{dt}\left(\frac{\partial L}{\partial \dot{y}_i}\right) - \frac{\partial L}{\partial y_i} = 0, \quad i = 1, 2, \ldots, n$$

lead to

$$m\frac{d^2\underline{y}}{dt^2} + m\omega_0^2 A\underline{y} = \underline{0}$$

The matrix A is self adjoint in the plain vanilla inner product and so has n orthogonal eigenvectors, denoted $\underline{x}_1$, $\underline{x}_2$, $\ldots$, $\underline{x}_n$, and real eigenvalues denoted λ_1, λ_2, $\ldots$, λ_n. The circle theorem tells us the eigenvalues cannot be negative.

Requiring $\langle\, \underline{x}_i\, \underline{x}_i\, \rangle = 1$, we can solve this second order differential equation by introducing generalized coordinates $q_1, q_2, \ldots, q_n$ via

$$\underline{y} = \underline{x}_1 q_1 + \underline{x}_2 q_2 + \cdots + \underline{x}_n q_n.$$

where $q_i = \langle \underline{x}_i,\, \underline{y}\rangle$

Find the equations satisfied by $q_1, q_2, \ldots, q_n$ and show that each generalized coordinate executes a purely harmonic motion at frequency ω_i where $\omega_i^2 = \lambda_i \omega_0^2$. Such a motion is called a normal mode of vibration.

Write L in terms of the generalized coordinates.

Let $n = 2$ and 3, determine the eigenvectors and eigenvalues of A and sketch the configuration of the particles in each normal mode of vibration.

10. Let $A_0, A_1, \ldots, A_m$ be a set of $n \times n$ matrices. Then

$$f(\lambda) = \lambda^m A_0 + \lambda^{m-1} A_1 + \cdots + A_m$$

is called a lambda matrix. Its elements are polynomials in λ of degree at most m. The latent roots of $f(\lambda)$ are the solutions of

$$\det f(\lambda) = 0$$

The corresponding latent vectors are the non-zero solutions of

$$f(\lambda)\underline{x} = \underline{0}$$

Show that the equations

$$\left\{ A_0 \frac{d^m}{dt^m} + A_1 \frac{d^{m-1}}{dt^{m-1}} + \cdots + A_m \right\} \underline{x}(t) = \underline{0}$$

and

$$\{A_0 S^m + A_1 S^{m-1} + \cdots + A_m\}\, \underline{x}(k) = \underline{0}$$

where $S^1 \underline{x}(k) = \underline{x}(k+1)$, $S^2 \underline{x}(k) = \underline{x}(k+2)$, etc., can be turned into first order equations, i.e., equations in only $\dfrac{d}{dt}$ and S^1.

Show that if λ_1 is a latent root of $f(\lambda)$ and $\underline{x}_1$ is a corresponding latent vector then

$$e^{\lambda_1 t} \underline{x}_1$$

satisfies the first equation while

$$\lambda_1^k \underline{x}_1$$

satisfies the second.

11. For a one shell pass, two tube pass heat exchanger, viz.,

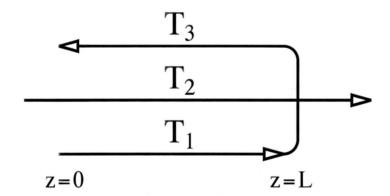

we have:

$$w_1\, c_{p_1} \frac{dT_1}{dz} = U_1 \pi D_1 \{T_2 - T_1\}$$

$$w_2\, c_{p_2} \frac{dT_2}{dz} = U_1 \pi D_1 \{T_1 + T_3 - 2T_2\}$$

$$-w_1 \, c_{p_1} \frac{dT_3}{dz} = U_1 \pi D_1 \{T_2 - T_3\}$$

where w_1 and w_2 denote the mass flow rates of the tube and the shell fluids, and where

$$T_1 \, (z = 0) = T_{1\text{in}}$$

and

$$T_2 \, (z = 0) = T_{2\text{in}}$$

Write this

$$\begin{pmatrix} w_1 \, c_{p_1} & 0 & 0 \\ 0 & w_2 \, c_{p_2} & 0 \\ 0 & 0 & -w_1 \, c_{p_1} \end{pmatrix} \frac{d}{dz} \begin{pmatrix} T_1 \\ T_2 \\ T_3 \end{pmatrix} = U_1 \pi D_1 \begin{pmatrix} -1 & 1 & 0 \\ 1 & -2 & 1 \\ 0 & 1 & -1 \end{pmatrix} \begin{pmatrix} T_1 \\ T_2 \\ T_3 \end{pmatrix}$$

and show that

$$\begin{pmatrix} 1 & 1 & 1 \end{pmatrix} \begin{pmatrix} -1 & 1 & 0 \\ 1 & -2 & 1 \\ 0 & 1 & -1 \end{pmatrix} = \begin{pmatrix} 0 & 0 & 0 \end{pmatrix}$$

is the conservation of energy equation.

To determine T_1, T_2 and T_3 write

$$\frac{d}{dz} \begin{pmatrix} T_1 \\ T_2 \\ T_3 \end{pmatrix} = \begin{pmatrix} -a & a & 0 \\ b & -2b & b \\ 0 & -a & a \end{pmatrix} \begin{pmatrix} T_1 \\ T_2 \\ T_3 \end{pmatrix}$$

and expand $\begin{pmatrix} T_1 \\ T_2 \\ T_3 \end{pmatrix}$ in terms of the eigenvalues and eigenvectors of the matrix

$$\begin{pmatrix} -a & a & 0 \\ b & -2b & b \\ 0 & -a & a \end{pmatrix}$$

where

$$a = \frac{U_1 \pi D_1}{w_1 c_{p_1}}$$

and

$$b = \frac{U_1 \pi D_1}{w_2 c_{p_2}}$$

Find $T_{3\text{out}}$ in terms of $T_{1\text{in}}$ and $T_{2\text{in}}$ by observing that

$$T_1 (z = L) = T_3 (z = L)$$

What happens in the limit as L grows large?

12. By reversing the direction of the shell flow, a second one shell pass, two tube pass heat exchanger configuration is obtained. Repeat the calculation in Problem 11 for this second configuration. Sketch the temperatures vs z in the two configurations if the shell side is hot and the tube side is cold. Show that the tube side temperature cannot cross the shell side temperature in the first configuration. In the second configuration no such restriction obtains and T_3 vs z need not be monotone.

13. For a two shell pass, four tube pass heat exchanger, viz.,

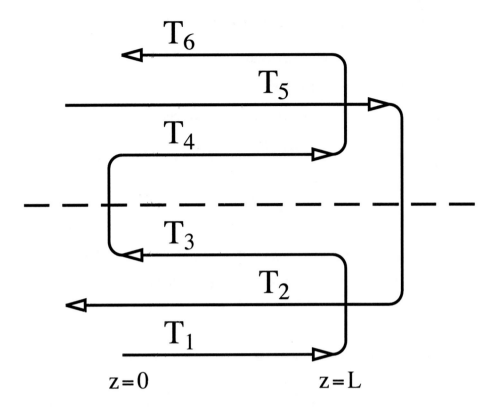

we have:

$$w_1 c_{p_1} \frac{dT_1}{dz} = U_1 \pi D_1 \{T_2 - T_1\}$$

$$-w_2 c_{p_2} \frac{dT_2}{dz} = U_1 \pi D_1 \{T_1 + T_3 - 2T_2\}$$

etc.

which can be written

$$\frac{d}{dz} \begin{pmatrix} T_1 \\ T_2 \\ \vdots \\ T_6 \end{pmatrix} = \begin{pmatrix} -a & a & 0 & 0 & 0 & 0 \\ -b & 2b & -b & 0 & 0 & 0 \\ 0 & -a & a & 0 & 0 & 0 \\ 0 & 0 & 0 & -a & a & 0 \\ 0 & 0 & 0 & b & -2b & b \\ 0 & 0 & 0 & 0 & -a & a \end{pmatrix} \begin{pmatrix} T_1 \\ T_2 \\ \vdots \\ T_6 \end{pmatrix}$$

Denote the matrix on the RHS by A. It is block diagonal and its diagonal blocks turn up in models of one shell pass, two tube pass heat exchangers.

Find the eigenvalues and eigenvectors of A and write a formula in terms of six undetermined constants for the dependence of T_1, T_2, ..., T_6 on z. It is only the values of these constants that make T_1, T_2, T_3 and T_4, T_5, T_6 interdependent.

Find T_{2out} and T_{6out} in terms of T_{1in} and T_{5in}. Sketch the temperatures vs z if the shell side is hot and the tube side is cold.

14. A one shell pass, one tube pass heat exchanger, i.e., a simple double pipe heat exchanger, is often built using n small diameter pipes in place of one large diameter pipe:

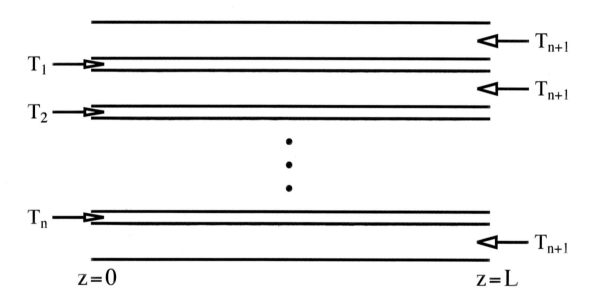

For this we have:

$$\frac{1}{n} w c_p \frac{dT_1}{dz} = U\pi D\{T_{n+1} - T_1\}$$

$$\vdots$$

$$\frac{1}{n} w c_p \frac{dT_n}{dz} = U\pi D\{T_{n+1} - T_n\}$$

$$-w_s c_{p_s} \frac{dT_{n+1}}{dz} = U\pi D\{T_1 + T_2 + \cdots + T_n - nT_{n+1}\}$$

The matrix that turns up here is

$$
\begin{pmatrix}
-a & 0 & 0 & \cdots & 0 & a \\
0 & -a & 0 & \cdots & 0 & a \\
\vdots & \vdots & \vdots & & \vdots & \vdots \\
0 & 0 & 0 & \cdots & -a & a \\
-b & -b & -b & \cdots & -b & nb
\end{pmatrix}
$$

where w denotes the total pipe side flow and a is n times as large as before.

Determine the eigenvalues and eigenvectors of this matrix.

The pipe side flow is already equally divided over the n pipes. If the pipe side inlet temperatures are also the same, show that all but two of the $n + 1$ constants in the solution must be zero. These are the constants corresponding to the eigenvalue $-a$ which is repeated $n - 1$ times,

This tells us that the temperature in each pipe is the same as it is in all other pipes. Under this condition the model reduces to the model of a simple double pipe heat exchanger if $\frac{1}{n}$ of the shell flow is assigned to each pipe.

15. For the simple $1 - 1$ heat exchanger, viz.,

we introduce the column vector $\begin{pmatrix} T_1 \\ T_2 \end{pmatrix}$. In the counterflow configuration we have

$$
\frac{d}{dz}\begin{pmatrix} T_1 \\ T_2 \end{pmatrix} = \begin{pmatrix} -a & a \\ -b & b \end{pmatrix}\begin{pmatrix} T_1 \\ T_2 \end{pmatrix}
$$

where

$$a = \frac{U\pi D}{w_1\, c_{p_1}} \quad \text{and} \quad b = \frac{U\pi D}{w_2\, c_{p_2}}$$

The eigenvalues and the eigenvectors of the matrix on the RHS are

$$0, \quad \begin{pmatrix} 1 \\ 1 \end{pmatrix} \quad \text{and} \quad -a+b, \quad \begin{pmatrix} a \\ b \end{pmatrix}$$

Using these produce the usual formulas for this heat exchanger.

The case where $a = b$ leads to a double eigenvalue to which corresponds only one independent eigenvector. Work out this case using a generalized eigenvector.

16. A cocurrent honeycomb heat exchanger is shown below.

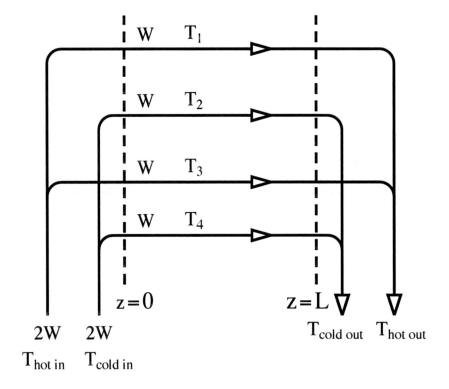

As a model, we write, assuming all c_p's are the same,

$$e^{-\frac{(UA)_{\text{eff}}}{wc_p}} = \frac{T_{\text{hot out}} - T_{\text{cold out}}}{T_{\text{hot in}} - T_{\text{cold in}}}$$

Suppose the internal heat transfer coefficient is U and the area of the plane wall separating each stream is A.

Show that

$$e^{-\frac{(UA)_{\text{eff}}}{wc_p}} = \frac{1}{4}\left(2 - \sqrt{2}\right) e^{\frac{(UA)}{wc_p}\left(-2+\sqrt{2}\right)} + \frac{1}{4}\left(2 + \sqrt{2}\right) e^{\frac{(UA)}{wc_p}\left(-2-\sqrt{2}\right)}$$

17. To see what happens when a condition such as $\sum x_i = 1$ must be satisfied in a problem where the stability of an equilibrium point is being investigated let

$$\frac{dx_1}{dt} = -f_1(x_1, x_2) + x_1$$

and

$$\frac{dx_2}{dt} = -f_2(x_1, x_2) + x_2$$

where $f_1 + f_2 = 1$ whenever $x_1 + x_2 = 1$. This requires

$$\frac{\partial f_1}{\partial x_1} - \frac{\partial f_1}{\partial x_2} + \frac{\partial f_2}{\partial x_1} - \frac{\partial f_2}{\partial x_2} = 0 \qquad (*)$$

whenever $x_1 + x_2 = 1$.

Let x_1^0, x_2^0 be an equilibrium point and let $x_1 = x_1^0 + \xi_1$, $x_2 = x_2^0 + \xi_2$ be a small excursion where $x_1^0 + x_2^0 = 1$ and $\xi_1 + \xi_2 = 0$

What does condition $(*)$ tell us about the eigenvalues and eigenvectors of the Jacobian matrix and about the solution to

$$\frac{d}{dt}\begin{pmatrix} \xi_1 \\ \xi_2 \end{pmatrix} = J \begin{pmatrix} \xi_1 \\ \xi_2 \end{pmatrix}$$

when

$$\begin{pmatrix} \xi_1\,(t=0) \\ \xi_2\,(t=0) \end{pmatrix} \propto \begin{pmatrix} 1 \\ -1 \end{pmatrix}$$

Does this agree with what is obtained on reducing the problem to

$$\frac{dx_1}{dt} = -f_1\,(x_1,\ 1-x_1) + x_1\ ?$$

18. A rider on a merry-go-round throws a ball at another rider directly opposite. The radius is R, the angular velocity is $\vec{w} = w\,\vec{k}$ and the speed of the ball is initially V.

The motion of the ball is viewed by an observer at the center of the merry-go-round and rigidly fixed to it. Under force free conditions the equation for the motion of the ball is

$$\vec{a} = -2\,\vec{w} \times \vec{v} - \vec{w} \times \vec{w} \times \vec{r}$$

where $\vec{r}\,(t=0) = R\,\vec{i}$ and $\vec{v}\,(t=0) = -V\,\vec{i}.$

Let $\vec{r} = x\,\vec{i} + y\,\vec{j}$, write the equations for the motion of the ball and put them in the form

$$\frac{d}{dt}\begin{pmatrix} x \\ \dfrac{dx}{dt} \\ y \\ \dfrac{dy}{dt} \end{pmatrix} = \begin{pmatrix} 0 & 1 & 0 & 0 \\ w^2 & 0 & 0 & 2w \\ 0 & 0 & 0 & 1 \\ 0 & -2w & w^2 & 0 \end{pmatrix}\begin{pmatrix} x \\ \dfrac{dx}{dt} \\ y \\ \dfrac{dy}{dt} \end{pmatrix}$$

Solve this, noticing that the matrix on the RHS has two double eigenvalues and to each there corresponds but one eigenvector.

Find the motion of the ball as seen by an observer fixed to the ground at the center of the merry-go-round. The equation of motion is then

$$\vec{a} = 0$$

where $\quad \vec{r}(t = 0) = R\,\vec{i} \quad$ and $\quad \vec{v}(t = 0) = -V\,\vec{i} + wR\,\vec{j}.$

This might explain why the physics of the problem requires that the observer fixed to the merry-go-round find a pair of double eigenvalues each corresponding to a single eigenvector.

19. Suppose we have a system of particles whose state is given by generalized coordinates $q_1, q_2, \ldots$ and generalized momenta $p_1, p_2, \ldots$. Denote the Hamiltonian for the system by $H = H(q_1, q_2, \ldots, p_1, p_2, \ldots)$. The equations of motion are

$$\frac{dq_i}{dt} = \frac{\partial H}{\partial p_i}$$

and

$$\frac{dp_i}{dt} = -\frac{\partial H}{\partial q_i}$$

Now for small oscillations about an equilibrium point, H can be approximated by

$$H = \frac{1}{2}\sum\sum Q_{ij}\,q_i\,q_j + \frac{1}{2}\sum\sum P_{ij}\,p_i\,p_j$$

where Q and P are real, symmetric and positive definite.

Then we have

$$\frac{dq_i}{dt} = \sum P_{ij}\,p_j$$

and hence

$$\frac{d^2 q_i}{dt^2} = \sum P_{ij} \frac{dp_j}{dt} = \sum P_{ij} \left(-\frac{\partial H}{\partial q_j} \right)$$

$$= -\sum \sum P_{ij} Q_{jk} q_k$$

which we can write

$$\frac{d^2 \underline{q}}{dt^2} = -W \underline{q}, \quad W = PQ$$

The eigenvalue problem for W is

$$W \underline{q}_i = w_i^2 \underline{q}_i$$

Prove $w_i^2 > 0$.

Then if w_i^2, $\underline{q}_i$ and w_j^2, $\underline{q}_j$ are two solutions to the eigenvalue problem, prove

$$\underline{q}_i^T Q \underline{q}_j = 0$$

To get going multiply the eigenvalue problem by Q whereupon it is:

$$Q P Q \underline{q} = w^2 Q \underline{q}$$

20. For a three stage stripping column, where we add H units of solution to the top every T units of time, we have

$$\frac{dx_3}{dt} = S (x_2 - x_3)$$

$$\frac{dx_2}{dt} = S (x_1 - x_2)$$

$$\frac{dx_1}{dt} = -Sx_1$$

where t is scaled by T, $S = \dfrac{mV}{L}$ and $L = \dfrac{H}{T}$. The x's are scaled by x_{in}.

Plot x_1, x_2 and x_3 vs t for the first few cycles where

$$\left.\begin{array}{l} x_3\,(t=0) = 1 \\[4pt] x_2\,(t=0) = 1 \\[4pt] x_1\,(t=0) = 1 \end{array}\right\} \quad \text{first cycle}$$

$$\left.\begin{array}{l} x_3\,(t=0) = 1 \\[4pt] x_2\,(t=0) = x_3\,(t=1), \quad \text{cycle before} \\[4pt] x_1\,(t=0) = x_2\,(t=1), \quad \text{cycle before} \end{array}\right\} \quad \text{thereafter}$$

After many cycles $x_1\,(t=1)$ should approach

$$\frac{1}{e^{3S} - 2Se^{2S} + \dfrac{1}{2}\,e^{S}}$$

Notice that the only eigenvalue of

$$\begin{pmatrix} -1 & 0 & 0 \\ 1 & -1 & 0 \\ 0 & 1 & -1 \end{pmatrix}$$

is $\lambda = -1$ to which there corresponds the eigenvector and generalized eigenvectors

$$\begin{pmatrix} 0 \\ 0 \\ 1 \end{pmatrix}, \quad \begin{pmatrix} 0 \\ 1 \\ 0 \end{pmatrix}, \quad \begin{pmatrix} 1 \\ 0 \\ 0 \end{pmatrix}$$

and notice that these vectors are orthogonal in the plain vanilla inner product.

21. If

$$A = \begin{pmatrix} -1 & -1 \\ 1 & -1 \end{pmatrix} \quad \text{and} \quad \langle\, \underline{y}, \underline{x} \,\rangle = \overline{\underline{y}}^T \underline{x}$$

show that the eigenvectors of A are orthogonal.

Observe that $A \neq A^*$ but $AA^* = A^*A$. If $AA^* = A^*A$ in one inner product, is $AA^* = A^*A$ in another inner product?

22. Cocurrent honeycomb heat exchanger.

We have a set of parallel channels of rectangular cross section through which hot and cold fluids flow, viz.,

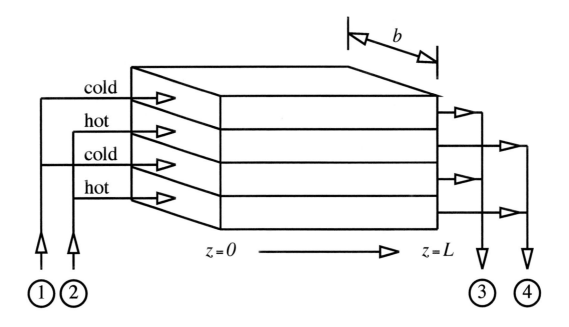

and our notation is

The cold flow rates are denoted W_{cold}, the hot flow rates are denoted W_{hot}. All the cold side heat transfer coefficients are the same, so too all the hot side heat transfer coefficients. Hence all the U's are the same.

Denoting $\dfrac{W_{cold}\, c_{p\,cold}}{W_{hot}\, c_{p\,hot}}$ by f and $\dfrac{Ub}{W_{cold}\, c_{p\,cold}}$ by β we have

$$\frac{d}{dz}\underline{T} = \beta A \underline{T}$$

where $\underline{T} = \begin{pmatrix} T_1 \\ T_2 \\ T_3 \\ T_4 \end{pmatrix}$

$$A = \begin{pmatrix} -1 & 1 & 0 & 0 \\ f & -2f & f & 0 \\ 0 & 1 & -2 & 1 \\ 0 & 0 & f & -f \end{pmatrix}$$

and

$$\underline{T}(z=0) = \begin{pmatrix} T_{\text{cold in}} \\ T_{\text{hot in}} \\ T_{\text{cold in}} \\ T_{\text{hot in}} \end{pmatrix}$$

Having access to inputs and outputs only at points ①, ②, ③ and ④ and knowing only that the flows are cocurrent, we define $(UA)_{\text{eff}}$ by

$$\frac{(T_{\text{hot}} - T_{\text{cold}})_{\text{out}}}{(T_{\text{hot}} - T_{\text{cold}})_{\text{in}}} = e^{-\dfrac{(UA)_{\text{eff}}\,(f+1)}{2W_{\text{cold}}\,c_{\text{p cold}}}}$$

and our job is to measure $(UA)_{\text{eff}}$ and estimate U, where

$$(T_{\text{hot}} - T_{\text{cold}})_{\text{out}} = T④ - T③ = \frac{1}{2}\begin{pmatrix} -1 \\ 1 \\ -1 \\ 1 \end{pmatrix}\underline{T}(z=L)$$

Define D, a diagonal matrix, by

$$D = \begin{pmatrix} 1 & & & \\ & f & & \\ & & 1 & \\ & & & f \end{pmatrix}, \quad \text{where} \quad D = \overline{D}^T$$

and denote $A(f=1)$ by A_1 where $A_1 = \overline{A_1}^T$, then observe that $A = DA_1$.

Show that the eigenvalues of A are real and not positive and that its eigenvectors are orthogonal in the inner product where $G = D^{-1}$.

Hence, write

$$\underline{T}(z = L) = \sum_G \left\langle \underline{x}_i, \underline{T}(z = 0) \right\rangle_G e^{\frac{UbL}{W_{\text{cold}} \, c_{\text{p cold}}} \lambda_i} \underline{x}_i$$

Derive the characteristic polynomial of A, viz.,

$$\lambda^4 + 3(1 + f)\lambda^3 + 2(1 + 3f + f^2)\lambda^2 + 2f(1 + f)\lambda + \text{zero} = 0$$

and conclude:

eigenvalues:

$$\lambda_1 = 0, \quad \lambda_2 = -f - 1 + \sqrt{f^2 + 1}, \quad \lambda_3 = -f - 1, \quad \lambda_4 = -f - 1 - \sqrt{f^2 + 1}$$

unnormalized eigenvectors:

$$\underline{x}_1 = \begin{pmatrix} 1 \\ 1 \\ 1 \\ 1 \end{pmatrix}, \quad \underline{x}_2 = \begin{pmatrix} 1 \\ -f + \sqrt{f^2 + 1} \\ \frac{1}{f}\left(1 - \sqrt{f^2 + 1}\right) \\ -1 \end{pmatrix},$$

$$\underline{x}_3 = \begin{pmatrix} 1 \\ -f \\ -f \\ f^2 \end{pmatrix}, \quad \underline{x}_4 = \begin{pmatrix} 1 \\ -f - \sqrt{f^2 + 1} \\ \frac{1}{f}\left(1 + \sqrt{f^2 + 1}\right) \\ -1 \end{pmatrix}$$

Then derive a formula for $(UA)_{\text{eff}}$ in terms of U. You worked out the case $f = 1$ in Problem 16.

23. You are going for a walk on a network of N points. A point is denoted $i, i = 1, \ldots, N$.

Each point is linked to others and we set $\ell_{ij} = 1$ if there is a one step path from point j to point i, otherwise $\ell_{ij} = 0$.

Denote by ℓ_j the number of one step paths from point j to the points of the network, viz.,

$$\ell_j = \sum_i \ell_{ij}, \quad \ell_j \neq 0$$

and assume on taking a step from point j the possible destinations are chosen with equal probability, $\dfrac{1}{\ell_j}$. Thus the probability of the step $j \to i$

$$p_{ij} = \frac{1}{\ell_j} > 0 \quad \text{if} \quad \ell_{ij} = 1$$

$$= 0 \qquad \text{if} \quad \ell_{ij} = 0$$

The probability of visiting point i at step $m + 1$ is

$$p_i^{(m+1)} = \sum p_{ij}\, p_j^{(m)}$$

and hence

$$\underline{p}^{(m+1)} = P\, \underline{p}^{(m)}$$

where

$$\begin{pmatrix} 1 \\ 1 \\ \vdots \\ 1 \end{pmatrix}^T \underline{p}^{(m)} = 1$$

What must be true of P in order that

$$\lim_{m \to \infty} \underline{p}^{(m)} = \underline{p}$$

independent of $\underline{p}^{(0)}$.

Given P how would you find $\underline{p}$?

Lecture 7

Simple Chemical Reactor Models

7.1 The Chemostat

A chemostat is a vessel in which a population of cells grows. We denote by V the volume of the vessel and assume the conditons therein to be spatially uniform. Then if n denotes the number density of cells in the vessel and W denotes the volume flow into and out of the vessel, the number of cells in the vessel satisfies

$$V\frac{dn}{dt} = -Wn + knV$$

where k is the growth constant and where the cells are not fed but grow from an intial injection which establishes $n\,(t=0)$. Because we have

$$\frac{dn}{dt} = \left\{ -\frac{W}{V} + k \right\} n,$$

if $k > \frac{W}{V}$ the cell culture grows without bound, whereas if $k < \frac{W}{V}$ it washes out.

To get a more interesting model we make the simple assumption that the value of k, which tells us the rate of cell multiplication, instead of being a constant, depends on the concentration of a single limiting nutrient. We let c denote this concentration and write $k = k\,(c)$. Assuming that the nutrient is consumed only when the cell population grows and that it must be fed to the chemostat to make up for this, we write

$$V\frac{dc}{dt} = Wc_{in} - Wc - \nu knV$$

and our model is

$$\frac{dn}{dt} = \left\{-\frac{W}{V} + k\right\} n$$

and

$$\frac{dc}{dt} = \frac{W}{V}\{c_{in} - c\} - \nu kn$$

where c_{in} is the nutrient concentration in the feed and ν is a stoichiometric coefficient.

The steady states are solutions to

$$0 = \left\{-\frac{W}{V} + k\right\} n$$

and

$$0 = \frac{W}{V}\{c_{in} - c\} - \nu kn$$

and these equations are statisfied by

$$n = 0, \ c = c_{in} \quad \text{(wash out)}$$

and by

$$n = \frac{c_{in} - c}{\nu}, \ k(c) = \frac{W}{V}.$$

Ordinarily k is a monotonically increasing function of c and assuming this to be so we require $k'(c) > 0$. As the values of c lie on the interval $[0, c_{in}]$, the largest value of k is $k(c_{in})$. We suppose that c_{in} and V are held at fixed values and that W is decreased from an arbitrarily large value. Then as long as $W > Vk\,(c_{in})$ we find only the washout solution, $n = 0$, $c = c_{in}$. The point $W = Vk\,(c_{in})$ is a branch point; as W passes through $Vk\,(c_{in})$ a new solution branches off the washout solution, and we have the following steady state diagram:

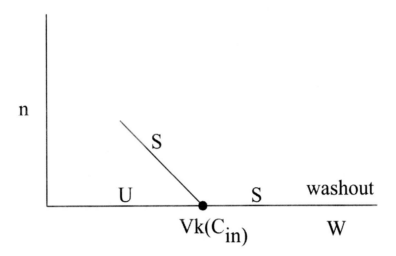

To establish the stability of these steady state solution branches, we investigate what happens to a small excursion from a steady solution denoted n_0, c_0. Writing our model

$$\frac{dn}{dt} = f\,(n, c) = \left\{ -\frac{W}{V} + k\,(c) \right\} n$$

and

$$\frac{dc}{dt} = g\,(n, c) = \frac{W}{V}\,(c_{in} - c) - \nu k\,(c)\, n$$

we find its linear approximation near n_0, c_0, in terms of the small displacements ξ and η, to be

$$\frac{d}{dt}\begin{pmatrix} \xi \\ \eta \end{pmatrix} = \begin{pmatrix} f_n & f_c \\ g_n & g_c \end{pmatrix}\begin{pmatrix} \xi \\ \eta \end{pmatrix} = \begin{pmatrix} -\frac{W}{V} + k\,(c_0) & k'\,(c_0)\, n_0 \\ -\nu k\,(c_0) & -\frac{W}{V} - \nu k'\,(c_0)\, n_0 \end{pmatrix}\begin{pmatrix} \xi \\ \eta \end{pmatrix} = J\begin{pmatrix} \xi \\ \eta \end{pmatrix}.$$

At a washout solution, where $n_0 = 0$, $c_0 = c_{in}$, we have

$$J = \begin{pmatrix} -\frac{W}{V} + k\,(c_{in}) & 0 \\ -\nu k\,(c_{in}) & -\frac{W}{V} \end{pmatrix}$$

whereas, at a non washout solution, where $\frac{W}{V} = k(c_0)$, our Jacobian matrix is

$$J = \begin{pmatrix} 0 & k'(c_0)\,n_0 \\ -\nu\frac{W}{V} & -\frac{W}{V} - \nu k'(c_0)\,n_0 \end{pmatrix}.$$

Hence if $W < Vk(c_{in})$ the new branch is stable whereas the washout branch is unstable. If $W > Vk(c_{in})$ the washout branch is stable. This is true because the eigenvalues of the Jacobian matrix on the new branch are $-\nu k'(c_0)\,n_0$ and $-\frac{W}{V}$ whereas on the washout branch they are $-\frac{W}{V} + k(c_0)$ and $-\frac{W}{V}$. Indeed as W decreases and passes through $Vk(c_{in})$ the washout solution loses its stability while the new solution picks up the lost stability of the washout solution. We observe that at the branch point an eigenvalue vanishes, i.e., the branch point is the point where the determinant of the Jacobian matrix vanishes. This corresponds to passing from the fourth to the third quadrant in the plane whose axes are the determinant and the trace of the Jacobian matrix.

The reader may wish to rework this problem after adding a term to account for cell metabolism i.e., $-\mu n$, $\mu > 0$, and specifying

$$k(c) = \beta c, \ \beta > 0,$$

which rules out cell growth at $c = 0$. Then the model is

$$\frac{dn}{dt} = \left(-\frac{W}{V} + \beta c \right) n$$

and

$$\frac{dc}{dt} = \frac{W}{V}(c_{in} - c) - (\nu\beta c + \mu)\,n.$$

7.2 The Stirred Tank Reactor

This is a model problem having a long history in chemistry and chemical engineering. There are many variations corresponding to many ways of making the reaction speed itself up. We assume the reaction is autothermal. The rate of a chemical reaction is ordinarily a strongly increasing function of the temperature at which the reaction takes place. This leads to a positive feedback when a reaction releases heat, for this heat then speeds up the reaction. This feedback makes the problem interesting, even in the simplest case, and carrying out the reaction in a stirred tank reactor produces a very simple problem. The model is plain vanilla, retaining only the Arrhenius temperature dependence of the chemical rate coefficient, and this in a simplified form. Our work is a part of what can be found in Poore's paper "*A Model Equation Arising from Chemical Reactor Theory*" (Arch. Rational Mech. Anal. $\underline{52}$, 358 (1973)).

Before we turn to this problem we presented a brief reminder, setting $n = 2$ so that

$$\underline{x} = \begin{pmatrix} x_1 \\ x_2 \end{pmatrix} \quad \text{and} \quad A = \begin{pmatrix} a_{11} & a_{12} \\ a_{21} & a_{22} \end{pmatrix}. \text{ The solution to}$$

$$\frac{d\underline{x}}{dt} = A\underline{x}$$

where $\underline{x}(t = 0)$ is assigned is

$$\underline{x} = \left\langle \underline{y}_1, \underline{x}(t=0) \right\rangle e^{\lambda_1 t} \underline{x}_1 + \left\langle \underline{y}_2, \underline{x}(t=0) \right\rangle e^{\lambda_2 t} \underline{x}_2$$

where $\{\underline{x}_1, \underline{x}_2\}$ and $\{\underline{y}_1, \underline{y}_2\}$ are biorthogonal sets of vectors in the inner product being used to write the solution, $\underline{x}_1$ and $\underline{x}_2$ being eigenvectors of A, $\underline{y}_1$ and $\underline{y}_2$ being eigenvectors of A^*. The corresponding eigenvalues, λ_1 and λ_2, satisfy

$$\lambda^2 - \mathrm{tr}A\,\lambda + \det A = 0$$

where $\mathrm{tr}A = a_{11} + a_{22}$ and $\det A = a_{11}a_{22} - a_{21}a_{12}$.

We assume that A is real and we observe that the qualitative behavior of $\underline{x}(t)$ differs according to where the point $(\det A, \mathrm{tr}A)$ lies in the $\det A - \mathrm{tr}A$ plane. The algebraic signs of the eigenvalues or their real parts are: $+, +$ in the first quadrant, $+, -$ in the second and third quadrants, $-, -$ in the fourth quadrant. The fourth quadrant is divided into two regions by the curve $(\mathrm{tr}A)^2 - 4\det A = 0$. Above the curve the two eigenvalues are complex conjugates having a negative real part, below they are negative real numbers. The path $\underline{x}(t)$ vs t differs in shape according to where the point $(\det A, \mathrm{tr}A)$ lies. If it lies below $(\mathrm{tr}A)^2 - 4\det A = 0$, $\underline{x}(t)$ is the sum of two vectors each remaining fixed in direction, their lengths shrinking exponentially. If $(\det A, \operatorname{tr} A)$ lies above $(\operatorname{tr} A)^2 - 4 \det A = 0$, $\underline{x}(t)$ returns to $\underline{0}$ on a spiral path. To see this we write $\lambda_2 = \overline{\lambda}_1$, $\underline{x}_2 = \overline{\underline{x}}_1$, $\underline{y}_2 = \overline{\underline{y}}_1$, then because $\underline{x}(t = 0)$ is real, $\underline{x}(t)$ is given by

$$x(t) = 2\operatorname{Re}\left\{ \left\langle \underline{y}_1, \underline{x}(t = 0) \right\rangle e^{\lambda_1 t} \underline{x}_1 \right\}$$

and, on writing $\left\langle \underline{y}_1, \underline{x}(t = 0) \right\rangle = \rho e^{i\phi}$, this is

$$\underline{x}(t) = 2\rho e^{\operatorname{Re}\lambda_1 t} \left\{ \cos\left(\operatorname{Im}\lambda_1 t + \phi\right) \operatorname{Re}\underline{x}_1 - \sin\left(\operatorname{Im}\lambda_1 t + \phi\right) \operatorname{Im}\underline{x}_1 \right\}$$

Hence $\operatorname{Re}\lambda_1$ tells us the rate of decay of the spiral, $\operatorname{Im}\lambda_1$ tells us its frequency of revolution and $\operatorname{Re}\underline{x}_1$ and $\operatorname{Im}\underline{x}_1$ determine its shape.

The point $(\det A, \operatorname{tr} A)$ can leave the fourth quadrant in two ways, either by crossing the line $\det A = 0$ or by crossing the line $\operatorname{tr} A = 0$. We call the first instance an exchange of stability, the second a Hopf bifurcation.

We turn now to the stirred tank reactor. Reactants are fed to the tank and products are removed along with some of the unused reactants and what we have is an autothermal process controlled by heat loss and reactant loss. We determine the steady states of the reactor and study their stability. We will find that the point $(\det A, \operatorname{tr} A)$ tells us all we want to know about the reactor close to a steady solution. The location of this point depends on the input variables to the problem and, as the values of these variables change, it may leave the fourth quadrant by crossing either the line $\det A = 0$ or the line $\operatorname{tr} A = 0$.

We write a simple model assuming that an exothermic, first order decomposition of a reactant in the feed stream takes place in a tank maintained spatially uniform by sufficient stirring. It is

$$V\frac{dc}{dt} = qc_{in} - qc - kcV$$

and

$$V\rho c_P \frac{dT}{dt} = q\rho c_P T_{in} - q\rho c_P T + \{-\Delta H\} kcV - UA(T - T_c)$$

where c denotes the concentration of the reactant and T denotes the temperature. The tank is equipped with a heat exchanger to remove the heat released by the reaction, and this explains the heat sink appearing as the fourth term on the right hand side of the second equation. The preceding term is the heat source, as $-\Delta H > 0$. The density, the heat capacity, etc. are taken to be constants while the chemical reaction rate coefficient is specified by the Arrhenius formula:

$$k = Ae^{-E/RT}.$$

This can be written in terms of k_{in} as

$$k = k_{in}e^{-\frac{E}{R}\left(\frac{1}{T} - \frac{1}{T_{in}}\right)} = k_{in}e^{\frac{y}{1 + \frac{RT_{in}}{E}y}},$$

where $y = \frac{E}{RT_{in}}\left(\frac{T - T_{in}}{T_{in}}\right)$, and then if $\frac{RT_{in}}{E}y = \frac{T - T_{in}}{T_{in}} << 1$ it is

$$k = k_{in}e^{y}.$$

This is what we use henceforth. It is called the Frank-Kamenetski approximation after D.A.Frank-Kamenetskii, a mining engineer interested in the problem of thermal explosions. The approximation makes sense as long as $\dfrac{T - T_{in}}{T_{in}} << 1$. It is explained in physical terms in his book, "*Diffusion and Heat Exchange in Chemical Kinetcs*." We use it for its mathematical convenience.

Then, letting x denote $1 - \dfrac{c}{c_{in}}$, the fractional conversion of the feed, scaling time by the holding time $\dfrac{V}{q}$ and writing it again as t, and introducing the dimensionless groups

$$\beta = \frac{UA}{\rho c_P V}\frac{V}{q} \geq 0$$

$$B = \frac{E}{RT_{in}}\frac{(-\Delta H)c_{in}}{\rho c_P T_{in}} > 0$$

and

$$D = \frac{V}{q}k_{in} > 0$$

we have

$$\frac{dx}{dt} = -x + De^y (1 - x) \equiv f(x, y)$$

and

$$\frac{dy}{dt} = -y + BDe^y (1 - x) - \beta y \equiv g(x, y)$$

where we have assumed $T_c = T_{in}$ and obtained a non-essential simplification. The input variables D, B and β, measure the strengths of the chemical reaction, the heat source and the heat sink. We will assume that they can be adjusted independently whereas that may not be so in a definite physical problem. Indeed to study the response of a system to the holding time τ, where $\tau = \dfrac{V}{q}$, it would be better to put $\beta_0 \tau$ and $k_0 \tau$ in place of β and D where $\dfrac{1}{\beta_0}$ and $\dfrac{1}{k_0}$ are the natural time scales for heat exchange and reaction.

The steady solutions satisfy

$$0 = -x + De^y (1 - x)$$

and

$$0 = -y + BDe^y (1 - x) - \beta y$$

or

$$y = \frac{B}{1 + \beta} x$$

and

$$D = \frac{x}{1 - x} e^{-y}$$

The dependence of x on D is then given implicity by

$$D = \frac{x}{1-x} e^{-\frac{B}{1+\beta}x}$$

To each x on $(0,1)$ there corresponds one value of D. As x increases from 0 to 1 the right hand side, called RHS henceforth, increases from 0 to ∞ and the question is: is this a monotonic increase? If it is, there will be one value of x corresponding to each value of D; otherwise there will be more than one value of x corresponding to some values of D. The answer depends on the size of $\frac{B}{1+\beta}$ for this determines whether the factor $e^{-\frac{B}{1+\beta}x}$ can turn around the strongly increasing factor $\frac{x}{1-x}$. To answer our question, the implicit function theorem instructs us to find where $\frac{dRHS}{dx}$ vanishes. Thus we calculate $\frac{dRHS}{dx}$ and find:

$$\frac{dRHS}{dx} = \frac{1}{(1-x)^2} \frac{1}{1+\beta} e^{-\frac{B}{1+\beta}x} \left\{ Bx^2 - Bx + (1+\beta) \right\}$$

The algebraic sign of $\frac{dRHS}{dx}$ is that of $\left\{ Bx^2 - Bx + (1+\beta) \right\}$ and hence is positive for $x = 0$ and for $x \to 1$. Our question then reduces to: does $Bx^2 - Bx + (1+\beta)$ vanish for intermediate values of $x : 0 < x < 1$? We let x_1 and x_2 denote the roots of $Bx^2 - Bx + (1+\beta) = 0$ and find

$$x_{1,2} = \frac{1}{2} \pm \frac{1}{2}\sqrt{1 - \frac{4(1+\beta)}{B}}$$

There are two possibilities: either $B < 4(1+\beta)$, x_1 and x_2 do not lie on $(0,1)$ and RHS is a monotonic increasing function of x, $\forall x \in (0,1)$ or $B > 4(1+\beta)$, x_1 and x_2 lie on $(0,1)$ and RHS exhibits turning points at x_1 and x_2.

The line $B = 4(1+\beta)$ divides the positive quadrant of the $\beta - B$ plane into two regions. In the lower region x vs D is monotonic, in the upper x vs D is S-shaped. Schematically then the steady state diagram looks as follows:

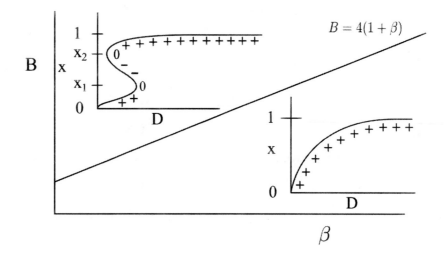

The steady state curve in the upper part of the sketch is not like the branching diagram in the chemostat problem. On this S-shaped curve jumps take place as D increases through the lower turning point corresponding to x_1 (ignition point) or decreases through the upper turning point corresponding to x_2 (extinction point). And we conclude that conversions between x_1 and x_2 may be difficult to achieve.

To establish the stability of the steady solutions we suppose the system to be in a steady state denoted (x, y) and ask whether a small excursion to $(x + \xi, y + \eta)$ does or does not return to (x, y). Because we have

$$\frac{d\xi}{dt} = f(x + \xi, y + \eta), \quad \frac{d\eta}{dt} = g(x + \xi, y + \eta)$$

and

$$f(x, y) = 0 = g(x, y)$$

we find, for small excursions from (x, y)

$$\frac{d}{dt}\begin{pmatrix} \xi \\ \eta \end{pmatrix} = A \begin{pmatrix} \xi \\ \eta \end{pmatrix}$$

where

$$A = \begin{pmatrix} f_x & f_y \\ g_x & g_y \end{pmatrix}$$

and where A is to be evaluated at the steady state under investigation. Now as

$$f_x = -1 - De^y$$

$$f_y = De^y (1 - x)$$

$$g_x = -BDe^y$$

and

$$g_y = -(1 + \beta) + BDe^y (1 - x)$$

and as $f(x, y) = 0$ implies

$$De^y = \frac{x}{1 - x}$$

we find

$$A = \begin{pmatrix} \frac{-1}{1-x} & x \\ \frac{-Bx}{1-x} & -(1+\beta) + Bx \end{pmatrix}$$

whence we have

$$\det A = \frac{1}{1 - x} \left\{ Bx^2 - Bx + (1 + \beta) \right\}$$

and

$$\mathrm{tr}\, A = -\frac{1}{1 - x} \left\{ Bx^2 - (B + 1 + \beta) x + 2 + \beta \right\}.$$

Looking at $\det A$ first, we see that as $x \to 0$ and as $x \to 1$ $\det A$ is positive and that $\det A$ vanishes at x_1 and x_2, the turning points of RHS. Hence $\det A$ is positive unless $B > 4(1 + \beta)$ and $x_1 \leq x \leq x_2$. The algebraic sign of $\det A$ is marked on the foregoing sketch and it indicates that the branch of the S-shaped steady state curve running between the turning points corresponding to x_1 and x_2 is unstable. Indeed as x increases through x_1 or decreases through x_2 an exchange of stabilities takes place which corresponds to passing from the fourth to the third quadrant in the $\det A - \text{tr} A$ plane by crossing the line $\det A = 0$. While the turning points at x_1 and x_2 correspond to $\det A = 0$ they do not look like the branch point, where also $\det A = 0$, discovered in the chemostat problem.

Thus at each point (β, B) where $B < 4(1 + \beta)$, we have $\det A > 0$ for all values of D but where $B > 4(1 + \beta)$ we have a bounded range of D, depending on β and B, where $\det A < 0$.

To see what is going on when $\det A$ is positive we must look at $\text{tr} A$. This is negative as $x \to 0$ and as $x \to 1$ so that all such states are stable to small upsets. The question is whether or not $\text{tr} A$ vanishes for intermediate values of x: $0 < x < 1$. Denoting by x_3 and x_4 the roots of $\text{tr} A = 0$ we find their values to be

$$x_{3,4} = \frac{B + (1 + \beta) \pm \sqrt{(B + (1 + \beta))^2 - 4B(2 + \beta)}}{2B}$$

The condition that $(B + (1 + \beta))^2 - 4B(2 + \beta)$ vanish places two curves on the $\beta - B$ diagram:

$$B = 3 + \beta \pm 2\sqrt{2 + \beta}$$

and between these curves x_3 and x_4 are complex conjugates and so do not lie on (0,1). Hence in the region between the two curves we have $\text{tr} A < 0$.

Now the roots, x_3 and x_4, of $\text{tr} A = 0$ satisfy $Bx^2 - (B + 1 + \beta)x + 2 + \beta = 0$ and because $B + 1 + \beta > 0$ and $2 + \beta > 0$ their real parts must be positive. If, then, the point (β, B) lies above the upper curve, i.e., $B > 3 + \beta + 2\sqrt{2 + \beta}$, the roots must be real and positive. In terms of z, where $z = x - 1$, the roots of $\text{tr} A = 0$ satisfy

$$Bx^2 - (B + 1 + \beta)x + 2 + \beta = Bz^2 - (-B + 1 + \beta)z + 1 = 0$$

but $-B + 1 + \beta < -2 - 2\sqrt{2 + \beta} < 0$ when $B > 3 + \beta + 2\sqrt{2 + \beta}$ hence the real parts of z_3 and z_4 must be negative and we see that x_3 and x_4 lie to the left of $x = 1$.

Likewise, if the point (β, B) lies below the lower curve, i.e., $B < 3 + \beta - 2\sqrt{2 + \beta}$, x_3 and x_4 must also be real and positive but now they lie to the right of $x = 1$.

So in terms of $\mathrm{tr}A$ what we find is this: for all points (β, B) lying below $B = 3 + \beta + 2\sqrt{2 + \beta}$ we have $\mathrm{tr}A < 0$, $\forall x \in (0, 1)$; for all points (β, B) lying above $B = 3 + \beta + 2\sqrt{2 + \beta}$ we have $\mathrm{tr}A < 0$ for $0 < x < x_3$, $\mathrm{tr}A > 0$ for $x_3 < x < x_4$ and $\mathrm{tr}A < 0$ for $x_4 < x < 1$. The correspondence between x and D is $D = \dfrac{x}{1 - x} e^{-\frac{B}{1 + \beta} x}$.

Thus at each point (β, B) where $B < 3 + \beta + 2\sqrt{2 + \beta}$, we have $\mathrm{tr}\, A < 0$ for all values of D but where $B > 3 + \beta + 2\sqrt{2 + \beta}$ we have a bounded range of D, depending on β and B, where $\mathrm{tr}\, A > 0$.

The two curves $B = 4(1 + \beta)$ and $B = 3 + \beta + 2\sqrt{2 + \beta}$ divide the $\beta - B$ plane into four regions as follows:

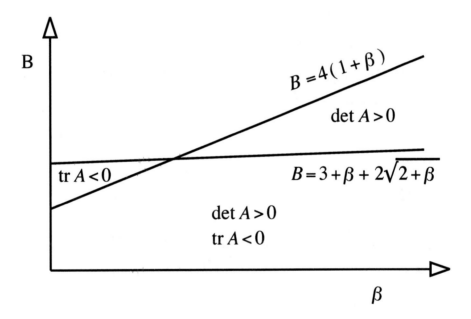

At each point where the sign of $\det A$ or $\mathrm{tr}A$ is not indicated, both signs are possible depending on the value of x.

The next figure presents one possibility

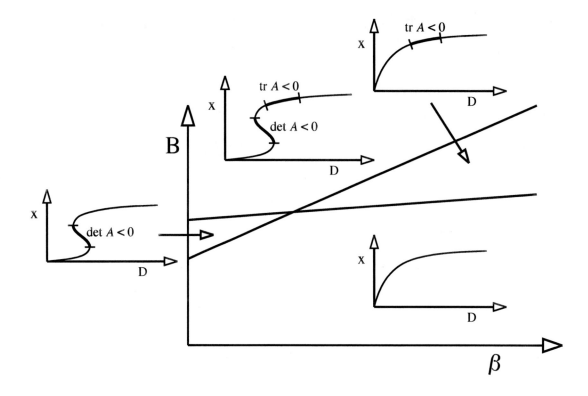

At each point of the upper most region x_1, x_2, x_3 and x_4 all fall on the interval (0,1) and so as x increases from 0 to 1 det A takes positive, negative then positive values while tr$A$ takes negative, positive then negative values. This region can be subdivided depending on how x_1, x_2, x_3 and x_4 are ordered. This is worked out in Poore's paper.

If we suppose that (β, B) is such that $x_1 < x_2 < x_3 < x_4$ then it is possible to find:

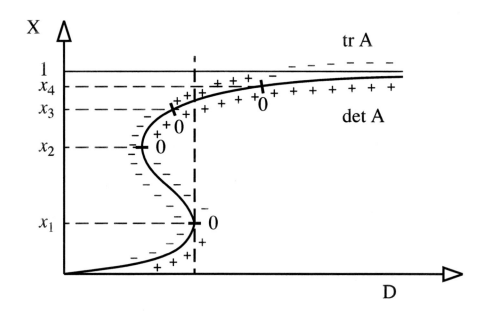

and where $\text{tr}A > 0$ or where $\det A < 0$ we have unstable steady states, the remaining steady states being stable to small disturbances. The sketch should convince the readers that they have no idea what happens as D increases and x passes through x_1. The points corresponding to x_3 and x_4, where $\text{tr}A = 0$ and $\det A > 0$ are called Hopf bifurcation points. For D and x such that x is just below x_3 or just above x_4 the corresponding steady state is stable but the return of a small perturbation to the steady state is not monotonic. As D increases and x increases through x_3 or as D decreases and x decreases through x_4 both of which correspond to passing from the fourth to the first quadrant in the $\det A - \text{tr}A$ plane by crossing the line $\text{tr}A = 0$, something new turns up that we can only guess at. In each instance it may be that a branch of stable periodic solutions grows from the bifurcation point, but there are other possibilities, and Poore deals with such questions. A simple way to get the required information can be found in Kuramoto's book "*Chemical Oscillations, Waves and Turbulence.*"

For example, as D increases in the sketch below, and the state crosses $\text{tr}\,A = 0$ from S to U,

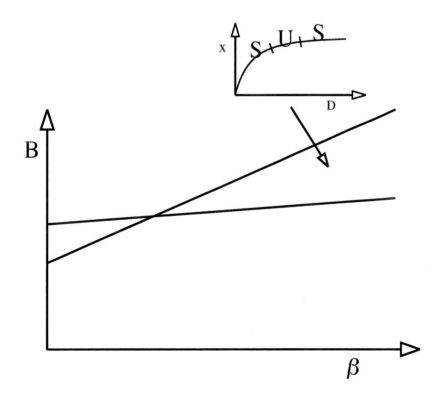

we may see

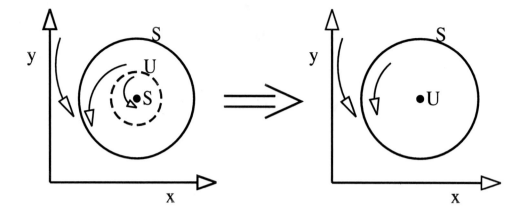

or

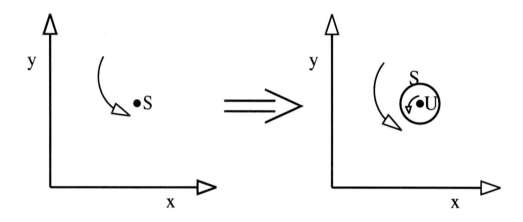

The reader may wish to work out the adiabatic case, viz., $\beta = 0$.

The two problems presented in this lecture illustrate the basic ideas of small amplitude stability studies and therefore serve our purposes very well. But they are too simple to represent real chemical reactor problems and even too simple to represent what is in the chemical reactor literature. The greatest simplification is in the use of two variables to define the state of the system and the consequent use of 2×2 matrices to determine its stability.

But even in two state variable problems there is a lot going on. To begin to learn about this the reader can consult chapter six in Gray and Scott's book "*Chemical Oscillations and Instabilities*."

More information on populations of microorganisms can be found in Waltman's book "*Competition Models in Population Biology*."

7.3 Home Problems

1. The autocatalytic reaction

$$A + P \xrightarrow{\;k\;} 2P$$

takes place in a spatially uniform reactor whose holding time is θ. If c_A and c_P denote the concentrations of A and P and $c_{P_{in}} = 0$, we can write

$$\frac{dc_A}{dt} = \frac{c_{A_{in}}}{\theta} - \frac{c_A}{\theta} - k\,c_A\,c_P$$

and

$$\frac{dc_P}{dt} = 0 - \frac{c_P}{\theta} - k\,c_A\,c_P$$

where $c_A(t=0)$ and $c_P(t=0) > 0$ are specified.

Find the steady solutions of this system of equations and their stability to small upsets, i.e., show that the steady state diagram in terms of $c_{A_{in}}$ is

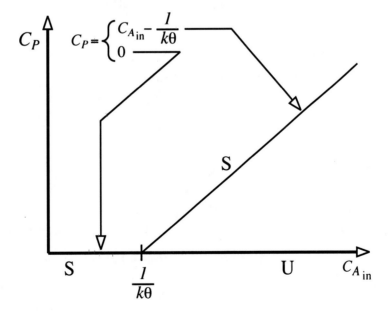

2. The reaction $A + B \longrightarrow C + D$ takes place by the mechanism

$$A \longrightarrow X$$

$$B + X \longrightarrow Y + C$$

$$2X + Y \longrightarrow 3X$$

$$X \longrightarrow D$$

Letting a denote the concentration of A, etc., assuming that a and b remain fixed and requiring all the elementary kinetic rate constants to be equal, we can write

$$\frac{dx}{dt} = a - bx + x^2 y - x$$

and

$$\frac{dy}{dt} = bx - x^2 y$$

Find the values of a and b for which these equations have a stable equilibrium point and show that the curve $b = 1 + a^2$ is a locus of Hopf bifurcations.

3. The predator-prey equations of Volterra and Lotka are

$$\frac{dx}{dt} = Ax - Bxy$$

$$\frac{dy}{dt} = Cxy - Dy$$

where A, B, C and D are positive constants and where y is the predator population while x is the prey population. Find the equilibrium points and determine their stability.

4. The Lorenz equations, "*Deterministic Nonperiodic Flow*," J. Atmos. Sci. <u>20</u>, 130 (1963),

viz.,

$$\frac{dx}{dt} = \sigma\,(y - x)$$

$$\frac{dy}{dt} = \rho x - y - xz$$

and

$$\frac{dz}{dt} = -\beta z + xy$$

where σ, ρ and β are positive constants, are well known in the theory of deterministic chaos. These equations represent a three mode truncation of the Boussinesq equations for natural convection in a fluid layer heated from below. The parameters σ, ρ and β denote the Prandtl number, the Rayleigh number and an aspect ratio. Lorenz set $\sigma = 10$ and $\beta = \frac{8}{3}$. For fixed σ and β investigate the equilibrium solutions as they depend on ρ and establish their stability. The point $\rho = 1$ is called a pitchfork as two new equilibrium solutions break off from the equilibrium solution $x = y = z = 0$. Be sure to find the Hopf bifurcation on each of the new equilibrium branches. To do this let I, II and III denote the principal invariants of the Jacobian matrix and observe that $I \times II = III$ at a Hopf bifurcation.

5. The reaction $A \longrightarrow B$, where A is a gas and B is a liquid, is ordinarily carried out by bubbling a gas stream containing A through a liquid containing B. Assuming that A dissolves in the liquid and that the reaction takes place there, we can write a simple model by supposing that the solubility of A is independent of temperature, that the rate of absorption of A is not enhanced by the chemical reaction and that the concentration of B is in large excess. The concentration of dissolved A and the temperature of the liquid then satisfy

$$V\frac{dc_A}{dt} = k_L^0\, a\, V\left(c_A^* - c_A\right) - Lc_A - k\, c_A\, c_{B_{\text{in}}} V$$

and

$$V \rho c_P \frac{dT}{dt} = \rho c_P L (T_{\text{in}} - T) + (-\Delta H) k c_A c_{B_{\text{in}}} V - U A (T - T_{\text{in}})$$

where L denotes the liquid feed rate and V denotes the liquid volume in the reactor and where we put $c_B = c_{B_{\text{in}}}$. The gas stream maintains A at partial pressure P_A and creates the surface area aV, diffusional resistance lies entirely on the liquid side and $c_A^* = \frac{P_A}{H}$.

Introducing

$$k = k_{\text{in}} e^{\frac{E}{RT_{\text{in}}} \frac{T - T_{\text{in}}}{T_{\text{in}}}}$$

$$\tau = \frac{V}{L}$$

$$D = \tau k_{\text{in}} C_{B_{\text{in}}}$$

$$D_m = \tau k_L^0 a$$

$$Y = \frac{E}{RT_{\text{in}}} \frac{T - T_{\text{in}}}{T_{\text{in}}}$$

$$X = \frac{c_A^* - c_A}{c_A^*}$$

we can write our model

$$-\tau \frac{dX}{dt} = D_m X - \left(1 + D e^Y\right) (1 - X)$$

and

$$\tau \frac{dY}{dt} = -Y + B D e^Y (1 - X) - \beta Y$$

Find the steady solutions of this system of equations and their stability to small upsets.

6. In problems where the model of a process is a system of differential and algebraic equations, e.g., p balance equations and q phase equilibrium equations to determine $p+q$ state variables,

the response of the system is inherently slow. To see this, suppose that x, y and v satisfy

$$\frac{dx}{dt} = f(x, y, v)$$

$$\frac{dy}{dt} = g(x, y, v)$$

and

$$x = y.$$

Then the equilibrium values of x, y and v satisfy

$$f(x, y, v) = 0$$

$$g(x, y, v) = 0$$

and

$$x - y = 0$$

and we assume that at an equilibrium point

$$\det \begin{pmatrix} f_x & f_y & f_v \\ g_x & g_y & g_v \\ 1 & -1 & 0 \end{pmatrix} \neq 0$$

The dynamic value of v is determined by

$$\frac{dx}{dt} = \frac{dy}{dt}$$

or

$$f(x, y, v) = g(x, y, v)$$

If x_0, y_0, v_0 denotes an equilibrium solution, we can construct a linear approximation to the nearby dynamics. To do this we obtain $v(x, y)$ via $f(x, y, v) = g(x, y, v)$ whence

$$v_x = -\frac{g_x - f_x}{g_v - f_v}$$

and

$$v_y = -\frac{g_y - f_y}{g_v - f_v}$$

Then the linear approximation is

$$\frac{d}{dt}\begin{pmatrix} \xi \\ \eta \end{pmatrix} = \begin{pmatrix} F_x & F_y \\ G_x & G_y \end{pmatrix}\begin{pmatrix} \xi \\ \eta \end{pmatrix}$$

where $F(x, y) = f\big(x, y, v(x, y)\big)$ and $G(x, y) = g\big(x, y, v(x, y)\big)$

Using

$$F_x = f_x + f_v v_x$$

$$F_y = f_y + f_v v_y$$

$$G_x = g_x + g_v v_x$$

and

$$G_y = g_y + g_v v_y$$

show that

$$\det\begin{pmatrix} F_x & F_y \\ G_x & G_y \end{pmatrix} = F_x G_y - G_x F_y = 0.$$

This tells us that the linear approximation always exhibits an eigenvalue that is zero. Hence the response of the system, when it is in an equilibrium state, to a small upset is slow. To see whether an upset strengthens or weakens, we need to go on and look at quadratic terms.

The boiling steady states of a constant pressure evaporator exhibit this problem. The model is:

$$M \frac{dx}{dt} = x_F F - x (F - V)$$

$$M c_P \frac{dT}{dt} = U A (T_S - T) + c_P (T_F - T) F - \lambda V$$

and

$$T = T^0 + \beta x$$

where the state variables are x, T and V, all else being fixed.

In terms of the holding time $\tau = \dfrac{M}{F}$, the heat transfer time $\tau_h = \dfrac{M c_P}{U A}$ and the dimensionless variables

$$y = \frac{T - T^0}{\beta}, \quad v = \frac{V}{F} \quad \text{and} \quad q = \frac{\lambda}{c_P \beta}$$

the model is

$$\frac{dx}{dt} = x_F - x (1 - v)$$

$$\frac{dy}{dt} = \frac{\tau}{\tau_h} \{ y_S - y \} + \{ y_F - y \} - q v$$

and

$$x = y$$

where t is written in place of t / τ.

To determine v in terms of x and y use $x = y$ and hence $\dfrac{dx}{dt} = \dfrac{dy}{dt}$ to conclude that

$$\frac{\tau}{\tau_h} \left(y_S - y \right) + \left(y_F - y \right) - q v = x_F - x + x v$$

This tells us that

$$\frac{\partial v}{\partial y} = -\frac{\dfrac{\tau}{\tau_h} + 1}{q + x}$$

and

$$\frac{\partial v}{\partial x} = \frac{1 - v}{q + x}$$

Then as long as boiling is taking place we have

$$\frac{dx}{dt} = x_F - x + xv\,(x, y)$$

and

$$\frac{dy}{dt} = \frac{\tau}{\tau_h}\left(y_S - y\right) + \left(y_F - y\right) - qv\,(x, y)$$

where $0 < v < 1$, and the linear approximation is determined by the matrix

$$\begin{pmatrix} -1 + v + x\dfrac{\partial v}{\partial x} & x\dfrac{\partial v}{\partial y} \\[2ex] -q\dfrac{\partial v}{\partial x} & -\dfrac{\tau}{\tau_h} - 1 - q\dfrac{\partial v}{\partial y} \end{pmatrix}$$

evaluated at a solution x_0, y_0, v_0 of the steady equations.

Show that the determinant of this matrix is zero and that its trace is

$$(-1 + v)\frac{q}{q + x} - \left(\frac{\tau}{\tau_h} + 1\right)\frac{x}{q + x} < 0$$

and as a result show that the solution to the linear equations describing a small upset of a boiling steady state is the sum of two terms, one constant in time, the other dying exponentially in time.

By carrying out an Euler approximation or otherwise, determine what in fact does happen when a boiling steady state experiences an upset.

This model takes the form

$$\frac{dx}{dt} = f(x, y, v)$$

$$\frac{dy}{dt} = g(x, y, v)$$

$$x = y$$

where v does not come into the third equation.

The boiling curve for a three component ideal solution is given by solving

$$\frac{dx_1}{dt} = \left\{ -\frac{P_1(T)}{P} + 1 \right\} x_1$$

$$\frac{dx_2}{dt} = \left\{ -\frac{P_2(T)}{P} + 1 \right\} x_2$$

and

$$\frac{P_1(T)}{P} x_1 + \frac{P_2(T)}{P} x_2 + \frac{P_3(T)}{P} (1 - x_1 - x_2) = 1$$

where $x_1 + x_2 + x_3 = 1$ is used to eliminate x_3.

Here T comes into the third equation and this system is not as sluggish as its look-a-like.

7. In the stirred tank reactor model put $\beta = 3$ and $B = 14$ (setting $\beta = 2$ and $B = 10$ leads to a better x vs D curve) then

$$3 + \beta + 2\sqrt{2 + \beta} < B < 4(1 + \beta)$$

and the conversion x is a monotonic increasing function of D. On this curve the determinant of the Jacobian matrix is always positive but its trace vanishes at $x_3 = 0.406$ and $x_4 = 0.880$.

Set $D = 0.162$, which corresponds to a steady conversion $x = 0.380$, and integrate the differential equations for the start up of the reactor from a variety of initial conditions, in particular $x = 0, y = 0$ which corresponds to starting the reactor up using its feed conditions as initial conditions. Do this using a machine having good graphics and plot the start up trajectories in the x, y plane.

You will find that only a small set of initial conditions lead to the one and only steady state. This steady state is stable to small upsets but it is close to the Hopf bifurcation point at $x_3 = 0.406$. As D increases through this bifurcation point, a small limit cycle breaks off. If the bifurcation is forward this cycle is stable and surrounds the steady state which turns unstable. But if it is backward, as it is here, it is unstable and for values of D short of the bifurcation point an unstable limit cycle surrounds the stable steady state. This in turn is surrounded by a large stable limit cycle that sets in at a lower value of D. As D passes through the bifurcation point the small unstable limit cycle vanishes, the stable steady state turns unstable while the large stable limit cycle is not sensitive to all of this. The picture is then:

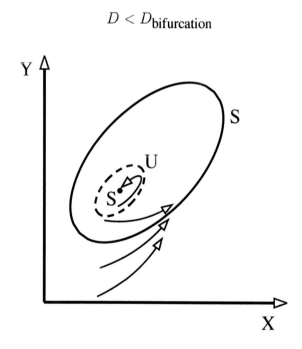

$$D < D_{\text{bifurcation}}$$

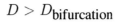

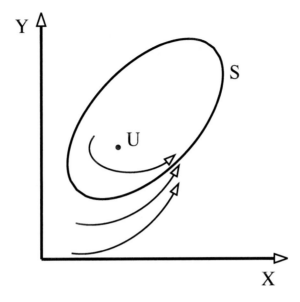

$D > D_{\text{bifurcation}}$

The top picture explains why your calculations turn out the way they do.

8. To put metabolism into the chemostat model in a simple way write

$$\frac{dn}{dt} = \left\{ -\frac{W}{V} + k \right\} n$$

and

$$\frac{dc}{dt} = \frac{W}{V} \left\{ c_{\text{in}} - c \right\} - \nu k n - \mu n$$

where $\mu > 0$.

Determine the steady solutions and their stability as it depends on $\dfrac{W}{V}$.

9. In the boiling curve problem for a three component ideal solution, assume $P_3(T) < P_2(T) < P_1(T)$ and $x_3(t = 0)$ is nearly zero. Show that as $t \to \infty$ we have $x_3 \to 1$.

Lecture 8

The Inverse Problem

In this lecture we present two problems where we must determine the values of the elements a_{ij} of a matrix A using measurements of the elements $x_i(t)$ of a vector $\underline{x}(t)$ satisfying $\dfrac{d\underline{x}}{dt} = A\underline{x}$. The first problem has been solved by J. Wei and C.D. Prater in "*The Structure and Analysis of Complex Reaction Systems*," Advances in Catalysis $\underline{13}$, 204 (1962).

In each problem A is self-adjoint in some inner product denoted $\langle\,,\,\rangle_G$ and we can discover this inner product while A itself remains unknown. Ordinarily this is not the plain vanilla inner product, but it does tell us that A has a complete set of eigenvectors corresponding to real eigenvalues. Denoting these $\underline{x}_1$, $\underline{x}_2$, $\ldots$, $\underline{x}_n$ and λ_1, λ_2, $\ldots$, λ_n, where $\langle\,\underline{x}_i,\,\underline{x}_j\,\rangle_G = \delta_{ij} = \langle\,\underline{y}_i,\,\underline{x}_i\,\rangle_I$, the idea is to obtain A by deriving from measurements of $\underline{x}(t)$ the terms that make up its spectral decomposition:

$$A = \sum \lambda_i\,\underline{x}_i\,\underline{y}_i^T.$$

To do this we expand the data, $\underline{x}(t)$, in the eigenvectors of A and write

$$\underline{x}(t) = \sum \langle\,\underline{y}_i,\underline{x}(t=0)\,\rangle_I\,e^{\lambda_i t}\underline{x}_i$$

where $\langle\,\underline{y}_i,\underline{x}(t=0)\,\rangle_I = \langle\,\underline{x}_i,\underline{x}(t=0)\,\rangle_G$ and where measurements give us the left hand side of this formula. The exponential separation of the terms on the right hand side as t grows large, together with the fact that $\underline{x}(t=0)$ is under our control leads to a plan for determining the eigenvectors of A and then the terms in its spectral decomposition. Indeed, if, for some value of i, we can guess $\underline{x}(t=0)$ so that $\langle\, \underline{y}_j,\, \underline{x}(t=0)\,\rangle_I = 0\ \forall j \neq i$, $\underline{x}(t)$ is just

$$\underline{x}(t) = \langle\, \underline{y}_i,\, \underline{x}(t=0)\,\rangle\, e^{\lambda_i t}\, \underline{x}_i$$

and measurements of $\underline{x}(t)$ determine $\underline{x}_i$.

If, in this way, $\underline{x}_1,\, \underline{x}_2,\, \ldots,\, \underline{x}_n$, can be determined in a sequence of experiments then $\underline{y}_1,\, \underline{y}_2,\, \ldots,\, \underline{y}_n$ can be calculated and $\lambda_1,\, \lambda_2,\, \ldots,\, \lambda_n$ can be obtained via

$$\ln \frac{\langle\, \underline{y}_i,\, \underline{x}(t)\,\rangle}{\langle\, \underline{y}_i,\, \underline{x}(t=0)\,\rangle} = \lambda_i t$$

This done, A can be recovered in terms of its eigenvalues and eigenvectors via

$$A = \sum \lambda_i\, \underline{x}_i\, \underline{y}_i^T.$$

The problem to be solved then is the selection of a useful sequence of $\underline{x}(t=0)$'s. What makes this possible is that the terms in the expansion of $\underline{x}(t)$ go to $\underline{0}$ at differing rates. Indeed if $\underline{0}$ is the unique equilibrium point, $\underline{x}_1$ can be obtained using the long time data from an arbitrary experiment, $\underline{x}_2$ can be obtained using the long time data from an experiment satisfying $\langle\, \underline{x}_1,\, \underline{x}(t=0)\,\rangle_G = 0$, etc.

8.1 A First Order Chemical Reaction Network.

Let $i = 1, \ldots, n$ denote the species in a chemically reacting system and suppose that each pair participates in a reversible chemical reaction:

where $k_{ji} > 0$ and $k_{ij} > 0$ are the forward and reverse chemical rate constants. A system of n chemical isomers provides the simplest physical realization of this. We carry out the reaction at constant temperature in a closed vessel. At time $t = 0$ we specify the number of moles of each species in the vessel and inquire as to how these numbers change as time runs on. As the total number of moles is fixed we can define the state of the system most easily in terms of the mole fractions of the species. Letting x_i, $i = 1, \ldots, n$, denote these we write

$$\frac{dx_i}{dt} = \sum_{\substack{j=1 \\ j \neq i}}^{n} k_{ij} x_j - \sum_{\substack{j=1 \\ j \neq i}}^{n} k_{ji} x_i$$

or in terms of $\underline{x} = \begin{pmatrix} x_1 \\ x_2 \\ \vdots \\ x_n \end{pmatrix}$

$$\frac{d\underline{x}}{dt} = K\underline{x}$$

where the off-diagonal elements of K are k_{ij}, the diagonal elements being the negative of the sums of the off-diagonal elements in the same column. This tells us that the Gerschgorin column circles are all centered on the negative real line with radius equal to the distance from the center to the origin. As a result the eigenvalues of K cannot have positive real parts, and if a real part is zero the eigenvalue itself must be zero. In the special case $n = 3$ the matrix K is

$$K = \begin{pmatrix} -k_{21} - k_{31} & k_{12} & k_{13} \\ k_{21} & -k_{12} - k_{32} & k_{23} \\ k_{31} & k_{32} & -k_{13} - k_{23} \end{pmatrix}$$

Now we have $\det K = 0$ because the rows of K add to $\underline{0}^T$, i.e.,

$$\begin{pmatrix} 1 \\ 1 \\ \vdots \\ 1 \end{pmatrix}^T K = \underline{0}^T,$$

so at least one eigenvalue of K is zero; and if $\sum x_i = 1$ when $t = 0$ then $\sum x_i = 1$ for all $t \geq 0$ due to

$$\frac{d}{dt} \sum x_i = \begin{pmatrix} 1 \\ 1 \\ \vdots \\ 1 \end{pmatrix}^T \frac{d\underline{x}}{dt} = \begin{pmatrix} 1 \\ 1 \\ \vdots \\ 1 \end{pmatrix}^T K\underline{x} = 0$$

Also it is not hard to see that if all $x_i \geq 0$ for $t = 0$ then each $x_i \geq 0$ for all $t \geq 0$. And so the motion of the state vector $\underline{x}$ takes place on the plane $\sum x_i = 1$ in the positive cone (quadrant, octant, ...) $x_i \geq 0$ of the composition space.

Because $\det K = 0$, the equation $K\underline{x} = \underline{0}$ has solutions other than $\underline{x} = \underline{0}$. We let

$$\underline{x}_{eq} = \begin{pmatrix} x_{1eq} \\ x_{2eq} \\ \vdots \\ x_{neq} \end{pmatrix}$$

denote one such solution. We make three assumptions about the problem:

first that x_{ieq} is positive, $i = 1, 2, \ldots, n$, and second that $\underline{x}_{eq}$ is the unique solution of $K\underline{x} = \underline{0}$ satisfying $\sum x_{ieq} = 1$. We call $\underline{x}_{eq}$ the equilibrium point for the reaction network and go on and assume that when the network is in equilibrium each reaction is in equilibrium. This is the principle of detailed balance and it tells us that

$$k_{ij}\, x_{jeq} = k_{ji}\, x_{ieq}$$

for all $i, j = 1, 2, \ldots, n$.

A readable explanation of the principle of detailed balance can be found in *"Treatise on Irreversible and Statistical Thermophysics"* by W. Yourgrau, A. van der Merwe and G. Raw. In H. Haken's book *"Synergetics"* the reader will find information on detailed balance as it has to do with what are called master equations.

The principle of detailed balance is sufficient that the first two assumptions hold. The requirement $k_{ij}\, x_{jeq} = k_{ji}\, x_{ieq}, \;\; k_{ij} > 0$, leads to the requirement that solutions of $K\underline{x} = \underline{0}$ have singly signed components. This tells us that the problem $K\underline{x} = \underline{0}$ can have but one independent solution, because two independent singly signed solutions have nonsingly signed linear combinations. Hence $\underline{x}_{eq}$ is unique up to a constant multiplier. It is unique and it lies in the positive cone as it is required to satisfy $\sum x_i = 1$.

The principle of detailed balance is also sufficient that K be self adjoint. To see this let $X_{eq} = \mathrm{diag}\left(x_{1eq}\; x_{2eq}\; \cdots\; x_{neq}\right)$ and observe that X_{eq} is a positive definite, Hermitian matrix. Then, by detailed balance, we can write $K X_{eq} = \left(\overline{K X_{eq}}\right)^T$, due to the fact that the jth column of $K X_{eq}$ is the jth column of K multiplied by x_{jeq}. Hence, in the inner product where $G = X_{eq}^{-1}$, we find that $K^* = K$. The readers should work this out for themselves using $K^* = G^{-1}\overline{K}^T G$.

We conclude therefore that K has a complete set of eigenvectors, denoted $\{\underline{x}_1,\, \underline{x}_2,\, \ldots,\, \underline{x}_n\}$, and we denote the corresponding eigenvalues $\lambda_1,\, \lambda_2,\, \ldots,\, \lambda_n$. The eigenvalues are real and not positive and we require that they be ordered: $\lambda_1 = 0 > \lambda_2 \geq \lambda_3 \geq \cdots \geq \lambda_n$. The eigenvectors are orthogonal in the inner product $G = X_{eq}^{-1}$, i. e.,

$$\underline{x}_i^T X_{eq}^{-1} \underline{x}_j = 0, \quad i \neq j$$

We denote by $\{\underline{y}_1,\, \underline{y}_2,\, \ldots,\, \underline{y}_n\}$ the set of vectors orthogonal to $\{\underline{x}_1,\, \underline{x}_2,\, \ldots,\, \underline{x}_n\}$ in the plain vanilla inner product and observe that, as $K^* = K^T$ in this inner product and $K^T \begin{pmatrix} 1 \\ 1 \\ \vdots \\ 1 \end{pmatrix} = \underline{0}$, we have $\underline{y}_1 = \begin{pmatrix} 1 \\ 1 \\ \vdots \\ 1 \end{pmatrix}$ if we set $\underline{x}_1 = \underline{x}_{eq}$. Then for any imtial composition we can write the solution

to $\dfrac{d\underline{x}}{dt} = K\underline{x}$ as

$$\underline{x}(t) = \sum_{i=1}^{n} \left\langle \underline{y}_i, \ \underline{x}(t=0) \right\rangle e^{\lambda_i t} \underline{x}_i$$

$$= \underline{x}_{eq} + \sum_{i=2}^{n} \left\langle \underline{y}_i, \ \underline{x}(t=0) \right\rangle e^{\lambda_i t} \underline{x}_i$$

To see how this can be used to find the matrix K and hence the $n(n-1)$ chemical rate constants we put $n=3$ and suppose $\lambda_3 < \lambda_2$. Then $\underline{x}(t) - \underline{x}_{eq}$ is the sum of two terms, $\left\langle \underline{y}_2, \ \underline{x}(t=0) \right\rangle e^{\lambda_2 t} \underline{x}_2$ and $\left\langle \underline{y}_3, \ \underline{x}(t=0) \right\rangle e^{\lambda_3 t} \underline{x}_3$ where the first dies out more slowly than the second. We call $\underline{x}_2$ the slow direction, $\underline{x}_3$ the fast direction. For large values of t the term in the slow direction approximates $\underline{x}(t) - \underline{x}_{eq}$ and so the long time data provide an estimate of $\underline{x}_2$ which we can refine in successive experiments. The idea is illustrated in the following sketch where the long time tangent direction is the estimate of $\underline{x}_2$. It is determined with increasing accuracy in successive experiments by using the latest estimate of $\underline{x}_2$ to derive a new initial condition wherein the magnitude of $\left\langle \underline{y}_2, \ \underline{x}(t=0) \right\rangle$ is increased vis-a-vis $\left\langle \underline{y}_3, \ \underline{x}(t=0) \right\rangle$. This can be done by extrapolating the long time tangent to the latest reaction path back to the edge of the triangle. The sequence $\underline{x}(t=0) - \underline{x}_{eq}$ then turns toward $\underline{x}_2$ and away from $\underline{x}_3$:

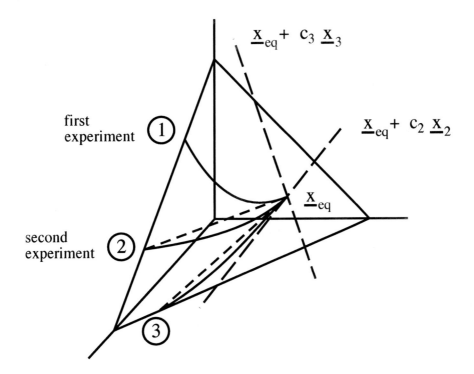

In the case $n = 3$ the subsequent work is especially simple for having obtained an estimate of $\underline{x}_2$ as indicated above, we can determine $\underline{x}_3$ in terms of $\underline{x}_1$ and $\underline{x}_2$ via orthogonality in the inner product $G = X_{eq}^{-1}$, i.e., $\underline{x}_3$ can be obtained as a solution to

$$\underline{x}_1^T X_{eq}^{-1} \underline{x}_3 = \begin{pmatrix} 1 \\ 1 \\ \vdots \\ 1 \end{pmatrix} \underline{x}_3 = 0$$

and

$$\underline{x}_2^T X_{eq}^{-1} \underline{x}_3 = 0$$

Then using the plane vanilla inner product, we can produce $\{\underline{y}_1, \underline{y}_2, \underline{y}_3\}$ via $\langle\, \underline{y}_i, \underline{x}_j \,\rangle = \delta_{ij}$ and return to a trajectory, such as ①, having, at least for short time, roughly equal contributions in the $\underline{x}_2$ and $\underline{x}_3$ directions and use it to obtain λ_2 and λ_3 via

$$\langle\, \underline{y}_2, \underline{x}(t) \,\rangle = \langle\, \underline{y}_2, \underline{x}(t = 0) \,\rangle e^{\lambda_2 t}$$

and

$$\langle \underline{y}_3, \underline{x}(t) \rangle = \langle \underline{y}_3, \underline{x}(t=0) \rangle e^{\lambda_3 t}$$

This information determines K via

$$K = \lambda_1 \underline{x}_1 \underline{y}_1^T + \lambda_2 \underline{x}_2 \underline{y}_2^T + \lambda_3 \underline{x}_3 \underline{y}_3^T.$$

This is what underlies the evaluation of K by the method of Wei and Prater.

And it obtains whatever the value of n. Indeed for any value of n we have $\underline{x}_1 = \underline{x}_{eq}$ and we can find $\underline{x}_2$ as above. To get $\underline{x}_3$ we select an initial condition at random and write

$$\underline{x}(t=0) = \underline{x}_{eq} + c_2 \underline{x}_2 + c_3 \underline{x}_3 + \cdots + c_n \underline{x}_n.$$

This cannot be used to determine $\underline{x}_3$ unless $c_2 = 0$, but because the eigenvectors $\underline{x}_1, \underline{x}_2, \ldots, \underline{x}_n$ are orthogonal in the inner product $G = X_{eq}^{-1}$ we can estimate c_2 as

$$c_2 = \frac{\underline{x}_2^T X_{eq}^{-1} \underline{x}(t=0)}{\underline{x}_2^T X_{eq}^{-1} \underline{x}_2}$$

and use the corrected initial condition $\underline{x}(t=0) - c_2 \underline{x}_2$ to generate a family of trajectories that will produce $\underline{x}_3$ in the same way that a random initial condition will produce $\underline{x}_2$. But there is a technical difficulty as the estimate of $\underline{x}_2$ we have is not perfect and neither are our composition measurements. Both factors make it impossible to completely free $\underline{x}(t=0)$ of its $\underline{x}_2$ component and hence $\underline{x}_2$ tends to reassert itself in any trajectory as time runs on. But this flaw is not fatal, it just makes the method somewhat more tedious than it might at first seem.

The Use of Flow Reactor Data

Instead of using data produced by a closed or batch reactor to determine K, we can investigate the possibility of using data produced by an open or flow reactor. By doing this we can avoid the problem of determining the time to which the composition measurements correspond.

Now the model for the steady operation of a well mixed reactor is

$$\underline{0} = \underline{x}_{\text{in}} - \underline{x}_{\text{out}} + \theta K \underline{x}_{\text{out}}$$

where θ is the holding time. The values of $\underline{x}_{\text{in}}$ and θ are under our control; the experiment produces the corresponding value of $\underline{x}_{\text{out}}$. Letting $n = 3$, denoting the eigenvectors of K as $\underline{x}_{\text{eq}}, \underline{x}_2$ and $\underline{x}_3$ and the corresponding eigenvalues as 0, λ_2 and λ_3, where $0 > \lambda_2 > \lambda_3$, and writing

$$\underline{x}_{\text{in}} = \underline{x}_{\text{eq}} + c_2 \underline{x}_2 + c_3 \underline{x}_3$$

and

$$\underline{x}_{\text{out}} = \underline{x}_{\text{eq}} + d_2 \underline{x}_2 + d_3 \underline{x}_3$$

we find

$$0 = c_2 - d_2 + \theta \lambda_2 d_2$$

and

$$0 = c_3 - d_3 + \theta \lambda_3 d_3$$

This tells us that $\underline{x}_{\text{in}}$, $\underline{x}_{\text{out}}$ and $\underline{x}_{\text{eq}}$ lie on a straight line iff that line is in the direction of an eigenvector of K. The reader can use this observation to devise a method for determining the matrix K via its spectral representation. In so doing it is useful to observe that an experiment turns $\underline{x}_{\text{in}} - \underline{x}_{\text{eq}}$ into the direction of the line through $\underline{x}_{\text{eq}}$ parallel to $\underline{x}_2$ and away from the line through $\underline{x}_{\text{eq}}$ in the direction of $\underline{x}_3$. Indeed as

$$\underline{x}_{\text{out}} - \underline{x}_{\text{eq}} = \left(I - \theta K \right)^{-1} \left(\underline{x}_{\text{in}} - \underline{x}_{\text{eq}} \right)$$

a sequence of experiments that might be worth some study is that in which the choice of $\underline{x}_{\text{in}}$ in any experiment is the value of $\underline{x}_{\text{out}}$ in the preceeding experiment. This is easily achieved by running a set of well mixed reactors in series.

8.2   Liquid Levels in a Set of Interconnected Tanks

We denote by $i = 1, 2, \ldots, n$ the tanks in a network of n pairwise connected tanks, by h_i the liquid level in tank i and by A_i its cross sectional area. Then the volume flow from tank i to tank j is

$$k_{ij}\big(h_i - h_j\big)$$

where k_{ij} is the conductivity of the pipe connecting tanks i and j and where $k_{ij} = k_{ji} > 0$, there being only one line connecting tanks i and j. Indeed under steady laminar flow conditions we would anticipate

$$k_{ij} = \frac{\pi}{8} \frac{R^4}{L} \frac{g}{\nu}$$

where R is the radius of the connecting line, L its length, etc.

The idea is to determine the values of the constants k_{ij} by studying the dynamics of the levels in a set of interconnected tanks as the levels go to equilibrium from an assigned set of initial values.

We can work in terms of the height, h_i, or the volume, V_i, of the liquid held in tank i. The heights make the equilibrium state simple but complicate the constant of the motion, whereas the reverse is true for the volumes. While this problem is more like the earlier problem when it is written in terms of volumes, we work in terms of heights and write

$$A_i \frac{dh_i}{dt} = \sum_{j \neq i} \big[- k_{ij}\big(h_i - h_j\big)\big] = \sum_{j \neq i} k_{ij} h_j - \sum_{j \neq i} k_{ij} h_i$$

or in terms of $\underline{h} = \begin{pmatrix} h_1 \\ h_2 \\ \vdots \\ h_n \end{pmatrix}$

$$\frac{d\underline{h}}{dt} = A^{-1}K\underline{h}$$

where $A = \mathrm{diag}\left(A_1\ A_2\ \ldots\ A_n\right)$ and where the off-diagonal elements of K are k_{ij}, the diagonal elements being the negative sums of the off-diagonal elements in the same column or row.

We first observe that $\overline{A}^T = A$ and $\overline{K}^T = K$ and then that $\begin{pmatrix} A_1 \\ A_2 \\ \vdots \\ A_n \end{pmatrix}^T A^{-1}K = \begin{pmatrix} 1 \\ 1 \\ \vdots \\ 1 \end{pmatrix}^T K = \underline{0}^T.$

Hence at least one eigenvalue of $A^{-1}K$ is zero and

$$\begin{pmatrix} A_1 \\ A_2 \\ \vdots \\ A_n \end{pmatrix}^T \frac{d\underline{h}}{dt} = \frac{d}{dt}\sum_i A_i h_i = \frac{dV}{dt} = 0$$

so that $\sum_i A_i h_i = V$ is constant and the motion takes place on a plane of constant volume, a plane whose normal is $\begin{pmatrix} A_1 \\ A_2 \\ \vdots \\ A_n \end{pmatrix}$ in the plane vanilla inner product. Also it is not hard to see that if all

$h_i \geq 0$ for $t = 0$ then each $h_i \geq 0$ for all $t \geq 0$. Therefore the curve mapped out by $\underline{h}(t)$, $t \geq 0$, lies on the plane $\sum_i A_i h_i = V$ in the positive cone of the vector space R^n where $\underline{h}$ resides.

The adjoint of $A^{-1}K$ in the inner product $\langle \underline{x}, \underline{y} \rangle_G = \overline{\underline{x}}^T G \underline{y}$ is

$$\left(A^{-1}K\right)^* = G^{-1}\left(\overline{A^{-1}K}\right)^T G$$

$$= G^{-1}\overline{K}^T \left(\overline{A}^{-1}\right)^T G = G^{-1}KA^{-1}G$$

and so, on taking $G = A$, we discover that $\left(A^{-1}K\right)^* = A^{-1}K$. Therefore in the inner product where $G = A$, $A^{-1}K$ is self adjoint and we conclude that $A^{-1}K$ has a complete set of eigenvectors and that the corresponding eigenvalues are real. We denote the eigenvectors $\underline{x}_1, \underline{x}_2, \ldots, \underline{x}_n$ and the corresponding eigenvalues $\lambda_1, \lambda_2, \ldots, \lambda_n$.

The rows of $A^{-1}K$ are multiples of the rows of K. Hence the Gerschgorin row circles for $A^{-1}K$, as for K itself, are all centered on the negative real line with radius equal to the distance from the center to the origin, whence the eigenvalues of $A^{-1}K$ cannot be positive. They can be ordered: $\lambda_n \leq \lambda_{n-1} \leq \cdots \leq \lambda_2 \leq \lambda_1 \leq 0$ where $\lambda_1 = 0$ and corresponding to λ_1 we have

$$\underline{x}_1 = \begin{pmatrix} 1 \\ 1 \\ \vdots \\ 1 \end{pmatrix} \quad \text{due to } K\underline{x}_1 = \underline{0}.$$

In the plain vanilla inner product, i.e., $G = I$, the adjoint of $A^{-1}K$ is

$$\left(A^{-1}K\right)^* = \left(\overline{A^{-1}K}\right)^T = KA^{-1}$$

and we denote its eigenvectors $\underline{y}_1, \underline{y}_2, \ldots, \underline{y}_n$ where $\langle \underline{x}_i, \underline{y}_j \rangle = \delta_{ij}$ and where

$$\underline{y}_1 = \frac{1}{A_1 + A_2 + \cdots + A_n} \begin{pmatrix} A_1 \\ A_2 \\ \vdots \\ A_n \end{pmatrix}$$

due to

$$KA^{-1} \begin{pmatrix} A_1 \\ A_2 \\ \vdots \\ A_n \end{pmatrix} = K \begin{pmatrix} 1 \\ 1 \\ \vdots \\ 1 \end{pmatrix} = \underline{0}$$

It is easy to see that the only solutions other than $\underline{x} = \underline{0}$ to $K\underline{x} = \underline{0}$ are multiples of $\begin{pmatrix} 1 \\ 1 \\ \vdots \\ 1 \end{pmatrix}$.

Indeed, as the off-diagonal elements of K satisfy $k_{ij} = k_{ji} > 0$ and the diagonal elements are the negative sums of the off-diagonal elements in the same column or row, we can eliminate x_1 from

$K\underline{x} = \underline{0}$ and discover that $\begin{pmatrix} x_2 \\ x_3 \\ \vdots \\ x_n \end{pmatrix} = \underline{x}^1$ satisfies $K^1\underline{x}^1 = \underline{0}^1$ where, like K, the off-diagonal

elements of K^1 satisfy $k_{ij}^1 = k_{ji}^1 > 0$ and the diagonal elements are the negative sums of the off-diagonal elements in the same column or row. Eliminating x_2, x_3, $\ldots$, x_{n-2} in the same way we find that

$$\begin{pmatrix} -a & a \\ a & -a \end{pmatrix} \begin{pmatrix} x_{n-1} \\ x_n \end{pmatrix} = \begin{pmatrix} 0 \\ 0 \end{pmatrix}, \quad a > 0,$$

whence $x_{n-1} = x_n$ and indeed we have $x_{n-2} = x_{n-1}, \ldots, x_1 = x_2$.

At equilibrium we have

$$A^{-1}K\underline{h}_{\text{eq}} = \underline{0}$$

and hence, due to $\det A \neq 0$,

$$K \underline{h}_{\text{eq}} = \underline{0}.$$

By this the equilibrium vector $\underline{h}_{\text{eq}}$ must be a multiple of $\begin{pmatrix} 1 \\ 1 \\ \vdots \\ 1 \end{pmatrix}$.

Detailed balance holds in this problem but this does not help us in this simple problem where K is symmetric as much as it did in the earlier problem where we had two independent paths connecting i and j.

Now zero is an eigenvalue of $A^{-1}K$ and it is simple as long as all $k_{ij} > 0$. If connecting lines are cut and the corresponding k_{ij} are set to zero, zero remains a simple eigenvalue as long as there remains at least one indirect flow path from each tank to each other tank. At the point where this is lost, zero cannot remain simple and our network splits into two disjoint subnetworks.

The solution to our problem is

$$\underline{h}(t) = \sum_{i=1}^{n} \left\langle \underline{y}_i,\ \underline{h}(t=0) \right\rangle e^{\lambda_i t} \underline{x}_i$$

and as

$$\left\langle \underline{y}_1,\ \underline{h}(t=0) \right\rangle = \frac{\sum A_i h_i(t=0)}{\sum A_i} = \frac{V}{\sum A_i} = h_{\text{eq}},$$

we can write this

$$\underline{h}(t) = h_{\text{eq}} \begin{pmatrix} 1 \\ 1 \\ \vdots \\ 1 \end{pmatrix} + \sum_{i=2}^{n} \left\langle \underline{y}_i,\ \underline{h}(t=0) \right\rangle e^{\lambda_i t} \underline{x}_i$$

and so in the case $n = 3$, which is sufficient to illustrate the main idea, we write

$$\underline{h}(t) = \underline{h}_{eq} + \langle\, \underline{y}_2,\, \underline{h}(t=0)\,\rangle\, e^{\lambda_2 t}\underline{x}_2 + \langle\, \underline{y}_3,\, \underline{h}(t=0)\,\rangle\, e^{\lambda_3 t}\underline{x}_3$$

where $\underline{h}_{eq} = h_{eq}\begin{pmatrix} 1 \\ 1 \\ 1 \end{pmatrix}$.

This formula determines $\underline{h}(t)$ in terms of $\underline{h}(t=0)$ and it can be used to recover K from experimental data for a variety of initial conditions in the way explained earlier in this lecture and used to determine chemical rate constants. In short when $\lambda_3 < \lambda_2$ we see that as t grows large $\underline{h}(t)$ approaches $\underline{h}_{eq}$ from the direction of $\underline{x}_2$. So using the direction of the tangent at $\underline{h}_{eq}$ to each of a sequence of experimental trajectories to determine an initial condition for the next trajectory, we step by step reduce the magnitude of $\langle\, \underline{y}_3,\, \underline{h}(t=0)\,\rangle$ in favor of $\langle\, \underline{y}_2,\, \underline{h}(t=0)\,\rangle$ and thereby determine $\underline{x}_2$ as accurately as we like. Using the orthogonality conditions $\underline{x}_3^T A \underline{x}_1 = 0 = \underline{x}_3^T A \underline{x}_2$ in the inner product where $G = A$ we can find $\underline{x}_3$ and then $\underline{y}_2$ and $\underline{y}_3$ in the plain vanilla inner product. As $\underline{y}_2, \underline{y}_3$, and an arbitrary trajectory, determine λ_2 and λ_3 via $\langle\, \underline{y}_2,\, \underline{h}(t)\,\rangle = \langle\, \underline{y}_2,\, \underline{h}(t=0)\,\rangle\, e^{\lambda_2 t}$ and $\langle\, \underline{y}_3,\, \underline{h}(t)\,\rangle = \langle\, \underline{y}_3,\, \underline{h}(t=0)\,\rangle\, e^{\lambda_3 t}$, the values of the k_{ij} are then recovered via

$$A^{-1}K = \lambda_1\, \underline{x}_1\, \underline{y}_1^T + \lambda_2\, \underline{x}_2\, \underline{y}_2^T + \lambda_3\, \underline{x}_3\, \underline{y}_3^T$$

The idea is to make a run at random where

$$\underline{h} = \underline{h}_{eq} + c_2\, e^{\lambda_2 t}\underline{x}_2 + c_3\, e^{\lambda_3 t}\underline{x}_3$$

and where $c_2 = \langle\, \underline{y}_2,\, \underline{h}(t=0)\,\rangle$ and $c_3 = \langle\, \underline{y}_3,\, \underline{h}(t=0)\,\rangle$ are not known. But the second and third terms may not separate until t is so large that $\underline{h} - \underline{h}_{eq}$ is inside experimental accuracy. This happens if $|c_2/c_3|$ is sufficiently small. The assumption is made that $|c_2/c_3|$ is large enough in the first run that separation takes place early enough in time that an estimate of $\underline{x}_2$ is obtained that can be used to increase $|c_2/c_3|$ for the second run. Then separation will take place even earlier in time leading to a better estimate of $\underline{x}_2$, etc. Such a sequence of experiments will produce an estimate of $\underline{x}_2$ limited only by the accuracy of liquid level measurements.

Because the curve $\underline{h}(t)$ vs t lies on the plane $\sum A_i h_i = V$ and $h_1 \geq 0, h_2 \geq 0, h_3 \geq 0$, it lies on a plane triangle. The readers can work out how to transfer this plane triangle to a piece of graph paper so that they can draw a graph of an experimental trajectory.

Ordinarily a set of experimental runs will be carried out at different volumes but as the eigenvectors and eigenvalues of $A^{-1}K$ do not depend on volume a sequence of runs can be brought to a common volume quite easily. Yet all this can be avoided by working in volume fractions. Then the matrix of interest is KA^{-1} and this also turns out to be self adjoint, but now $G = A^{-1}$.

All the reader needs are three fifty-five gallon drums, measuring sticks and some pipe to build a nice unit operations lab experiment.

8.3 Home Problems

1. The reactions

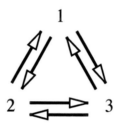

are carried out in a spatially uniform flow reactor whose holding time is denoted θ. We have

$$\theta \frac{d\underline{x}}{dt} = \underline{x}_F - \underline{x} + \theta K \underline{x}$$

where $\underline{x}$ is the column vector of species mole fractions and $\underline{x}_F$ denotes the feed composition.

The steady solutions, denoted $\underline{x}_S$ satisfy

$$\underline{x}_F = \{I - \theta K\}\underline{x}_S$$

Determine the eigenvectors and eigenvalues of $I - \theta K$ in terms of the eigenvectors and eigenvalues of K. Show that there is one and only one value of $\underline{x}_S$ corresponding to each value of $\underline{x}_F$. Show that if $\underline{x}_F$ is physically meaningful then so too $\underline{x}_S$.

Let $\underline{y} = \underline{x} - \underline{x}_S$ then $\underline{y}$ satisfies

$$\frac{d\underline{y}}{dt} = \left\{ -\frac{1}{\theta} I + K \right\} \underline{y}$$

where $\underline{y}(t = 0) = \underline{x}(t = 0) - \underline{x}_S$.

Show that as t grows large $\underline{y}(t)$ converges to $\underline{0}$ for all values of $\underline{y}(t = 0)$.

2. Let n solutes be dissolved in a solvent which is confined to a layer of thickness L. The solvent layer is in contact with a reservoir at $x = 0$ which maintains the solute concentration there at the value $\underline{c}_0$. The edge of the layer at $x = L$ is impermeable to solute. The solutes undergo an isomerization reaction

$$i \underset{k_{ij}}{\overset{k_{ji}}{\rightleftarrows}} j$$

in the solvent and for their distribution there we have

$$D \frac{d^2\underline{c}}{dx^2} + K \underline{c} = \underline{0}, \qquad 0 < x < L$$

$$\underline{c}(x = 0) = \underline{c}_0$$

and

$$\frac{d\underline{c}}{dx}(x = L) = \underline{0}$$

$$
\text{where } \underline{c} = \begin{pmatrix} c_1 \\ c_2 \\ \vdots \\ c_n \end{pmatrix} \text{ and } D = \begin{pmatrix} D_1 & 0 & 0 & \cdots & 0 \\ 0 & D_2 & 0 & \cdots & 0 \\ \vdots & \vdots & \vdots & & \vdots \\ 0 & 0 & 0 & \cdots & D_n \end{pmatrix}, \; D_i > 0. \text{ Determine the total}
$$

rate of reaction in the film and express this in terms of an effectiveness factor matrix.

Because D and K do not ordinarily have a complete set of eigenvectors in common this might seem like a new problem. Multiplication by D^{-1} shows that it is not. But the facts about $D^{-1}K$ have yet to be established. This can be done by observing that $D^{-1}K$ is similar to $D^{-\frac{1}{2}}KD^{-\frac{1}{2}}$ and to KD^{-1}. The first is symmetrizable due to $D^{-\frac{1}{2}}KD^{-\frac{1}{2}}C_{\text{eq}} = \left(D^{-\frac{1}{2}}KD^{-\frac{1}{2}}C_{\text{eq}}\right)^T$. To see this requires $KC_{\text{eq}} = \left(KC_{\text{eq}}\right)^T$ and use of the symmetry and commutativity of diagonal matrices. The Gerschgorin column circles of the second lie in the left half plane because its columns are multiples of the corresponding columns of K, the multiplying factors being positive, i.e., $\dfrac{1}{D_1}, \dfrac{1}{D_2}, \ldots$.

The matrix K is self adjoint in the inner product $G = C_{\text{eq}}^{-1}$. Show that the matrix $D^{-1}K$ is self adjoint in the inner product $G = C_{\text{eq}}^{-1}D$.

3. For reactions taking place in a solvent layer in contact with a reservoir supplying the reactants, the effectiveness factor matrix multiplies the rate of production vector evaluated at reservoir conditions to determine the true rate of production vector. When the reactions are

$$
i \; \underset{k_{ij}}{\overset{k_{ji}}{\rightleftarrows}} \; j
$$

the effectiveness factor matrix is

$$
D \sum_{i=1}^{n} \frac{\tanh\sqrt{-\lambda_i L}}{\sqrt{-\lambda_i L}} \, \underline{x}_i \, \underline{y}_i^T D^{-1}
$$

where λ_i and $\underline{x}_i$, $i = 1, 2, \ldots, n$, denote the eigenvalues and eigenvectors of $D^{-1}K$ and $\underline{y}_i^T \underline{x}_j = \delta_{ij}$

When the simple reversible reaction

$$1 \quad \underset{k'}{\overset{k}{\rightleftarrows}} \quad 2$$

takes place in the solvent show that the effectiveness factor matrix is

$$\frac{1}{D_1 k' + D_2 k} \begin{pmatrix} D_1 k' & D_1 k' \\ D_2 k & D_2 k \end{pmatrix} + \frac{\tanh\sqrt{-\lambda_2 L}}{\sqrt{-\lambda_2 L}} \frac{1}{D_1 k' + D_2 k} \begin{pmatrix} D_2 k & -D_1 k' \\ -D_2 k & D_1 k' \end{pmatrix}$$

where

$$-\lambda_2 = \frac{D_1 k' + D_2 k}{D_1 D_2}$$

Observe that the rate of production of either species depends on the rate of production of both species at reservoir conditions.

4. Let

$$A = \begin{pmatrix} -1 & 1 \\ -1 & -1 \end{pmatrix}.$$

Then show that the eigenvalue problem

$$A\underline{x} = \lambda \underline{x}$$

is satisfied by

$$\lambda_1 = -1 + i, \quad \underline{x}_1 = \begin{pmatrix} i \\ -1 \end{pmatrix}$$

and

$$\lambda_2 = -1 - i, \quad \underline{x}_2 = \begin{pmatrix} -i \\ -1 \end{pmatrix}$$

and that $\{\underline{x}_1, \underline{x}_2\}$ and $\{\underline{y}_1, \underline{y}_2\}$ are biorthogonal sets if

$$\underline{y}_1 = -\frac{1}{2} \begin{pmatrix} -i \\ 1 \end{pmatrix}, \quad \underline{y}_2 = -\frac{1}{2} \begin{pmatrix} i \\ 1 \end{pmatrix}$$

Show that the solution to

$$\frac{d\underline{x}}{dt} = A\underline{x}$$

where $\underline{x}\,(t = 0)$ is assigned is

$$\underline{x}\,(t) = \left\langle \underline{y}_1, \underline{x}\,(t = 0) \right\rangle e^{(-1+i)\,t} \begin{pmatrix} i \\ -1 \end{pmatrix} + \left\langle \underline{y}_2, \underline{x}\,(t = 0) \right\rangle e^{(-1-i)\,t} \begin{pmatrix} -i \\ -1 \end{pmatrix}$$

Sketch the solution in the x_1, x_2 plane when $\underline{x}\,(t = 0) = \begin{pmatrix} 1 \\ -1 \end{pmatrix}$. Because the decay constant is Re $\lambda_1 = -1$ and the period of the revolution is $\dfrac{2\pi}{\text{Im } \lambda_1} = 2\pi$ it may be difficult to see the spiral as $e^{-2\pi}$ is small, where $e^{-2\pi}$ is the factor by which the length of $\underline{x}\,(t)$ is shortened each period.

An experiment on a system where either Re λ_1 is much less than zero or Im λ_1 is close to zero may look like it identifies a straight line path as the spiral will be difficult to detect and a long time tangent direction may appear to be defined. But the direction of the apparent approach to $\underline{0}$ will depend on $\underline{x}\,(t = 0)$. To see this sketch the solution in the x_1, x_2 plane when $\underline{x}\,(t = 0) = \begin{pmatrix} 1 \\ 1 \end{pmatrix}$ and then look at the earlier sketch.

5. Two experiments run on three tanks having cross-sectional areas $A_1 = 1$, $A_2 = 2$ and $A_3 = 3$ produce the following data:

t	h_1	h_2	h_3	h_1	h_2	h_3
0	1.0	4.0	2.0	1.0	2.0	10/3
1	2.1226	2.7030	2.4905	1.8900	2.4323	2.7484
2	2.3848	2.5275	2.5201	2.2677	2.4908	2.5835
4	2.4860	2.5005	2.5043	2.4680	2.4998	2.5108
8				2.4994	2.5000	2.5002
∞	2.5	2.5	2.5	2.5	2.5	2.5

Use the data on the right hand side to estimate the conductivities of the connecting pipes. Then use the estimates to predict the data on the left hand side.

It is of some interest to see how the "experimental data" were determined. As $A_1 = 1$, $A_2 = 2$ and $A_3 = 3$ the geometric conditions of the problem require $\underline{x}_1 = \begin{pmatrix} 1 \\ 1 \\ 1 \end{pmatrix}$,

$\underline{y}_1 = \dfrac{1}{A_1 + A_2 + A_3} \begin{pmatrix} A_1 \\ A_2 \\ A_3 \end{pmatrix} = \dfrac{1}{6} \begin{pmatrix} 1 \\ 2 \\ 3 \end{pmatrix}$. Then as $\underline{x}_2$ and $\underline{x}_3$ can be determined only up to a constant multiple one degree of freedom is used in setting $\underline{x}_2$ and this also determines $\underline{x}_3$ via

$$\underline{x}_1^T A \underline{x}_3 = \underline{0} = \underline{x}_2^T A \underline{x}_3$$

Then $\underline{x}_1$, $\underline{x}_2$ and $\underline{x}_3$ determine $\underline{y}_1$, $\underline{y}_2$ and $\underline{y}_3$. The remaining two degrees of freedom are used in setting λ_2 and λ_3 $\{\lambda_3 < \lambda_2 < 0\}$ as $\lambda_1 = 0$. However this cannot be done arbitrarily as k_{12}, k_{23} and k_{31} must be positive.

6. An experiment on the isomerization of species A_1, A_2 and A_3 produces the data

t	x_1	x_2	x_3
0	1	0	0
1/4	0.7910	0.1967	0.0122
1/2	0.6451	0.3161	0.0389
1	0.4677	0.4324	0.0999
2	0.3222	0.4908	0.1869
4	0.2592	0.4998	0.2409
∞	0.2500	0.5000	0.2500

Determine the chemical rate coefficients k_{ij}. Improve the estimates of the rate coefficients given that the point $x_1 = 0.5$, $x_2 = 0.5$, $x_3 = 0$ lies on the slow straight line path.

7. Straight Line Paths in Diffusion

Denote $\begin{pmatrix} c_1 \\ c_2 \end{pmatrix}$ by $\underline{c}$ and $\begin{pmatrix} D_{11} & D_{12} \\ D_{21} & D_{22} \end{pmatrix}$ by D where c_1 and c_2 are the composition of two dilute solutes diffusing in a solvent.

At $t = 0$ we have $\underline{c} = \underline{c}_0$ for $0 \leq x < \infty$ and as $x \to \infty$ we have $\underline{c} = \underline{c}_0$ for all $t \geq 0$. At $x = 0$ we have $\underline{c} = \underline{c}_0 + \underline{\Delta}$ for $t > 0$.

Our model is

$$\frac{\partial \underline{c}}{\partial t} = D \frac{\partial^2 \underline{c}}{\partial x^2}$$

and its solution is

$$\underline{c}(\xi) = \underline{c}_0 + \left\{ I - \mathrm{erf}\,\frac{1}{2}\xi D^{-1/2} \right\} \underline{\Delta}$$

where $\xi = xt^{-1/2}$ and where $D_1 > D_2 > 0$ denote the eigenvalues of D corresponding to the eigenvectors $\underline{x}_1$ and $\underline{x}_2$.

Sketch $\underline{c}(\xi) - \underline{c}_0$ vs ξ for several $\underline{\Delta}$'s. You should see the curves turning toward the slow straight line path, away from the fast straight line path. Use your favorite matrix D.

To run an experiment, we would try to bring two solutions of different compositions together at $t = 0$ and run for a short time. The reader ought to derive a formula for $\underline{c}$ in this experiment.

Lecture 9

More Uses of Gerschgorin's Circle Theorem

9.1 Difference Approximations

As we explained in Lecture 6 Gerschgorin's circle theorem establishes a set of circles in the complex plane inside of which the eigenvalues of a matrix must lie, outside of which they cannot lie. The theorem leads to estimates of the eigenvalues and while the estimates may not be sharp, neither are they difficult to obtain.

Now stability conditions and, what is the same thing, convergence conditions tell us where the eigenvalues of a matrix must not lie. So when convergence is the problem, what the circle theorem tells us about where the eigenvalues of a matrix do not lie is interesting. This is our emphasis in this lecture.

Because we will study the diffusion equation in subsequent lectures, we investigate here some simple approximations to its solution.

The diffusion equation acts to smooth out solute irregularities. Hence it acts to damp concentration excursions from equilibrium in a region whose boundary is held at equilibrium, viz.,

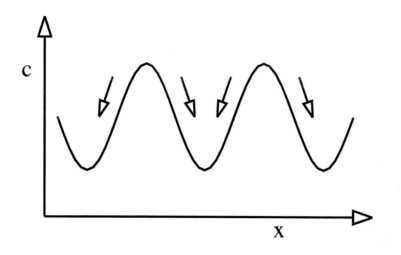

solute diffusion filling up troughs at the expense of crests

We also should see this in approximations to its solution.

For one dimensional diffusion across a layer of thickness L, we write

$$\frac{\partial c}{\partial t} = \frac{\partial^2 c}{\partial x^2}, \ 0 < x < 1$$

and

$$c\left(x = 0\right) = 0 = c\left(x = 1\right), \ t > 0$$

where $c\left(t = 0\right)$ is assigned. Here distance is scaled by L and time is scaled by $\frac{L^2}{D}$. We subdivide the interval (0,1) into $n+1$ subintervals of length h and approximate c at each of the points $x_i = ih$, $i = 1, \ldots, n$, by c_i so that at any time t the function $c\left(x, t\right)$ is approximated by the vector

$$\underline{c}\left(t\right) = \begin{pmatrix} c_1\left(t\right) \\ c_2\left(t\right) \\ \vdots \\ c_n\left(t\right) \end{pmatrix}.$$

Using a second central difference to approximate the spatial derivative, we require c_i to satisfy

$$\frac{dc_i}{dt} = \frac{c_{i+1} - 2c_i + c_{i-1}}{h^2}$$

which we can write

$$\frac{d\underline{c}}{dt} = \frac{1}{h^2} A \underline{c}$$

where

$$A = \begin{pmatrix} -2 & 1 & 0 & 0 & \cdots \\ 1 & -2 & 1 & 0 & \cdots \\ 0 & 1 & -2 & 1 & \cdots \\ & & \cdots & & \end{pmatrix}$$

As $\overline{A}^T = A$ we see that A is self adjoint in the plain vanilla inner product, viz., $G = I$. Hence its eigenvalues, denoted λ_i, must be real and its eigenvectors, denoted $\underline{x}_i$, can be scaled so that $\langle \underline{x}_i, \underline{x}_j \rangle = \delta_{ij}$ in the plain vanilla inner product. Then our approximation is

$$\underline{c}(t) = \sum_{i=1}^{n} \langle \underline{x}_i, \underline{c}(t=0) \rangle \, e^{\frac{1}{h^2}\lambda_i t} \underline{x}_i$$

The circle theorem tells us that $-4 \leq \lambda_i \leq 0$ and as $\det A \neq 0$, $\lambda_i \neq 0$. Ordering the eigenvalues as $-4 \leq \lambda_n \leq \lambda_{n-1} \leq \cdots \lambda_1 < 0$ we see that $\underline{c}(t)$ dies out to $\underline{0}$ exponentially as t grows large, the last gasp being $\langle \underline{x}_1, \underline{c}(t=0) \rangle \, e^{\frac{1}{h^2}\lambda_1 t} \underline{x}_1$. So, no matter how ragged the initial solute concentration, $\underline{c}(t=0)$, it is finally only as ragged as $\underline{x}_1$.

In this simple case the solutions to the eigenvalue problem $A\underline{x} = \lambda \underline{x}$ are

$$\underline{x}_i \propto \begin{pmatrix} \sin i\pi \dfrac{1}{n+1} \\ \vdots \\ \sin i\pi \dfrac{n}{n+1} \end{pmatrix}$$

and

$$\lambda_i = 2\cos \frac{i\pi}{n+1} - 2$$

so $\underline{x}_1$ is smooth and singly signed.

We can go on and ask for another approximation defined only at equally spaced values of t, viz., $t = kT$, $k = 1, 2, \ldots$. The approximation resulting on replacing the time derivative by a forward difference satisfies

$$c_i^{k+1} - c_i^k = \frac{T}{h^2} \left\{ c_{i+1}^k - 2c_i^k + c_{i-1}^k \right\}$$

or

$$\underline{c}^{k+1} = \left\{ I + \frac{T}{h^2} A \right\} \underline{c}^k.$$

The eigenvectors of $I + \frac{T}{h^2} A$ are those of A itself, the corresponding eigenvalues being $1 + \frac{T}{h^2} \lambda_i$, and as $-4 \leq \lambda_i < 0$, we find $1 - 4\frac{T}{h^2} \leq 1 + \frac{T}{h^2} \lambda_i < 1$. The approximation then is

$$\underline{c}^k = \sum_{i=1}^n \langle \underline{x}_i, \underline{c}\,(k = 0) \rangle \left(1 + \frac{T}{h^2} \lambda_i \right)^k \underline{x}_i \qquad .$$

and $\underline{c}^k \to \underline{0}$ as $k \to \infty$ iff $\left| 1 + \frac{T}{h^2} \lambda_i \right| < 1$. To be sure that this is so we must have $1 - 4\frac{T}{h^2} > -1$ or $T < \frac{1}{2} h^2$. In fact we require $1 - 4\frac{T}{h^2} > 0$ or $T < \frac{1}{4} h^2$ to be certain that $0 < 1 + \frac{T}{h^2} \lambda_i < 1$, thereby eliminating the possibility that terms in the approximation alternate in sign step by step.

We see that in this second approximation, having set h, we cannot set T freely and be certain that the approximation is well behaved. We also see that the two approximations differ, the factor $\left(e^{\frac{1}{h^2} \lambda_i} \right)^t$ in the first being replaced in the second by $\left(1 + \frac{T}{h^2} \lambda_i \right)^{\frac{t}{T}}$ where $t = kT$. In fact the second converges to the first if we fix t and let $k \to \infty$ and $T \to 0$ so that $kT = t$.

Instead of replacing the time derivative by a forward difference, we can get another approximation if a backward difference is used. It satisfies

$$c_i^k - c_i^{k-1} = \frac{T}{h^2} \left\{ c_{i+1}^k - 2c_i^k + c_{i-1}^k \right\}$$

or

$$\underline{c}^{k+1} = \left\{ I - \frac{T}{h^2} A \right\}^{-1} \underline{c}^k.$$

The eigenvectors of $\left\{I - \frac{T}{h^2}A\right\}^{-1}$, like those of $\left\{I + \frac{T}{h^2}A\right\}$, are those of A; the corresponding eigenvalues are now $\frac{1}{1 - \frac{T}{h^2}\lambda_i}$. Because $\lambda_i < 0$, these eigenvalues are all positive and lie to the left of $+1$, and so this approximation,

$$\underline{c}^k = \sum_{i=1}^{n} \langle \underline{x}_i, \underline{c}(k=0) \rangle \left(\frac{1}{1 - \frac{T}{h^2}\lambda_i} \right)^k \underline{x}_i,$$

goes to $\underline{0}$ as k goes to ∞, each term maintaining a fixed sign for all values of k, and we do not need a condition on T to make this happen. If we fix t and let $k \to \infty$ and $T \to 0$ so that $kT = t$, this approximation, like the second, converges to the first.

Anticipating higher accuracy, we can replace $\dfrac{\partial c}{\partial t}$ by the average of the two one-sided differences. Indeed as

$$e^{\frac{T}{h^2}\lambda_i} = 1 + \frac{T}{h^2}\lambda_i + \frac{1}{2}\left(\frac{T}{h^2}\lambda_i\right)^2 + \cdots$$

$$1 + \frac{T}{h^2}\lambda_i = 1 + \frac{T}{h^2}\lambda_i$$

and

$$\frac{1}{1 - \frac{T}{h^2}\lambda_i} = 1 + \frac{T}{h^2}\lambda_i + \left(\frac{T}{h^2}\lambda_i\right)^2 + \cdots$$

the average of the second and third expansions agrees with the first to three terms while each by itself agrees with the first to only two terms. Using this idea, this, then, we get

$$c_i^{k+1} - c_i^{k-1} = \frac{2T}{h^2}\left\{c_{i+1}^k - 2c_i^k + c_{i-1}^k\right\}$$

or

$$\underline{c}^{k+1} = 2\frac{T}{h^2}A\underline{c}^k + I\underline{c}^{k-1}$$

This is a second order difference equation; but it can be rewritten as a first order difference equation, viz.,

$$\begin{pmatrix} \underline{c}^{k+1} \\ \underline{c}^k \end{pmatrix} = \begin{pmatrix} 2\frac{T}{h^2}A & I \\ I & 0 \end{pmatrix} \begin{pmatrix} \underline{c}^k \\ \underline{c}^{k-1} \end{pmatrix}$$

and solved as above. The readers can use Gerschgorin's theorem to see if they can learn anything about this problem.

Ordinarily a simple second order difference equation looks like this

$$\underline{c}^{k+1} = A\underline{c}^k + B\underline{c}^{k-1}$$

and it may be introduced in an attempt to stabilize an unstable first order difference equation by introducing a delay. The second order equation can be written in first order form as

$$\begin{pmatrix} \underline{c}^{k+1} \\ \underline{c}^k \end{pmatrix} = \begin{pmatrix} A & B \\ I & 0 \end{pmatrix} \begin{pmatrix} \underline{c}^k \\ \underline{c}^{k-1} \end{pmatrix}$$

whereupon the corresponding eigenvalue problem,

$$\begin{pmatrix} A & B \\ I & 0 \end{pmatrix} \begin{pmatrix} \underline{x} \\ \underline{y} \end{pmatrix} = \lambda \begin{pmatrix} \underline{x} \\ \underline{y} \end{pmatrix},$$

is then

$$\left(\lambda^2 I - \lambda A - B\right)\underline{y} = \underline{0}$$

The condition that there be solutions $\underline{y} \neq \underline{0}$ is that $\det\left(\lambda^2 I - \lambda A - B\right) = 0$. A matrix whose elements are polynomials in a scalar λ is called a lambda matrix. Information about lambda matrices, their latent roots and latent vectors can be found in Lancaster's book "Theory of Matrices" and in the book "Matrix Polynomials" by Gohberg, Lancaster and Rodman, as well as in Frazer, Duncan and Collar's book "Elementary Matrices". If, however, A and B have a common set of eigenvectors, elementary methods can be used.

Returning then to our problem

$$\underline{c}^{k+1} = 2\frac{T}{h^2} A\underline{c}^k + I\underline{c}^{k-1}$$

we expand $\underline{c}^k$ in the eigenvectors of A as

$$\underline{c}^k = \sum_{i=1}^{n} \langle \underline{x}_i, \underline{c}^k \rangle \, \underline{x}_i,$$

and find the equation satisfied by $\langle \underline{x}_i, \underline{c}^k \rangle$ to be

$$\langle \underline{x}_i, \underline{c}^{k+1} \rangle = 2\frac{T}{h^2}\lambda_i \langle \underline{x}_i, \underline{c}^k \rangle + \langle \underline{x}_i, \underline{c}^{k-1} \rangle$$

This is a second order constant coefficient difference equation to be solved for each value of i. Its solution is

$$\langle \underline{x}_i, \underline{c}^k \rangle = a_i \mu_{i1}^k + b_i \mu_{i2}^k$$

where μ_{i1} and μ_{i2} are the roots of $\mu^2 - 2\frac{T}{h^2}\lambda_i\mu - 1 = 0$ and where a_i and b_i satisfy

$$\langle \underline{x}_i, \underline{c}\,(k = 0) \rangle = a_i + b_i$$

and

$$\langle \underline{x}_i, \underline{c}\,(k = 1) \rangle = a_i \mu_{i1} + b_i \mu_{i2}$$

The values of μ_{i1} and μ_{i2} are

$$\mu_{i1,2} = \frac{T}{h^2}\lambda_i \pm \sqrt{\left(\frac{T}{h^2}\lambda_i\right)^2 + 1}$$

and hence, because $\lambda_i < 0$, half of these values lie to the left of -1 and so the approximation

$$\underline{c}^k = \sum_{i=1}^n \left(a_i \mu_{i1}^k + b_i \mu_{i2}^k \right) \underline{x}_i$$

grows in magnitude and, sooner or later, oscillates in sign each time k increases by 1 whatever values of T and h are used.

This fourth approximation is named after L. Richardson and is an example of a good idea that did not work out.

The forward difference approximation leads to the iteration matrix $I + \frac{T}{h^2} A$ whereas the backward difference approximation leads to $\left(I - \frac{T}{h^2} A \right)^{-1}$. Stability places a condition on T in the first but not in the second. But to expand the second in powers of $\frac{T}{h^2} A$ we must have $\left| \frac{T}{h^2} \lambda_i \right| < 1$ and for this it is sufficient that $T < \frac{1}{4} h^2$.

The readers may wish to investigate the stability of the Crank-Nicholson approximation which leads to the iteration matrix $\left(2I - \frac{T}{h^2} A \right)^{-1} \left(2I + \frac{T}{h^2} A \right)$.

9.2 Home Problems

1. Two simple difference approximations to the convective diffusion equation

$$\frac{\partial c}{\partial t} = D \frac{\partial^2 c}{\partial x^2} - v \frac{\partial c}{\partial x}$$

result on making a forward or a backward difference approximation to $\dfrac{\partial c}{\partial x}$, viz.,

$$\frac{\partial c_i}{\partial t} = D \frac{c_{i+1} - 2c_i + c_{i-1}}{h^2} - \frac{v}{h} \begin{cases} c_{i+1} - c_i \\ \quad \text{or} \\ c_i - c_{i+1} \end{cases}$$

Let v be positive and use Gerschgorin circle theorem to obtain stability conditions on the size of h in each approximation. When $v = 0$ stability is obtained for all values of h.

2. Determine where the eigenvalues of the Crank-Nicholson iteration matrix

$$\left(2I - \frac{T}{h^2} A\right)^{-1} \left(2I + \frac{T}{h^2} A\right)$$

lie as a function of $\dfrac{T}{h^2}$.

3. Every practical method of solving $A\underline{x} = \underline{b}$ stops at some point, producing an estimate of $\underline{x}$. Often this estimate needs to be improved. Two simple iterations used to do this are called Jacobi's and Gauss' iterations.

To present these improvement methods we write $A = L + D + U$, where L and U are strictly lower and upper triangular. Then in Jacobi's method, we rearrange $A\underline{x} = \underline{b}$ as

$$\underline{x} = -D^{-1}(L + U)\underline{x} + D^{-1}\underline{b}$$

and use the iteration formula

$$\underline{x}_{k+1} = -D^{-1}(L + U)\underline{x}_k + D^{-1}\underline{b}$$

to generate a sequence of estimates $\underline{x}_1$, $\underline{x}_2$, ... from an initial guess $\underline{x}_0$. The error satisfies

$$\underline{e}_{k+1} = J\underline{e}_k$$

where $J = -D^{-1}(L + U)$ and convergence obtains iff the eigenvalues of J lie inside the unit circle. You can obtain a sufficient condition for this, viz.,

$$\sum_{\substack{j=1 \\ \neq i}} |a_{ij}| \leq |a_{ii}|, \quad i = 1, 2, \ldots, n$$

by using Gerschgorins' theorem, and it tells you to carry out elementary row and column operations on your problem so that when you write it $A\underline{x} = \underline{b}$, A is as diagonally dominant as possible.

In Gauss' iteration we rearrange $A\underline{x} = \underline{b}$ as

$$\underline{x} = -(L+D)^{-1}U\underline{x} + (L+D)^{-1}\underline{b}$$

whence we have

$$\underline{e}_{k+1} = G\underline{e}_k$$

where $G = -(L+D)^{-1}U$.

4. In an iteration process where $\underline{x}_k$ determines $\underline{x}_{k+1}$, the components of $\underline{x}_{k+1}$ are not obtained simultaneously but one at a time. Show that in a Jacobi iteration, the components of $\underline{x}_{k+1}$ obtained in the step $k \to k+1$ are not used in the calculation until the step $k+1 \to k+2$. Show that in a Gauss iteration, the components obtained in the step $k \to k+1$ are used in the calculation as soon as they are obtained, i.e., the $i+1$ component of $\underline{x}_{k+1}$ is determined using the first i components of $\underline{x}_{k+1}$ and the last $n-i$ components of $\underline{x}_k$.

Use Gerschgorin's circle theorem to derive sufficient conditions for a Gauss iteration to converge.

Part II

Elementary Differential Equations

Lecture 10

A Word To The Reader Upon Leaving Finite Dimensional Vector Spaces

In Part II we are going to study linear differential equations in which the differential operator is, for the most part, ∇^2. The solutions to our problems will depend on position as well as on time and the spaces where they reside will be called function spaces. The solutions to the eigenvalue problem for ∇^2 will be called eigenfunctions and ordinarily there will be infinitely many independent solutions. The function spaces will be infinite dimensional and our solutions will be in the form of infinite series. Many difficult questions will then arise that did not arise in Part I and many of these difficulties can be reduced to the question: what interpretation we can place on an infinite sum of eigenfunctions?

Before we take up such questions, if we do so at all, we explain the way to go about constructing the solution to a linear differential equation. We do this by expanding the solution in a series of eigenfunctions. To obtain the eigenfunctions we need to explain how to solve the eigenvalue problem. We do this by separation of variables. So we have two aims: first to explain how eigenfunction expansions are used to solve linear differential equations, second to explain the method of separation of variables as it is used to solve the corresponding eigenvalue problem.

As ∇^2 is the focus of our work, we first establish the elementary facts about ∇^2. We then use separation of variables to reduce the eigenvalue problem for ∇^2 to a set of three one-dimensional eigenvalue problems and we use Frobenius' method to solve these eigenvalue problems.

Earlier, in Part I, the solutions to our problems were finite sums and we might, but did not, have substituted such a sum directly into an equation to be solved in order to determine the coefficients of the eigenvectors in the sum. Indeed to see how easily this works the reader needs to determine the equations for the $c_i(t)$ by substituting

$$\underline{x}(t) = \sum c_i(t)\,\underline{x}_i$$

into the equation

$$\frac{d\underline{x}}{dt} = A\underline{x}$$

In Part II this may work, but it may not. The problem is that the solutions are infinite sums and it may not be known before the problem is solved whether or not the derivative of a sum is in fact the sum of the derivatives of its terms. There is at least one simple example in what follows to illustrate this point.

What we do does not differ from what we did in Part I and does not require that an assumed solution be substituted into the equation being solved. Indeed the coefficients to be determined are found by integrating this equation, after it is multiplied by suitable weighting functions.

The lectures are on the diffusion equation, its solution in bounded regions in terms of eigenfunctions, the solution of the eigenvalue problem by separation of variables and some problems in Cartesian, cylindrical and spherical coordinate systems to fill in the details.

Lecture 11

The Differential Operator ∇^2

11.1 The Differential Operator ∇

Part II is about the differential operator ∇^2. To begin we need to learn how to write ∇ and $\nabla^2 = \nabla \cdot \nabla$ in coordinate systems of interest and for us this means only orthogonal coordinate systems. The simplest of these and our starting point is a system of Cartesian coordinates denoted x, y, z.

A point in space, say P, can be located with respect to an origin O by the vector $\vec{r} = x\vec{i} + y\vec{j} + z\vec{k} = \overrightarrow{OP}$ where (x, y, z) denotes the Cartesian coordinates of the point P and where $\vec{i}, \vec{j}$ and $\vec{k}$ are unit vectors along the axes Ox, Oy and Oz. By design we have

$$\vec{i} \cdot \vec{j} = 0, \qquad \vec{j} \cdot \vec{k} = 0, \qquad \vec{k} \cdot \vec{i} = 0$$

i. e., a Cartesian coordinate system is constructed to be an orthogonal coordinate system, and we have

$$\vec{i} \times \vec{j} = \vec{k}, \qquad \vec{j} \times \vec{k} = \vec{i}, \qquad \vec{k} \times \vec{i} = \vec{j}$$

Then, at the point P, we have tangents to the coordinate curves passing through P, viz., the tangent to the coordinate curve where x is increasing at constant y and z is $\dfrac{\partial \vec{r}}{\partial x} = \vec{i}$. Likewise we

245

have $\dfrac{\partial \vec{r}}{\partial y} = \vec{j}$ and $\dfrac{\partial \vec{r}}{\partial z} = \vec{k}$. Thus, at P, the set of vectors $\{\vec{i},\, \vec{j},\, \vec{k}\}$ forms an orthogonal basis of unit length vectors, the same basis at any point P as at any other point. Hence the nine derivatives $\dfrac{\partial \vec{i}}{\partial x}$, etc., all vanish. This is what makes Cartesian coordinates the simplest coordinate system.

Now, suppose that we have a smooth scalar or vector or tensor valued function defined throughout a region of space and that we wish to introduce a notation that allows us to differentiate this function. To do this let C denote a curve lying in this region and let s denote arc length along this curve. The positions of points on the curve are denoted $\vec{r}(s)$ or $x(s), y(s), z(s)$, and the tangent to the curve at a point P on the curve is denoted by $\vec{t}$ where $\vec{t} = \dfrac{d\vec{r}}{ds}$ and $\vec{t} \cdot \vec{t} = 1$ due to $ds^2 = d\vec{r} \cdot d\vec{r}$.

We introduce the differential operator ∇ so that at a point P of C the derivative of f with respect to arc length along C is given by

$$\frac{df}{ds} = \vec{t} \cdot \nabla f$$

where ∇f depends on the point P and $\vec{t}$, at the point P, depends on the curve C.

We now select three curves passing through P having unit tangents at P denoted $\vec{t}_1, \vec{t}_2$ and $\vec{t}_3$ and we introduce the set of vectors $\{\vec{a}_1,\, \vec{a}_2,\, \vec{a}_3\} \perp \{\vec{t}_1,\, \vec{t}_2,\, \vec{t}_3\}$. Then we have

$$\vec{t}_1 \cdot \nabla f = \frac{df}{ds_1}, \qquad \vec{t}_2 \cdot \nabla f = \frac{df}{ds_2}, \qquad \vec{t}_3 \cdot \nabla f = \frac{df}{ds_3}$$

whereupon

$$\nabla f = \{\vec{a}_1 \vec{t}_1 + \vec{a}_2 \vec{t}_2 + \vec{a}_3 \vec{t}_3\} \cdot \nabla f = \vec{a}_1 \frac{df}{ds_1} + \vec{a}_2 \frac{df}{ds_2} + \vec{a}_3 \frac{df}{ds_3}$$

and therefore we have ∇f in terms of three derivatives of f, viz., $\dfrac{df}{ds_1}, \dfrac{df}{ds_2}$ and $\dfrac{df}{ds_3}$, along three curves C_1, C_2 and C_3 passing through P.

Thus on any curve C the derivative of f with respect to arc length along the curve is given by

$$\frac{df}{ds} = \vec{t} \cdot \vec{a}_1 \frac{df}{ds_1} + \vec{t} \cdot \vec{a}_2 \frac{df}{ds_2} + \vec{t} \cdot \vec{a}_3 \frac{df}{ds_3}$$

Now if we have a coordinate system, we will choose the three curves through P to be the three coordinate curves passing through P. Hence in Cartesian coordinates we choose $\vec{t}_1 = \vec{i}, \vec{t}_2 = \vec{j}$, and $\vec{t}_3 = \vec{k}$ whereupon we have

$$\frac{df}{ds_1} = \frac{\partial f}{\partial x}, \qquad \frac{df}{ds_2} = \frac{\partial f}{\partial y}, \quad \text{and} \quad \frac{df}{ds_3} = \frac{\partial f}{\partial z}$$

and, therefore,

$$\nabla f = \vec{i}\,\frac{\partial f}{\partial x} + \vec{j}\,\frac{\partial f}{\partial y} + \vec{k}\,\frac{\partial f}{\partial z}$$

Hence, in Cartesian coordinates, we denote by ∇ the differential operator

$$\vec{i}\,\frac{\partial}{\partial x} + \vec{j}\,\frac{\partial}{\partial y} + \vec{k}\,\frac{\partial}{\partial z}$$

whereupon we have

$$\nabla^2 = \nabla \cdot \nabla = \frac{\partial^2}{\partial x^2} + \frac{\partial^2}{\partial y^2} + \frac{\partial^2}{\partial z^2}$$

making use of the fact that $\vec{i}, \vec{j}$ and $\vec{k}$ are independent of x, y and z.

11.2   New Coordinate Systems

Now, we may introduce a new coordinate system where the new coordinates, (u, v, w), like (x, y, z), are coordinates of a point P. We do this by writing

$$x = f(u, v, w)$$

$$y = g(u, v, w)$$

$$z = h(u, v, w)$$

where the vectors

$$\vec{r}_u = \frac{\partial \vec{r}}{\partial u}, \qquad \vec{r}_v = \frac{\partial \vec{r}}{\partial v} \qquad \text{and} \qquad \vec{r}_w = \frac{\partial \vec{r}}{\partial w}$$

are tangent to the coordinate curves, viz., the curves u increasing at constant v and w, v increasing at constant w and u, etc.

Then our new coordinate system is an orthogonal coordinate system iff

$$\vec{r}_u \cdot \vec{r}_v = 0, \qquad \vec{r}_v \cdot \vec{r}_w = 0 \qquad \text{and} \qquad \vec{r}_w \cdot \vec{r}_u = 0$$

This being the case, and it is the only case of interest to us, we introduce unit length vectors along the three coordinate curves by scaling $\vec{r}_u, \vec{r}_v$ and $\vec{r}_w$, viz.,

$$\vec{i}_u = \frac{\vec{r}_u}{|\vec{r}_u|}, \qquad \vec{i}_v = \frac{\vec{r}_v}{|\vec{r}_v|} \qquad \text{and} \qquad \vec{i}_w = \frac{\vec{r}_w}{|\vec{r}_w|}$$

And now at each point of space we have an orthogonal basis of unit vectors that, algebraically, acts just like $\{\vec{i}, \vec{j}, \vec{k}\}$, viz.,

$$\vec{i}_u \cdot \vec{i}_v = 0, \qquad \vec{i}_v \cdot \vec{i}_w = 0 \qquad \text{and} \qquad \vec{i}_w \cdot \vec{i}_u = 0$$

and

$$\vec{i}_u \times \vec{i}_v = \vec{i}_w, \qquad \vec{i}_v \times \vec{i}_w = \vec{i}_u \qquad \text{and} \qquad \vec{i}_w \times \vec{i}_u = \vec{i}_v$$

However, the vectors $\vec{r}_u, \vec{r}_v$ and $\vec{r}_w$ ordinarily do not remain fixed in direction as we move from a point P to a nearby point.

To write a formula for ∇ in our new coordinate system, we introduce a curve C defined by

$$u = u(s), \qquad v = v(s) \qquad \text{and} \qquad w = w(s)$$

and then we differentiate f with respect to arc length along this curve, viz.,

$$\frac{df}{ds} = \frac{\partial f}{\partial u}\frac{du}{ds} + \frac{\partial f}{\partial v}\frac{dv}{ds} + \frac{\partial f}{\partial w}\frac{dw}{ds}$$

Observing that the tangent to the curve is given by

$$\vec{t} = \frac{d\vec{r}}{ds} = \vec{r}_u\frac{du}{ds} + \vec{r}_v\frac{dv}{ds} + \vec{r}_w\frac{dw}{ds}$$

we have

$$\frac{du}{ds} = \frac{\vec{r}_u \cdot \vec{t}}{|\vec{r}_u|^2} = \frac{\vec{i}_u}{|\vec{r}_u|} \cdot \vec{t}$$

$$\frac{dv}{ds} = \frac{\vec{r}_v \cdot \vec{t}}{|\vec{r}_v|^2} = \frac{\vec{i}_v}{|\vec{r}_v|} \cdot \vec{t}$$

and

$$\frac{dw}{ds} = \frac{\vec{r}_w \cdot \vec{t}}{|\vec{r}_w|^2} = \frac{\vec{i}_w}{|\vec{r}_w|} \cdot \vec{t}$$

whence we obtain

$$\frac{df}{ds} = \vec{t} \cdot \left\{ \frac{\vec{i}_u}{|\vec{r}_u|}\frac{\partial}{\partial u} + \frac{\vec{i}_v}{|\vec{r}_v|}\frac{\partial}{\partial v} + \frac{\vec{i}_w}{|\vec{r}_w|}\frac{\partial}{\partial w} \right\} f = \vec{t} \cdot \nabla f$$

where, in our new coordinate system,

$$\nabla = \frac{\vec{i}_u}{h_u}\frac{\partial}{\partial u} + \frac{\vec{i}_v}{h_v}\frac{\partial}{\partial v} + \frac{\vec{i}_w}{h_w}\frac{\partial}{\partial w}$$

and where

$$h_u = |\vec{r}_u|, \qquad h_v = |\vec{r}_v| \qquad \text{and} \qquad h_w = |\vec{r}_w|$$

Now a displacement $d\vec{r}$ is given in terms of du, dv and dw by

$$d\vec{r} = \vec{r}_u du + \vec{r}_v dv + \vec{r}_w dw$$

whereupon we see that

$$ds^2 = d\vec{r} \cdot d\vec{r}$$

$$= |\vec{r}_u|^2 du^2 + |\vec{r}_v|^2 dv^2 + |\vec{r}_w|^2 dw^2$$

$$= h_u{}^2 du^2 + h_v{}^2 dv^2 + h_w{}^2 dw^2$$

Hence if we can write a formula for ds^2 in our coordinate system, we can read off the formulas for h_u, h_v and h_w. For example, in cylindrical coordinates we have

$$ds^2 = dr^2 + r^2 d\theta^2 + dz^2$$

and therefore $h_r = 1, h_\theta = r, h_z = 1$. And in spherical coordinates we have

$$ds^2 = dr^2 + r^2 d\theta^2 + r^2 \sin^2 \theta \, d\phi^2$$

and therefore $h_r = 1, h_\theta = r, h_\phi = r \sin \theta$.

We now have a formula for ∇ in any orthogonal coordinate system and we can proceed to derive a formula for $\nabla^2 = \nabla \cdot \nabla$.

11.3   The Surface Gradient

Before we do this, we introduce the surface gradient, denoted ∇_S in order that we may differentiate functions defined on a surface.

Points (x, y, z), such that

$$x = f(\alpha, \beta)$$

$$y = g(\alpha, \beta)$$

and

$$z = h(\alpha, \beta)$$

lie on a surface, denoted S, on which the curves α constant, β increasing, and α increasing, β constant, are coordinate curves. The vectors $\vec{r}_\alpha$ and $\vec{r}_\beta$, tangent to these curves at a point P of S, ordinarily are not perpendicular. They can be used to determine the unit normal to S at P via

$$\vec{n} = \frac{\vec{r}_\alpha \times \vec{r}_\beta}{|\vec{r}_\alpha \times \vec{r}_\beta|}$$

where

$$|\vec{r}_\alpha \times \vec{r}_\beta|^2 = |\vec{r}_\alpha|^2 |\vec{r}_\beta|^2 - \left(\vec{r}_\alpha \cdot \vec{r}_\beta\right)^2$$

The two sets of vectors $\{\vec{a}, \vec{b}, \vec{n}\}$ and $\{\vec{r}_\alpha, \vec{r}_\beta, \vec{n}\}$ are biorthogonal if $\vec{a}$ and $\vec{b}$ are given by

$$\vec{a} = \frac{\vec{r}_\beta \times \vec{n}}{\vec{r}_\alpha \cdot \vec{r}_\beta \times \vec{n}}$$

and

$$\vec{b} = \frac{\vec{n} \times \vec{r}_\alpha}{\vec{r}_\alpha \cdot \vec{r}_\beta \times \vec{n}}$$

Then if f is defined on S as a function of α and β and $\vec{t}$ is tangent to a curve C on S passing through P, we have

$$\frac{df}{ds} = \frac{df}{d\alpha}\frac{d\alpha}{ds} + \frac{df}{d\beta}\frac{d\beta}{ds}$$

where $\alpha = \alpha(s)$ and $\beta = \beta(s)$ define the curve and s denotes arc length along the curve.

Due to

$$\vec{t} = \frac{d\vec{r}}{ds} = \vec{r}_\alpha \frac{d\alpha}{ds} + \vec{r}_\beta \frac{d\beta}{ds}$$

we have

$$\frac{d\alpha}{ds} = \vec{a} \cdot \vec{t}, \qquad \frac{d\beta}{ds} = \vec{b} \cdot \vec{t}$$

and hence

$$\frac{df}{ds} = \vec{t} \cdot \left\{ \vec{a} \frac{\partial}{\partial \alpha} + \vec{b} \frac{\partial}{\partial \beta} \right\} f$$

which we write as

$$\frac{df}{ds} = \vec{t} \cdot \nabla_S f$$

where

$$\nabla_S = \vec{a} \frac{\partial}{\partial \alpha} + \vec{b} \frac{\partial}{\partial \beta}$$

and at points along the surface we have

$$\nabla = \nabla_S + \vec{n} \frac{d}{ds}$$

The mean curvature of a surface, denoted by H, is important in some of our later examples and we record the fact that it can be obtained via

$$2H = -\nabla_S \cdot \vec{n}$$

More simply, defining

$$g_{\alpha\alpha} = \vec{r}_\alpha \cdot \vec{r}_\alpha, \qquad g_{\alpha\beta} = \vec{r}_\alpha \cdot \vec{r}_\beta, \qquad \text{and} \qquad g_{\beta\beta} = \vec{r}_\beta \cdot \vec{r}_\beta,$$

$$b_{\alpha\alpha} = \vec{n} \cdot \vec{r}_{\alpha\alpha}, \qquad b_{\alpha\beta} = \vec{n} \cdot \vec{r}_{\alpha\beta}, \qquad \text{and} \qquad b_{\beta\beta} = \vec{n} \cdot \vec{r}_{\beta\beta}$$

we have

$$2H = \operatorname{tr}\left\{ \begin{pmatrix} g_{\alpha\alpha} & g_{\alpha\beta} \\ g_{\alpha\beta} & g_{\beta\beta} \end{pmatrix}^{-1} \begin{pmatrix} b_{\alpha\alpha} & b_{\alpha\beta} \\ b_{\alpha\beta} & b_{\beta\beta} \end{pmatrix} \right\}$$

For example if the surface is defined by

$$z = Z(x, y)$$

where x and y are surface coordinates, we then have

$$\vec{r}_x = \vec{i} + Z_x \vec{k}, \qquad \vec{r}_{xx} = Z_{xx} \vec{k}$$

$$\vec{r}_y = \vec{j} + Z_y \vec{k}, \qquad \vec{r}_{yy} = Z_{yy} \vec{k}$$

$$\vec{n} = \frac{\vec{k} - Z_x \vec{i} - Z_y \vec{j}}{\sqrt{1 + Z_x^2 + Z_y^2}}$$

and

$$\vec{r}_{xy} = Z_{xy} \vec{k}$$

whereupon we find

$$2H = \frac{(1 + Z_y^2)Z_{xx} - 2 Z_x Z_y Z_{xy} + (1 + Z_x^2)Z_{yy}}{\left(1 + Z_x^2 + Z_y^2\right)^{3/2}}$$

11.4   A Formula for ∇^2

To derive a formula for ∇^2 in the u, v, w coordinate system we first notice that if f is a vector valued function, say $\vec{v}$, where

$$\vec{v} = v_u \vec{i}_u + v_v \vec{i}_v + v_w \vec{i}_w$$

then to calculate the tensor $\nabla \vec{v}$ where $\nabla \vec{v}$ can be used to find the derivative of $\vec{v}$ with respect to arc length along a curve, say a particle path, we write

$$\nabla \vec{v} = \left\{ \frac{1}{h_u} \vec{i}_u \frac{\partial}{\partial u} + \cdots \right\} \left\{ v_u \vec{i}_u + \cdots \right\}$$

The first term is

$$\frac{1}{h_u} \vec{i}_u \frac{\partial}{\partial u} \left\{ v_u \vec{i}_u \right\} = \frac{1}{h_u} \frac{\partial v_u}{\partial u} \vec{i}_u \vec{i}_u + \frac{v_u}{h_u} \vec{i}_u \frac{\partial \vec{i}_u}{\partial u}$$

and we see that to get $\nabla \vec{v}$ we are going to need three derivatives of each of the three base vectors, twenty seven components in all. These components are of the form

$$\vec{i}_\xi \cdot \frac{\partial \vec{i}_\eta}{\partial \zeta}$$

where ξ, η and ζ are any of u, v and w.

First we have

$$\vec{i}_\xi \cdot \frac{\partial \vec{i}_\xi}{\partial \eta} = 0$$

due to $\vec{i}_\xi \cdot \vec{i}_\xi = 1$ and then because

$$\frac{\partial^2 \vec{r}}{\partial \xi \, \partial \eta} = \frac{\partial^2 \vec{r}}{\partial \eta \, \partial \xi}$$

we have

$$\frac{\partial}{\partial \xi}\left(h_\eta \, \vec{i}_\eta\right) = \frac{\partial}{\partial \eta}\left(h_\xi \, \vec{i}_\xi\right)$$

hence

$$\frac{\partial h_\eta}{\partial \xi}\, \vec{i}_\eta + h_\eta \frac{\partial \vec{i}_\eta}{\partial \xi} = \frac{\partial h_\xi}{\partial \eta}\, \vec{i}_\xi + h_\xi \frac{\partial \vec{i}_\xi}{\partial \eta}$$

whereupon we obtain

$$\vec{i}_\eta \cdot \frac{\partial \vec{i}_\xi}{\partial \eta} = \frac{1}{h_\xi}\frac{\partial h_\eta}{\partial \xi}$$

and this is all that we need to derive a formula for $\nabla \cdot \vec{v}$ or for $\nabla^2 = \nabla \cdot \nabla$, i. e., many of the terms that appear in ∇v or in $\nabla\nabla$ are eliminated by the dot product.

To obtain a formula for ∇^2 we write

$$\nabla^2 = \left(\underline{\frac{1}{h_u}\, \vec{i}_u \frac{\partial}{\partial u}} + \frac{1}{h_v}\, \vec{i}_v \frac{\partial}{\partial v} + \frac{1}{h_w}\, \vec{i}_w \frac{\partial}{\partial w}\right) \cdot \left(\frac{1}{h_u}\, \vec{i}_u \frac{\partial}{\partial u} + \frac{1}{h_v}\, \vec{i}_v \frac{\partial}{\partial v} + \frac{1}{h_w}\, \vec{i}_w \frac{\partial}{\partial w}\right)$$

where the underlined term leads to

$$\frac{1}{h_u}\, \vec{i}_u \cdot \left\{ \frac{\partial}{\partial u}\left(\frac{1}{h_u}\right)\vec{i}_u \frac{\partial}{\partial u} + \frac{1}{h_u}\frac{\partial \vec{i}_u}{\partial u}\frac{\partial}{\partial u} + \frac{1}{h_u}\, \vec{i}_u \frac{\partial^2}{\partial u^2} \right.$$

$$+ \frac{\partial}{\partial u}\left(\frac{1}{h_v}\right)\vec{i}_v \frac{\partial}{\partial v} + \frac{1}{h_v}\frac{\partial \vec{i}_v}{\partial u}\frac{\partial}{\partial v} + \frac{1}{h_v}\, \vec{i}_v \frac{\partial^2}{\partial u\,\partial v}$$

$$\left. + \frac{\partial}{\partial u}\left(\frac{1}{h_w}\right)\vec{i}_w \frac{\partial}{\partial w} + \frac{1}{h_w}\frac{\partial \vec{i}_w}{\partial u}\frac{\partial}{\partial w} + \frac{1}{h_w}\, \vec{i}_w \frac{\partial^2}{\partial u\,\partial w} \right\}$$

$$= \frac{1}{h_u}\frac{\partial}{\partial u}\left(\frac{1}{h_u}\right)\frac{\partial}{\partial u} + \frac{1}{h_u{}^2}\frac{\partial^2}{\partial u^2} + \frac{1}{h_u h_v}\, \vec{i}_u \cdot \frac{\partial \vec{i}_v}{\partial u}\frac{\partial}{\partial v} + \frac{1}{h_u h_w}\, \vec{i}_u \cdot \frac{\partial \vec{i}_w}{\partial u}\frac{\partial}{\partial w}$$

$$= \frac{1}{h_u}\frac{\partial}{\partial u}\left(\frac{1}{h_u}\right)\frac{\partial}{\partial u} + \frac{1}{h_u{}^2}\frac{\partial^2}{\partial u^2} + \frac{1}{h_u h_v{}^2}\frac{\partial h_u}{\partial v}\frac{\partial}{\partial v} + \frac{1}{h_u h_w{}^2}\frac{\partial h_u}{\partial w}\frac{\partial}{\partial w}$$

The terms coming from $\dfrac{1}{h_v}\, \vec{i}_v \cdot \dfrac{\partial}{\partial v}$ and $\dfrac{1}{h_w}\, \vec{i}_w \cdot \dfrac{\partial}{\partial w}$ can be written by replacing u, v and w by v, w and u, etc., in this formula.

More simplifications are possible. In fact the reader can derive the formula:

$$\nabla^2 = \frac{1}{h_u\, h_v\, h_w}\left\{ \frac{\partial}{\partial u}\left(\frac{h_v\, h_w}{h_u}\frac{\partial}{\partial u}\right) + \frac{\partial}{\partial v}\left(\frac{h_w\, h_u}{h_v}\frac{\partial}{\partial v}\right) + \frac{\partial}{\partial w}\left(\frac{h_u\, h_v}{h_w}\frac{\partial}{\partial w}\right)\right\}$$

But we turn our attention to some examples which indicate a direct way to write ∇^2.

First we introduce cylindrical coordinates defined by

$$x = r\cos\theta$$

$$y = r\sin\theta$$

$$z = z$$

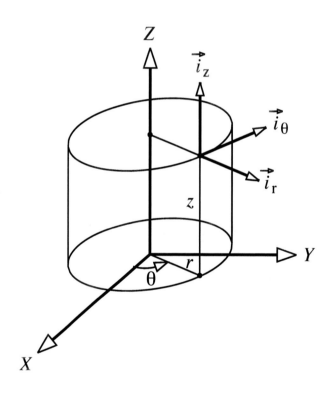

whereupon we have

$$\vec{i}_r = \cos\theta\,\vec{i} + \sin\theta\,\vec{j}, \qquad h_r = 1$$

$$\vec{i}_\theta = -\sin\theta\,\vec{i} + \cos\theta\,\vec{j}, \qquad h_\theta = r$$

$$\vec{i}_z = \vec{k}, \qquad\qquad\qquad h_z = 1$$

and hence we learn that

$$\frac{\partial\vec{i}_r}{\partial\theta} = \vec{i}_\theta, \qquad \frac{\partial\vec{i}_\theta}{\partial\theta} = -\vec{i}_r$$

the other seven derivatives being zero. Therefore we have

$$\nabla = \vec{i}_r\frac{\partial}{\partial r} + \frac{1}{r}\vec{i}_\theta\frac{\partial}{\partial\theta} + \vec{i}_z\frac{\partial}{\partial z}$$

and

$$\nabla^2 = \frac{\partial^2}{\partial r^2} + \frac{1}{r}\frac{\partial}{\partial r} + \frac{1}{r^2}\frac{\partial^2}{\partial\theta^2} + \frac{\partial^2}{\partial z^2}$$

where the term $\dfrac{1}{r}\dfrac{\partial}{\partial r}$ appears due to $\dfrac{1}{r}\vec{i}_\theta\dfrac{\partial}{\partial\theta}\cdot\left(\vec{i}_r\dfrac{\partial}{\partial r}\right)$ and where

$$\frac{\partial^2}{\partial r^2} + \frac{1}{r}\frac{\partial}{\partial r} = \frac{1}{r}\frac{\partial}{\partial r}\left(r\frac{\partial}{\partial r}\right)$$

Second, spherical coordinates are defined by

$$x = r\sin\theta\cos\phi, \qquad 0 \le \phi \le 2\pi, \quad 0 \le \theta \le \pi$$

$$y = r\sin\theta\sin\phi$$

$$z = r\cos\theta$$

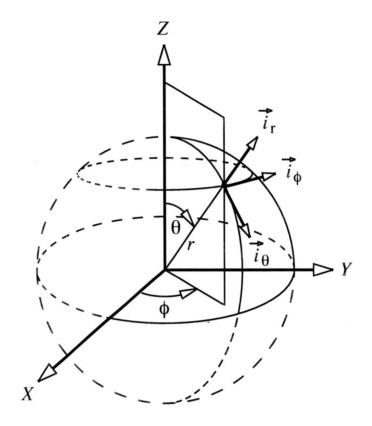

and we have

$$\vec{i}_r = \sin\theta\,\cos\phi\,\vec{i} + \sin\theta\,\sin\phi\,\vec{j} + \cos\theta\,\vec{k}, \qquad h_r = 1$$

$$\vec{i}_\theta = \cos\theta\,\cos\phi\,\vec{i} + \cos\theta\,\sin\phi\,\vec{j} - \sin\theta\,\vec{k}, \qquad h_\theta = r$$

$$\vec{i}_\phi = -\sin\phi\,\vec{i} + \cos\phi\,\vec{j}, \qquad\qquad h_\phi = r\sin\theta$$

and we see that

$$\frac{\partial\vec{i}_r}{\partial\theta} = \vec{i}_\theta, \qquad \frac{\partial\vec{i}_r}{\partial\phi} = \sin\theta\,\vec{i}_\phi,$$

$$\frac{\partial\vec{i}_\theta}{\partial\theta} = -\vec{i}_r, \qquad \frac{\partial\vec{i}_\theta}{\partial\phi} = \cos\theta\,\vec{i}_\phi,$$

$$\frac{\partial\vec{i}_\phi}{\partial\phi} = -\sin\theta\,\vec{i}_r - \cos\theta\,\vec{i}_\theta$$

the remaining four derivatives being zero.

Hence we have

$$\nabla = \vec{i}_r \frac{\partial}{\partial r} + \frac{1}{r} \vec{i}_\theta \frac{\partial}{\partial \theta} + \frac{1}{r \sin \theta} \vec{i}_\phi \frac{\partial}{\partial \phi}$$

whereupon

$$\vec{i}_r \frac{\partial}{\partial r} \cdot \nabla = \frac{\partial^2}{\partial r^2}$$

and

$$\frac{1}{r} \vec{i}_\theta \frac{\partial}{\partial \theta} \cdot \nabla = \frac{1}{r} \frac{\partial}{\partial r} + \frac{1}{r^2} \frac{\partial^2}{\partial \theta^2}$$

All of the interesting terms come out of

$$\frac{1}{r \sin \theta} \vec{i}_\phi \frac{\partial}{\partial \phi} \cdot \nabla = \frac{1}{r \sin \theta} \sin \theta \frac{\partial}{\partial r} + \frac{1}{r \sin \theta} \frac{1}{r} \cos \theta \frac{\partial}{\partial \theta} + \frac{1}{r^2 \sin^2 \theta} \frac{\partial^2}{\partial \phi^2}$$

and we find

$$\nabla^2 = \frac{1}{r^2} \frac{\partial}{\partial r} \left(r^2 \frac{\partial}{\partial r} \right) + \frac{1}{r^2 \sin \theta} \frac{\partial}{\partial \theta} \left(\sin \theta \frac{\partial}{\partial \theta} \right) + \frac{1}{r^2 \sin^2 \theta} \frac{\partial^2}{\partial \phi^2}$$

Now we have formulas for ∇^2 in three coordinate systems and enough information to write ∇^2 in any other orthogonal coordinate system. Given a coordinate system it is often easier to proceed to ∇^2 directly via the base vectors and their derivatives than it is to use general formulas.

For axisymmetric flow of an incompressible fluid, i.e., a flow where the velocity components are the same in every plane through the z-axis we can write

$$\vec{v} = \frac{1}{r} \vec{i}_\theta \times \nabla \psi (r, z)$$

in cylindrical coordinates and

$$\vec{v} = \frac{1}{r \sin \theta} \, \vec{i}_\phi \times \nabla \psi \, (r, \theta)$$

in spherical coordinates.

Hence in cylindrical coordinates we have

$$\nabla \times \vec{v} = \frac{1}{r} \left\{ r \frac{\partial}{\partial r} \left(\frac{1}{r} \frac{\partial \psi}{\partial r} \right) + \frac{\partial^2 \psi}{\partial z^2} \right\} \vec{i}_\theta = \frac{1}{r} \, E^2 \psi \, \vec{i}_\theta$$

and the reader ought to work out the corresponding formula in spherical coordinates, viz.,

$$\nabla \times \vec{v} = \frac{1}{r \sin \theta} \, E^2 \psi \, \vec{i}_\phi$$

learning, in the course of the calculation, a formula for E^2 in spherical coordinates.

11.5 Domain Pertubations

Our aim is to solve problems on a domain $\mathcal{D}$ and to do this in terms of the eigenfunctions of ∇^2 on the domain we must solve the eigenvalue problem

$$\nabla^2 \psi + \lambda^2 \psi = 0 \qquad \text{on} \quad \mathcal{D}$$

where, say, $\psi = 0$ on the boundary of $\mathcal{D}$. We now know how to write ∇^2 in a variety of orthogonal coordinate systems and if one of these coordinate systems fits our needs we will try to solve our eigenvalue problem in this coordinate system.

But there are many domains where we can not do this and for some of these, those close to a domain where our methods will work, we have an option.

Suppose our domain $\mathcal{D}$ lies close to a domain $\mathcal{D}_0$ on which we are able to solve our problem. We call $\mathcal{D}_0$ the reference domain and we wish to know what problems to solve on $\mathcal{D}_0$ in order to estimate the solution to our problem on $\mathcal{D}$.

To sketch the main idea, we assume $\mathcal{D}$ and $\mathcal{D}_0$ are two dimensional and in a Cartesian coordinate system we denote the points of $\mathcal{D}$ by (x, y) and those of $\mathcal{D}_0$ by (x_0, y_0).

We imagine a family of domains, $\mathcal{D}_\varepsilon$, growing out of $\mathcal{D}_0$, one of these being $\mathcal{D}$. Then points $(x, y, \varepsilon) \in \mathcal{D}_\varepsilon$ grow out of points $(x_0, y_0) \in \mathcal{D}_0$ via

$$x = f(x_0, y_0, \varepsilon), \qquad y = g(x_0, y_0, \varepsilon)$$

where the boundary points of $\mathcal{D}_\varepsilon$, viz., $y = Y(x, \varepsilon) = g(x_0, Y_0(x_0), \varepsilon)$ grow out of the corresponding boundary points of $\mathcal{D}_0$, viz., $y_0 = Y_0(x_0)$

We simplify our mapping of $\mathcal{D}_0$ into $\mathcal{D}_\varepsilon$ to

$$x = x_0, \qquad y = g(x_0, y_0, \varepsilon)$$

and we first expand $y = g(x_0, y_0, \varepsilon)$, viz.,

$$g(x_0, y_0, \varepsilon) = g(x_0, y_0, \varepsilon = 0) + \varepsilon \frac{dg}{d\varepsilon}(x_0, y_0, \varepsilon = 0) + \frac{1}{2}\varepsilon^2 \frac{d^2g}{d\varepsilon^2}(x_0, y_0, \varepsilon = 0) + \cdots$$

where

$$g(x_0, y_0, \varepsilon = 0) = y_0$$

and we define

$$y_1 = \frac{dg}{d\varepsilon}(x_0, y_0, \varepsilon = 0)$$

$$y_2 = \frac{d^2g}{d\varepsilon^2}(x_0, y_0, \varepsilon = 0)$$

etc.

Thus our mapping of $\mathcal{D}_0$ into $\mathcal{D}_\varepsilon$ can be written

$$x = x_0$$

$$y = y_0 + \varepsilon \, y_1 \, (x_0, y_0) + \frac{1}{2} \, \varepsilon^2 \, y_2 \, (x_0, y_0) + \cdots$$

on the domain and

$$x = x_0$$

$$y = Y \, (x, \varepsilon) = g \, (x_0, Y_0 \, (x), \varepsilon) = Y_0 \, (x_0) + \varepsilon \, Y_1 \, (x_0) + \frac{1}{2} \, \varepsilon^2 \, Y_2 \, (x_0) + \cdots$$

on its boundary where $Y_0 \, (x_0)$ defines the boundary of $\mathcal{D}_0$, where

$$Y_1 \, (x_0) = y_1 \big(x_0, Y \, (x_0) \big), \qquad Y_2 \, (x_0) = y_2 \big(x_0, Y \, (x_0) \big), \qquad \text{etc.}$$

and where $Y_0, Y_1, Y_2, \cdots$ define the boundary of $\mathcal{D}_\varepsilon$.

Now our problem is to find a function, denoted u, defined on a domain $\mathcal{D}_\varepsilon$. Assuming the equations satisfied by u on $\mathcal{D}_\varepsilon$ have the same form for all values of ε, we begin by expanding $u \, (x, y, \varepsilon)$ along the mapping where all derivatives along the mapping hold x_0 and y_0 fixed.

Thus expanding u in powers of ε along the mapping of $\mathcal{D}_0$ into $\mathcal{D}_\varepsilon$ we have:

$$u \, (x, y, \varepsilon) = u\big(x = x_0, y = y_0, \varepsilon = 0 \big) + \varepsilon \, \frac{du}{d\varepsilon} \, \big(x = x_0, y = y_0, \varepsilon = 0 \big) +$$

$$\frac{1}{2} \, \varepsilon^2 \, \frac{d^2 u}{d\varepsilon^2} \, \big(x = x_0, y = y_0, \varepsilon = 0 \big) + \cdots$$

where u depends on x, y, ε and $\dfrac{d}{d\varepsilon}$ holds x_0 and y_0 fixed.

Hence we have, using the chain rule,

$$\frac{du}{d\varepsilon} \, (x, y, \varepsilon) = \frac{\partial u}{\partial \varepsilon} \, (x, y, \varepsilon) + \frac{\partial u}{\partial y} \, (x, y, \varepsilon) \frac{dy}{d\varepsilon} \, (x_0, y_0, \varepsilon)$$

and therefore

$$\frac{du}{d\varepsilon} \, (x = x_0, y = y_0, \varepsilon = 0) = u_1 \, (x_0, y_0) + y_1 \, (x_0, y_0) \frac{\partial u_0}{\partial y_0} \, (x_0, y_0)$$

where

$$u_1(x_0, y_0) = \frac{\partial u}{\partial \varepsilon}(x = x_0, y = y_0, \varepsilon = 0)$$

$$y_1(x_0, y_0) = \frac{dg}{d\varepsilon}(x_0, y_0, \varepsilon = 0)$$

and

$$\frac{\partial u_0}{\partial y_0}(x_0, y_0) = \frac{\partial u}{\partial y}(x = x_0, y = y_0, \varepsilon = 0)$$

Likewise we have

$$\frac{d^2 u}{d\varepsilon^2}(x = x_0, y = y_0, \varepsilon = 0) = u_2(x_0, y_0) + 2y_1(x_0, y_0)\frac{\partial u_1(x_0, y_0)}{\partial y_0} +$$

$$y_1^2(x_0, y_0)\frac{\partial^2 u_0(x_0, y_0)}{\partial y_0^2} + y_2(x_0, y_0)\frac{\partial u_0(x_0, y_0)}{\partial y_0}$$

where

$$u_2(x_0, y_0) = \frac{\partial^2 u}{\partial \varepsilon^2}(x = x_0, y = y_0, \varepsilon = 0)$$

Thus we write our expansion of u as

$$u(x, y, \varepsilon) = u_0 + \varepsilon\left(u_1 + y_1\frac{\partial u_0}{\partial y_0}\right) + \frac{1}{2}\varepsilon^2\left(u_2 + 2y_1\frac{\partial u_1}{\partial y_0} + y_1^2\frac{\partial^2 u_0}{\partial y_0^2} + y_2\frac{\partial u_0}{\partial y_0}\right) + \cdots$$

where u_0, u_1, y_1, etc. are all evaluated at (x_0, y_0) in $\mathcal{D}_0$.

To obtain the expansion of $\frac{\partial u}{\partial y}$ we differentiate the RHS of the expansion of u with respect to y holding x and ε fixed, where x_0 and y_0 are obtained in terms of x, y and ε via

$$x = x_0, \qquad y = y_0 + \varepsilon\, y_1(x_0, y_0) + \frac{1}{2}\varepsilon^2 y_2(x_0, y_0) + \cdots$$

and we find

$$\frac{\partial u\,(x,y,\varepsilon)}{\partial y} = \frac{\partial u_0}{\partial y_0} + \varepsilon\left(\frac{\partial u_1}{\partial y_0} + y_1\frac{\partial^2 u_0}{\partial y_0^2}\right) +$$

$$\frac{1}{2}\,\varepsilon^2\left(\frac{\partial u_2}{\partial y_0} + 2y_1\frac{\partial^2 u_1}{\partial y_0^2} + y_1^2\frac{\partial^3 u_0}{\partial y_0^3} + y_2\frac{\partial^2 u_0}{\partial y_0^2}\right) + \cdots$$

Likewise we have

$$\frac{\partial u\,(x,y,\varepsilon)}{\partial x} = \frac{\partial u_0}{\partial x_0} + \varepsilon\left(\frac{\partial u_1}{\partial x_0} + y_1\frac{\partial^2 u_0}{\partial y_0\partial x_0}\right) +$$

$$\frac{1}{2}\,\varepsilon^2\left(\frac{\partial u_2}{\partial x_0} + 2y_1\frac{\partial^2 u_1}{\partial y_0\partial x_0} + y_1^2\frac{\partial^3 u_0}{\partial y_0^2\partial x_0} + y_2\frac{\partial^2 u_0}{\partial y_0\partial x_0}\right) + \cdots$$

The reader may notice that only $y_1, y_2, \ldots$ appear in these two formulas, not their derivatives. To see how the derivatives are lost, the algebra must be worked out. The main idea is to replace $\frac{\partial y_0}{\partial y}$ with $1 - \varepsilon\frac{\partial y_1}{\partial y} - \frac{1}{2}\,\varepsilon^2\frac{\partial y_2}{\partial y} - \cdots$ in the derivation of the formula for $\frac{\partial u}{\partial y}$.

Now to derive the equations for $u_0, u_1, u_2, \ldots$ on the reference domain, we substitute our expansions for u and its derivatives into the equation satisfied by u. Doing this we discover that the mappings do not survive. For example, if our problem is

$$\nabla^2 u = f \quad \text{on} \quad \mathcal{D}_\varepsilon$$

we substitute

$$\frac{\partial^2 u}{\partial x^2} = \frac{\partial^2 u_0}{\partial x_0^2} + \varepsilon\left(\frac{\partial^2 u_1}{\partial x_0^2} + y_1\frac{\partial}{\partial y_0}\frac{\partial^2 u_1}{\partial x_0^2}\right) + \cdots$$

$$\frac{\partial^2 u}{\partial y^2} = \frac{\partial^2 u_0}{\partial y_0^2} + \varepsilon\left(\frac{\partial^2 u_1}{\partial y_0^2} + y_1\frac{\partial}{\partial y_0}\frac{\partial^2 u_1}{\partial y_0^2}\right) + \cdots$$

and

$$f = f_0 + \varepsilon\left(f_1 + y_1\frac{\partial}{\partial y_0}\,f_0\right) + \cdots$$

to obtain

$$\nabla_0^2 u_0 + \varepsilon \left(\nabla_0^2 u_1 + y_1 \frac{\partial}{\partial y_0} \nabla_0^2 u_0 \right) + \cdots = f_0 + \varepsilon \left(f_1 + y_1 \frac{\partial}{\partial y_0} f_0 \right) + \cdots$$

whereupon we have, on $\mathcal{D}_0$,

$$\nabla_0^2 u_0 = f_0$$

$$\nabla_0^2 u_1 = f_1$$

etc.

and the conclusion is this: the equations for u_0, u_1, u_2 on $\mathcal{D}_0$, i.e., on the reference domain, can be derived by ordinary methods from the equation for u on $\mathcal{D}_\varepsilon$. The mapping of $\mathcal{D}_0$ into $\mathcal{D}_\varepsilon$ does not appear, and we are grateful, because we can never know what it is.

Thus we can substitute $u = u_0 + \varepsilon u_1 + \frac{1}{2} \varepsilon^2 u_2 + \cdots$ into the equation for u and set to zero terms of order zero, one, two, etc. while paying no attention to the fact that $\mathcal{D}_\varepsilon$ is not $\mathcal{D}_0$. By doing this we obtain the equations for $u_0, u_1, u_2, \ldots$ on $\mathcal{D}_0$.

The boundary is different. For example suppose $u = 0$ must be satisfied at $y = Y(x)$ in the forgoing problem. Then using the expansion

$$u(x, Y(x)) = u_0 + \varepsilon \left(u_1 + Y_1 \frac{\partial u_0}{\partial y_0} + \cdots \right) + \cdots$$

we obtain at $y_0 = Y_0(x_0)$

$$u_0 = 0$$

$$u_1 + Y_1 \frac{\partial u_0}{\partial y_0} = 0$$

etc.

And we see that in the boundary conditions the displacement of $\mathcal{D}_0$ into $\mathcal{D}_\varepsilon$ appears.

To obtain $u(x, y, \varepsilon)$ from $u_0(x_0, y_0), u_1(x_0, y_0)$, etc., not knowing the mapping, we rearrange the expansion

$$u\left(x, y, \varepsilon\right) = u_0\left(x_0, y_0\right) + \varepsilon\left(u_1 + y_1 \frac{\partial u_0}{\partial y_0}\right) + \frac{1}{2}\varepsilon^2\left(u_2 + 2y_1 \frac{\partial u_1}{\partial y_0} + y_1^2 \frac{\partial^2 u_0}{\partial y_0^2} + y_2 \frac{\partial u_0}{\partial y_0}\right) + \cdots$$

using

$$y - y_0 = \varepsilon y_1 + \frac{1}{2}\varepsilon^2 y_2 + \cdots$$

and

$$u_0\left(x, y\right) = u_0\left(x_0, y_0\right) + \frac{\partial u_0}{\partial y_0}\left(x_0, y_0\right)\left(y - y_0\right) + \frac{1}{2}\frac{\partial^2 u_0}{\partial y_0^2}\left(x_0, y_0\right)\left(y - y_0\right)^2 + \cdots$$

etc.

and conclude

$$u\left(x, y, \varepsilon\right) = u_0\left(x, y\right) + \varepsilon u_1\left(x, y\right) + \frac{1}{2}\varepsilon^2 u_2\left(x, y\right) + \cdots$$

and by this we obtain u at all points (x, y) of $\mathcal{D}_\varepsilon$ which are also points of $\mathcal{D}_0$.

Now to solve our eigenvalue problem

$$\nabla^2\psi + \lambda^2\psi = 0 \qquad \text{on} \quad \mathcal{D}$$

$$\psi = 0 \qquad \text{on} \quad y = Y\left(x\right)$$

we solve

$$\nabla^2\psi_0 + \lambda_0^2\psi_0 = 0 \qquad \text{on} \quad \mathcal{D}_0$$

$$\psi_0 = 0 \qquad \text{on} \quad y_0 = Y_0\left(x_0\right)$$

$$\nabla^2 \psi_1 + \lambda_0^2 \psi_1 = -\lambda_1^2 \psi_0 \qquad \text{on} \quad \mathcal{D}_0$$

$$\psi_1 + Y_1 \frac{\partial \psi_0}{\partial y_0} = 0 \qquad \text{on} \quad y_0 = Y_0(x_0)$$

$$\nabla^2 \psi_2 + \lambda_0^2 \psi_2 = -\lambda_2^2 \psi_0 - 2\lambda_1^2 \psi_1 \qquad \text{on} \quad \mathcal{D}_0$$

$$\psi_2 + 2Y_1 \frac{\partial \psi_1}{\partial y_0} + Y_1^2 \frac{\partial^2 \psi_0}{\partial y_0^2} + Y_2 \frac{\partial \psi_0}{\partial y_0} = 0 \qquad \text{on} \quad y_0 = Y_0(x_0)$$

etc.

And we notice that the homogeneous part of every problem has a solution, not zero. Hence a solvability condition must be satisfied at every order and it will determine $\lambda_1^2, \lambda_2^2, \ldots$ the corrections to λ_0^2. The displacement of the boundary , given by $Y_1, Y_2, \ldots$, will appear in the solvability conditions.

Domain perturbations would be useful if we have a heavy fluid lying above a light fluid in a container of arbitrary cross section and we wish to learn if the interface is stable to a small displacement. Hopefuly, the arbitrary cross section is close to, say, a circle, and the displaced interface is close to, say, a plane.

More the details can be found in the book *"Interfacial Instability"* by L.E. Johns and R. Narayanan.

11.6   Home Problems

1. Suppose $\vec{v}$ satisfies $\nabla \cdot \vec{v} = 0$ and $\vec{v}$

   (a) is tangent to the family of planes $z = const$,

      or

   (b) is tangent to the family of planes passing through the z–axis

In (a) we can write

$$\vec{v} = \vec{k} \times \nabla \psi \, (x, y)$$

whereas in (b) we can write

$$\vec{v} = \frac{1}{r} \, \vec{i}_\theta \times \nabla \psi \, (r, z) \qquad \left(\text{cylindrical}, \quad r^2 = x^2 + y^2\right)$$

or

$$\vec{v} = \frac{1}{r \sin \theta} \, \vec{i}_\phi \times \nabla \psi \, (r, \theta) \qquad \left(\text{spherical}, \quad r^2 = x^2 + y^2 + z^2\right)$$

Introducing orthogonal coordinates ξ, η, derive formulas for v_ξ and v_η in terms of $\psi \, (\xi, \eta)$.

In (a) you have

$$x = f \, (\xi, \eta), \quad y = g \, (\xi, \eta)$$

whereas in (b) you have either

$$r = f \, (\xi, \eta), \quad z = g \, (\xi, \eta)$$

or

$$r = f \, (\xi, \eta), \quad \theta = g \, (\xi, \eta)$$

Derive $\vec{v} \cdot \nabla \psi = 0$ and conclude that $\vec{v}$ is tangent to surfaces $\psi = const.$

2. You are to write out the terms appearing in the Navier-Stokes equation. Do this in cylindrical coordinates where

$$\vec{v} = v_r \vec{i}_r + v_\theta \vec{i}_\theta + v_z \vec{i}_z$$

and where

$$\frac{\partial \vec{i}_r}{\partial \theta} = \vec{i}_\theta \quad \text{and} \quad \frac{\partial \vec{i}_\theta}{\partial \theta} = -\vec{i}_r$$

the other seven derivatives being zero. Write out

$$\nabla \vec{v}$$

$$\nabla \cdot \vec{v}$$

$$\nabla \times \vec{v}$$

$$\vec{v} \cdot \nabla \vec{v}$$

$$\nabla \cdot \nabla \vec{v} = \nabla^2 \vec{v}$$

$$\nabla \cdot (\nabla \vec{v})^T = \nabla (\nabla \cdot \vec{v})$$

3. Denote by r, θ and z cylindrical coordinates, viz.,

$$x = r \cos \theta$$

$$y = r \sin \theta$$

where $r^2 = x^2 + y^2$ and define coordinate systems of revolution by

$$r = r(\xi, \eta), \quad z = z(\xi, \eta)$$

so that

$$x = r\left(\xi, \eta\right)\cos\theta$$

$$y = r\left(\xi, \eta\right)\sin\theta$$

$$z = z\left(\xi, \eta\right)$$

For example, spherical coordinates obtain if θ is replaced by ϕ and

$$r\left(\xi, \eta\right) = \xi \sin\eta$$

$$z\left(\xi, \eta\right) = \xi \cos\eta$$

where $\xi^2 = x^2 + y^2 + z^2$, $\eta = \theta$ and $\cos\theta = \vec{k} \cdot \vec{r}$

Prolate and oblate spheroidal coordinates are defined by

$$r = c \sinh\xi \sin\eta$$

$$z = c \cosh\xi \cos\eta$$

and

$$r = c \cosh\xi \sin\eta$$

$$z = c \sinh\xi \cos\eta$$

where $0 \leq \xi \leq \infty, 0 \leq \eta \leq \pi$.

Show that prolate and oblate spheroidal coordinates are orthogonal, write ∇^2 and reduce $\left(\nabla^2 + \lambda^2\right)\psi = 0$ to ordinary differential equations by separation of variables.

4. Show that $\vec{v} = \dfrac{1}{T}\left(x\vec{i} - y\vec{j}\right)$ and $p = -\dfrac{1}{2}\rho\,\dfrac{x^2 + y^2}{T^2}$ satisfy

$$\rho\vec{v}\cdot\nabla\vec{v} = -\nabla p + \mu\nabla^2\vec{v}$$

and

$$\nabla\cdot\vec{v} = 0$$

Notice that:

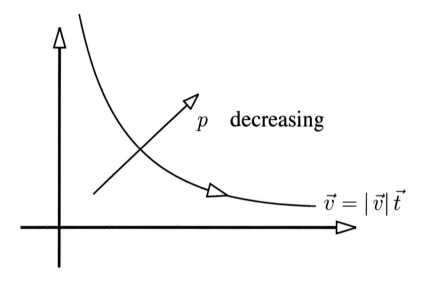

where, denoting by κ the curvature of a particle path, we have

$$\vec{a} = \vec{v}\cdot\nabla\vec{v} = |\vec{v}|\,\vec{t}\cdot\nabla\left\{|\vec{v}|\,\vec{t}\right\} = |\vec{v}|\,\frac{d}{ds}\left\{|\vec{v}|\,\vec{t}\right\} = |\vec{v}|\,\frac{d|\vec{v}|}{ds}\,\vec{t} + |\vec{v}|\,\frac{d\vec{t}}{ds}$$

and where $\dfrac{d\vec{t}}{ds} = \kappa\vec{p}$ and $\vec{p}\cdot\vec{t} = 0$.

5. Denote by $\vec{m}$ a magnetic dipole. The vector potential and the magnetic induction due to $\vec{m}$ are:

$$\vec{A} = \frac{\mu_0}{4\pi}\,\vec{m}\times\vec{r}$$

and

$$\vec{B} = \nabla \times \vec{A}$$

Assume $\vec{m} = m\vec{k}$ and find $\vec{B}$.

Lecture 12

Diffusion in Unbounded Domains

To learn something about diffusion without doing very much, we present some diffusion problems in unbounded domains.

12.1 Power Moments of an Evolving Solute Concentration Field: The Formula $D = \dfrac{1}{2}\dfrac{d\sigma^2}{dt}$

An easy way to get information about the solution to the diffusion equation without actually solving the equation is to determine the spatial power moments of the concentration field. As spatial derivatives in the diffusion equation lower the order of the power moments, a sequence of ordinary differential equations results that can be solved recursively and the first few moments tell us some interesting things about diffusion. To see how this works we begin with a one dimensional problem. Suppose that a solute is distributed at $t = 0$ over $-\infty < x < \infty$ according to the smooth function $c(t = 0)$ where $c(t = 0) = 0$ for $|x| > a$, $a < \infty$. Then, for $t > 0$, the solute distribution satisfies

$$\frac{\partial c}{\partial t} = D\frac{\partial^2 c}{\partial x^2}, \quad -\infty < x < \infty$$

and we only need c to vanish strongly enough as $|x| \to \infty$ so that all of its moments are finite.

Defining the power moments of c via

$$c_m(t) = \int_{-\infty}^{+\infty} x^m c(x,t)dx, \quad m = 0,1,2,\ldots$$

we can derive the equation satisfied by $c_m(t)$ by multiplying the diffusion equation by x^m and integrating the result over all x. Simplifying the right hand side by integration by parts and setting all the terms evaluated at $\pm\infty$ to zero we get

$$\frac{dc_m}{dt} = Dm(m-1)c_{m-2}, \quad m = 0,1,2,\ldots$$

Indeed, for $m = 0,1$ and 2, we have

$$\frac{dc_0}{dt} = 0$$

$$\frac{dc_1}{dt} = 0$$

and

$$\frac{dc_2}{dt} = 2Dc_0$$

whence, for $t > 0$, we obtain

$$c_0 = c_0(t = 0)$$

$$c_1 = c_1(t = 0)$$

and

$$c_2 = c_2(t = 0) + 2Dtc_0(t = 0)$$

The first result expresses the fact that solute is neither gained nor lost in diffusion, it is just redistributed. The second and third results tell us something about this redistribution. If we think of c at any time t as the result of an experiment designed to determine the spatial positions of the solute molecules at that time then $\frac{c(x,t)}{c_0}dx$ is the probability that the measurement of a molecule's position falls between x and $x + dx$ at time t and hence we have

$$\langle x \rangle = \int_{-\infty}^{+\infty} x \frac{c(x,t)}{c_0} dx = \frac{c_1}{c_0}$$

and

$$\langle x^2 \rangle = \int_{-\infty}^{+\infty} x^2 \frac{c(x,t)}{c_0} dx = \frac{c_2}{c_0}$$

where $\langle \ \rangle$ denotes the average or expected value of a function of x, weighted by the solute density. So $c_1(t)$ and $c_2(t)$ determine the expected values of x and x^2 and in a simple diffusion experiment and we see $\langle x \rangle$ is fixed while $\langle x^2 \rangle$ increases linearly in time.

The variance of the solute distribution, i.e., the average of $(x - \langle x \rangle)^2$, tells us about the spreading of the solute, and as this is

$$\sigma^2 = \left\langle \left(x - \langle x \rangle \right)^2 \right\rangle$$
$$= \langle x^2 \rangle - \langle x \rangle^2$$
$$= \frac{c_2}{c_0} - \left(\frac{c_1}{c_0} \right)^2$$

we find that

$$\sigma^2 = \sigma^2(t = 0) + 2Dt$$

This is a formula for D in terms of the variance of the solute concentration as independent solute molecules spread out in a solvent, viz.,

$$D = \frac{1}{2} \frac{d\sigma^2}{dt} = \frac{1}{2c_0} \frac{dc_2}{dt}$$

It can be used to determine the value of D if measurements of $c_2(t)$ can be made.

All this carries over to three dimensional problems where a solute distribution is assigned at

$t = 0$ according to $c(t = 0)$ and then spreads out for $t > 0$ according to

$$\frac{\partial c}{\partial t} = D \left\{ \frac{\partial^2 c}{\partial x^2} + \frac{\partial^2 c}{\partial y^2} + \frac{\partial^2 c}{\partial z^2} \right\} = D \nabla^2 c$$

The power moments of c defined by

$$c_{\ell m n} = \int_{-\infty}^{+\infty} \int_{-\infty}^{+\infty} \int_{-\infty}^{+\infty} x^\ell y^m z^n c(x, y, z, t) \, dx \, dy \, dz$$

then satisfy

$$\frac{dc_{\ell m n}}{dt} = D\ell(\ell - 1)c_{(\ell-2)mn} + Dm(m - 1)c_{\ell(m-2)n} + Dn(n - 1)c_{\ell m(n-2)}$$

and again this set of equations can be solved recursively. And a formula for D in three dimensional diffusion in terms of

$$\sigma^2 = \left(\langle x^2 \rangle - \langle x \rangle^2 \right) + \left(\langle y^2 \rangle - \langle y \rangle^2 \right) + \left(\langle z^2 \rangle - \langle z \rangle^2 \right)$$

can be obtained.

Suppose now that an initial solute distribution not only spreads out by diffusion but is also displaced by a flow field. Here too, we can get useful information about what is going on by looking at the moments of the solute distribution. To do the simplest example we suppose that an initial solute distribution in two dimensions is being displaced by a stagnation flow such as

$$\vec{v} = \frac{x}{T} \vec{i} - \frac{y}{T} \vec{j}$$

viz.,

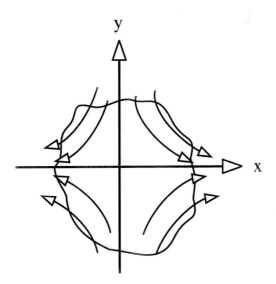

whose tendency is to stretch the solute out in the x direction while compressing it in the y direction so that as t increases the solute might pass through the following stages:

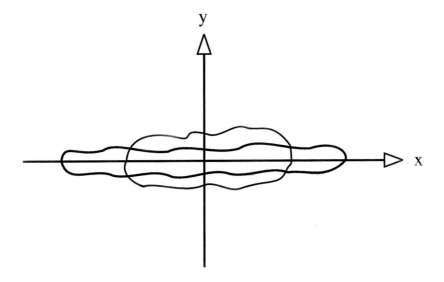

Then, assuming $c(t = 0)$ is assigned, we have for $c(t > 0)$

$$\frac{\partial c}{\partial t} = D\nabla^2 c - \vec{v} \cdot \nabla c$$

and using $\vec{v} = \dfrac{x}{T}\,\vec{i} - \dfrac{y}{T}\,\vec{j}$ this is

$$\frac{\partial c}{\partial t} = D\frac{\partial^2 c}{\partial x^2} + D\frac{\partial^2 c}{\partial y^2} - \frac{x}{T}\frac{\partial c}{\partial x} + \frac{y}{T}\frac{\partial c}{\partial y}$$

where T is a time constant of the flow.

Defining the power moments of c via

$$c_{mn} = \int_{-\infty}^{+\infty} \int_{-\infty}^{+\infty} x^m y^n c(x, y, t) \, dx dy$$

we can determine the equations satisfied by the moments just as we did before and here we find

$$\frac{dc_{mn}}{dt} = Dm(m-1)c_{(m-2)n} + Dn(n-1)C_{m(n-2)} + \frac{m-n}{T} c_{mn}$$

Again the equations can be solved recursively and doing this we find that the moment equations

$$\frac{dc_{00}}{dt} = 0$$

$$\frac{dc_{10}}{dt} = \frac{1}{T} c_{10}$$

$$\frac{dc_{01}}{dt} = -\frac{1}{T} c_{01}$$

$$\frac{dc_{11}}{dt} = 0$$

$$\frac{dc_{20}}{dt} = 2Dc_{00} + \frac{2}{T} c_{20}$$

and

$$\frac{dc_{02}}{dt} = 2Dc_{00} - \frac{2}{T} c_{02}$$

lead to

$$c_{00} = c_{00}(t = 0)$$

$$c_{10} = c_{10}(t = 0)e^{t/T}$$

$$c_{01} = c_{01}(t = 0)e^{-t/T}$$

$$c_{11} = c_{11}(t = 0)$$

$$c_{20} = e^{2t/T}c_{20}(t=0) + TDc_{00}(t=0)\left\{e^{2t/T} - 1\right\}$$

and

$$c_{02} = e^{-2t/T}c_{02}(t=0) - TDc_{00}(t=0)\left\{e^{-2t/T} - 1\right\}$$

Using these formulas we can determine averages and variances, viz.,

$$\langle x \rangle = \frac{c_{10}}{c_{00}}, \qquad \langle x^2 \rangle = \frac{c_{20}}{c_{00}}, \qquad \sigma_x^2 = \langle x^2 \rangle - \langle x \rangle^2 = \frac{c_{20}}{c_{00}} - \left(\frac{c_{10}}{c_{00}}\right)^2, \qquad \text{etc.}$$

and the results are

$$\langle x \rangle = \langle x \rangle (t=0)e^{t/T}$$

$$\langle y \rangle = \langle y \rangle (t=0)e^{-t/T}$$

$$\sigma_x^2 = \sigma_x^2(t=0)e^{2t/T} + TD\left\{e^{2t/T} - 1\right\}$$

and

$$\sigma_y^2 = \sigma_y^2(t=0)e^{-2t/T} - TD\left\{e^{-2t/T} - 1\right\}$$

This tells us that as time grows large σ_y^2 achieves a constant value TD. This value expresses the balance between the tendency of diffusion to increase σ_y^2 and the tendency of the flow to decrease it. The tendency of the flow is to carry the solute toward the line $y = 0$ building concentration gradients there until ultimately diffusion in the y direction can just offset this.

The reader can use this method to investigate the displacement of a solute concentration field by a simple shearing flow: $v_x = \dfrac{1}{T}y, v_y = 0$.

The reader can also do two simple calculations in the spherically symmetric case where

$$\frac{\partial c}{\partial t} = D\frac{1}{r^2}\frac{\partial}{\partial r}\left(r^2\frac{\partial c}{\partial r}\right)$$

Define c_0 and c_2 via

$$c_0 = \int_0^\infty c\, 4\pi r^2 \, dr$$

and

$$c_2 = \int_0^\infty r^2\, c\, 4\pi r^2 \, dr$$

and derive

$$\frac{dc_0}{dt} = 0$$

and

$$\frac{dc_2}{dt} = 6Dc_0$$

Now suppose we have one dimensional diffusion in an inhomogeneous region so that $D = D(x)$, then c, the solute concentration, satisfies

$$\frac{\partial c}{\partial t} = \frac{\partial}{\partial x}\left\{ D(x)\frac{\partial c}{\partial x}\right\}$$

whereupon $c_0(t) = \int_{-\infty}^{+\infty} c\, dx$ satisfies

$$\frac{dc_0}{dt} = 0$$

and $c_1(t) = \int_{-\infty}^{+\infty} xc\, dx$ satisfies

$$\frac{dc_1}{dt} = \int_{-\infty}^{+\infty} D'(x)c\, dx$$

whence we have

$$\frac{d}{dt}\langle x\rangle = \langle D'(x)\rangle$$

where $\langle x \rangle = \dfrac{c_1}{c_0}$

Thus we see that solute no longer redistributes about a fixed point. And moving on to solute spreading we observe that

$$c_2 = \int_{-\infty}^{+\infty} x^2 c \, dx$$

satisfies

$$\frac{dc_2}{dt} = 2 \int_{-\infty}^{+\infty} (xD(x))' \, c \, dx$$

whereupon

$$\frac{d\sigma^2}{dt} = \frac{1}{c_0} \frac{dc_2}{dt} - 2 \frac{c_1}{c_0} \frac{dc_1}{dt}$$

$$= 2\langle (xD(x))' \rangle - 2\langle x \rangle \frac{d\langle x \rangle}{dt}$$

$$= 2\langle D(x) \rangle + 2\langle xD'(x) - \langle x \rangle \langle D'(x) \rangle \rangle$$

Now these formulas certainly have the expected form and we could use them if we could write $c(x, t)$ in terms of $c_0(t), c_1(t)$, etc. but that is not so easy.

12.2 Chromatographic Separations

Carrying on with our plan to try to learn something about diffusion without doing much, we turn to a problem in which we discover diffusion where at first there would appear to be none.

A solute is injected into a carrier gas flowing through a packed column where the solute is adsorbed by the packing. As the solute moves through our column it is adsorbed at its leading edge, desorbed at its trailing edge. The solute concentration in the carrier phase and in the solid phase are denoted by c and by a and the dilute solute equilibrium isotherm is assumed to be

$$c = ma \qquad \text{where } m \text{ is constant}$$

The smaller the value of m, the more strongly the solute is bound to the solid.

We denote the volumetric flow rate of the carrier by G, the cross sectional area of the empty column by A and the porosity of the bed by ε.

Then $v_0 = \dfrac{G}{A}$ denotes the superficial velocity of the carrier and $v = \dfrac{v_0}{\varepsilon}$ denotes its interstitial velocity.

We can write a simple model, assuming no variation of c and a on the cross section. It is

$$\varepsilon \frac{\partial c}{\partial t} + v_0 \frac{\partial c}{\partial z} = -K\left(c - ma\right)$$

and

$$\left(1 - \varepsilon\right) \frac{\partial a}{\partial t} = K\left(c - ma\right)$$

where the volumetric mass transfer coefficient is denoted by K and where $[K] = \dfrac{1}{\text{time}}$.

The bed is assumed to be infinitely long and at time zero a finite amount of solute is injected into the carrier near $z = 0$.

Our model equations can be solved, but we do not do so. Instead, we will see what we can learn by solving the moment equations.

The zeroth moments of c and a, viz.,

$$c_0 = \int_{-\infty}^{+\infty} c \, dz, \qquad a_0 = \int_{-\infty}^{+\infty} a \, dz$$

satisfy

$$\varepsilon \frac{dc_0}{dt} = -K\left(c_0 - ma_0\right)$$

and

$$\left(1 - \varepsilon\right) \frac{da_0}{dt} = K\left(c_0 - ma_0\right)$$

By adding these equations, differentiating the first and eliminating a_0 we obtain

$$\frac{d^2 c_0}{dt^2} + K \left(\frac{1}{\varepsilon} + \frac{m}{1-\varepsilon} \right) \frac{dc_0}{dt} = 0$$

whence we have

$$c_0 = A_0 + B_0 \, e^{-tK \left(\frac{1}{\varepsilon} + \frac{m}{1-\varepsilon} \right)}$$

and, as $t \to \infty$,

$$c_0 \to A_0$$

$$\frac{dc_0}{dt} \to 0$$

and

$$a_0 \to \frac{c_0}{m}$$

The first order moments of c and a, viz.,

$$c_1 = \int_{-\infty}^{+\infty} zc \, dz, \qquad a_1 = \int_{-\infty}^{+\infty} za \, dz$$

satisfy

$$\varepsilon \frac{dc_1}{dt} - v_0 c_0 = -K \left(c_1 - ma_1 \right)$$

and

$$(1 - \varepsilon) \frac{da_1}{dt} = K \left(c_1 - ma_1 \right)$$

Again, adding these equations, differentiating the first and eliminating a_1, we have

$$\frac{d^2 c_1}{dt^2} + K\left(\frac{1}{\varepsilon} + \frac{m}{1-\varepsilon}\right)\frac{dc_1}{dt} = K\frac{v_0}{\varepsilon}\frac{m}{1-\varepsilon}c_0 + \frac{v_0}{\varepsilon}\frac{dc_0}{dt}$$

Hence, at times long after solute injection, we must solve

$$\frac{d^2 c_1}{dt^2} + K\left(\frac{1}{\varepsilon} + \frac{m}{1-\varepsilon}\right)\frac{dc_1}{dt} = K\frac{v_0}{\varepsilon}\frac{m}{1-\varepsilon}A_0$$

whereupon we obtain

$$c_1 = A_1 + \underset{\text{zero}}{\cancel{B_1 e^{-tK\left(\frac{1}{\varepsilon} + \frac{m}{1-\varepsilon}\right)}}} + \frac{v_0}{\varepsilon}\frac{\frac{m}{1-\varepsilon}}{\frac{1}{\varepsilon} + \frac{m}{1-\varepsilon}}A_0 t$$

The average distance transversed by the solute in the carrier phase, measured from the injection point, $z = 0$, and denoted $\langle z\rangle$, viz.,

$$\langle z\rangle = \frac{\int_{-\infty}^{+\infty} zc\,dz}{\int_{-\infty}^{+\infty} c\,dz} = \frac{c_1}{c_0}$$

advances, at times long after solute injection, according to

$$\frac{d\langle z\rangle}{dt} = \frac{1}{c_0}\frac{dc_1}{dt} = \frac{v_0}{\varepsilon}\frac{\frac{m}{1-\varepsilon}}{\frac{1}{\varepsilon} + \frac{m}{1-\varepsilon}}$$

and we call $\dfrac{d\langle z\rangle}{dt}$ the speed at which the solute in the carrier phase proceeds through the bed.

For large m, weakly bound solute, we have

$$\frac{d\langle z\rangle}{dt} = \frac{v_0}{\varepsilon},$$

where $\dfrac{v_0}{\varepsilon}$ is the carrier speed, whereas for small m, strongly bound solute, we have

$$\frac{d\langle z \rangle}{dt} = \frac{v_0}{\varepsilon} \frac{\varepsilon}{1-\varepsilon} m$$

Thus the ratio of the speeds of two strongly bound solutes is the ratio of their m's. The mass transfer coefficient does not appear in these formulas.

The reader may wish to find the speed of the solute in the solid phase.

What is interesting about this problem is that we have a transverse distribution of longitudinal speeds, simple though it may be, viz., $\dfrac{v_0}{\varepsilon}$ in the carrier, zero in the solid, and we have solute that can move between the regions of different speed. This is a recipe for solute spreading in the flow direction and, therefore, a hint that a diffusion model may be appropriate.

To see this we look at second moments of c and a defined by

$$c_2 = \int_{-\infty}^{+\infty} z^2 c\, dz, \qquad a_2 = \int_{-\infty}^{+\infty} z^2 a\, dz$$

These moments satisfy

$$\varepsilon \frac{dc_2}{dt} - 2v_0 c_1 = -K\left(c_2 - m a_2\right)$$

and

$$(1-\varepsilon)\frac{da_2}{dt} = K\left(c_2 - m a_2\right)$$

By eliminating a_2 we obtain

$$\frac{d^2 c_2}{dt^2} + K\left(\frac{1}{\varepsilon} + \frac{m}{1-\varepsilon}\right)\frac{dc_2}{dt} = 2\frac{v_0}{\varepsilon}\frac{dc_1}{dt} + 2\frac{v_0}{\epsilon} K \frac{m}{1-\varepsilon} c_1$$

whence, upon using

$$c_1 = A_1 + \frac{v_0}{\varepsilon} \frac{\dfrac{m}{1-\varepsilon}}{\dfrac{1}{\varepsilon} + \dfrac{m}{1-\varepsilon}} A_0 t = A_1 + \text{const } A_0 t$$

we have, for times long after solute injection,

$$\frac{d^2 c_2}{dt^2} + K\left(\frac{1}{\varepsilon} + \frac{m}{1-\varepsilon}\right)\frac{dc_2}{dt} = 2\frac{v_0}{\varepsilon}\,\text{const}\,A_0 + 2\frac{v_0}{\varepsilon}\,K\,\frac{m}{1-\varepsilon}\{A_1 + \text{const}\,A_0 t\}$$

Now we introduce σ^2, the longitudinal variance of the solute distribution in the carrier, where

$$\sigma^2 = \frac{\int_{-\infty}^{+\infty} (z - \langle z \rangle)^2\, c\, dz}{\int_{-\infty}^{+\infty} c\, dz} = \frac{c_2}{c_0} - \left(\frac{c_1}{c_0}\right)^2$$

and, therefore, we have, at long time,

$$\frac{d\sigma^2}{dt} = \frac{1}{c_0}\frac{dc_2}{dt} - \frac{1}{c_0^2}\,2c_1\,\frac{dc_1}{dt}$$

where $\dfrac{d\sigma^2}{dt}$ tells us the rate of spreading of the solute. If σ^2 is a multiple of t or $\dfrac{d\sigma^2}{dt}$ is a constant, this constant defines the longitudinal diffusion coefficient.

The long time value of $\dfrac{d\sigma^2}{dt}$ can be worked out and after some algebra we obtain

$$\frac{d\sigma^2}{dt} = 2\,\frac{(v_0/\varepsilon)^2}{K\varepsilon}\,\frac{\dfrac{m}{1-\varepsilon}}{\left(\dfrac{1}{\varepsilon} + \dfrac{m}{1-\varepsilon}\right)^2}$$

which, for strongly bound solute, is

$$\frac{d\sigma^2}{dt} = 2\,\frac{(v_0/\varepsilon)^2}{K}\,\varepsilon^2\,\frac{m}{1-\varepsilon}$$

Hence, the larger the value of K and the smaller the value of m, the more the solute hangs together. Indeed, the larger the value of K, the less the effect of the velocity gradient, and the smaller the value of m, the stronger the solute is bound to the solid.

The reader ought to derive these formulas. Only a particular solution of the c_2 equation is needed.

In order for the spreading process to be diffusive, the terms in $\dfrac{1}{c_0}\dfrac{dc_2}{dt}$ and in $\dfrac{1}{c_0^2}\,2c_1\,\dfrac{dc_1}{dt}$ that are multiples of t must cancel, otherwise $\dfrac{d\sigma^2}{dt}$ will be a multiple of t and in that case the spreading is called ballistic, not diffusive.

This problem serves as an introduction to the problem called Taylor dispersion presented in Lecture 20.

12.3   Random Walk Model

The diffusion equation tells us that

$$\langle x \rangle = 0$$

and

$$\langle x^2 \rangle = 2Dt$$

assuming all the solute particles start at $x = 0$ and diffuse in the $\pm x$ directions.

The probability in a simple random walk on a one-dimensional lattice satisfies the difference equation

$$P\left(N_1, N_2; N\right) = pP\left(N_1 - 1, N_2; N - 1\right) + qP\left(N_1, N_2 - 1 : N - 1\right)$$

where $P\left(N_1, N_2; N\right)$ is the probability that in N steps, N_1 steps are to the right, N_2 steps are to the left and where, at each step, p is the probability that it is to the right, q the probability that it is to the left. We must have $N_1 + N_2 = N$ and $p + q = 1$. After N steps the possible values of N_1 are $0, 1, 2, \ldots, N$.

The solution to this difference equation is

$$P\left(N_1, N_2; N\right) = \frac{N!}{N_1! N_2!} p^{N_1} q^{N_2}$$

where $N_1 + N_2 = N$ and $p + q = 1$.

This formula is called the binormal distribution. It tells us the probability of N_1 successes in N trials where the trials are independent and where p denotes the probability of a success in any trial.

Now, due to

$$\sum_{N_1}\sum_{N_2} P\left(N_1, N_2 : N\right) = (p+q)^N, \qquad N_1 + N_2 = N$$

we can determine the average values of N_1 and N_1^2 to be

$$\langle N_1 \rangle = Np$$

and

$$\langle N_1^2 \rangle = Np\left(1 - p\right) + N^2 p^2$$

Hence if we denote $N_1 - N_2$ by Δ, i.e., the net number of steps to the right, then the average values of Δ and Δ^2 are

$$\langle \Delta \rangle = N\left(2p - 1\right)$$

and

$$\langle \Delta^2 \rangle = 4Np\left(1 - p\right) + N^2\left(2p - 1\right)^2$$

If the lattice spacing is ℓ and a particle is released at $x = 0$ then its position on the lattice after N steps is $x = \ell\Delta$. So if all particles are released at $x = 0$ and if $p = q = \dfrac{1}{2}$ then

$$\langle x \rangle = 0$$

and

$$\langle x^2 \rangle = \ell^2 N$$

Assuming the time required to take a step is τ, we see that N steps corresponds to time $t = N\tau$ and

$$\langle x^2 \rangle = \ell v t$$

where $v = \ell/\tau$ is the speed of a particle. Our two formulas for $\langle x^2 \rangle$ tell us that random walk acts statistically like a diffusion process where

$$D = \frac{1}{2} \ell v$$

We can repeat this calculation assuming that diffusion takes place, not in one, but in three dimensions. Then if all the particles start at the origin the diffusion equation tells us directly that

$$\langle r^2 \rangle = \langle x^2 \rangle + \langle y^2 \rangle + \langle z^2 \rangle = 6Dt$$

To introduce the corresponding simple random walk on a three dimensional cubic lattice we let $P(N_1, N_2, N_3, N_4, N_5, N_6; N)$ be the probabiity that in N steps, N_1 steps are in the $+x$ direction, N_2 steps are in the $-x$ direction, N_3 steps are in the $+y$ direction, etc. And at each step we let p be the probability that it is in the $+x$ direction, q the probability that it is in the $-x$ direction, r the probability that it is in the $+y$ direction, etc. Then we find

$$P(N_1, N_2, \ldots, N_6; N) = \frac{N!}{N_1! N_2! N_3! N_4! N_5! N_6!} \, p^{N_1} q^{N_2} r^{N_3} s^{N_4} u^{N_5} v^{N_6}$$

where $N_1 + N_2 + \cdots + N_6 = N$ and $p + q + \cdots + v = 1$ and where

$$\sum_{N_1} \sum_{N_2} \cdots \sum_{N_6} P(N_1, N_2, \ldots, N_6; N) = (p + q + r + s + u + v)^N = 1, \quad N_1 + N_2 + \cdots + N_6 = N$$

We can determine the averages we need as follows

$$\langle N_1 \rangle = \sum_{N_1} \sum_{N_2} \cdots \sum_{N_6} N_1 P(N_1, N_2, \ldots, N_6; N)$$

$$= p \frac{\partial}{\partial p} \sum_{N_1} \sum_{N_2} \cdots \sum_{N_6} P(N_1, N_2, \ldots, N_6; N)$$

$$= pN$$

and

$$\langle N_1^2 \rangle = \sum_{N_1} \sum_{N_2} \cdots \sum_{N_6} N_1^2 P(N_1, N_2, \ldots, N_6; N)$$

$$= p \frac{\partial}{\partial p} p \frac{\partial}{\partial p} \sum_{N_1} \sum_{N_2} \cdots \sum_{N_6} P(N_1, N_2, \ldots, N_6; N)$$

$$= pN + p^2 N(N - 1)$$

Likewise we find

$$\langle N_2 \rangle = qN$$

$$\langle N_1 N_2 \rangle = pqN(N - 1)$$

and

$$\langle N_2^2 \rangle = qN + q^2 N(N - 1)$$

So, if we let $\Delta_x = N_1 - N_2$, $\Delta_y = N_3 - N_4$ and $\Delta_z = N_5 - N_6$ then Δ_x is the net number of steps in the positive x direction. Its average is

$$\langle \Delta_x \rangle = \langle N_1 - N_2 \rangle = (p - q)N$$

while the average of its square is

$$\langle \Delta_x^2 \rangle = \langle N_1^2 - 2N_1 N_2 + N_2^2 \rangle$$

$$= \{p+q\}N + \{p-q\}^2 N(N-1)$$

If the lattice spacing is ℓ and a particle is released at $x = y = z = 0$ then its position on the lattice after N steps is $x = \ell \Delta_x$, $y = \ell \Delta_y$ and $z = \ell \Delta_z$. So if all particles are released at $x = y = z = 0$ and if $p = q = r = s = u = v = \dfrac{1}{6}$ then

$$\langle x \rangle = 0$$

and

$$\langle x^2 \rangle = \ell^2 \frac{1}{3} N$$

Likewise we have $\langle y \rangle = 0 = \langle z \rangle$ and $\langle y^2 \rangle = \ell^2 \dfrac{1}{3} N = \langle z^2 \rangle$.

Again if the time required to take a step is τ then $t = N\tau$ and

$$\langle r^2 \rangle = \langle x^2 \rangle + \langle y^2 \rangle + \langle z^2 \rangle = \ell v t$$

whence

$$D = \frac{1}{6} \ell v$$

So when we hold the length of a free path fixed, require the free path to lie on a lattice and hold the speed of a particle fixed we get

$$D = \frac{1}{2} \ell v$$

and

$$D = \frac{1}{6}\ell v$$

depending on whether the random walk is one or in three dimensions.

This may be a surprising result. Even more surprising is the way the lattice calculations turn out. After N steps the average of the square of the distance from the origin is the same whether the random walk is on a one dimensional lattice where $\langle x^2 \rangle = \ell^2 N$ or on a three dimensional lattice where $\langle r^2 \rangle = \ell^2 N$.

It is worth recording the elementary mean free path result that

$$D = \frac{1}{3}\ell v$$

when a dilute gas is diffusing in a gas of fixed atoms.

12.4   The Point Source Solution to the Diffusion Equation

A solute is distributed throughout all space at time $t = 0$. The amount of solute is finite. There are no natural length and time scales. Our job is to predict the solute concentration, c, at $t > 0$. To do this we must solve

$$\frac{\partial c}{\partial t} = D\nabla^2 c, \qquad \forall \vec{r}$$

where we must have $c \to 0$ as $|\vec{r}| \to \infty$ and where $c(t = 0)$ is assigned. Our first step is to find out how a point source of solute spreads out in time in an unbounded domain. The result will be the Green's function for our problem and we obtain it by assuming we know the solution to a step function initial condition. Then we come back to Green's functions in Lecture 19.

To begin we first observe that

$$A + B \operatorname{erf} \frac{x}{\sqrt{4t}}$$

satisfies

$$\frac{\partial c}{\partial t} = \frac{\partial^2 c}{\partial x^2}$$

where erf z, called the error function, is defined by

$$\text{erf } z = \frac{2}{\pi} \int_0^z e^{-\zeta^2} d\zeta$$

and where for real values of z it looks like:

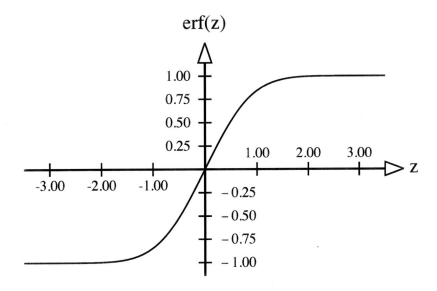

While this would not seem to be a very useful solution to the diffusion equation due to the limited class of boundary conditions that can be satisfied by setting two constants, nonetheless it plays an important role in modeling diffusion processes as can be seen by looking in Bird, Stewart and Lightfoot's book "*Transport Phenomena.*" To get the solution when the diffusivity is other than $D = 1$, write Dt in place of t.

If $c(t = 0)$ is zero for $x < \xi$ and $x > \xi + \Delta\xi$ and uniform but not zero for $\xi < x < \xi + \Delta\xi$ then, for $t > 0$ we have

$$c = c_0 \left\{ \frac{1}{2} \text{ erf} \left(\frac{x - \xi}{\sqrt{4t}} \right) - \frac{1}{2} \text{ erf} \left(\frac{x - (\xi + \Delta\xi)}{\sqrt{4t}} \right) \right\}$$

and to the first order in $\Delta\xi$ this is

$$c_0 \frac{1}{\sqrt{4\pi t}} e^{-\dfrac{(x-\xi)^2}{4\pi}} \Delta\xi$$

To go to three dimensions observe that if $f(x,t)$ satisfies

$$\frac{\partial c}{\partial t} = \frac{\partial^2 c}{\partial x^2}$$

then $f(x,t)f(y,t)f(z,t)$ satisfies

$$\frac{\partial c}{\partial t} = \nabla^2 c$$

and hence to first order in $\Delta\xi$, $\Delta\eta$ and $\Delta\zeta$

$$c = c_0 \,\Delta\xi\Delta\eta\Delta\zeta \left\{ \frac{1}{\sqrt{4\pi t}} \right\}^3 e^{-\dfrac{(x-\xi)^2 + (x-\eta)^2 + (x-\zeta)^2}{4t}}$$

satisfies

$$\frac{\partial c}{\partial t} = \nabla^2 c$$

where

$$c(t=0) = c_0 \begin{cases} \xi < x < \xi + \Delta\xi \\ \eta < y < \eta + \Delta\eta \\ \zeta < z < \zeta + \Delta\zeta \end{cases}$$

$$= 0 \quad \text{otherwise}$$

The product $c_0\Delta\xi\Delta\eta\Delta\zeta$ is the total amount of solute initially present. If we hold this fixed, equal to one unit of mass, and let $\Delta\xi$, $\Delta\eta$ and $\Delta\zeta$ go to zero, we find

$$c = \frac{1}{\left\{\sqrt{4\pi t}\right\}^3} \, e^{-\frac{(x-\xi)^2 + (x-\eta)^2 + (x-\zeta)^2}{4t}}$$

and c satisfies

$$\frac{\partial c}{\partial t} = \nabla^2 c, \qquad t > 0$$

$$c(t=0) = 0, \quad \forall\, (x,y,z) \neq (\xi, \eta, \zeta)$$

and

$$\int_{-\infty}^{+\infty} \int_{-\infty}^{+\infty} \int_{-\infty}^{+\infty} c \, dV = 1, \quad \forall t \geq 0$$

This is called the point source solution to the diffusion equation as it tells us the solute density at the point (x, y, z) and at the time t if at $t = 0$ a unit mass of solute is injected into a region of vanishingly small extent at the point (ξ, η, ζ). It is called the Green's function for the diffusion equation in an unbounded region.

If the diffusivities are other that $D = 1$ it is a simple matter to decide that the point source solution to

$$\frac{\partial c}{\partial t} = D_{xx}\frac{\partial^2 c}{\partial x^2} + D_{yy}\frac{\partial^2 c}{\partial y^2} + D_{zz}\frac{\partial^2 c}{\partial z^2}$$

is

$$c = \frac{1}{\left\{\sqrt{4\pi t}\right\}^3} \frac{1}{\sqrt{D_{xx}}\sqrt{D_{yy}}\sqrt{D_{zz}}} \, e^{-\frac{D_{xx}^{-1}(x-\xi)^2 + D_{yy}^{-1}(x-\eta)^2 + D_{zz}^{-1}(x-\zeta)^2}{4t}}$$

and this is the Green's function for diffusion in an anisotropic solvent.

Thus suppose c satisfies

$$\frac{\partial c}{\partial t} = \nabla \cdot \vec{\vec{D}} \cdot \nabla c = \vec{\vec{D}} : \nabla\nabla c$$

where we have assumed $\vec{\vec{D}}$ to be symmetric. Thus being so there is an orthogonal basis in which $\vec{\vec{D}}$ is diagonal. Hence requiring $\vec{i}, \vec{j}$ and $\vec{k}$ to lie along the eigenvectors of $\vec{\vec{D}}$, we can write

$$\vec{\vec{D}} : \nabla\nabla c = D_{xx}\frac{\partial^2 c}{\partial x^2} + D_{yy}\frac{\partial^2 c}{\partial y^2} + D_{zz}\frac{\partial^2 c}{\partial z^2},$$

where D_{xx}, D_{yy} and D_{zz} are the eigenvalues of $\vec{\vec{D}}$. And in the eigenbasis of $\vec{\vec{D}}$ we can write the point source solution to the anisotropic diffusion equation as above. In coordinate free form it is

$$c\left(\vec{r}, t\right) = \frac{1}{\left\{\sqrt{4\pi t}\right\}^3 \left\{\det \vec{\vec{D}}\right\}^{1/2}} e^{-\frac{\vec{\vec{D}}^{-1} : (\vec{r} - \vec{r_0})(\vec{r} - \vec{r_0})}{4t}}$$

where $\vec{r_0} = \xi\vec{i} + \eta\vec{j} + \zeta\vec{k}$.

We can use the Green's function to write the solution to the diffusion equation in a way that does not require us to decompose the sources but accepts them as they stand. To do this observe that the point source solution tells us the solute concentration at the point (x, y, z) at a time t following the introduction of a unit mass of solute at the point (ξ, η, ζ). Knowing this we can write the solution to

$$\frac{\partial c}{\partial t} = \vec{\vec{D}} : \nabla\nabla c + Q\left(\vec{r}, t\right)$$

where

$$c(t = 0) = c_0\left(\vec{r}\right)$$

in terms of a superposition of point source solutions. The result is

$$c\left(\vec{r},t\right) = \int_{-\infty}^{+\infty}\int_{-\infty}^{+\infty}\int_{-\infty}^{+\infty}\frac{1}{\left\{\sqrt{4\pi t}\right\}^3\left\{\sqrt{\det\vec{\vec{D}}}\right\}}e^{-\dfrac{\vec{\vec{D}}^{-1}:(\vec{r}-\vec{r_0})(\vec{r}-\vec{r_0})}{4t}}c_0\left(\vec{r_0}\right)\,dV_0$$

$$+\int_{-\infty}^{+\infty}\int_{-\infty}^{+\infty}\int_{-\infty}^{+\infty}\int_{0}^{t}\frac{1}{\left\{\sqrt{4\pi(t-t_0)}\right\}^3\left\{\sqrt{\det\vec{\vec{D}}}\right\}}e^{-\dfrac{\vec{\vec{D}}^{-1}:(\vec{r}-\vec{r_0})(\vec{r}-\vec{r_0})}{4(t-t_0)}}Q\left(\vec{r_0},t_0\right)\,dV_0\,dt_0$$

inasmuch as $c_0(\vec{r_0})dV_0$ is the amount of solute introduced into dV_0 at $\vec{r_0}$ at $t=0$ and $Q\left(\vec{r_0},t_0\right)dV_0\,dt_0$ is the amount of solute introduced into dV_0 at $\vec{r_0}$ during dt_0 at t_0. Again the solution is a sum of terms, each term itself being the solution corresponding to one of the sources when the others vanish. If this result is used to determine $\vec{\vec{D}}$, only symmetric $\vec{\vec{D}}$'s can be found.

The point source solution for diffusion in three dimensions leads easily to the point source solution for diffusion in two dimensions and then in one dimension.

If we introduce M''' mass units of solute into an isotropic solvent at the point $\vec{r}=\vec{r_0}$ and at time $t=0$ then the resulting solute concentration is

$$c\left(\vec{r},t\right) = M'''\left\{\frac{1}{\sqrt{4\pi Dt}}\right\}^3 e^{-\dfrac{(x-x_0)^2+(y-y_0)^2+(z-z_0)^2}{4Dt}}$$

and this is the point source solution in three dimensions. The dimensions of c are M/L^3 and the dimension of $\sqrt{Dt}$ is L.

If instead we introduce M'' mass units of solute per unit of length uniformly on the line $x=x_0, y=y_0, -\infty < z < \infty$, at time $t=0$, then the resulting solute concentration is

$$c\left(\vec{r},t\right) = \int_{-\infty}^{+\infty}M''dz_0\left\{\frac{1}{\sqrt{4\pi Dt}}\right\}^3 e^{-\dfrac{(x-x_0)^2+(y-y_0)^2+(z-z_0)^2}{4Dt}}$$

$$= M''\left\{\frac{1}{\sqrt{4\pi Dt}}\right\}^2 e^{-\dfrac{(x-x_0)^2+(y-y_0)^2}{4Dt}}$$

and this is called the line source solution in three dimensions. It is uniform in z and does not vanish as $|z| \to \infty$. It can be called the point source solution in two dimensions, taking the dimensions of M'' to be M instead of M/L and then the dimensions of c to be M/L^2. Likewise if we introduce M' mass units of solute per unit of area uniformly over the plane $x = x_0, -\infty < y < \infty, -\infty < z < \infty$, at time $t = 0$, then the resulting solute concentration is

$$c\left(\vec{r}, t\right) = M' \frac{1}{\sqrt{4\pi Dt}} \, e^{-\frac{\left(x - x_0\right)^2}{4Dt}}$$

and this is called the plane source solution in three dimensions. It is uniform in y and z and does not vanish as $|y| \to \infty$ or as $|z| \to \infty$. It can be called the point source solution in one dimension, taking the dimensions of M' to be M instead of M/L^2 and then the dimensions of c to be M/L.

12.5   Home Problems

1. Suppose c satisfies

$$\frac{\partial c}{\partial t} = \vec{\vec{D}} : \nabla \nabla c$$

where $c\left(t = 0\right)$ is assigned and c vanishes strongly as $|\vec{r}| \to \infty$. Derive equations for the power moments of c, viz.,

$$c_0 = \iiint c \, dV$$

$$\vec{c}_1 = \iiint \vec{r} c \, dV$$

$$\vec{\vec{c}}_2 = \iiint \vec{r} \vec{r} c \, dV$$

   etc.

Derive a formula for $\vec{\vec{D}}$ in terms of the power moments of c.

You ought to begin by deriving:

$$\vec{r}\,\nabla^2 c = \nabla^2\left(\vec{r}c\right) - 2\nabla c$$

$$\vec{r}\,\vec{r}\,\nabla^2 c = \nabla^2\left(\vec{r}\,\vec{r}\,\nabla^2 c\right) - 2\vec{\vec{I}}c - 2\nabla c\,\vec{r} - 2\vec{r}\nabla c$$

and

$$\left(\nabla c\right)\vec{r} = \nabla\left(\vec{r}c\right) - \vec{\vec{I}}c$$

It helps to know:

$$\iiint\limits_{V} \nabla f \, dV = \iint\limits_{S} dA\,\vec{n}f$$

where f can be scalar, vector, etc. valued.

2. A thermometer, your finger, senses the temperature of its tip shortly after it touches another body. Let α_B, k_B and T_B denote the thermal diffusivity, conductivity and temperature of a body while α_F, k_F and T_F denote the thermal diffusivity, conductivity and temperature of the thermometer.

   Use the special solution of the diffusion equation, viz.,

   $$A + B \,\mathrm{erf}\left(\sqrt{\frac{x}{4\alpha t}}\right)$$

   to decide that two bodies at the same temperature ordinarily do not feel like they are at the same temperature. What determines which feels cooler?

3. Denote by σ a surface heat source and assume σ is constant over a sphere of radius R. Derive a formula for T inside and for T outside, assuming $T \to 0$ as $r \to \infty$. The thermal conductivities inside and outside may differ.

Here, $\nabla^2 T$ is simply

$$\frac{1}{r^2}\frac{d}{dr^2}\left(r^2\frac{dT}{dr}\right)$$

Write your result in terms of $\Sigma = 4\pi R^2\sigma$ and then let $R \to 0$ holding Σ constant.

4. Leveling a storm surge.

Water lies in a porous rock above an impermeable plane, viz.,

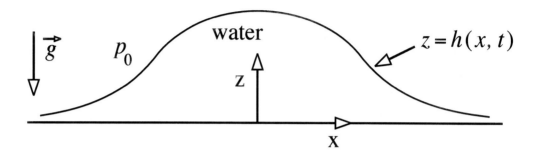

Writing Darcy's law for v_x, viz.,

$$\frac{\mu}{K}v_x = -\frac{\partial p}{\partial x}$$

assuming the pressure is hydrostatic, viz.,

$$p = p_0 + \rho g\,(h - z)$$

and using

$$\frac{\partial v_x}{\partial x} + \frac{\partial v_z}{\partial z} = 0, \qquad v_z = 0 \quad \text{at} \quad z = 0$$

derive

$$\left\{K\frac{\rho}{\mu}g\right\}\frac{\partial}{\partial x}\left(h\frac{\partial h}{\partial x}\right) = \frac{\partial h}{\partial t}$$

starting from the balance satisfied by the water at the surface $z = h(x, t)$, viz.,

$$v_z - v_x \frac{\partial h}{\partial x} = \frac{\partial h}{\partial t}$$

Notice that

$$\left[\frac{K \rho g}{\mu} h \right] = \frac{L^2}{T}$$

and that what you have is a nonlinear diffusion equation where the diffusivity at (x, t) depends on how much h is there.

Define

$$h_m = \int_{-\infty}^{+\infty} x^m h(x, t) \, dx$$

and show that

$$\frac{dh_0}{dt} = 0 = \frac{dh_1}{dt}$$

and

$$\frac{dh_2}{dt} > 0$$

Thus the amount of water does not change nor does its mean position as it spreads out.

5. Chromatography

Solve for

$$\begin{pmatrix} c_0 \\ a_0 \end{pmatrix}, \quad \begin{pmatrix} c_1 \\ a_1 \end{pmatrix}, \quad \cdots$$

in terms of the eigenvalues and eigenvectors of

$$
\begin{pmatrix}
-\dfrac{K}{\varepsilon} & \dfrac{Km}{\varepsilon} \\[2mm]
\dfrac{K}{1-\varepsilon} & -\dfrac{Km}{1-\varepsilon}
\end{pmatrix}
$$

Define $\bar{c}$, V_{eff} and D_{eff} via

$$
\bar{c} = \varepsilon\, c + (1 - \varepsilon)\, a
$$

$$
V_{\text{eff}} = \frac{1}{\bar{c}_0} \frac{d\bar{c}_1}{dt}
$$

and

$$
D_{\text{eff}} = \frac{1}{2\bar{c}_0} \frac{d\bar{c}_2}{dt} - \frac{\bar{c}_1}{\bar{c}_0^2} \frac{d\bar{c}_1}{dt}
$$

and derive formulas for V_{eff} and D_{eff} as $t \to \infty$.

Lecture 13

Multipole Expansions

13.1 Steady Diffusion in an Unbounded Domain

In this lecture we deal with the problem of steady solute diffusion or heat conduction from a source of solute or heat near the origin to a sink at zero concentration or temperature, infinitely far away. Again we derive a point source solution.

Suppose heat is generated uniformly inside a sphere of radius R centered on the origin O. Then, assuming the heat is conducted steadily to $T = 0$ at $r = \infty$, we have

$$\nabla^2 T + \frac{Q}{k} = 0 \ , \ \ [Q] = \frac{\text{heat}}{\text{volume time}}$$

inside the sphere and

$$\nabla^2 T = 0$$

outside the sphere where

$$\nabla^2 = \frac{1}{r^2} \frac{d}{dr} r^2 \frac{d}{dr}$$

and therefore $\nabla^2 T = 0$ implies

$$T = A + \frac{B}{r}$$

Before going on we ought to notice that there would be no steady solution if the sphere were replaced by an infinitely long cylinder.

We find

$$T = A - \frac{1}{6}r^2\frac{Q}{k}$$

inside,

$$T = \frac{B}{r}$$

outside and, denoting by k the outside thermal conductivity, we have

$$B = \frac{1}{3}R^3\frac{Q}{k}$$

and therefore, outside,

$$T = \frac{1}{3}R^3\frac{Q}{k}\frac{1}{r} = \frac{VQ}{4\pi k}\frac{1}{r}, \quad r > R$$

where VQ is the heat supplied per unit time. Denoting VQ by Q, we have, for a point source of heat at the origin:

$$T = \frac{Q}{4\pi k}\frac{1}{r}, \quad r > 0.$$

Now that we have the temperature at a point $\vec{r}$ due to a point source of heat at a point $\vec{r_0}$, viz.,

$$T(\vec{r}) = \frac{Q}{4\pi k}\frac{1}{|\vec{r} - \vec{r_0}|}, \qquad [Q] = \frac{\text{heat}}{\text{time}}$$

we can find the temperature at the point $\vec{r}$ due to a continuously distributed source of heat specified by the density ρ, vanishing outside a region of space denoted V_0, viz.,

$$T(\vec{r}) = \frac{1}{4\pi k} \int\int\int_{V_0} \frac{\rho(\vec{r_0})\, dV_0}{|\vec{r} - \vec{r_0}|}, \qquad [\rho] = \frac{\text{heat}}{\text{time volume}}$$

The readers ought to convince themselves that the integral makes sense, whether $\vec{r}$ lies inside or outside the sphere, by looking at the special case where heat is generated uniformly in a finite sphere centered on O

Thus, we have the solution to

$$\nabla^2 T + \frac{\rho(\vec{r})}{k} = 0, \quad T \to 0 \text{ as } |\vec{r}| \to \infty$$

in an infinite region. This equation is called Poisson's equation.

Now suppose we have a system of point sources in the neighbourhood of a point O and we wish to know T at a point P some distance away, viz.,

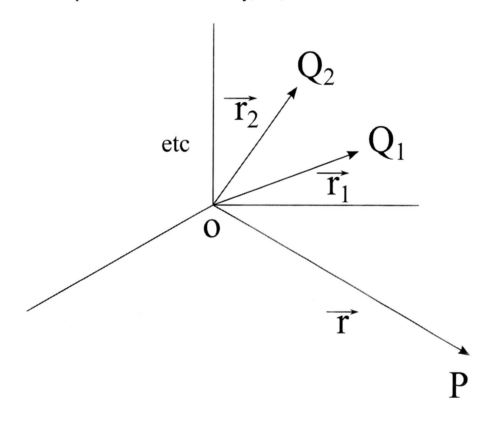

At P we have

$$T\left(P\right) = \sum \frac{Q_i}{4\pi k} \frac{1}{|\vec{r} - \vec{r_i}|}$$

where

$$|\vec{r} - \vec{r_i}|^2 = (\vec{r} - \vec{r_i}) \cdot (\vec{r} - \vec{r_i}) = r^2 \left(1 - 2\frac{r_i}{r} \cos \theta_i + \frac{r_i^2}{r^2}\right)$$

and therefore

$$|\vec{r} - \vec{r_i}|^{-1} = \frac{1}{r} \left(1 - \frac{1}{2} \left(-2\frac{r_i}{r} \cos \theta_i + \frac{r_i^2}{r^2}\right) + \frac{3}{8} \left(4\frac{r_i^2}{r^2} \cos^2 \theta_i + \cdots\right) + \cdots\right)$$

and we find that $T\left(P\right)$ is given by

$$T\left(P\right) = \frac{1}{4\pi k} \frac{1}{r} \sum Q_i \left\{1 + \frac{r_i}{r} \cos \theta_i + \frac{3}{2} \frac{\left(r_i \cos \theta_i\right)^2}{r^2} - \frac{1}{2} \frac{r_i^2}{r^2} + \cdots\right\}$$

Then using

$$r_i \cos \theta_i = \frac{\vec{r} \cdot \vec{r_i}}{r},$$

we have

$$T\left(P\right) = \frac{1}{4\pi k} \left\{\frac{1}{r} \sum Q_i + \frac{1}{r^3} \vec{r} \cdot \sum Q_i \vec{r_i} + \frac{1}{r^5} \vec{r} \vec{r} : \sum Q_i \left(\frac{3}{2} \vec{r_i} \vec{r_i} - \frac{1}{2} \vec{r_i} \cdot \vec{r_i} \overset{=}{I}\right) + \cdots\right\}$$

The factors $\sum Q_i, \sum Q_i \vec{r_i}, \sum Q_i \left(\frac{3}{2} \vec{r_i} \vec{r_i} - \frac{1}{2} \vec{r_i} \cdot \vec{r_i} \overset{=}{I}\right), \cdots$ depend only on the distribution of the point heat sources near O and are called the monopole, dipole, quadrupole, $\cdots$ moments of the point source distribution. Denoting these $M, \vec{D}, \overset{=}{Q}, \cdots$ our formula can be written

$$4\pi k T\left(P\right) = \frac{1}{r} M + \frac{1}{r^3} \vec{r} \cdot \vec{D} + \frac{1}{r^5} \vec{r} \vec{r} : \overset{=}{Q} + \cdots$$

where $\vec{r}$ denotes the position of P with respect to O and $M, \vec{D}, \overset{=}{Q}, \cdots$ do not depend on the field point $\vec{r}$.

If, instead of a heat conduction problem, we have an electrostatic problem, where the electrical potential through out space is created by a system of point charges near the origin, we find for the electrostatic potential at P, via Coulomb's law or, as above, via $\nabla^2 \phi = 0$:

$$4\pi\epsilon_0 \phi(P) = \frac{1}{r}M_O + \frac{1}{r^3}\vec{r} \cdot \vec{D}_O + \frac{1}{r^5}\vec{r}\vec{r} : \overset{\Rightarrow}{Q}_O + \cdots$$

where $M_O, \vec{D}_O, \overset{\Rightarrow}{Q}_O, \cdots$ denote the monopole, dipole, quadrupole, $\cdots$ moments of the charge distribution about O.

We might then have a second charge distribution in the neighbourhood of P and ask for the potential energy of this second set of charges due to the electrical potential created by the first set of charges, i.e., the potential energy of one molecule due to the electrical potential created by another molecule. The result is

$$PE = \sum \phi(P_i) Q_i = \sum \phi(\vec{r} + \vec{r}_i) Q_i$$

corresponding to the sketch

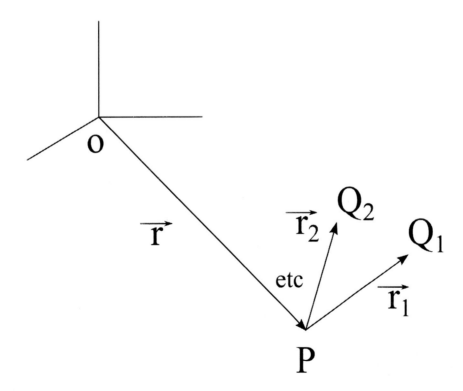

where $\vec{r_i'}$ denotes the position with respect to P of the ith charge in the second charge distribution.

The potential energy can be written in terms of the moments of the two charge distributions by expanding $\phi(P_i)$ via

$$\phi(\vec{r} + \vec{r_i'}) = \phi(\vec{r}) + \vec{r_i'} \cdot \nabla\phi(\vec{r}) + \frac{1}{2}\vec{r_i'}\,\vec{r_i'} : \nabla\nabla\phi(\vec{r}) + \cdots$$

Hence we have

$$PE = \sum Q_i \left\{ \phi(\vec{r}) + \vec{r_i'} \cdot \nabla\phi(\vec{r}) + \frac{1}{2}\vec{r_i'}\,\vec{r_i'} : \nabla\nabla\phi(\vec{r}) + \cdots \right\}$$

and using

$$\nabla\left(\frac{1}{r}\right) = -\frac{\vec{r}}{r^3}$$

$$\nabla\nabla\left(\frac{1}{r}\right) = -\nabla\left(\frac{\vec{r}}{r^3}\right) = \frac{3\vec{r}\,\vec{r}}{r^5} - \frac{1}{r^3}\overset{\Rightarrow}{I}$$

and

$$\nabla\left(\frac{\vec{r}}{r^3}\right) = -\frac{3\vec{r}\,\vec{r}}{r^5} + \frac{1}{r^3}\overset{\Rightarrow}{I}$$

we find, to terms of order $\dfrac{1}{r^3}$,

$$
\begin{aligned}
4\pi\epsilon_0 PE \;=\; & M_P\left\{ \frac{1}{r}M_O + \frac{1}{r^3}\vec{r} \cdot \vec{D}_O + \frac{1}{r^5}\vec{r}\,\vec{r} : \overset{\Rightarrow}{Q}_O \right\} \\
& + \vec{D}_P \cdot \left\{ -\frac{\vec{r}}{r^3}M_O + \left(-\frac{3\vec{r}\,\vec{r}}{r^5} + \frac{1}{r^3}\overset{\Rightarrow}{I} \right) \cdot \vec{D}_O \right\} \\
& + \sum Q_i \frac{1}{2}\vec{r}_i\,\vec{r}_i : \left\{ \frac{3\vec{r}\,\vec{r}}{r^5} - \frac{1}{r^3}\overset{\Rightarrow}{I} \right\} M_O
\end{aligned}
$$

Then, using

$$\operatorname{tr}\left\{ \frac{3\vec{r}\,\vec{r}}{r^5} - \frac{1}{r^3}\overset{\Rightarrow}{I} \right\} = \overset{\Rightarrow}{I} : \left\{ \frac{3\vec{r}\,\vec{r}}{r^5} - \frac{1}{r^3}\overset{\Rightarrow}{I} \right\} = 0$$

and

$$\operatorname{tr} \vec{\vec{Q}}_P = 0$$

where

$$\vec{\vec{Q}}_P = \sum Q_i \left\{ \frac{3}{2} \vec{r}_i \vec{r}_i - \frac{1}{2} \vec{r}_i \cdot \vec{r}_i \vec{\vec{I}} \right\}$$

we have

$$4\pi\epsilon_0 PE = \frac{M_O M_P}{r} + \frac{\vec{r}}{r^3} \cdot \left\{ \vec{D}_O M_P - \vec{D}_P M_O \right\} + \left(\frac{1}{r^3} \vec{\vec{I}} - \frac{1}{r^5} \vec{r} \vec{r} \right) : \vec{D}_O \vec{D}_P$$
$$+ \frac{1}{r^5} \vec{r} \vec{r} : \left\{ \vec{\vec{Q}}_O M_P + \vec{\vec{Q}}_P M_O \right\}$$

where $\vec{r} = \overrightarrow{OP}$.

We see that if the two charge distributions are neutral; i.e., $M_O = 0 = M_P$, the leading term is the dipole-dipole term, falling off as $\dfrac{1}{r^3}$.

If the point charges were replaced by point masses and ε_0 by G, and if O and P were taken to lie at the centers of mass so that $\vec{D}_O = \vec{0} = \vec{D}_P$, then the first correction to the leading term $\dfrac{M_O M_P}{r}$ would be the quadrupole term, where the quadrupole moment of a mass distribution can be expressed in terms of its inertia tensor.

13.2   Home Problems

1. A source of heat is distributed over the surface of a sphere of radius R according to

$$\sigma = \sigma_0 + \sigma_1 \cos \theta$$

Your job is to solve $\nabla^2 T = 0$ for $r > R$ and $r < R$ where $T \to 0$ as $r \to \infty$. The thermal conductivities inside and outside may differ.

You have

$$\nabla^2 = \frac{1}{r^2} \frac{\partial}{\partial r} \left(r^2 \frac{\partial}{\partial r} \right) + \frac{1}{r^2 \sin\theta} \frac{\partial}{\partial\theta} \left(\sin\theta \frac{\partial}{\partial\theta} \right)$$

and to solve your problem write, inside and outside,

$$T = T_0 (r) + T_1 (r) \cos\theta$$

and derive

$$\frac{1}{r^2} \frac{d}{dr} \left(r^2 \frac{dT_0}{dr} \right) = 0$$

and

$$\frac{1}{r^2} \frac{d}{dr} \left(r^2 \frac{dT_1}{dr} \right) - \frac{2}{r^2} T_1 = 0$$

whereupon

$$T_0 = A_0 + \frac{B_0}{r}$$

and

$$T_1 = A_1 r + \frac{B_1}{r^2}$$

What if the inside of the sphere is non conducting?

2. A heat source, denoted Q, is distributed over a region V_0. The temperature at a point $\vec{r}$ is given by

$$T(\vec{r}) = \frac{1}{4\pi k} \iiint_{V_0} \frac{Q(\vec{r}_0)}{|\vec{r} - \vec{r}_0|} \, dV_0$$

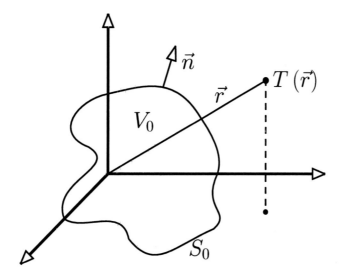

show that if $|\vec{r}| >> |\vec{r}_0|$ we have

$$T(\vec{r}) = \frac{1}{4\pi k} \left\{ \frac{1}{r} \iiint\limits_{V_0} Q(\vec{r}_0)\, dV_0 + \frac{\vec{r}}{r^3} \cdot \iiint\limits_{V_0} \vec{r}_0\, Q(\vec{r}_0)\, dV_0 + \cdots \right\}$$

Now if $\vec{q}$ denotes $-k\nabla T$ we have

$$\nabla \cdot \vec{q} = Q$$

and hence we see that

$$\iint\limits_{S_0} \vec{n} \cdot \vec{q}\, dA_0 = \iiint\limits_{V_0} Q(\vec{r}_0)\, dV_0$$

This suggests that perhaps we can replace all the moments of Q over V_0 by moments of $\vec{n} \cdot \vec{q}$ over S_0 in our equation for $T(\vec{r})$.

Show that

$$\vec{r}\ \nabla \cdot \vec{q} = \nabla \cdot (\vec{q}\,\vec{r}) - \vec{q}$$

and hence derive

$$\iiint\limits_{V_0} \vec{r}_0\, Q\, dV_0 = \iint\limits_{S_0} \vec{r}_0\, \vec{n} \cdot \vec{q}\, dA_0 - \iiint\limits_{V_0} \vec{q}\, dV_0$$

and conclude that we cannot replace first moments of Q over V_0 by first moments of $\vec{n} \cdot \vec{q}$ over S_0.

However, we have

$$\iiint\limits_{V_0} \vec{q}\, dV_0 = -k \iiint\limits_{V_0} \nabla T\, dV_0 = -k \iint\limits_{S_0} \vec{n}\, T\, dA_0$$

whence if T is constant on S_0 we have

$$\iiint\limits_{V_0} \vec{q}\, dV_0 = \vec{0}$$

Turn to second moments and evaluate

$$\iiint\limits_{V_0} \vec{r}_0\, \vec{r}_0\, Q\, dV_0 - \iint\limits_{S_0} \vec{n} \cdot \vec{q}\, \vec{r}_0\, \vec{r}_0\, dA_0$$

3. Derive a formula for the potential energy due to systems of charges near O, P and Q.

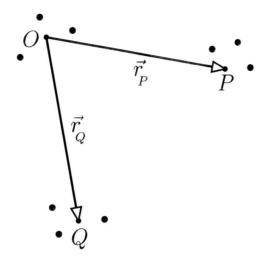

where $|\vec{r}_P|, |\vec{r}_Q|$ and $|\vec{r}_P - \vec{r}_Q|$ are all much greater than the distances of the charges from O, P and Q.

Assume the monopole moments are all zero, and account only for the dipole moments.

4. The temperature at $\vec{r}$ due to a point source at $\vec{r}_0$ is

$$T\left(\vec{r}\right) = \frac{\dfrac{Q}{4\pi k}}{|\vec{r} - \vec{r}_0|}$$

Assume $\vec{r}_0 = z_0 \vec{k}$, $z_0 > 0$ and show that the temperature at the points (R, θ, ϕ) is

$$T\left(R, \theta, \phi\right) = \frac{\dfrac{Q}{4\pi k}}{\sqrt{z_0^2 - 2z_0 R \cos\theta + R^2}}$$

Assume $z_0 >> R$ and write out the first three terms.

Setting this aside, find the temperature at $\vec{r}$ if the temperature, $T\left(R, \theta, \phi\right)$, is specified on the surface of a sphere of radius R, $r > R$ and $T \to 0$ as $r \to \infty$.

Now we have a point source of heat at $z = z_0$, $x = 0 = y$ and we wish to derive a formula for $T\left(\vec{r}\right)$ due to this point source, given that a sphere of radius R is centered at the origin and given that its surface temperature is held constant at a temperature $T_S > 0$.

The result is the sum of two contributions, that of the point source in the absence of the sphere and that of a sphere whose surface temperature is

$$T_S = \frac{\frac{Q}{4\pi k}}{\sqrt{z_0^2 - 2z_0 R \cos\theta + R^2}}$$

in the absence of the point source.

How should we think about finding the temperature at $\vec{r}$ due to two spheres of radius R, one centered at $\vec{r}_1$ having surface temperature T_1, the other centered at $\vec{r}_2$ having surface temperature T_2, T_1 and T_2 constants?

5. Suppose our charge distribution is composed of two charges, q_1 at $\vec{r}_1$ and q_2 at $\vec{r}_2$, where $q_1 = q = -q_2$ and $\vec{r}_1 - \vec{r}_2 = d\vec{u}$ and where $\vec{u}$ is a vector of unit length in the direction $\vec{r}_1 - \vec{r}_2$.

Derive the monopole, dipole and quadrupole moments of this charge distribution. What is the limiting form of each of these moments as $d \to 0$ holding qd fixed.

6. The solute concentration at a point $\vec{r}$ due to point sources at $\vec{r}_1, \vec{r}_2, \ldots$, near the origin is

$$4\pi Dc(\vec{r}) = \sum \frac{m_i}{|\vec{r} - \vec{r}_i|}$$

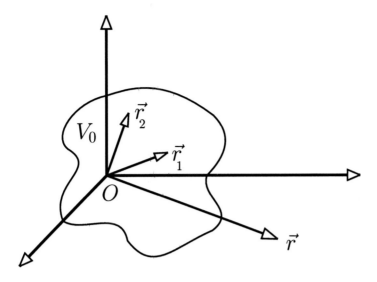

which we can write

$$4\pi D\, c\,(\vec{r}) = \iiint\limits_{V_0} \frac{\rho\,(\vec{r}_0)}{|\vec{r} - \vec{r}_0|}\, dV_0$$

if the discrete sources are replaced by a continuous source density ρ.

If $\vec{r}$ lies inside V_0 we might wonder about the integral on the right hand side as $\vec{r}_0$ passes through $\vec{r}$. Suppose $\vec{r} = 0$ and write

$$\iiint\limits_{V_0} = \iiint\limits_{V_0 - V_\varepsilon} + \iiint\limits_{V_\varepsilon}$$

where V_ε is the volume of a sphere of radius ε centered on O.

Show that

$$\iiint\limits_{V_\varepsilon} \frac{\rho\,(r_0)}{|\vec{r}_0|}\, dV_0$$

is not singular if ρ is bounded.

Now for $\vec{r}$ outside V_0 derive

$$4\pi D c(\vec{r}) = \frac{1}{r} \iiint_{V_0} \rho(\vec{r}_0) \; dV_0 + \frac{1}{r^2}\frac{\vec{r}}{r} \cdot \iiint_{V_0} \vec{r}_0 \rho(\vec{r}_0) \; dV_0$$

$$+ \frac{1}{r^3}\left\{\frac{3}{2}\frac{\vec{r}}{r}\frac{\vec{r}}{r} - \frac{1}{2}\vec{\vec{I}}\right\} : \iiint_{V_0} \vec{r}_0 \vec{r}_0 \rho(\vec{r}_0) \; dV_0 + \text{etc.}$$

Suppose the solute source is distributed over a surface S_0, its density being denoted by σ. Then decide

$$4\pi D c(\vec{r}) = \iint_{S_0} \frac{\sigma(\vec{r}_0)}{|\vec{r} - \vec{r}_0|} \; dA_0$$

If $\vec{r}$ is a point of S_0 does the right hand side present a technical difficulty?

Suppose S_0 is the surface of a sphere of radius R_0 centered at the point $\vec{r}*$ and define

$$G(\vec{r}, \vec{r}_0) = \frac{1}{|\vec{r} - \vec{r}_0|}, \quad \vec{r}_0 \in S_0$$

Then expand G about $\vec{r}*$ via

$$G(\vec{r}, \vec{r}_0) = G(\vec{r}, \vec{r}*) + (\vec{r}_0 - \vec{r}*) \cdot \nabla_0 G(\vec{r}, \vec{r}*)$$

$$+ \frac{1}{2}(\vec{r}_0 - \vec{r}*)(\vec{r}_0 - \vec{r}*) : \nabla_0\nabla_0 G(\vec{r}, \vec{r}*) + \cdots$$

where $\vec{r}_0 - \vec{r}* = R_0 \vec{n}(\vec{r}_0)$ and $\vec{n}$ is the unit normal to the sphere at $\vec{r}_0$.

Then we can write

$$4\pi D\, c\,(\vec{r}) = \iint\limits_{S_0} \sigma\,(\vec{r}_0)\; dA_0\; G\,(\vec{r},\,\vec{r}\,*)$$

$$+ \iint\limits_{S_0} \sigma\,(\vec{r}_0)\,(\vec{r}_0 - \vec{r}\,*)\; dA_0 \cdot \nabla_0\, G\,(\vec{r},\,\vec{r}\,*)$$

$$+ \iint\limits_{S_0} \sigma\,(\vec{r}_0)\,(\vec{r}_0 - \vec{r}\,*)\left(\vec{r}_0 - \vec{r}\,*\right) dA_0 : \nabla_0\nabla_0\, G\,(\vec{r},\,\vec{r}\,*)$$

$$+ \cdots$$

Suppose the interior of the sphere is not permeable to solute. How must the foregoing be corrected?

7. The temperature T at a point P due to a heat source distributed over a volume V_0 is

$$T\,(P) = \frac{1}{4\pi k}\iint\limits_{V_0} \frac{\rho\,(\vec{r}_0)}{|\vec{r} - \vec{r}_0|}\, dV_0$$

Assume ρ is constant inside a sphere of radius R_0 centered on the origin and zero outside. Derive

$$T\,(\text{origin}) = \frac{\rho}{k}\frac{1}{2}\, R_0^2$$

Check this by solving

$$\frac{1}{r^2}\frac{d}{dr}\left(r^2 \frac{dT}{dr}\right) = \begin{cases} 0, & r > R_0 \\[2mm] -\dfrac{\rho}{k}, & r < R_0 \end{cases}$$

Lecture 14

One Dimensional Diffusion in Bounded Domains

14.1   Introduction

We begin with a simple problem, solute diffusion in one dimension. The diffusion takes place in a solvent layer separating two solute reservoirs where we control what is going on in the reservoirs.

The differential operator ∇^2 is then $\dfrac{\partial^2}{\partial x^2}$ and our problem is to solve

$$\frac{\partial c}{\partial t} = D \frac{\partial^2 c}{\partial x^2} \ , \ \ 0 < x < L$$

to determine $c\,(t > 0)$ whenever $c\,(t = 0)$ is assigned.

In our first problem the solvent layer is in perfect contact with large reservoirs maintained solute free. Hence, in scaled variables $\dfrac{x}{L} \Rightarrow x, \dfrac{Dt}{L^2} \Rightarrow t$ we have

$$\frac{\partial c}{\partial t} = \frac{\partial^2 c}{\partial x^2} \ , \ \ 0 < x < 1$$

and

$$c\,(x = 0) = 0 = c\,(x = 1)$$

where $c(t = 0)$ is assigned and it is the only source of solute in the problem.

We are going to make several assumptions as we go along, one being that our functions are smooth enough that integration by parts formulas can be used.

Earlier we solved

$$\frac{d\underline{x}}{dt} = A\underline{x} \ , \ \underline{x}(t = 0) \ \text{specified}$$

by expanding $\underline{x}(t)$ in the eigenvectors of A and solving for the time dependent coefficients in the expansion. In fact, in one of our problems A was derived from difference approximations to $\frac{\partial^2 c}{\partial x^2}$.

To maintain continuity with our earlier work, we are going to introduce the eigenvalue problem for the differential operator $\frac{d^2}{dx^2}$, viz.

$$\frac{d^2}{dx^2}\psi = -\lambda^2\psi \ , \ 0 < x < 1$$

Solutions, ψ, to this problem are called eigenfunctions, and we will call the corresponding values of λ^2 the eigenvalues.

This problem has been stated too generally and we need to introduce restrictions on its solutions. At this point we can only say that

$$\psi = A \cos \lambda x + B \sin \lambda x$$

where A, B and λ are unknown. The problem is homogeneous and will remain homogeneous as we introduce additional conditions, hence if ψ is a solution, so too any multiple of ψ.

What we need to do is to define the domain of the differential operator $\frac{d^2}{dx^2}$ in a way that is specific to the problem at hand. Here, as our solutions satisfy $c(x = 0) = 0 = c(x = 1)$ it is natural to require $\psi(x = 0) = 0 = \psi(x = 1)$. This is the best choice and we will see why this is so as we go along. We might simply argue that if we make each term in a sum vanish at $x = 0$, then the sum vanishes at $x = 0$.

So, the eigenvalue problem is

$$\frac{d^2\psi}{dx^2} = -\lambda^2\psi \ , \ \ 0 < x < 1$$

and

$$\psi\left(x = 0\right) = 0 = \psi\left(x = 1\right)$$

and its solutions are

$$\left.\begin{array}{rcl}\psi &=& \sin n\pi x \\ \lambda^2 &=& n^2\pi^2\end{array}\right\} n = 1, 2, \dots$$

where $A = 0$ and we set $B = 1$.

The eigenvalues are determined by the condition at $x = 1$, viz.,

$$\sin \lambda = 0$$

Thus, $\lambda = n\pi$, $n = 0, \pm 1, \pm 2, \dots$. The value $n = 0$ leads to $\psi = 0$ whereas the values $n = -1, -2, \dots$ lead to eigenfunctions that are multiples of those corresponding to $n = 1, 2, \dots$ Thus, we find an infinite set of eigenfunctions, $\psi = \sin n\pi x$, corresponding to an infinite set of eigenvalues.

$$\lambda^2 = n^2\pi^2 \ , \ \ n = 1, 2, \dots$$

The main question that we would like to answer about an infinite set of functions, such as these eigenfunctions, is whether or not it can be used as a basis for the expansion of a fairly arbitrary function, viz., the function $c\left(t = 0\right)$. This is what the theory of Fourier series is about and a good elementary account of this theory can be found in Weinberger's book (*A First Course in Partial Differential Equations with Complex Variables and Transform Methods*). The expansion we require is an infinite series, not a finite sum, and the question is not easy to answer. Still there are conditions on the arbitrary function and on the basis functions sufficient for the series representation of the function to mean something. On their part, the set of basis functions $\sin n\pi x$, $n = 1, 2, \ldots$ satisfy the conditions required, viz., first, they are a set of orthogonal functions. We can establish this directly by using the functions themselves or indirectly by using the eigenvalue problem defining the functions.

14.2 Orthogonality of the Functions ψ_n

We show that orthogonality is implied by the differential operator $\dfrac{d^2}{dx^2}$ and the boundary conditions and we go on and draw some conclusions about the eigenvalues. To do this we use two integration formulas. If the functions ϕ and ψ are smooth enough so that we can integrate by parts, we get two formulas

$$\int_0^1 \phi \frac{d^2\psi}{dx^2} dx = \left[\phi \frac{d\psi}{dx}\right]_0^1 - \int_0^1 \frac{d\phi}{dx} \frac{d\psi}{dx} dx$$

and

$$\int_0^1 \phi \frac{d^2\psi}{dx^2} dx = \left[\phi \frac{d\psi}{dx} - \frac{d\phi}{dx} \psi\right]_0^1 + \int_0^1 \frac{d^2\phi}{dx^2} \psi dx$$

where

$$[\phi]_0^1 = \phi(x = 1) - \phi(x = 0)$$

and where these formulas hold for complex valued functions as well as for real valued functions.

Then as $\overline{\psi}$ and $\overline{\lambda^2}$ satisfy the eigenvalue problem for $\dfrac{d^2}{dx^2}$ whenever it is satisfied by ψ and λ^2, we let ψ be an eigenfunction and ϕ be its complex conjugate in the second formula and determine that $\overline{\lambda^2} = \lambda^2$, and hence that λ^2 is real. It follows that if ψ is an eigenfunction corresponding to the eigenvalue λ^2 so too is its complex conjugate and its real and imaginary parts.

Again if ψ and ϕ denote an eigenfunction and its complex conjugate, the first formula tells us that:

$$-\lambda^2 \int_0^1 |\psi|^2 \, dx = \left[\overline{\psi} \frac{d\psi}{dx} \right]_0^1 - \int_0^1 \left| \frac{d\psi}{dx} \right|^2 dx = -\int_0^1 \left| \frac{d\psi}{dx} \right|^2 dx$$

and hence that the corresponding eigenvalue, λ^2, is strictly positive. If λ^2 were zero, this formula would require that ψ be a constant and as $\psi(x = 0) = 0 = \psi(x = 1)$ this constant would then be zero.

Returning to the second formula and taking ϕ and ψ to be eigenfunctions corresponding to different eigenvalues, we discover that

$$\int_0^1 \overline{\phi}\psi \, dx = 0$$

and we call ϕ and ψ orthogonal functions. We can introduce the inner product

$$\langle \phi, \psi \rangle = \int_0^1 \overline{\phi}\psi \, dx$$

and restate this as

$$\langle \phi, \psi \rangle = 0.$$

In terms of this inner product the two integration by parts formulas can be rewritten as

$$\left\langle \phi, \frac{d^2}{dx^2} \psi \right\rangle = \left[\overline{\phi} \frac{d\psi}{dx} \right]_0^1 - \int_0^1 \frac{d\overline{\phi}}{dx} \frac{d\psi}{dx} dx$$

and

$$\left\langle \phi, \frac{d^2}{dx^2} \psi \right\rangle = \left[\overline{\phi} \frac{d\psi}{dx} - \frac{d\overline{\phi}}{dx} \psi \right]_0^1 + \left\langle \frac{d^2}{dx^2} \phi, \psi \right\rangle.$$

What we have established then is just what we can show to be true by direct calculation:

the eigenvalues $n^2\pi^2$, $n = 1, 2, \ldots$, are real and positive and the eigenfunctions, $\sin n\pi x$, $n = 1, 2, \ldots$, are orthogonal, viz.,

$$\int_0^1 \sin m\pi x \, \sin n\pi x \, dx = 0 \,, \; m \neq n$$

What is important is not the confirmation of results we already have, but the possibility of obtaining new results. To see this we observe the important role played by the boundary conditions in eliminating the term $\left[\phi\dfrac{d\psi}{dx}\right]_0^1$ in the first formula and the term $\left[\phi\dfrac{d\psi}{dx} - \dfrac{d\phi}{dx}\psi\right]_0^1$ in the second. Indeed if ϕ and ψ, and therefore their complex conjugates, satisfy the boundary conditions $\psi(x = 0) = 0$, $\dfrac{d\psi}{dx}(x = 1) = 0$ or the boundary conditions $\dfrac{d\psi}{dx}(x = 0) = 0$, $\psi(x = 1) = 0$ then the conclusions are again as above. So too if the boundary conditions are $\dfrac{d\psi}{dx}(x = 0) = 0$, $\dfrac{d\psi}{dx}(x = 1) = 0$, except that now we can conclude only that $\lambda^2 \geq 0$ as $\psi = \text{constant} \neq 0$ satisfies the boundary conditions. This also obtains for periodic boundary conditions, where $\psi(x = 0) = \psi(x = 1)$ and $\dfrac{d\psi}{dx}(x = 0) = \dfrac{d\psi}{dx}(x = 1)$, as again the terms $\left[\phi\dfrac{d\psi}{dx}\right]_0^1$ and $\left[\phi\dfrac{d\psi}{dx} - \dfrac{d\phi}{dx}\psi\right]_0^1$ vanish and $\psi = \text{constant} \neq 0$ satisfies the boundary conditions.

Because the specification of a linear combination of ψ and $\dfrac{d\psi}{dx}$ at a boundary is of physical interest, we look also at the boundary conditions $\dfrac{d\psi}{dx}(x = 0) + \beta_0\psi(x = 0) = 0$ and $\dfrac{d\psi}{dx}(x = 1) + \beta_1\psi(x = 1) = 0$ where the constants β_0 and β_1 take real values. Then because $\left[\phi\dfrac{d\psi}{dx} - \dfrac{d\phi}{dx}\psi\right]_0^1 = 0$ all conclusions drawn from the second integration by parts formula are as above whereas the first now tells us that

$$-\lambda^2 \int_0^1 |\psi|^2 \, dx = -\beta_1 |\psi(x = 1)|^2 + \beta_0 |\psi(x = 0)|^2 - \int_0^1 \left|\frac{d\psi}{dx}\right|^2 dx.$$

If β_0 and β_1 are not both zero then $\beta_1 \geq 0$, $\beta_0 \leq 0$ are sufficient, but, as we shall see, not necessary, that $\lambda^2 > 0$.

What we conclude then is that the operator $\dfrac{d^2}{dx^2}$, restricted to a variety of domains by a variety of boundary conditions of differing physical interpretation, leads via the solution of its eigenvalue problem to a variety of sets of orthogonal functions. Denote one such set $\psi_1, \psi_2, \ldots$ and require that $\int_0^1 \overline{\psi_i}\psi_i dx = 1$ so that $\langle \psi_i, \psi_j \rangle = \delta_{ij}$. Then sums of terms like $e^{-\lambda_i^2 t}\psi_i(x)$ satisfy the diffusion equation and the corresponding homogeneous boundary conditions. It remains only to determine the weight of each such term in a series solution to a diffusion problem. And this must be decided by the assigned solute distribution at the initial instant, $t = 0$. Thus we need to learn how to determine the coefficients in an expansion such as

$$c(t = 0) = c_1 \psi_1(x) + c_2 \psi_2(x) + \cdots$$

where $c(t = 0)$ is a fairly arbitrary function of $x, 0 \le x \le 1$.

14.3 Least Mean Square Error

Writing $c(t = 0)$ as $f(x)$ we find the error in an n term approximation to $f(x)$ to be $f(x) - S_n(x)$ where $S_n(x) = \sum_{i=1}^n c_i \psi_i(x)$. The mean square error is then

$$\int_0^1 \left\{ \overline{f(x) - S_n(x)} \right\} \left\{ f(x) - S_n(x) \right\} dx$$

which, on using $\int_0^1 \overline{\psi_i}\psi_j dx = \delta_{ij}$, can be rewritten

$$\int_0^1 |f|^2 dx + \sum_{i=1}^n \left| c_i - \int_0^1 \overline{\psi_i} f dx \right|^2 - \sum_{i=1}^n \left| \int_0^1 \overline{\psi_i} f dx \right|^2$$

where only the second term, $\sum_{i=1}^n \left| c_i - \int_0^1 \overline{\psi_i} f dx \right|^2$, depends on the values of the coefficients $c_1, c_2, \ldots, c_n$. Because the second term is not negative we can make the mean square error least by making this term zero. To do this we assign the coefficients in the expansion the values

$$c_i = \int_0^1 \overline{\psi_i} f dx = \langle \psi_i, f \rangle$$

whence our approximation to $f(x)$ is

$$\sum_{i=1}^{n} \langle \psi_i, f \rangle \, \psi_i(x)$$

and its mean square error is

$$\int_{0}^{1} |f|^2 \, dx - \sum_{i=1}^{n} |c_i|^2 \, .$$

What is going on is indicated in the picture below where the best approximation of f using only ψ_1 is shown

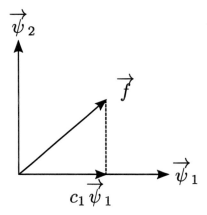

and where we see that the length of $\vec{f} - c_1 \vec{\psi}_1$ is least if $\vec{f} - c_1 \vec{\psi}_1$ is $\perp$ to $\vec{\psi}_1$ or if $c_1 = \left\langle \vec{\psi}_1, \vec{f} \right\rangle$. Indeed on requiring $f - \sum_{i=1}^{n} c_i \psi_i$ to be $\perp$ to $\psi_1, \psi_2, \cdots, \psi_n$ we get immediately $c_i = \langle \psi_i, f \rangle$. And we see that the values of the coefficients $c_1, c_2, \cdots, c_n$ do not depend on n, remaining fixed as n increases once determined for some value of n.

As the mean square error is non-negative we have

$$\sum_{i=1}^{n} |c_i|^2 \le \int_{0}^{1} |f|^2 \, dx$$

and for $n \longrightarrow \infty$

$$\sum_{i=1}^{\infty} |c_i|^2 \le \int_{0}^{1} |f|^2 \, dx.$$

This tells us that the series $\sum_{i=1}^{\infty} |c_i|^2$ converges and therefore that $|c_i|^2 \longrightarrow 0$ as $i \longrightarrow \infty$. The coefficients $c_i = \int_0^1 \overline{\psi_i} f \, dx = \langle \psi_i, f \rangle$ are called the Fourier coefficients of f and the set of orthogonal functions $\psi_1, \psi_2, \cdots$ is said to be complete if for every function f in a class of functions of interest we have

$$\sum_{i=1}^{\infty} |c_i|^2 = \int_0^1 |f|^2 \, dx.$$

Then $f(x)$ is approximated by $S_n(x) = \sum_{i=1}^{n} \langle \psi_i, f \rangle \psi_i$ to a mean square error that vanishes as $n \longrightarrow \infty$ and the sequence $S_n(x)$ is said to converge to $f(x)$ in the mean or in the norm. This does not imply that the sequence $S_n(x)$ converges to $f(x)$ for an arbitrary value of x on the interval $[0, 1]$. But it does imply that the sequence $S_n(x)$ converges to $f(x)$ pointwise almost everywhere. A discussion of pointwise and norm convergence is given in Weinberger's book in terms of conditions on the functions being approximated.

We will assume that the sets of orthogonal functions that we generate by solving the eigenvalue problem for $\dfrac{d^2}{dx^2}$, and indeed for ∇^2 itself, are complete vis-a-vis functions of interest in physical problems. But whereas completeness implies only convergence in norm, we will go on and write

$$f(x) = \sum_{i=1}^{\infty} \langle \psi_i, f \rangle \psi_i(x)$$

mindful of the warning that this might not be true for all values of x. Indeed the series obtained by differentiating $\sum_{i=1}^{\infty} \langle \psi_i, f \rangle \psi_i$ termwise might not converge for any value of x and therefore not have a meaning in any ordinary sense.

An infinite series is nearly as useful as a finite sum if the function $f(x)$ is smooth enough and satisfies the same boundary conditions as do the functions $\psi_1(x), \psi_2(x), \cdots$. For suppose $\dfrac{d^2 f}{dx^2}$ has the Fourier series $\sum_{i=1}^{\infty} d_i \psi_i(x)$ where $d_i = \left\langle \psi_i, \dfrac{d^2 f}{dx^2} \right\rangle$ and $d_i \longrightarrow 0$ as $i \longrightarrow \infty$. Then we have

$$d_i = \int_0^1 \overline{\psi_i} \frac{d^2 f}{dx^2} dx = \left[\overline{\psi_i} \frac{df}{dx} - \frac{d\overline{\psi_i}}{dx} f \right]_0^1 + \int_0^1 \frac{d^2 \overline{\psi_i}}{dx^2} f \, dx$$

or

$$d_i = -\lambda_i^2 \int_0^1 \overline{\psi}_i f \, dx = -\lambda_i^2 \langle \psi_i, f \rangle$$

and hence

$$c_i = \langle \psi_i, f \rangle = -\frac{d_i}{\lambda_i^2}$$

where $\sum_{i=1}^{\infty} c_i \psi_i(x)$ is the Fourier series corresponding to f. Assuming $\lambda_i^2 \propto i^2$, we see that $c_i \longrightarrow 0$ faster than $\frac{1}{i^2}$ as $i \longrightarrow \infty$ and hence that the Fourier series for f may then be a useful representation.

14.4   Example Problems

With the foregoing as background we investigate a sequence of problems where diffusion takes place in a solvent layer separating two reservoirs. If the reservoirs are very large, well mixed and solute free we put $c(x = 0) = 0 = c(x = 1)$. If the reservoirs are impermeable to solute, we put $\frac{\partial c}{\partial x}(x = 0) = 0 = \frac{\partial c}{\partial x}(x = 1)$. If the solute diffusing to the right hand boundary is dissipated by a first order reaction taking place there we put $-\frac{D}{l}\frac{\partial c}{\partial x}(x = 1) = kc(x = 1)$ or $\frac{\partial c}{\partial x}(x = 1) + \beta c(x = 1) = 0$ where $\beta = \frac{kl}{D}$. As we have written it, k is positive for an ordinary decomposition of solute, but negative for an autocatalytic reaction wherein, at the wall, solute catalyzes the production of more solute.

We can also assume that the solute diffusing to the right hand boundary accumulates in a finite, well mixed reservoir whose composition is in equilibrium with the composition at the right-hand edge of the film. Then we put

$$-\frac{D}{l}\frac{\partial c}{\partial x}(x = 1) = \frac{l_1 m}{l^2/D}\frac{\partial c}{\partial t}(x = 1)$$

or

$$\frac{\partial c}{\partial t}(x=1) + \alpha \frac{\partial c}{\partial t}(x=1) = 0$$

where $\alpha = \frac{l_1}{l} m$, l_1 denoting the volume of the reservoir divided by the cross sectional area of the diffusion layer, m denoting the equilibrium distribution ratio. This problem comes up in separations by chromatography and its solutions can be used to explain why retention times depend on initial solute distributions even if there is no competition for the adsorbent.

In each of these problems, differing only in the conditions specified at the boundary of the diffusion layer, we have

$$\frac{\partial c}{\partial t} = \frac{\partial^2 c}{\partial x^2} \ , \ \ 0 < x < 1$$

where $c(t=0)$ is an asssigned function of x, $0 \le x \le 1$. To solve our problem, we expand the solution we seek in the eigenfunctions of $\frac{d^2}{dx^2}$ and so we write

$$c(x,t) = \sum_{i=1}^{\infty} c_i(t) \psi_i(x)$$

where the coefficients $c_i(t) = \langle \psi_i, c \rangle$ remain to be determined. To find the equations satisfied by the c_i we multiply the diffusion equation by $\overline{\psi}_i$ and integrate over the domain, viz.,

$$\int_0^1 \overline{\psi}_i \frac{\partial c}{\partial t} dx = \int_0^1 \overline{\psi}_i \frac{\partial^2 c}{\partial x^2} dx$$

and observe, by our second integration by parts formula, that this is

$$\frac{d}{dt} \langle \psi_i, c \rangle = \left[\overline{\psi}_i \frac{\partial c}{\partial x} - \frac{d\overline{\psi}_i}{dx} c \right]_0^1 - \lambda_i^2 \langle \psi_i, c \rangle .$$

Now we have what appears to be a technical difficulty: we do not know both c and $\frac{\partial c}{\partial x}$ at both $x = 0$ and $x = 1$.

If the first term on the right hand side vanishes, and we set the boundary conditions in the eigenvalue problem to make this happen, if we can, this is simply

$$\frac{d}{dt} \langle \psi_i, c \rangle = -\lambda_i^2 \langle \psi_i, c \rangle$$

whence

$$\langle \psi_i, c \rangle = e^{-\lambda_i^2 t} \langle \psi_i, c(t = 0) \rangle$$

and our solution is

$$c(x, t) = \sum_{i=1}^{\infty} \langle \psi_i, c(t = 0) \rangle e^{-\lambda_i^2 t} \psi_i(x).$$

We order the eigenvalues so that $0 \le \lambda_1^2 \le \lambda_2^2 \le \cdots$.

This series is an increasingly useful representation of $c(x, t)$ as t increases. How useful it is for small values of t depends on what $c(t = 0)$ is. For very large values of t, $c(x, t)$ is approximately

$$\langle \psi_1, c(t = 0) \rangle e^{-\lambda_1^2 t} \psi_1(x)$$

if $\lambda_1^2 > 0$ and $\lambda_2^2 > \lambda_1^2$ or

$$\langle \psi_1, c(t = 0) \rangle \psi_1(x) + \langle \psi_2, c(t = 0) \rangle e^{-\lambda_2^2 t} \psi_2(x)$$

if $\lambda_1^2 = 0$, $\lambda_2^2 > 0$ and $\lambda_3^2 > \lambda_2^2$.

It is important to notice that we have the solution to our problem, even though it is an infinite series, and that we have not differentiated the series to obtain it.

We now look at the solutions to the eigenvalue problem corresponding to a variety of boundary conditions where in every case $\psi = A \cos \lambda x + B \sin \lambda x$ satisfies

$$\frac{\partial^2 \psi}{\partial x^2} + \lambda^2 \psi = 0.$$

Example (1): $c\,(x = 0) = 0 = c\,(x = 1)$

In the first problem a solute initially distributed over a solvent layer according to $c\,(t = 0)$, is lost to the adjacent reservoirs which maintain the edges of the diffusion layer solute free, i.e., $c\,(x = 0) = 0 = c\,(x = 1)$. Hence we choose the boundary conditions satisfied by ψ to be $\psi\,(x = 0) = 0 = \psi\,(x = 1)$ whereupon the term $\left[\overline{\psi}_i\dfrac{\partial c}{\partial x} - \dfrac{\partial\overline{\psi}_i}{\partial x}c\right]_0^1$ on the right hand side of the equation for $\langle\psi_i, c\rangle$ vanishes.

Then $\psi\,(x = 0) = 0$ implies $A = 0$ and $\psi\,(x = 1) = 0$ implies $\sin\lambda = 0$ whence the eigenfunctions and the corresponding eigenvalues are

$$\psi_i = \sqrt{2}\sin i\pi x\ ,\quad i = 1, 2, \cdots$$

and

$$\lambda_i^2 = i^2\pi^2\ ,\quad i = 1, 2, \cdots$$

Example (2): $\dfrac{\partial c}{\partial x}\,(x = 0) = 0,\ c\,(x = 1) = 0$

The solute sink at $x = 0$ in Example (1) is replaced by a barrier impermeable to solute, all else remaining the same.

To solve this problem we need a new set of eigenfunctions which we obtain by introducing new boundary conditions, viz.,

$$\frac{d\psi}{dx}\,(x = 0) = 0,\ \psi\,(x = 1) = 0$$

for this is what is required to make the term $\left[\overline{\psi}_i\dfrac{\partial c}{\partial x} - c\dfrac{d\overline{\psi}_i}{dx}\right]_0^1$ on the right hand side of the equation for $\langle\psi_i, c\rangle$ vanish.

Then $\dfrac{d\psi}{dx}\,(x = 0) = 0$ implies $B = 0$ and $\psi\,(x = 0) = 0$ implies $\cos\lambda = 0, A \neq 0$, whence we have

$$\psi_i = \sqrt{2} \cos \left(i - \frac{1}{2} \right) \pi x, \quad i = 1, 2, \cdots$$

and

$$\lambda_i^2 = \left(i - \frac{1}{2} \right)^2 \pi^2, \quad i = 1, 2, \cdots$$

The loss of solute is slowed by imposing the barrier on the left hand side of the layer. For long times, after the details of the initial solute distribution are forgotten, the second layer is effectively twice as thick as the first and to see this we need only observe that λ_1^2 in the second problem is one fourth λ_1^2 in the first.

Example (3): $\dfrac{\partial c}{\partial x} (x = 0) = 0, \ \dfrac{\partial c}{\partial x} (x = 1) = 0$

Here both edges of the solute layer are isolated from the bounding reservoirs and the initial solute distribution in the film is simply rearranged by diffusion, no solute being lost. The boundary conditions on ψ are now

$$\frac{\partial \psi}{\partial x} (x = 0) = 0, \ \frac{\partial \psi}{\partial x} (x = 1) = 0$$

as this leads to

$$\left[\overline{\psi}_i \frac{\partial c}{\partial x} - \frac{d\overline{\psi}_i}{dx} c \right]_0^1 = 0$$

Because

$$\frac{d\psi}{dx} = -A\lambda \sin \lambda + B\lambda \cos \lambda x$$

we see that $\dfrac{d\psi}{dx} (x = 0) = 0$ implies $B\lambda = 0$. Thus we have either

$$\lambda = 0, \ \psi = A$$

or

$$B = 0, \quad \psi = A \cos \lambda x, \quad A \neq 0$$

where $\dfrac{d\psi}{dx}(x = 1) = 0$ implies

$$A\lambda \sin \lambda = 0$$

and our eigenfunctions and eigenvalues are

$$\psi_i = \begin{cases} 1 & , \quad i = 1 \\ \sqrt{2} \cos (i - 1) \pi x & , \quad i = 2, \cdots \end{cases}$$

and

$$\lambda_i^2 = (i - 1)^2 \pi^2 \ , \quad i = 1, 2, \cdots .$$

The average solute concentration in the solvent layer is

$$c_{avg}(t) = \int_0^1 c(x, t)\, dx = \langle \psi_1, c \rangle$$

and in terms of this the solution can be written

$$c(x, t) = c_{avg}(t = 0) + \sum_{i=2}^{\infty} \langle \psi_i, c(t = 0) \rangle\, e^{-\lambda_i^2 t} \psi_i(x)$$

Due to $\langle \psi_1, \psi_i \rangle = 0, i = 2, \cdots$, we have

$$c_{avg}(t) = c_{avg}(t = 0).$$

Example (4): $\dfrac{\partial c}{\partial x}(x=0)=0, \ \dfrac{\partial c}{\partial x}(x=1)+\beta c(x=1)=0, \ \beta \geq 0$

Again the left hand edge of the solute layer is impermeable to solute but at the right hand edge solute is lost by a first order process. If $\beta = 0$ we get Example (3), if $\beta \longrightarrow \infty$ we get Example (2).

The boundary conditions are now

$$\frac{d\psi}{dx}(x=0)=0, \ \frac{\partial \psi}{\partial x}(x=1)+\beta \psi(x=1)=0$$

because this choice makes the term

$$\left[\overline{\psi}_i \frac{\partial c}{\partial x} - \frac{d\overline{\psi}_i}{dx}c\right]_0^1$$

vanish.

First we notice that if $\lambda = 0$, then ψ must be constant. But that constant must be zero if $\beta \neq 0$. The boundary condition $\dfrac{d\psi}{dx}(x=0)=0$ implies $B=0$ hence we have

$$\psi = A\cos \lambda x \ , \ \ A \neq 0$$

whereupon the $\lambda's$ are the solutions of

$$\lambda \sin \lambda - \beta \cos \lambda = 0$$

where if λ is a solution so too $-\lambda$, both implying the same λ^2 and ψ. Hence we write our equation

$$\frac{\lambda}{\beta} = \cot \lambda$$

and look for solutions $\lambda > 0$. These can be obtained graphically. The figure illustrates their dependence on β.

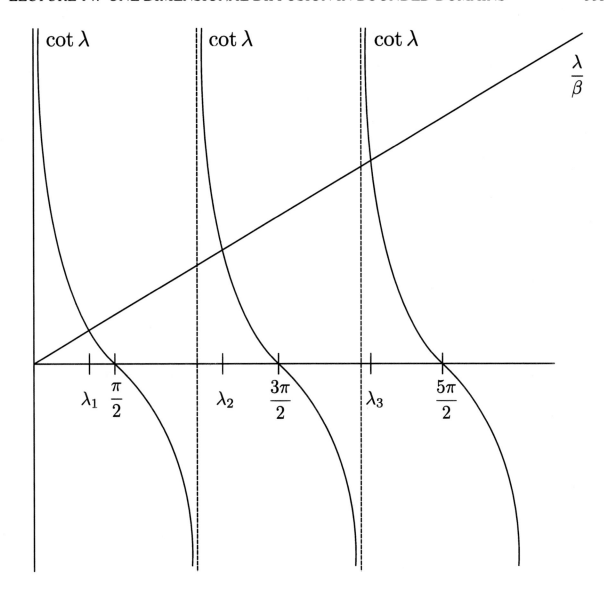

We see that $0 \leq \lambda_1 \leq \frac{1}{2}\pi$, $\pi \leq \lambda_2 \leq \frac{3}{2}\pi$ and indeed that $(i-1)\pi \leq \lambda_i \leq \left(i-1+\frac{1}{2}\right)\pi = \left(i-\frac{1}{2}\right)\pi$, $i = 1, 2, \cdots$. We also see that $\lambda_i \longrightarrow (i-1)\pi$ as $\beta \longrightarrow 0$, the result for Example (3), and $\lambda_i \longrightarrow \left(i - \frac{1}{2}\right)\pi$ as $\beta \longrightarrow \infty$, the result for Example (2). As β increases from 0 to ∞, λ_i increases monotonically from $(i-1)\pi$ to $\left(i - \frac{1}{2}\right)\pi$ and this corresponds to an increasing rate of loss of solute. Some information on how the eigenvalues depend on β is in Appendix 1.

While each eigenvalue is a smooth monotonic function of β, moving from its $\beta \longrightarrow 0$ limit (chemical reaction control) to its $\beta \longrightarrow \infty$ limit (diffusion control) as β increases, we observe that if we hold β fixed, at any value other than ∞, then as $i \longrightarrow \infty$, $\lambda_i \longrightarrow (i-1)\pi$, and this is its $\beta = 0$ value. The larger the value of β the larger the value of i before λ_i can be closely approximated by its $\beta = 0$ value, viz., $\lim_{i \to \infty} \lim_{\beta \to \infty} \lambda_i(\beta) \neq \lim_{\beta \to \infty} \lim_{i \to \infty} \lambda_i(\beta)$. Ordinarily it is only the first few λ_i's that depart greatly from their $\beta = 0$ values but these are the most important eigenvalues for all but the shortest times.

This is the first problem where we might not have been able to use trigonometric identities to discover that

$$\int_0^1 \overline{\psi}_i \psi_j dx = 0 \ , \ \ i \neq j$$

and yet this condition holds here just as it does in the earlier problems.

The eigenvalues then are the squares of the values $\lambda_1, \lambda_2, \cdots$ determined as above and the corresponding eigenfunctions are $\cos \lambda_i x$ divided by $\sqrt{\frac{1}{2\lambda_i} \sin \lambda_i \cos \lambda_i + \frac{1}{2}}$ as $\int_0^1 \cos^2 \lambda x dx = \frac{1}{2\lambda} \sin \lambda \cos \lambda + \frac{1}{2}$.

It may be worthwhile to observe that for each value of β we generate an infinite set of orthogonal eigenfunctions:

$$\cos \lambda_1 x, \ \cos \lambda_2 x, \ \cos \lambda_3 x, \ \cdots$$

where $\lambda_1, \lambda_2, \lambda_3, \cdots$ depend on β. The eigenfunctions for one value of β are not particularly useful in writing the solution for another value of β. The readers may wish to satisfy themselves that this is so. An example is presented in Appendix 3.

Example (5): $c\left(x = 0\right) = 0, \ \dfrac{\partial c}{\partial x}\left(x = 1\right) + \beta c\left(x = 1\right) = 0, \ \beta < 0$

Here we have a source of solute at $x = 1$, not a sink as in the earlier case where $\beta > 0$, and for β near zero we might imagine that all λ^2's are positive. Solute is produced at the right hand boundary, lost at the left hand boundary and we wish to know: at what value of β, as the source becomes increasingly stronger, can diffusion across the layer no longer control the source.

Our eigenvalue problem is

$$\frac{d^2\psi}{dx^2} + \lambda^2 \psi = 0 \ , \ \ 0 < x < 1$$

and

$$\psi = 0 \text{ at } x = 0; \quad \frac{d\psi}{dx} + \beta\psi = 0 \text{ at } x = 1$$

and we observe, assuming λ^2 and ψ to be real, that

$$-\lambda^2 \int_0^1 \psi^2 dx = \left[\psi \frac{d\psi}{dx} \right]_0^1 - \int_0^1 \frac{d\psi}{dx} \frac{d\psi}{dx} dx$$

where the first term on the right is

$$-\beta\psi^2 \left(x = 1 \right).$$

Hence if $\beta > 0$ we have $\lambda^2 > 0$, but if $\beta < 0$ we cannot tell the sign of λ^2 without a calculation.

The first term depends explicitly on β, the second implicitly. The signs of both are known and opposite if $\beta < 0$. We therefore anticipate that stability will be lost if β becomes sufficiently negative (i.e., at least one value of λ^2 will become negative). Indeed our formula for λ^2 continues to hold if $\psi\left(x = 0\right) = 0$, a sink, is replaced by $\frac{d\psi}{dx}\left(x = 0\right) = 0$, a barrier. In that case the critical value of β is certainly zero; i.e., every negative value of β leads to growth.

By choosing ψ to satisfy

$$\psi\left(x = 0\right) = 0, \quad \frac{d\psi}{dx}\left(x = 1\right) + \beta\psi\left(x = 1\right) = 0$$

we have

$$\left[\overline{\psi_i} \frac{\partial c}{\partial x} - \frac{d\overline{\psi_i}}{dx} c \right]_0^1 = 0$$

and the equation for $\langle \psi_i, c \rangle$ is the same as in the earlier examples.

The boundary condition $\psi\left(x = 0\right) = 0$ implies $A = 0$ and $\lambda \neq 0$, whence $B \neq 0$. Then the condition $\frac{\partial\psi}{\partial x}\left(x = 1\right) + \beta\psi\left(x = 1\right) = 0$ tells us that λ must satisfy

$$\lambda \cos \lambda + \beta \sin \lambda = 0.$$

This is an equation for λ^2. If λ is a solution so also is $-\lambda$ and both $\pm\lambda$ lead to the same eigenvalue and eigenfunction. However, if we anticipate a solution $\lambda = 0$, we ought to write

$$\psi = A + Bx$$

and then we see that $\lambda = 0$ is indeed a solution corresponding to $\psi = \beta x$, iff $\beta = -1$.

First we look for positive real values of λ leading to positive real values of λ^2. To do this we write our equation

$$-\frac{\lambda}{\beta} = \tan \lambda$$

and indicate its solutions graphically as:

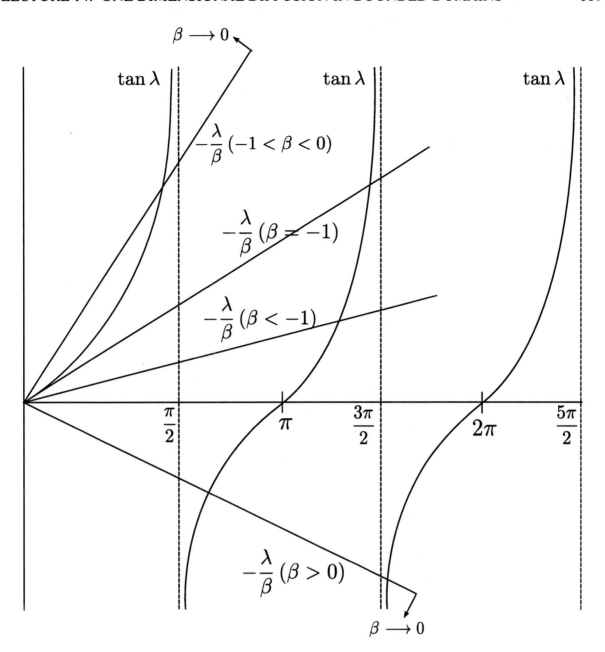

For $\beta > 0$ and indeed for $\beta > -1$ all is well. The value of λ_1, the smallest positive root, decreases from π to $\dfrac{\pi}{2}$ as β decreases from ∞ to 0 and then decreases from $\dfrac{\pi}{2}$ to 0 as β decreases from 0 to -1. This makes physical sense as it tells us that an initial solute distribution dies out more and more slowly as a solute sink loses strength and turns into a weak source. But as β passes through -1 a root is lost and something new seems to happen. Indeed if we were to inquire as to whether a steady concentration field were possible, wherein diffusion to the left hand reservoir just balances production at the right hand boundary, we would find that such a condition obtains only for $\beta = -1$.

Now λ^2 must be real and, when $\beta \geq 0$, it must also be positive. This is what directed our earlier attention to real and positive values of λ; but if we admit purely imaginary values of λ, i.e., $\lambda = i\omega$ where ω is real, then $\lambda^2 = -\omega^2$ and now λ^2 is real, as it must be, but it is negative and this is new. To see if this might be what is happening we put $\lambda = i\omega, \omega > 0$, into $\lambda \cos \lambda + \beta \sin \lambda = 0$ and then use $\cos i\omega = \cosh \omega$ and $\sin i\omega = i \sinh \omega$ to get

$$\omega \cosh \omega + \beta \sinh \omega = 0.$$

If β is not negative, this equation is not satisfied by any real values of ω. If β is negative we write $\beta = -|\beta|$ whence ω satisfies

$$\frac{\omega}{|\beta|} = \tanh \omega$$

But $\tanh \omega$ increases monotonically from 0 to 1 while its derivative decreases monotonically from 1 to 0 as ω increases from 0 to ∞ and $\lim\limits_{\omega \to 0} \dfrac{\tanh \omega}{\omega} = 1$. Hence there are no solutions to this equation for $0 < |\beta| < 1$; but for $|\beta| > 1$ there is a solution as we can see on the graph:

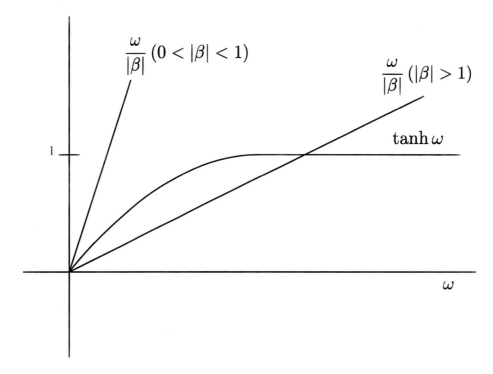

So the eigenvalue λ_1^2 decreases from π^2 to 0 as β decreases from ∞ to -1 and then decreases from 0 to $-\infty$ as β decreases from -1 to $-\infty$. Hence for $\beta < -1$ an initial solute distribution, no matter what its shape, runs away, while for $\beta > -1$ an initial solute distribution, again no matter what its shape, dies out.

What we have found then is this. The diffusion equation, under the conditions $c(x = 0) = 0$, $\dfrac{\partial c}{\partial x}(x = 1) + \beta c(x = 1) = 0$, acts to dissipate imposed solute fields (of any size) so long as $-1 < \beta < \infty$. The parameter β is $\dfrac{kl}{D}$, where k depends on temperature and l is determined by the proximity of the source and the sink, i.e., the diffusion path length. Calculations of this sort are of interest in the design of cylinders for the storage of acetylene, but there the autocatalytic reaction is homogeneous and is controlled by diffusion to the wall where deactivation takes place. Again the diameter of the tank is important as is the temperature.

We might have expected to see first one negative value of λ^2, then two, then three, etc., as β decreases below -1. But we do not.

The Steady Solution When $\beta = -1$

The problem

$$\frac{\partial c}{\partial t} = \frac{\partial^2 c}{\partial x^2}, \quad 0 < x < 1$$

and

$$c(x = 0) = 0, \quad \frac{\partial c}{\partial x}(x = 1) + \beta c(x = 1) = 0$$

where $c(t = 0)$ is assigned has the solution $c(x, t) = 0$ for all values of β if $c(t = 0) = 0$; likewise the corresponding steady problem has the solution $c(x) = 0$ for all values of β. If $\beta > -1$ this is the long time limit of all unsteady solutions.

The steady solution is

$$c = A + Bx$$

and as $c(x = 0) = 0$ we have $A = 0$. Then B must satisfy $B + \beta B = 0$ and hence $B = 0$ for all values of β except $\beta = -1$ where B is indeterminate. If $\beta = -1$ we must solve the dynamic problem to discover the value of B, for the steady solution depends on how much solute is in the film at the outset.

The eigenvalue problem is

$$\frac{d^2\psi}{dx^2} + \lambda^2\psi = 0, \ \ 0 < x < 1$$

and

$$\psi\,(x = 0) = 0, \ \frac{d\psi}{dx}\,(x = 1) + \beta\psi\,(x = 1) = 0$$

and its solutions are

$$\psi = A\cos\lambda x + \frac{B}{\lambda}\sin\lambda x$$

where $A = 0$ as $\psi\,(x = 0) = 0$. The eigenvalues corresponding to $\psi = \dfrac{B}{\lambda}\sin\lambda x$ are determined by the boundary condition at $x = 1$ and hence by the solutions to

$$\cos\lambda + \frac{\beta}{\lambda}\sin\lambda = 0.$$

This has the solution $\lambda = 0$ iff $\beta = -1$. For small λ we can write this equation

$$\left(1 - \frac{1}{2}\lambda^2 \cdots\right) + \frac{\beta}{\lambda}\left(\lambda - \frac{1}{6}\lambda^3 \cdots\right) = 0$$

or

$$(1 + \beta) - \lambda^2\left(\frac{1}{2} + \frac{1}{6}\beta\right) + \cdots = 0$$

whence $\lambda^2 = 0$ is a root and a simple root, iff $\beta = -1$. When λ^2 is zero it corresponds to the eigenfunction $\psi = Bx$.

So if $\beta = -1$, the eigenvalues are the squares of the roots of $\lambda\cos\lambda - \sin\lambda = 0$ and the corresponding eigenfunctions are $\psi_1 = \sqrt{3}x$, $\psi_2 = B_2 \sin \lambda_2 x$, $\cdots$. And we can demonstrate by direct calculation that $\int_0^1 x \sin \lambda_i x \, dx = 0, i = 2, \cdots$.

The solution to our problem when $\beta = -1$ is then

$$c = \left\{ \int_0^1 \sqrt{3}xc\,(t = 0)\, dx \right\} \sqrt{3}x + \sum_{i=2}^{\infty} \langle \psi_i, c\,(t = 0) \rangle \, e^{-\lambda_i^2 t} \psi_i\,(x)$$

the first term being the steady solution.

Example (6): $\dfrac{\partial c}{\partial x}\,(x = 0) = 0$, $\dfrac{\partial c}{\partial x}\,(x = 1) + \alpha \dfrac{\partial c}{\partial t}\,(x = 1) = 0$, $\alpha > 0$

Now the diffusion layer is isolated from the left hand reservoir but exchanges solute with the right hand reservoir, assumed to be of finite extent so that its concentration responds to this solute exchange. To simplify the problem we assume that the right hand edge of the diffusion layer and the reservoir remain in phase equilibrium for all time.

Here we see something new: time derivatives appear in our problem in two places and hence we anticipate that our eigenvalue problem will not be the same as it was in the earlier examples. But imagining that our time dependence will remain exponential we propose that ψ and λ^2 must satisfy

$$\frac{d^2\psi}{dx^2} + \lambda^2 \psi = 0$$

$$\frac{d\psi}{dx}\,(x = 0) = 0 \text{ and } \frac{d\psi}{dx}\,(x = 1) - \alpha\lambda^2\psi\,(x = 1) = 0$$

where the eigenvalue λ^2 now also appears in the boundary condition.

Before we solve this problem we ought to use our two integration formulas to learn something about it.

Thus, as $\overline{\lambda^2}$ and $\overline{\psi}$ is a solution to our problem whenever λ^2 and ψ is a solution, we substitute $\phi = \overline{\psi}$ in the second formula to see that $\lambda^2 = \overline{\lambda^2}$, hence the eigenvalues must be real. Then substituting $\phi = \overline{\psi}$ in the first formula, we see that λ^2 must be non negative.

Something new appears on substituting $\phi = \overline{\psi}_1$ and $\psi = \psi_2$, eigenfunctions corresponding to distinct eigenvalues, in the second formula. We see that ϕ and ψ are not orthogonal in the plane vanilla inner product. Instead they are orthogonal in the inner product

$$\langle a, b \rangle = \int_0^1 \bar{a} b \, dx + \alpha \bar{a} \left(x = 1 \right) b \left(x = 1 \right)$$

Hence, to solve our diffusion problem, we write

$$c = \sum c_i \left(t \right) \psi_i$$

and find

$$\frac{d}{dt} \langle \psi_i, c \rangle = -\lambda_i^2 \langle \psi_i, c \rangle$$

just as before, but now in a new inner product, and our solution is as before, but now in a new inner product, viz.,

$$c = \sum \langle \psi_i, c \left(t = 0 \right) \rangle e^{-\lambda_i^2 t} \psi_i$$

where

$$\langle \psi_i, \, c \left(t = 0 \right) \rangle = \int_0^1 \overline{\psi_i} c \left(t = 0 \right) dx + \alpha \overline{\psi_i} \left(x = 1 \right) c \left(t = 0 \right) \Big|_{x=1}$$

and we can see how the initial states of the diffusion layer and the reservoir come into the solution.

It remains only to solve the eigenvalue problem. First we observe that zero is an eigenvalue corresponding to $\psi = 1$. This is not surprising because we expect the system to come to rest as $t \longrightarrow \infty$ with a uniform solute concentration in the diffusion layer, in equilibrium with whatever solute concentration winds up in the reservoir. So writing

$$\psi = A \cos \lambda x + B \sin \lambda x$$

we see that $\dfrac{d\psi}{dx} \left(x = 0 \right) = 0$ implies that $B = 0$, hence A must not be zero. The values of λ then satisfy

$$\frac{d\psi}{dx}(x=1) - \lambda^2 \alpha \psi(x=1) = 0$$

and this tells us that

$$\lambda \sin \lambda + \lambda^2 \alpha \cos \lambda = 0.$$

This is an equation for λ^2 and we see that $\lambda^2 = 0$ is a simple root. We can find the remaining values of λ^2 by graphical means, by solving

$$\alpha \lambda = -\tan \lambda$$

for $\lambda > 0$. Because α is positive there is no solution $\lambda = i\omega$ and hence λ^2 is not negative.

Example (7): Periodic Conditions

Suppose we have

$$c(x=0) = c(x=1)$$

and

$$\frac{\partial c}{\partial x}(x=0) = \frac{\partial c}{\partial x}(x=1)$$

then we find that

$$\left\{ \overline{\psi_i} \frac{\partial c}{\partial x} - \frac{d\overline{\psi_i}}{dx} c \right\} \Big|_0^1$$

vanishes if we choose

$$\psi(x=0) = \psi(x=1)$$

and

$$\frac{d\psi}{dx}(x=0) = \frac{d\psi}{dx}(x=1).$$

Hence writing the solution to

$$\frac{d^2\psi}{dx^2} + \lambda^2\psi = 0$$

as

$$\psi = A\cos\lambda x + B\sin\lambda x$$

we see that A, B and λ must satisfy

$$A = A\cos\lambda + B\sin\lambda$$

and

$$B\lambda = -A\lambda\sin\lambda + B\lambda\cos\lambda$$

and hence

$$\begin{pmatrix} \cos\lambda - 1 & \sin\lambda \\ -\lambda\sin\lambda & \lambda(\cos\lambda - 1) \end{pmatrix} \begin{pmatrix} A \\ B \end{pmatrix} = \begin{pmatrix} 0 \\ 0 \end{pmatrix}$$

To have a solution to these homogeneous equations such A and B are not both zero, we must have

$$\det\begin{pmatrix} \cos\lambda - 1 & \sin\lambda \\ -\lambda\sin\lambda & \lambda(\cos\lambda - 1) \end{pmatrix} = 0$$

First, $\lambda = 0$ is a solution (and at $\lambda = 0$ we have $\psi = A + Bx$) and it implies $\psi = A \neq 0$, i.e., there is only one periodic solution at $\lambda = 0$. Then, for $\lambda \neq 0$, we have

$$\cos \lambda = 1$$

or

$$\lambda = 2\pi,\ 4\pi,\ \ldots$$

Hence our eigenvalues are

$$0,\ (2\pi)^2,\ (4\pi)^2,\ \ldots$$

and the corresponding eigenfunctions (not normalized) are

$$1,\ \begin{Bmatrix} \cos 2\pi x \\ \sin 2\pi x \end{Bmatrix},\ \cdots,\ \begin{Bmatrix} \cos 2\pi n x \\ \sin 2\pi n x \end{Bmatrix},\ \cdots$$

$n = 1,\ 2,\ \cdots$ Thus to every eigenvalue not zero, we have two periodic eigenfunctions, viz., $A = 1$, $B = 0$ and $A = 0$, $B = 1$. But corresponding to $\lambda = 0$, we have only one periodic solution.

The reader may observe that the expansion of a periodic function in these eigenfunctions is what is ordinarily called a Fourier series expansion, where the coefficient of the first eigenfunction is the average value of the function being expanded.

14.5 An Eigenvalue Problem Arising in Frictional Heating

We present here an eigenvalue problem that is a little out of the ordinary. First, suppose ϕ and μ^2 satisfy

$$\frac{d^2\phi}{dx^2} + \mu^2\phi = 0,\quad -1 < x < 1$$

where

$$\phi\left(x=-1\right)=0=\phi\left(x=1\right)$$

The solutions are

$$\cos\frac{1}{2}\pi x \qquad \mu^2=\left(\frac{1}{2}\pi\right)^2$$

$$\sin\pi x \qquad \mu^2=\pi^2$$

$$\cos\frac{3}{2}\pi x \qquad \mu^2=\left(\frac{3}{2}\pi\right)^2$$

$$\text{etc.}$$

Now the eigenvalue problem of interest in this example is

$$\frac{d^2\psi}{dx^2}+\nu^2\left(\psi-\int_{-1}^{1}\psi dx\right)=0$$

where

$$\psi\left(x=-1\right)=0=\psi\left(x=1\right).$$

This is a model for an eigenvalue problem arising in frictional heating. It appears if a small pertur-bation is imposed on the solution to a problem in plane Couette flow.

The solutions odd about $x=0$ are as above, viz.,

$$\sin\pi x \qquad \nu^2=\pi^2$$

$$\sin 2\pi x \qquad \nu^2=\left(2\pi\right)^2$$

$$\text{etc.}$$

Then setting

$$C = \int_{-1}^{1} \psi \, dx$$

we have

$$\frac{d^2\psi}{dx^2} + \nu^2\psi = \nu^2 C$$

whereupon

$$\psi = C + A\cos\nu x + B\sin\nu x$$

and we see that A, B, C and ν satisfy

$$\left.\begin{array}{l} 0 = C + A\cos\nu - B\sin\nu \\ 0 = C + A\cos\nu + B\sin\nu \end{array}\right\} \Rightarrow \left\{\begin{array}{l} 0 = C + A\cos\nu \\ 0 = B\sin\nu \end{array}\right.$$

and

$$C = 2C + \frac{2A}{\nu}\sin\nu$$

The case $B \neq 0$, $A = 0$, $\nu = \pi, 2\pi, \cdots$ corresponds to the odd solutions about $x = 0$ and hence to $C = 0$. In the remaining case, $C \neq 0$, we have $A \neq 0$, $B = 0$ and

$$\frac{1}{2}\nu = \frac{\sin\nu}{\cos\nu} = \tan\nu$$

This equation has many positive solutions, which can be found graphically, as well as one negative solution $\nu = i\omega$, whence $\nu^2 = -\omega^2 < 0$.

There are similarities here to two earlier problems in this lecture, one of which also has a negative eigenvalue.

The reader can use our two integration by parts formulas, now on the interval $-1 \leq x \leq 1$, to derive general conclusions about the solutions to this new eigenvalue problem. For example setting $\phi = \overline{\psi}$ in our second integration by parts formula and observing that $\overline{\psi}, \overline{\lambda^2}$ is a solution if ψ, λ^2 is a solution we have

$$-\nu^2 \left\{ \int_{-1}^{1} \overline{\psi}\psi dx - \int_{-1}^{1} \overline{\psi}dx \int_{-1}^{1} \psi dx \right\} = -\overline{\nu^2} \left\{ \int_{-1}^{1} \overline{\psi}\psi dx - \int_{-1}^{1} \overline{\psi}dx \int_{-1}^{1} \psi dx \right\}$$

whereupon we conclude ν^2 must be real.

14.6   More on Examples (5) and (6)

Our problem in Example (5) is to solve the eigenvalue problem

$$\frac{d^2\psi}{dx^2} + \lambda^2\psi = 0, \quad \psi = 0 \quad \text{at} \quad x = 0$$

and

$$\frac{d\psi}{dx} = -\beta\psi, \quad \text{at} \quad x = 1$$

whereas in Example (6) the boundary condition at $x = 1$ is

$$\frac{d\psi}{dx} = \alpha\lambda^2\psi$$

All β's make physical sense but $\beta \geq 0$ is the ordinary case. Only $\alpha \geq 0$ makes sense but here we suppose $\alpha < 0$ is possible, i.e., we have an antireservoir.

In both problems

$$\psi = A \sin \lambda x$$

where in the first we have

$$\frac{\lambda}{-\beta} = \tan \lambda$$

whereas in the second

$$\frac{1}{\alpha \lambda} = \tan \lambda$$

The readers can satisfy themselves that all λ^2's are real.

If we ask: is there a value of α or β where $\lambda^2 = 0$, in the first problem we find one and only one value of $\beta : \beta = -1$. In the second problem there are no values of α such that $\lambda^2 = 0$. For $\beta > -1$ all λ^2's are positive, at $\beta = -1$, one becomes zero and at $\beta < -1$, one is negative, the others remaining positive because there is no value of β other than -1 where $\lambda^2 = 0$. All λ^2's are smooth functions of β.

This is not the way the second problem works. For $\alpha \geq 0$ all λ^2's are positive, $\alpha = 0$ corresponding to an impermeable wall at $x = 1$. However, if we admit the possibility $\alpha < 0$ we have

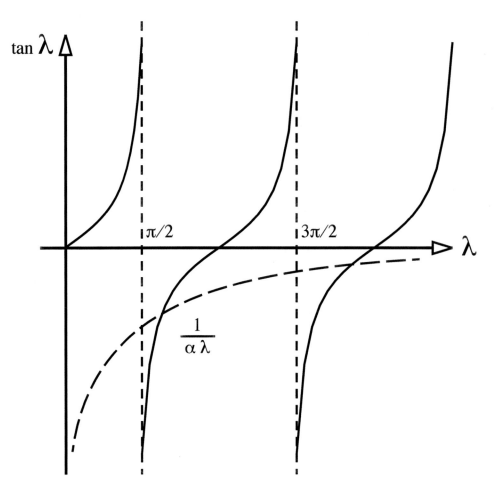

and the root $\pi/2$ at $\alpha = 0$ appears to have been lost. However, at $\alpha < 0$ by setting $\lambda = i\omega$ we have

$$\frac{1}{(-\alpha)\,\omega} = \tan\omega$$

and we recover our lost root. Now we have: $\alpha \to 0^- \implies \omega \to \infty$ and $\alpha \to -\infty \implies \omega \to 0$

and we see

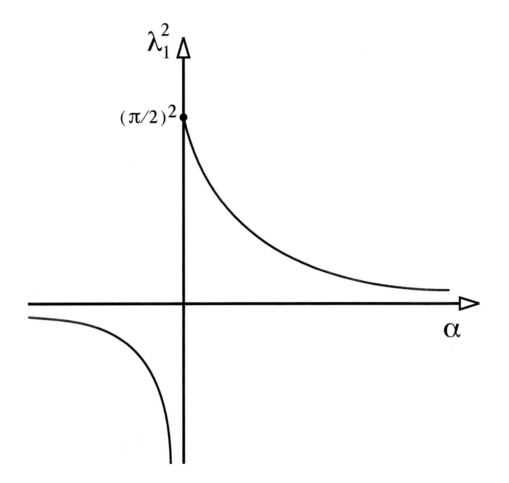

Our pictures show that in the second problem all λ^2's are positive for $\alpha \geq 0$, one λ^2 is negative, all others remaining positive, for $\alpha < 0$ and now no λ^2 is ever zero.

Our frictional heating problem is like the second problem if we write it

$$\frac{d^2\psi}{dx^2} + \nu^2 \left(\psi - \gamma \int_{-1}^{+1} \psi \, dx \right) = 0$$

$$\psi \left(x = \pm 1 \right) = 0$$

The crisis here occurs at $\gamma = \dfrac{1}{2}$. The two earlier cases correspond to $\gamma = 0$ and $\gamma = 1$.

If we now put our autocatalytic reaction on the domain, our eigenvalue problem is

$$\frac{d^2\psi}{dx^2} + \lambda^2\psi - \beta\psi = 0, \quad \beta < 0$$

$$\psi = 0 \quad \text{at} \quad x = 0, 1$$

And we have

$$\psi = A \sin \sqrt{\lambda^2 - \beta}\, x$$

where

$$\lambda^2 = \beta + n^2 \pi^2$$

And if we ask: at what values of β can λ^2 be zero we find

$$\beta = -n^2 \pi^2, \quad n = 1, 2, \ldots$$

Every λ^2 can turn negative and we see that the larger $-\beta$ the more spatial variation is needed to control growth. But once $-\beta$ exceeds π^2 we have lost stability.

14.7   Differentiating the Eigenvalue Problem

In example (4) we have

$$\left(\frac{d^2}{dx^2} + \lambda^2 \right) \psi = 0$$

and

$$\frac{d}{dx} \psi\, (x = 0) = 0, \quad \left(\frac{d}{dx} + \beta \right) \psi\, (x = 1) = 0$$

and our aim here is to determine the dependence of $\lambda_1^2, \lambda_2^2, \cdots$ on β. To do this we find, by differentiating the problem, that $\dfrac{d\psi}{d\beta}$ satisfies

$$\left(\frac{d^2}{dx^2} + \lambda^2 \right) \frac{d\psi}{d\beta} = -\frac{d\lambda^2}{d\beta} \psi$$

and

$$\frac{d}{dx}\frac{d\psi}{d\beta}(x=0)=0, \quad \left(\frac{d}{dx}+\beta\right)\frac{d\psi}{d\beta}(x=1)=-\psi(x=1)$$

The corresponding homogeneous problem has a nonzero solution, hence a solvability condition must be satisfied. It is satisfied because $\frac{d\psi}{d\beta}$ can be found by differentiating ψ, and solvability determines $\frac{d\lambda^2}{d\beta}$. Thus we use our second integration by parts formula to write

$$\int_0^1 \psi\left(\frac{d^2}{dx^2}+\lambda^2\right)\frac{d\psi}{d\beta}dx = \left[\psi\frac{d}{dx}\frac{d\psi}{d\beta}-\frac{d\psi}{dx}\frac{d\psi}{d\beta}\right]_0^1 + \int_0^1\left(\frac{d^2}{dx^2}+\lambda^2\right)\psi\frac{d\psi}{d\beta}dx$$

and then substitute $\left(\dfrac{d^2}{dx^2}+\lambda^2\right)\dfrac{d\psi}{d\beta}=-\dfrac{d\lambda^2}{d\beta}\psi$ and $\left(\dfrac{d^2}{dx^2}+\lambda^2\right)\psi=0$ into this to get

$$\frac{d\lambda^2}{d\beta}\int_0^1 \psi\psi dx = -\psi\psi(x=1)$$

Using the fact that ψ is a multiple of $\cos\lambda x$ we can write this

$$\frac{d\lambda^2}{d\beta}=\frac{\cos^2\lambda}{\frac{1}{2\lambda}\sin\lambda\cos\lambda+\frac{1}{2}}$$

and hence, using $\lambda\sin\lambda-\beta\cos\lambda=0$ and $\sin^2\lambda+\cos^2\lambda=1$, we get

$$\frac{d\lambda^2}{d\beta}=\frac{2\lambda^2}{\lambda^2+\beta^2+\beta}$$

This differential equation determines λ_i^2 vs β given that $\lambda_i^2(\beta=0)=(i-1)^2\pi^2$, $i=1,2,\cdots$. Indeed we see that

$$\frac{d\lambda_i^2}{d\beta}(\beta=0)=2, \quad i=2,\cdots$$

but $\dfrac{d\lambda_1^2}{d\beta}(\beta=0)$ is indeterminate because $\lambda_1^2(\beta=0)=0$.

To see what λ_1^2 is doing when β is small we first observe that λ_1^2 as a function of β satisfies

$$\lambda_1 \sin \lambda_1 - \beta \cos \lambda_1 = 0.$$

Then when λ_1^2 is near zero, as it is when β is small, we can approximate $\lambda_1 \sin \lambda_1$ and $\cos \lambda_1$ by

$$\lambda_1 \sin \lambda_1 = \lambda_1^2 - \frac{1}{6}\lambda_1^4 + \frac{1}{120}\lambda_1^6 - \cdots$$

and

$$\cos \lambda_1 = 1 - \frac{1}{2}\lambda_1^2 + \frac{1}{24}\lambda_1^4 - \cdots$$

and write

$$\lambda_1^2 = c_1\beta + c_2\beta^2 + \cdots$$

Substituting this in

$$\lambda_1^2 - \frac{1}{6}\lambda_1^4 + \frac{1}{120}\lambda_1^6 - \cdots - \beta\left(1 - \frac{1}{2}\lambda_1^2 + \frac{1}{24}\lambda_1^4 - \cdots\right) = 0$$

we find that

$$c_1 = 1, \ c_2 = -\frac{1}{3}, \ \cdots$$

and hence as $\beta \longrightarrow 0$ that

$$\lambda_1^2 = \beta - \frac{1}{3}\beta^2$$

or

$$\lambda_1^2 = \frac{\beta}{1 + \frac{1}{3}\beta}.$$

This approximation is useful in many ways; indeed it shows that

$$\frac{d\lambda_1^2}{d\beta}(\beta = 0) = 1.$$

14.8 The Use of a Nondiagonalizing Basis.

To get a clear idea how much help a diagonalizing basis is in writing the solution to the diffusion equation we carry out a calculation in a nondiagonalizing basis.

Let c satisfy

$$\frac{\partial c}{\partial t} = \frac{\partial^2 c}{\partial x^2}, \quad 0 < x < 1$$

$$\frac{\partial c}{\partial x}(x = 0) = 0$$

and

$$\frac{\partial c}{\partial x}(x = 1) + \beta c(x = 1) = 0$$

where $c(t = 0)$ is assigned. Then the eigenvalue problem is

$$\frac{d^2\psi}{dx^2} + \lambda^2\psi = 0, \quad 0 < x < 1$$

$$\frac{d\psi}{dx}(x = 0) = 0$$

and

$$\frac{d\psi}{dx}(x = 1) + \beta\psi(x = 1) = 0$$

and it is satisfied by

$$\psi = A\cos\lambda x + \frac{B}{\lambda}\sin\lambda x$$

where A, B and λ remain to be determined.

As $\dfrac{d\psi}{dx}(x = 0) = 0$ we find $B = 0$, and the eigenfunctions are then

$$\psi_i = A \cos \lambda_i x$$

where the λ_i are the non-negative solutions to

$$\lambda \sin \lambda - \beta \cos \lambda = 0$$

and the $A's$ are normalization constants. For each value of β we get a set of eigenfunctions and these eigenfunctions depend on the value of β.

Earlier, Example (4) page 334, we solved this problem. Whatever the value of β we expanded the solution in the corresponding eigenfunctions. Here we try something else. Let $\beta > 0$ be fixed. Then to determine the solution for this value of β we expand it in the eigenfunctions corresponding to $\beta = 0$ as they are easy to determine.

When $\beta = 0$ the eigenvalues satisfy

$$\lambda^{(0)} \sin \lambda^{(0)} = 0$$

which is an equation in $\left(\lambda^{(0)}\right)^2$ having simple roots, viz.,

$$\left(\lambda^{(0)}\right)^2 = 0,\ \pi^2,\ 2^2\pi^2,\ \cdots$$

We normalize the corresponding eigenfunctions and denote them ψ_i^0, $i = 0, 1, 2, \cdots$, where

$$\psi_0^0 = 1$$

$$\psi_1^0 = \sqrt{2} \cos \pi x$$

$$\psi_2^0 = \sqrt{2} \cos 2\pi x$$

etc

To solve the problem corresponding to a fixed value of $\beta > 0$ in terms of the eigenfunctions corresponding to $\beta = 0$ we write

$$c = \sum_{i=0}^{\infty} c_i \psi_i^0$$

and our job is to find the coefficients c_i in this series, where

$$c_i = \langle \psi_i^0, c \rangle = \int_0^1 \psi_i^0 \, c \, dx$$

To do this we multiply the equation for c by ψ_i^0, integrate the result from 0 to 1 and use the integration by parts formula

$$\int_0^1 u \frac{d^2 v}{dx^2} dx = \int_0^1 \frac{d^2 u}{dx^2} v \, dx + \left[u \frac{dv}{dx} - \frac{du}{dx} v \right]_0^1$$

to get

$$\left\langle \psi_i^0, \frac{\partial c}{\partial t} \right\rangle = \left\langle \psi_i^0, \frac{\partial^2 c}{\partial x^2} \right\rangle = \left\langle \frac{d^2 \psi_i^0}{dx^2}, c \right\rangle + \left[\psi_i^0 \frac{\partial c}{\partial x} - \frac{d\psi_i^0}{dx} c \right]_0^1$$

and this is

$$\frac{d}{dt} c_i = - \left(\lambda_i^0 \right)^2 c_i + \psi_i^0 (x = 1) \frac{\partial c}{\partial x} (x = 1).$$

The technical difficulty here is that $\frac{\partial c}{\partial x} (x = 1)$ is not zero. In fact $\frac{\partial c}{\partial x} (x = 1) = -\beta c (x = 1) = -\beta \sum_{j=0} c_j \psi_j^0 (x = 1)$ and using this we get

$$\frac{d}{dt} c_i = - \left(\lambda_i^0 \right)^2 c_i - \beta \psi_i^0 (x = 1) \sum_{j=0} c_j \psi_j^0 (x = 1)$$

This illustrates what happens when we do not use a diagonalizing basis; the equations satisfied by the c_i are not uncoupled.

Using $\psi_0^0 (x = 1) = 1$, $\psi_j^0 (x = 1) = \sqrt{2} (-1)^j$, $j = 1, 2, \cdots$, $(\lambda_0^0)^2 = 0$ and $(\lambda_j^0)^2 = j^2 \pi^2$, $j = 1, 2, \cdots$ we must solve

$$\frac{dc_i}{dt} = - (\lambda_i^0)^2 c_i - \beta \psi_i^0 (x = 0) c_0 - \sqrt{2} \beta \psi_i^0 (x = 0) \sum_{j=1} (-1)^j c_j$$

or

$$\frac{dc_0}{dt} = 0 - \beta c_0 - \sqrt{2} \beta \sum_{j=1} (-1)^j c_j$$

and

$$\frac{dc_i}{dt} = -i^2 \pi^2 c_i - \sqrt{2} \beta (-1)^i c_0 - 2\beta (-1)^i \sum_{j=1} (-1)^j c_j, \quad i = 1, 2, \cdots .$$

To try to learn something about our solution, we truncate the first two equations to

$$\frac{dc_0}{dt} = -\beta c_0 + \sqrt{2} \beta c_1$$

and

$$\frac{dc_1}{dt} = -\pi^2 c_1 + \sqrt{2} \beta c_0 - 2\beta c_1$$

and hence to determine c_0 and c_1 in this approximation we need the eigenvalues of the matrix

$$\begin{pmatrix} -\beta & \sqrt{2}\beta \\ \sqrt{2}\beta & -\pi^2 - 2\beta \end{pmatrix}$$

For small β these are $-\beta$ and $-\pi^2$ and so for long time and small β the solute is dissipated as $e^{-\beta t}$ and this is correct. But more information than this is difficult to obtain in this basis.

14.9 A Warning About Series Solutions

Before we go on, we can get an idea of what is to come and at the same time make an observation about Fourier series. The eigenvalue problem

$$\frac{d^2\psi}{dx^2} + \lambda^2\psi = 0 \ , \ \ 0 < x < 1$$

and

$$\psi\left(x=0\right) = 0, \ \psi\left(x=1\right) = 0$$

has solutons $\lambda_i^2 = i^2\pi^2$ and $\psi_i = \sqrt{2}\sin i\pi x$, $i = 1, 2, \cdots$. This orthogonal set of eigenfunctions can be used to solve problems such as

$$0 = \frac{d^2c}{dx^2} + Q\left(x\right)$$

where

$$c\left(x=0\right) = c_0, \ c\left(x=1\right) = c_1.$$

Indeed, writing

$$c\left(x\right) = \sum_{i=1}^{\infty} \langle \psi_i, c \rangle \, \psi_i$$

we can find $\langle \psi_i, c \rangle$ by multiplying $0 = \dfrac{d^2c}{dx^2} + Q$ by $\overline{\psi_i}$ and integrating over $0 \le x \le 1$. Doing this we get

$$0 = \left\langle \psi_i, \frac{d^2c}{dx^2} \right\rangle + \langle \psi_i, Q \rangle$$

and using our second integration by parts formula, we find

$$0 = \left[\overline{\psi_i} \frac{dc}{dx} - \frac{d\overline{\psi_i}}{dx} c \right]_0^1 + \left\langle \frac{d^2\psi_i}{dx^2}, c \right\rangle + \langle \psi_i, Q \rangle$$

Something new happens here. The term $\left[\overline{\psi_i} \frac{dc}{dx} - \frac{d\overline{\psi_i}}{dx} c \right]_0^1$ does not vanish, but it can be evaluated

because ψ_i vanishes on the boundary, eliminating the piece $\left[\overline{\psi_i} \frac{dc}{dx} \right]_0^1$, while c is assigned there,

establishing the value of the piece $\left[-\frac{d\overline{\psi_i}}{dx} c \right]_0^1$. Using this we get

$$\langle \psi_i, c \rangle = \frac{\langle \psi_i, Q \rangle}{\lambda_i^2} - \frac{1}{\lambda_i^2} \frac{d\overline{\psi_i}}{dx} (x = 1) c_1 + \frac{1}{\lambda_i^2} \frac{d\overline{\psi_i}}{dx} (x = 0) c_0$$

and hence

$$c(x) = \sum \frac{\langle \psi_i, Q \rangle}{\lambda_i^2} \psi_i - c_1 \sum \frac{1}{\lambda_i^2} \frac{d\overline{\psi_i}}{dx} (x = 1) \psi_i + c_0 \sum \frac{1}{\lambda_i^2} \frac{d\overline{\psi_i}}{dx} (x = 0) \psi_i$$

We see, then, that $c(x)$ is the sum of three terms each accounting for the contribution of one of the three sources. The boundary sources introduce a special problem. To see this let $Q = 0$, $c_0 = 0$ and $c_1 = 1$, then

$$c(x) = -\sum \frac{1}{\lambda_i^2} \frac{d\overline{\psi_i}}{dx} (x = 1) \psi_i = -\sum \frac{1}{i^2\pi^2} \sqrt{2} i\pi \cos i\pi \sqrt{2} \sin i\pi x = \sum_{i=1}^{\infty} -\frac{2}{i\pi} (-1)^i \sin i\pi x$$

Because $c(x) = x$ satisfies this problem, this expansion must be the Fourier series for x and indeed it is, viz.,

$$x = \sum_{i=1}^{\infty} -\frac{2}{i\pi} (-1)^i \sin i\pi x$$

The terms in this series fall off as $\frac{1}{i}$ and so convergence depends on the alternating sign, $(-1)^i$, i.e., $\frac{1}{i} - \frac{1}{i+1} \sim \frac{1}{i^2}$, and gets a little help from the sign pattern of $\sin i\pi x$.

What we seem to have here is the function $f(x) = x$ on $0 \le x \le 1$ expanded in a series of odd functions of period 2. But in fact what we really have is the function $f(x) = x$ on $0 \le x \le 1$, extended first to $-1 \le x \le 1$ as an odd function and then extended to all x as a function of period 2, expanded in a series of functions of period 2. That is, what we have expanded is the function shown here:

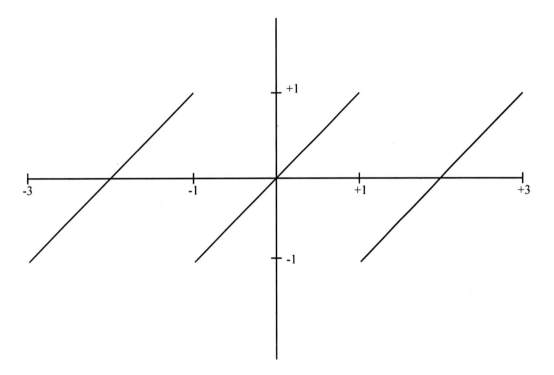

The convergence is slow because this function is not smooth, having a jump at $x = 1$. But the series converges to x for all $x : 0 \le x < 1$. It converges to 0 at $x = 1$, where 0 is the average of 1 and -1, the limits of the value of the function as x goes to 1 from the left and the right.

The series that results on termwise differentiation,

$$\sum -2(-1)^i \cos i\pi x,$$

does not converge anywhere and has no ordinary meaning.

The solution to this boundary source problem is not a superposition of terms that satisfy the original differential equation as we found earlier for initial value problems nor is it a superposition of terms that satisfy special problems of the same kind as we find for interior sources where

$$c = \frac{\langle \psi_i, Q \rangle \, \psi_i}{\lambda_i^2}$$

satisfies

$$0 = \frac{d^2 c}{dx^2} + \langle \psi_i, Q \rangle \, \psi_i$$

and

$$c(x = 0) = 0, \; c(x = 1) = 0.$$

What we have then is a series expansion of the solution to a problem driven by a boundary source which is correct but which cannot be verified by direct substitution into the problem. Indeed what can be learned about the solution to the problem using its series expansion is limited to operations on the series that exclude differentiation.

14.10   Home Problems

1. Let c_1 and c_2 denote the concentrations of two solute species dissolved in a solvent which is confined to a thin layer. The two solute species are distributed across the solvent layer at $t = 0$ in some assigned way. The edges of the layer are impermeable to both species and an estimate of the time for the layer to equilibrate is required. In terms of

$$\underline{c} = \begin{pmatrix} c_1 \\ c_2 \end{pmatrix}$$

we have for the diffusive homogenization of the solvent layer

$$\frac{\partial \underline{c}}{\partial t} = D \frac{\partial^2 \underline{c}}{\partial x^2}, \qquad 0 < x < 1$$

and

$$\frac{\partial \underline{c}}{\partial x}(x = 0) = \underline{0}, \qquad \frac{\partial \underline{c}}{\partial x}(x = 1) = \underline{0},$$

where, in units of the layer thickness,

$$D = \begin{pmatrix} D_{11} & D_{12} \\ D_{21} & D_{22} \end{pmatrix} = \begin{pmatrix} 3/2 & 1/2 \\ 1/2 & 3/2 \end{pmatrix}$$

and where at $t = 0$:

$$c_1 = \begin{cases} 1, & 0 \leq x < \dfrac{1}{2} \\ 0, & \dfrac{1}{2} < x \leq 1 \end{cases}$$

$$c_2 = \begin{cases} 0, & 0 \leq x < \dfrac{1}{2} \\ 1, & \dfrac{1}{2} < x \leq 1 \end{cases}$$

How long before the maximum deviation from uniformity is less than 0.000001?

2. Let $c_1, c_2, \ldots, c_n$ denote the concentrations of n solute species dissolved in a solvent and suppose that the species participate in a set of reversible first order chemical reactions, viz.,

$$i \underset{k_{ij}}{\overset{k_{ji}}{\rightleftarrows}} j$$

Then the rate of production of solute is $K\underline{c}$ where K is introduced in Lecture 8.

Let the solute be in chemical equilibrium and be distributed uniformly in the solvent which is confined to a thin layer, $0 < x < 1$. At $t = 0$ the edges of the solvent layer contact large solute free reservoirs that hold the solute concentration there at zero for all $t > 0$. Then for the loss of solute to the reservoir we have

$$\frac{\partial \underline{c}}{\partial t} = D \frac{\partial^2 \underline{c}}{\partial x^2} + K\underline{c}, \quad 0 < x < 1$$

$$\underline{c}(x = 0) = \underline{0}, \quad \underline{c}(x = 1) = \underline{0}$$

and

$$\underline{c}\,(t=0) = \underline{c}\,_{\text{eq}}$$

where

$$\underline{c} = \begin{pmatrix} c_1 \\ c_2 \\ \vdots \\ c_n \end{pmatrix}$$

$$K\underline{c}\,_{\text{eq}} = \underline{0}$$

and

$$D = \begin{pmatrix} D_{11} & D_{12} & \cdots \\ D_{21} & D_{22} & \cdots \\ \vdots & \vdots & \vdots \end{pmatrix}$$

Determine $\underline{c}\,(x,t)$ and hence the time required for the solvent to be cleared of solute. In problem 1 expanding $\underline{c}$ in the eigenvectors of D leads to two familiar diffusion equations. That idea will not work here unless D and K have a complete set of eigenvectors in common. But this is not ordinarily so, even if D is diagonal.

Yet the problem is special in another way: the boundary conditions are Dirichlet conditions for all solute species. And so its solution can be obtained by expanding $\underline{c}$ in the eigenfunctions determined by the ordinary eigenvalue problem

$$\frac{d^2\Psi}{dx^2} + \lambda^2\Psi = 0$$

and

$$\Psi\,(x=0) = 0, \qquad \Psi\,(x=1) = 0$$

Denote the solutions to this Ψ_i, $i = 1, 2, \ldots$, and write

$$\underline{c} = \sum \underline{c}_i(t) \, \Psi_i(x)$$

where $\underline{c}_i(t) = \langle \Psi_i, \underline{c} \rangle$ and $\langle \ , \ \rangle = \displaystyle\int_0^1 dx$. Then as

$$\int_0^1 \left\{ \Psi_i \frac{d^2\underline{c}}{dx^2} - \frac{d^2\Psi_i}{dx^2} \underline{c} \right\} dx = \left[\Psi_i \frac{d\underline{c}}{dx} - \frac{d\Psi_i}{dx} \underline{c} \right]_0^1 = \underline{0}$$

our expansion works out here just as it does in more familiar problems.

How much time must elapse before only 1% of the initial equilibrium solute remains in the solvent, if $n = 2$, if, in units of film thickness,

$$D = \begin{pmatrix} 1 & 0 \\ 0 & \dfrac{1}{2} \end{pmatrix}$$

and if

$$K = \begin{pmatrix} -1 & 1 \\ 1 & -1 \end{pmatrix} ?$$

3. Let c, the concentration of a solute in a solvent, satisfy

$$\frac{\partial c}{\partial t} = \frac{\partial^2 c}{\partial x^2}, \quad 0 < x < 1$$

where $c\,(t = 0)$ is assigned. Study the homogenization of the solute in three experiments:

(i) $c\,(x = 0) = 0 = c\,(x = 1)$

(ii) $c\,(x = 0) = 0 = \dfrac{\partial c}{\partial x}\,(x = 1)$

(iii) $\dfrac{\partial c}{\partial x}\,(x = 0) = 0 = \dfrac{\partial c}{\partial x}\,(x = 1)$

Order the rates at which the solute goes to its long time uniform state in the three experiments. Observe that the uniform state in (iii) cannot be determined by solving the steady equation but depends on $c(t = 0)$. This is not true in (i) and (ii).

4. Free radicals are created in acetylene whereupon they catalyse the production of more free radicals.

The growth of free radicals is controlled by diffusion to the wall whereupon the free radicals are destroyed upon collision with the solid surface. A tank of acetylene must not be too large in diameter if a runaway is to be averted.

Suppose acetylene is stored between two plane walls a distance L apart. In terms of k and D, how large can L be before a runaway occurs?

The model is

$$\frac{1}{D}\frac{\partial c}{\partial t} = \frac{\partial^2 c}{\partial x^2} + \frac{k}{D}c, \quad k > 0$$

$c = 0$ at $x = 0$ and $x = L$ where c denotes the free radical concentration, $[D] = \dfrac{L^2}{T}$ and $[k] = \dfrac{1}{T}$.

Does the value of L depend on $c(t = 0)$?

The steady problem

$$0 = \frac{\partial^2 c}{\partial x^2} + \frac{k}{D}c, \quad 0 < x < L$$

and

$$c(x = 0) = 0 = c(x = L)$$

has the solution $c = 0$ for all values of $\dfrac{kL^2}{D} > 0$. But for special values of $\dfrac{kL^2}{D}$ it has solutions other than $c = 0$. Find these values.

The steady problem does not have only non-negative solutions. But you can show that if $c(t = 0) > 0$, the unsteady problem must have non-negative solutions. Do this.

Write the solution to the unsteady problem if $c(t = 0) > 0$ and $\dfrac{kL^2}{D}$ is less than, equal to or greater than π^2.

If $\dfrac{kL^2}{D} = \pi^2$, the solution to the steady problem can only be obtained by solving the unsteady problem and then letting t grow large. This steady solution depends on $c(t = 0)$.

5. Two species having concentrations a and b are distributed over a one dimensional domain, $0 < x < 1$. At the ends we have $a = 0$ and $\dfrac{\partial b}{\partial x} = 0$ and on the domain, where the reaction

$$a \; \underset{\longleftarrow}{\overset{\longrightarrow}{}} \; b$$

takes place, we have

$$\frac{\partial a}{\partial t} = \frac{\partial^2 a}{\partial x^2} + b - a$$

and

$$\frac{\partial b}{\partial t} = \frac{\partial^2 b}{\partial x^2} - b + a$$

Estimate the rate at which a and b go to zero.

This simple looking problem is not so simple.

6. Write the solution to the problem

$$\frac{\partial c}{\partial t} = \frac{\partial^2 c}{\partial x^2}, \quad 0 < x < 1$$

$$c(x = 0) = 0, \qquad c(x = 1) = 1$$

and

$$c(t = 0) = 0$$

Do this by expanding $c(x, t)$ in the solutions to the eigenvalue problem

$$\frac{d^2\Psi}{dx^2} + \lambda^2\Psi = 0$$

and

$$\Psi(x = 0) = 0, \qquad \Psi(x = 1) = 0$$

Take the limit of the solution as $t \to \infty$ and verify that this is the Fourier series for $f(x) = x$ on the interval $-1 \le x \le 1$. This series converges for all values of x and defines a periodic function of period 2. Its value when $x = 1$ is zero, its limit as $x \to 1^-$ is one. It is the solution to the problem

$$0 = \frac{d^2c}{dx^2}, \quad 0 < x < 1$$

and

$$c(x = 0) = 0, \qquad c(x = 1) = 1$$

This can be verified by construction but not by direct substitution, as the series derived by termwise differentiation does not converge.

7. Let D be the linear differential operator

$$D = \frac{d^2}{dx^2}, \quad 0 < x < \varepsilon$$

$$D = \beta\frac{d^2}{dx^2}, \quad \varepsilon < x < 1$$

Then using the integration by parts formulas

$$\int_0^1 u Dv\, dx = \left[u \frac{dv}{dx} \right]_0^\varepsilon + \left[u\beta \frac{dv}{dx} \right]_\varepsilon^1 - \int_0^\varepsilon \frac{du}{dx} \frac{dv}{dx}\, dx - \int_\varepsilon^1 \frac{du}{dx} \beta \frac{dv}{dx}\, dx$$

$$= \left[u \frac{dv}{dx} - \frac{du}{dx} v \right]_0^\varepsilon + \left[u\beta \frac{dv}{dx} - \beta\frac{du}{dx} v \right]_\varepsilon^1 + \int_0^1 Duv\, dx$$

show that the solutions to the eigenvalue problem

$$D\Psi + \lambda^2 \Psi = 0, \quad 0 < x < 1$$

$$\Psi(x = 0) = 0, \quad \Psi(x = 1) = 0$$

$$\Psi(x = \varepsilon^-) = \Psi(x = \varepsilon^+)$$

and

$$\frac{d\Psi}{dx}(x = \varepsilon^-) = \beta \frac{d\Psi}{dx}(x = \varepsilon^+)$$

satisfy the usual orthogonality, etc., conditions

Use the solutions to this eigenvalue problem to obtain a formula for the solution to a diffusion problem where a solute, initially distributed over a layer composed of two immiscible solvents, diffuses out of the layer when it is placed in contact at $t = 0$ with two solute free reservoirs that maintain the solute concentration at its edges at $c = 0$. The solute concentration then satisfies

$$\frac{\partial c}{\partial t} = \begin{cases} D_1 \dfrac{\partial^2 c}{\partial x^2}, & 0 < x < x_{12} \\[4mm] D_2 \dfrac{\partial^2 c}{\partial x^2}, & x_{12} < x < \ell \end{cases}$$

and

$$c(x = 0) = 0, \qquad c(x = \ell) = 0$$

where $c(t = 0)$ is assigned. At $x = x_{12}$, the solvent-solvent interface, phase equilibrium obtains and is given by the equilibrium distribution ratio α. The rate of solute diffusion is continuous there.

8. Solve the eigenvalue problem

$$\frac{d^2\psi}{dx^2} + \lambda^2\psi = c, \quad \psi = 0 \quad \text{at} \quad x = \pm 1, \qquad \int_{-1}^{+1} \psi\, dx = 0$$

9. Solve the eigenvalue problem

$$\frac{d^2\psi}{dx^2} + \lambda^2\psi = \int_{-1}^{+1} \psi\, dx, \quad \psi = 0 \quad \text{at} \quad x = \pm 1$$

In this and the preceding problem, by using our integration by parts formula, you can prove that λ^2 must be real, etc.

Lecture 15

Two Examples of Diffusion in One Dimension

15.1 Instability due to Diffusion

Diffusion is always smoothing and ordinarily it is stabilizing; nonetheless there is a paper by Segel and Jackson (L. A. Segel, J. L. Jackson, *J. Theoretical Biology*, (1972), <u>37</u>, 545) in which it is proposed that diffusion is destabilizing, causing non uniformities to appear in an otherwise stable, spatially uniform system.

We present the model. Two chemical species occupy the real line, $-\infty < x < \infty$. We denote their concentrations c_1 and c_2 and refer to the first as the activator, the second as the inhibitor.

The model is

$$\frac{\partial c_1}{\partial t} = D_1 \frac{\partial^2 c_1}{\partial x^2} + R_1 (c_1, c_2)$$

$$\frac{\partial c_2}{\partial t} = D_2 \frac{\partial^2 c_2}{\partial x^2} + R_2 (c_1, c_2)$$

where the equations

$$R_1 (c_1, c_2) = 0 = R_2 (c_1, c_2)$$

have a solution $c_1 = c_1^{(0)}, c_2 = c_2^{(0)}$ and hence $c_1 = c_1^{(0)}, c_2 = c_2^{(0)}$ is a spatially uniform, time independent solution of our model problem.

We would like to know if we can see this solution in an experiment.

Interest in this stems from the fact that an activator and an inhibitor might be in balance in a cell wall where there also reside receptors picking up signals that the cell must respond to. Our uniform activator-inhibitor state may not persist in the face of perturbations due to such signals and our aim might be to find conditions that cause the cell to spring into action.

To do this we introduce small perturbations to $c_1^{(0)}$ and $c_2^{(0)}$, denoted ξ_1 and ξ_2, and derive

$$
\frac{\partial}{\partial t}
\begin{pmatrix} \xi_1 \\ \xi_2 \end{pmatrix}
=
\begin{pmatrix} D_1 \dfrac{\partial^2}{\partial x^2} & 0 \\ 0 & D_2 \dfrac{\partial^2}{\partial x^2} \end{pmatrix}
\begin{pmatrix} \xi_1 \\ \xi_2 \end{pmatrix}
+
\begin{pmatrix} a_{11} & a_{12} \\ a_{21} & a_{12} \end{pmatrix}
\begin{pmatrix} \xi_1 \\ \xi_2 \end{pmatrix}
$$

where the algebraic signs of the a_{ij} have a physical meaning.

Because species 1 is an activator, it causes both species to grow, hence we have

$$a_{11} > 0, \qquad a_{21} > 0$$

Likewise, because species 2 is an inhibitor, we have

$$a_{12} < 0, \qquad a_{22} < 0$$

In the absence of diffusion, the uniform state is assumed to be stable. Thus we have

$$a_{11} + a_{22} < 0 \qquad \text{(trace condition)}$$

and

$$a_{11} a_{22} - a_{21} a_{12} > 0 \qquad \text{(det condition)}$$

and this implies

$$(-a_{22}) > a_{11}$$

and

$$a_{21}(-a_{12}) > a_{11}(-a_{22})$$

Then, for a perturbation of wave number k we write

$$\xi_1 = a_1 \cos kx e^{\sigma t}$$

$$\xi_2 = a_2 \cos kx e^{\sigma t}$$

whereupon we have

$$\sigma \begin{pmatrix} a_1 \\ a_2 \end{pmatrix} = \left(-k^2 \begin{pmatrix} D_1 & 0 \\ 0 & D_2 \end{pmatrix} + A \right) \begin{pmatrix} a_1 \\ a_2 \end{pmatrix}, \qquad A = \begin{pmatrix} a_{11} & a_{12} \\ a_{21} & a_{22} \end{pmatrix}$$

and we see that the σ's, the growth constants, are eigenvalues of the matrix

$$\begin{pmatrix} a_{11} - k^2 D_1 & a_{12} \\ a_{21} & a_{22} - k^2 D_2 \end{pmatrix}$$

where, because x runs from $-\infty$ to $+\infty$ all values of k are admissible. Had x a finite range the admissible k's would be limited by the end conditions, eg., Neumann conditions, periodic conditions, etc.

Our system is stable to long wave length perturbations ($k^2 = 0$), by construction, and to small wave length perturbations ($k^2 \to \infty$) due to diffusive smoothing.

If there is an intermediate range of wave numbers where stability is lost, we say that the instability is brought about by diffusion, though both A and $-k^2 \begin{pmatrix} D_1 & 0 \\ 0 & D_2 \end{pmatrix}$ are stable matrices.

Upon setting

$$\det\left(-k^2\begin{pmatrix} D_1 & 0 \\ 0 & D_2 \end{pmatrix} + A - \sigma I\right) = 0$$

we obtain $\mathrm{Re}\,\sigma < 0$ iff

$$-k^2 D_1 + a_{11} - k^2 D_2 + a_{22} < 0 \qquad \text{(trace condition)}$$

and

$$\left(-k^2 D_1 + a_{11}\right)\left(-k^2 D_2 + a_{22}\right) - a_{21}a_{12} > 0 \qquad \text{(det condition)}$$

Because we have stability at $k^2 = 0$, viz., $a_{11} + a_{22} < 0$, we see that

$$-k^2 D_1 + a_{11} - k^2 D_2 + a_{22} < 0$$

and the trace condition is satisfied for all values of k^2.

Turning to the det condition then, we have

$$\det\left(k^2\right) = \underbrace{D_1 D_2 k^4}_{(+)} - \left(D_1\,a_{22} + D_2\,a_{11}\right)k^2 + \underbrace{a_{11}a_{22} - a_{21}a_{12}}_{(+)}$$

and we observe that if $\det\left(k^2\right)$ is increasing at $k^2 = 0$, it will continue to increase as k^2 increases and it will always be positive. Thus to have a chance of finding an instability caused by the presence of diffusion we must have $\dfrac{d}{dk^2}\det\left(k^2\right)$ negative at $k^2 = 0$, i.e., we must have

$$D_1 a_{22} + D_2 a_{21} > 0$$

whence we need

$$D_2 > D_1$$

due to $(-a_{22}) > a_{11}$.

A graph of $\det (k^2)$ vs k^2 appears as shown in the sketch:

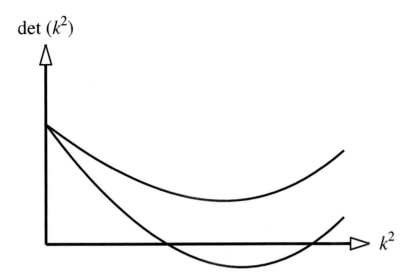

and we see that to have an instability $\det (k^2)$ must be negative at its least value. The least value occurs at

$$2D_1 D_2 k^2 - (D_1 a_{22} + D_2 a_{11}) = 0$$

where the value of $\det (k^2)$ is

$$-\frac{1}{4} \frac{(D_1 a_{22} + D_2 a_{11})^2}{D_1 D_2} + a_{11} a_{22} - a_{21} a_{12}$$

and hence we must have

$$\Big(\underbrace{D_1 a_{22}}_{(-)} + \underbrace{D_2 a_{11}}_{(+)} \Big)^2 > 4 D_1 D_2 \underbrace{\Big(a_{11} a_{22} - a_{21} a_{12} \Big)}_{(+)}$$

and this can be satisfied if D_1 is small enough or D_2 is large enough.

Thus we can arrange an instability and we can understand it: At a site where a perturbation due to an outside signal increases the activator concentration, and hence the rates of production of the activator and the inhibitor are both increased, instability obtains if the activator remains in place (low D_1) and the inhibitor diffuses away (high D_2), leaving the activator to reinforce itself. This is referred to as diffusion induced symmetry breaking.

Given an unstable range of k's, we could go on and get an estimate of the non uniform state that appears. This is done in Grindrod's book "*The Theory and Applications of Reaction-Diffusion Equations: Patterns and Waves*." We turn instead to the Petri Dish problem where this is easier to do.

15.2 Petri Dish

A steady solution to our Petri Dish problem (Lecture 1) satisfies, in scaled variables,

$$0 = \frac{d^2c}{dx^2} + \lambda F(c)$$

where $F(0) = 0$, $F'(0) > 0$ and where $c = 0$ at $x = 0, 1$.

We already know that $c = c_0 = 0$ is a solution for all values of λ, and we know that this solution is stable to small perturbations for all values of $\lambda < \lambda_{\text{crit}}$ (see also §15.3); beyond $\lambda = \lambda_{\text{crit}}$ it is unstable and we wish to find out what the new solution looks like for λ just beyond λ_{crit}.

To find the non zero steady solution branch emerging from λ_{crit} as λ increases we write

$$c = c_0 + \varepsilon\, c_1 + \frac{1}{2}\varepsilon^2 c_2 + \cdots$$

where $c_0 = 0$, and we try to find how λ depends on ε or vice versa.

There is a method, called *dominant balance*, for doing this and it is explained in the books by Bender and Orzag and by Grindrod. ("*Advanced Mathematical Methods for Scientists and Engineers*" and "*The Theory and Application of Reaction-Diffusion Equations*.") But we can illustrate the main ideas by trying two possibilities:

(1) $\lambda = \lambda_{\text{crit}} + \varepsilon$

and

(2) $\lambda = \lambda_{\text{crit}} + \frac{1}{2}\varepsilon^2$

First we expand the nonlinear part of the problem in powers of ε, viz.,

$$F(c) = F(c_0) + \varepsilon F'(c_0) c_1 + \frac{1}{2} \varepsilon^2 \left(F'(c_0) c_2 + F''(c_0) c_1^2 \right)$$

$$+ \frac{1}{6} \varepsilon^3 \left(F'(c_0) c_3 + 3F''(c_0) c_1 c_2 + F'''(c_0) c_1^3 \right) + \cdots$$

and then, assuming expansion (1), we have, using $F(c_0) = 0$,

$$\frac{d^2 c_1}{dx^2} + \lambda_{\text{crit}} F'(c_0) c_1 = 0, \qquad c_1 = 0 \quad \text{at} \quad x = 0, 1$$

and

$$\frac{d^2 c_2}{dx^2} + \lambda_{\text{crit}} F'(c_0) c_2 = -2F'(c_0) c_1 - \lambda_{\text{crit}} F''(c_0) c_1^2, \qquad c_2 = 0 \quad \text{at} \quad x = 0, 1$$

at order ε and at order $\dfrac{1}{2} \varepsilon^2$

The first problem is the eigenvalue problem that we solved earlier, Lecture 1, to obtain λ_{crit}, and we found $\lambda_{\text{crit}} F'(c_0) = \pi^2$. Whence we have

$$c_1 = A \sin \pi x$$

and our job is to find the factor A.

For our expansion, here (1), to be correct we need to be able to find $c_1, c_2, \ldots$ and hence we move on to the second order problem and we notice that the homogeneous part of this problem is the eigenvalue problem and we already know that it has a non zero solution, viz., c_1. Thus a solvability condition must be satisfied in order for the calculation to continue. and we find this condition by multiplying the second problem by c_1 the first by c_2, subtracting and integrating over $0 < x < 1$.

The result is

$$\int_0^1 c_1 \left(2F'(c_0) c_1 + \lambda_{\text{crit}} F''(c_0) c_1^2 \right) dx = 0$$

and we look at two functions $F(c)$:

First we suppose

$$F(c) = c - c^2 \begin{cases} F'(0) = 1 \\ F''(0) = -2 \\ F'''(0) = 0 \end{cases}$$

whereupon solvability at second order tells us that

$$2F'(0) A^2 \int_0^1 \sin^2 \pi x \, dx + \lambda_{\text{crit}} F''(0) A^3 \int_0^1 \sin^3 \pi x \, dx = 0$$

This determines A, hence our solution just beyond λ_{crit} is

$$c = (\lambda - \lambda_{\text{crit}}) A \sin \pi x$$

and expansion (1) appears to be correct.

Second we suppose

$$F(c) = c - c^3 \begin{cases} F'(0) = 1 \\ F''(0) = 0 \\ F'''(0) = -6 \end{cases}$$

and now we find, at second order, that $A = 0$. Hence we conclude that expansion (1) fails in this case.

Turning to expansion (2), and continuing with our second case, we have, at first and second orders,

$$\frac{d^2 c_1}{dx^2} + \lambda_{\text{crit}} F'(c_0) c_1 = 0, \qquad c_1 = 0 \quad \text{at} \quad x = 0, 1$$

and

$$\frac{d^2 c_2}{dx^2} + \lambda_{\text{crit}} F'(c_0) c_2 = -\lambda_{\text{crit}} F''(c_0) c_1^2, \qquad c_2 = 0 \quad \text{at} \quad x = 0, 1$$

whence

$$c_1 = A \sin \pi x$$

as before, but now solvability at second order, viz.,

$$\int_0^1 c_1 \left(-\lambda_{\text{crit}} F''(c_0) c_1^2 \right) dx = 0$$

is satisfied for all values of A due to $F''(c_0) = 0$. Hence we must go to third order where we have

$$\frac{d^2 c_3}{dx^2} + \lambda_{\text{crit}} F'(c_0) c_3 = -3\lambda_{\text{crit}} F''(c_0) c_1 c_2 - \lambda_{\text{crit}} F'''(c_0) c_1^3 - 3F'(c_0) c_1$$

$$c_3 = 0 \quad \text{at} \quad x = 0, 1$$

and solvability must be satisfied if we are to be able to find c_3 and continue our calculations. Ordinarily c_2 would be needed, and it can be found, but it is not needed here because $F''(c_0) = 0$. (It is not usually true that the condition needed to satisfy solvability at second order eliminates the need for c_2 at third order.) The solvability condition at third order is

$$-\lambda_{\text{crit}} F'''(c_0) \int_0^1 c_1^4 \, dx - 3F'(c_0) \int_0^1 c_1^2 \, dx = 0$$

and this determines A^2 as

$$A^2 = \frac{1}{2\pi^2} \frac{\displaystyle\int_0^1 \sin^2 \pi x \, dx}{\displaystyle\int_0^1 \sin^4 \pi x \, dx}$$

where we have used $\lambda_{\text{crit}} F'(c_0) = \pi^2$, $F'(c_0) = 1$ and $F'''(c_0) = -6$.

Our solution branch as λ passes through λ_{crit} is then

$$c = \sqrt{2(\lambda - \lambda_{\text{crit}})} A \sin \pi x$$

and we see that how our solution depends on λ, for λ just beyond λ_{crit}, differs as the nonlinearity differs.

15.3   A Lecture 14 Problem

We wish to find the critical value of λ at which the solution $c = 0$, which holds for all λ, becomes unstable. The model is

$$\frac{\partial c}{\partial t} = \frac{\partial^2 c}{\partial x^2} + \lambda F(c)$$

where $c = 0$ at $x = 0$ and at $x = 1$, where $F(0) = 0$ and where $F'(0) > 0$.

The perturbation equation is

$$\frac{\partial c_1}{\partial t} = \frac{\partial^2 c_1}{\partial x^2} + \lambda F'(0) c_1$$

where $c_1 = 0$ at $x = 0, 1$.

The eigenvalue problem is

$$\frac{\partial^2 \psi}{\partial x^2} + \mu^2 \psi = 0$$

where $\psi = 0$ at $x = 0, 1$.

Its solutions are

$$\psi = A \sin \mu x \qquad \mu^2 = \pi^2, \ (2\pi)^2, \ \dots$$

Expanding c_1 in the eigenfunctions of $\dfrac{d^2}{dx^2}$ we have

$$c_1 = \sum \langle \psi, c_1 \rangle \psi$$

where

$$\frac{d}{dt} \langle \psi, c_1 \rangle = \left(-\mu^2 + \lambda F'(0) \right) \langle \psi, c_1 \rangle$$

As λ increases from zero, $-\mu^2 + \lambda F'(0)$ starts out negative. It first becomes zero at $\lambda = \lambda_{\text{crit}}$, corresponding to $\mu^2 = \mu_1^2$, and thereafter remains positive, viz.,

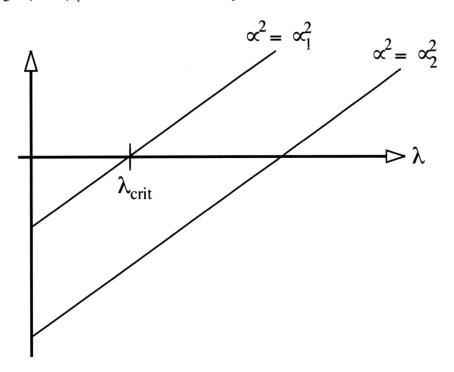

15.4   Home Problems

1. Solve the Segel and Jackson problem on a bounded, one dimensional domain, assuming homogeneous Neumann conditions at the ends.

Lecture 16

Diffusion in Bounded, Three Dimensional Domains

16.1 The Use of the Eigenfunctions of ∇^2 to Solve Inhomogeneous Problems

In this lecture we do not assume the boundary conditions are homogeneous and we do not assume the domain is one dimensional. Thus, we replace the differential operator $\dfrac{d^2}{dx^2}$ by ∇^2 corresponding to diffusion in more than one dimension. Our emphasis will be on ∇^2 and its eigenvalue problem.

We suppose that at time $t = 0$ a solute is distributed in some specified way throughout a solvent occupying a bounded region of three dimensional space. We let V denote the region as well as its volume and we suppose that V is separated from the remainder of physical space, over which we have control, by a piecewise smooth surface S.

Our interest is in determining how our initial distribution of solute changes as time goes on and we assume that its concentration satisfies

$$\frac{\partial c}{\partial t} = \nabla^2 c + Q\left(\overrightarrow{r}, t\right), \ \ t > 0, \ \ \overrightarrow{r} \in V$$

where $c(t=0)$ is assigned $\forall \; \vec{r} \in V$ and where we measure lengths in units of a length L, say $L = V^{\frac{1}{3}}$, and measure time in units of L^2/D. This sets the value of the diffusion coefficient to one in the scaled units. The assigned functions $Q(\vec{r},t)$ and $c(t=0)$ specify the source of solute in the region and the initial distribution of solute there, but the problem is indeterminate until we go on and specify the conditions on S, the boundary of V, which define the effect of the surroundings on what is going on inside V. To do this we divide S into three parts S_1, S_2 and S_3 and on each of these we specify a definite condition:

$$c(\vec{r},t) = g_1(\vec{r},t), \quad \vec{r} \in S_1, \; t > 0$$

$$\vec{n} \cdot \nabla c(\vec{r},t) = g_2(\vec{r},t), \quad \vec{r} \in S_2, \; t > 0$$

and

$$\vec{n} \cdot \nabla c(\vec{r},t) + \beta c(\vec{r},t) = g_3(\vec{r},t), \quad \vec{r} \in S_3, \; t > 0$$

where $\vec{n}$ is the outward unit normal vector to S and where β is assigned on S_3. It is assumed to be real and it may not be constant. Ordinarily it is a positive constant. Thus we specify the solute concentration itself on S_1, the rate of solute diffusion across S_2 and a linear combination of these on S_3 by specifying the functions g_1, g_2, and g_3 defined on S_1, S_2 and S_3 $\forall \, t > 0$. The conditions on S_1, S_2 and S_3 are called Dirichlet, Neuman and Robin conditions and if $S_1 = S$ the problem is called a Dirichlet problem, etc.

Our goal here is to learn how to write the solution to this problem. To do this we introduce the eigenvalue problem

$$\nabla^2 \psi = -\lambda^2 \psi, \quad \vec{r} \in V$$

and denote its solutions, the eigenfunctions and the eigenvalues, $\psi_1, \psi_2, \ldots$ corresponding to $\lambda_1^2, \lambda_2^2, \ldots$

We face two problems. The first is to specify the boundary conditions on S_1, S_2 and S_3 that the eigenfunctions must satisfy in order that they can be used in solving for c. The second is to prove that eigenfunctions are an orthogonal set of functions so that the coefficients in the expansion of c

in the eigenfunctions are the Fourier coefficients of c.

Thus we complete the statement of the eigenvalue problem and then prove orthogonality of the eigenfunctions in the inner product

$$\langle \phi, \psi \rangle = \iiint\limits_{V} \overline{\phi} \psi \, dV$$

We assume our set of eigenfunctions is complete and we introduce the normalization

$$\langle \psi_i, \psi_i \rangle = 1$$

Integration by Parts Formulas

By asking what we must do to determine the coefficients in the expansion of the solution to our problem in the eigenfunctions of ∇^2, we will discover the conditions that the eigenfunctions must satisfy on S_1, S_2 and S_3 and to do this we need two integration by parts formulas for functions defined on V, then the argument is much as it was in Lecture 14.

In Brand's book, "*Vector and Tensor Analysis*," there are many very general integration theorems. But all we need are the three dimensional forms of our earlier integration by parts formulas. If ϕ and ψ are sufficiently smooth these are:

$$\iiint\limits_{V} \phi \nabla^2 \psi \, dV = \iint\limits_{S} \phi \vec{n} \cdot \nabla \psi \, dA - \iiint\limits_{V} \nabla \phi \cdot \nabla \psi \, dV$$

and

$$\iiint\limits_{V} \phi \nabla^2 \psi \, dV = \iint\limits_{S} \{ \phi \vec{n} \cdot \nabla \psi - \psi \vec{n} \cdot \nabla \phi \} \, dA + \iiint\limits_{V} \psi \nabla^2 \phi \, dV$$

where ϕ and ψ are real or complex valued functions of $\vec{r}$ defined throughout V and on its boundary. These formulas are called Green's first and second theorems and we can write the solution to our diffusion problem using them.

To begin we assume that the solution to the diffusion equation can be expanded in the eigenfunctions of ∇^2 and that the coefficients in the expansion are the Fourier coefficients of $c\left(\vec{r}, t\right)$. Thus we write

$$c\left(\vec{r}, t\right) = \sum_{i=1}^{\infty} c_i\left(t\right)\psi_i\left(\vec{r}\right)$$

and our job is to find the coefficients $c_i\left(t\right)$, viz., to find

$$c_i\left(t\right) = \langle\psi_i, c\rangle = \iiint_V \overline{\psi}_i c\, dV.$$

To derive the equation satisfied by $c_i\left(t\right)$ we multiply the diffusion equation by $\overline{\psi}_i$ and integrate over V obtaining

$$\iiint_V \overline{\psi}_i \frac{\partial c}{\partial t} dV = \iiint_V \overline{\psi}_i \nabla^2 c\, dV + \iiint_V \overline{\psi}_i Q dV.$$

Then, using Green's second theorem to turn the first term on the right hand side into terms we can evaluate, we have

$$\frac{d}{dt}\langle\psi_i, c\rangle = \iint_S dA\left\{\overline{\psi}_i \vec{n}\cdot\nabla c - c\vec{n}\cdot\nabla\overline{\psi}_i\right\} + \langle\nabla^2\psi_i, c\rangle + \langle\psi_i, Q\rangle$$

and we write this

$$\frac{d}{dt}\langle\psi_i, c\rangle = -\overline{\lambda}_i^2\langle\psi_i, c\rangle + \langle\psi_i, Q\rangle$$

$$+ \iint_{S_1} dA\left\{\overline{\psi}_i\vec{n}\cdot\nabla c - c\vec{n}\cdot\nabla\overline{\psi}_i\right\}$$

$$+ \iint_{S_2} dA\left\{\overline{\psi}_i\vec{n}\cdot\nabla c - c\vec{n}\cdot\nabla\overline{\psi}_i\right\}$$

$$+ \iint_{S_3} dA\left\{\overline{\psi}_i\vec{n}\cdot\nabla c - c\vec{n}\cdot\nabla\overline{\psi}_i\right\}$$

The Boundary Conditions Satisfied by the Eigenfunctions

Assuming that we have the eigenfunctions and the eigenvalues, this is an equation by which we can determine $c_i(t) = \langle \psi_i, c \rangle$, i.e., the coefficient of ψ_i in the solution to our problem. The first two terms on the right hand side present no problem. But in the third term $\vec{n} \cdot \nabla c$ is not specified on S_1, in the fourth term c is not specified on S_2 while in the fifth term neither $\vec{n} \cdot \nabla c$ nor c are specified on S_3. The equation then is indeterminate and it is our choice of the boundary conditions satisfied by ψ_i, which completes the definition of the eigenvalue problem, that removes this indeterminancy. So to make this a determinate equation for $\langle \psi_i, c \rangle$, we put $\psi_i = 0$ on the part of the boundary where c is specified but $\vec{n} \cdot \nabla c$ is not; while on the part of the boundary where $\vec{n} \cdot \nabla c$ is specified but c is not we put $\vec{n} \cdot \nabla \psi_i = 0$; on the remaining part of the boundary where $\vec{n} \cdot \nabla c + \beta c$ is specified we put $\vec{n} \cdot \nabla \psi_i + \beta \psi_i = 0$. Then the differential equation

$$
\begin{aligned}
\frac{d}{dt} \langle \psi_i, c \rangle &= -\overline{\lambda_i^2} \langle \psi_i, c \rangle + \langle \psi_i, Q \rangle \\
&+ \iint\limits_{S_1} -g_1 \vec{n} \cdot \nabla \overline{\psi}_i dA \\
&+ \iint\limits_{S_2} \overline{\psi}_i g_2 dA \\
&+ \iint\limits_{S_3} \overline{\psi}_i g_3 dA
\end{aligned}
$$

and the initial condition

$$
\langle \psi_i, c \rangle (t = 0) = \langle \psi_i, c(t = 0) \rangle
$$

determine the coefficient $\langle \psi_i, c \rangle$ in the expansion

$$
c(\vec{r}, t) = \sum \langle \psi_i, c \rangle \psi_i(\vec{r}).
$$

Each coefficient, $\langle \psi_i, c \rangle$, can be written as the sum of five terms, each one corresponding to one of the sources: $c(t = 0)$, Q, g_1, g_2 and g_3. If , as an example, S_1 is all of S then the sources are

$c\left(t=0\right), Q$ and g_1 and $\langle \psi_i, c \rangle$ is

$$
\begin{aligned}
\langle \psi_i, c \rangle \;=\;& \langle \psi_i, c\left(t=0\right) \rangle\, e^{-\overline{\lambda_i^2}t} \\[1em]
&+ \int_0^t e^{-\overline{\lambda_i^2}(t-\tau)} \langle \psi_i, Q\left(t=\tau\right) \rangle\, d\tau \\[1em]
&- \int_0^t e^{-\overline{\lambda_i^2}(t-\tau)} \iint_S g_1\left(t=\tau\right) \overrightarrow{n} \cdot \nabla \overline{\psi}_i\, dA d\tau
\end{aligned}
$$

This, when multiplied by ψ_i and summed over i, is the solution to our problem as it depends on the three sources of solute: $c\left(t=0\right), Q$ and g_1. Each term, in fact, produces a solution to the diffusion equation corresponding to one of the sources when the other two vanish.

This method of solving for c requires that an eigenvalue problem be solved. Doing this determines a set of eigenfunctions and a way of doing this will be presented in Lecture 17. Then to produce a solution to the diffusion equation acting under a specified set of sources, each eigenfunction must be multiplied by a coefficient $\langle \psi_i, c \rangle$ and the product summed over the set of eigenfunctions. Each coefficient is the solution of a linear, constant coefficient, first order differential equation, independent of every other coefficient. Each coefficient depends in its own way on the sources of the solute , i.e., on $c\left(t=0\right), Q, g_1, g_2$ and g_3, and its dependence on the sources is additive. The coefficient $\langle \psi_i, c \rangle$ depends on t and is a sum of terms each depending on t and each corresponding to one of the sources of the field. This separation of the contributions of the sources carries over to the solution itself and is one form of the principle of superposition satisfied by the solutions to the diffusion equation. It is also satisfied by the solutions to all linear equations.

We now know what our method is and how the sources of the field make their contribution to the solution. We also know what the eigenvalue problem is; it is the homogeneous problem

$$
\nabla^2 \psi = -\lambda^2 \psi, \quad \overrightarrow{r} \in V
$$

$$
\psi = 0, \quad \overrightarrow{r} \in S_1
$$

$$
\overrightarrow{n} \cdot \nabla \psi = 0, \quad \overrightarrow{r} \in S_2
$$

$$\vec{n} \cdot \nabla \psi + \beta \psi = 0, \quad \vec{r} \in S_1$$

The nonzero solutions ψ are called eigenfunctions while the corresponding values of λ^2 are called eigenvalues.

Expanding a Function in a Series of Eigenfunctions

Suppose we wish to approximate a function defined in a region V by a sum

$$f = \sum_{i=1}^{n} c_i \psi_i$$

The question is: how should we assign values to the coefficients c_i?

Ordinarily we expect to have an orthogonal set of functions $\psi_1, \psi_2, \ldots$, viz.,

$$\langle \psi_i, \psi_j \rangle = \iiint\limits_V \overline{\psi}_i \psi_j \, dV = \delta_{ij}$$

and we do not expect this set of functions to be finite in number.

We denote by S_n the n term approximation to f,

$$S_n = \sum_{i=1}^{n} c_i \psi_i$$

hence the error is $f - S_n$ and the mean square error, MSE, viz.,

$$\iiint\limits_V \overline{(f - S_n)} \, (f - S_n) \, dV$$

is positive.

Now we have

$$\overline{(f - S_n)} \, (f - S_n) = \overline{f} f - \overline{S}_n f - \overline{f} S_n + \overline{S}_n S_n$$

$$= \overline{f} f - \sum \overline{c}_i \overline{\psi}_i f - \sum c_i \psi_i \overline{f} + \sum \overline{c}_i \overline{\psi}_i \sum c_j \psi_j$$

and therefore

$$MSE = \iiint_V |f|^2 \, dV - \sum \bar{c}_i \iiint_V \overline{\psi}_i \, f \, dV - \sum c_i \iiint_V \psi_i \, \overline{f} \, dV + \sum \bar{c}_i \, c_i$$

$$= \iiint_V |f|^2 \, dV + \sum_{i=1}^{n} \left| c_i - \iiint_V \overline{\psi}_i f \, dV \right|^2 - \sum_{i=1}^{n} \left| \iiint_V \overline{\psi}_i f \, dV \right|^2$$

due to

$$\left| c_i - \iiint_V \overline{\psi}_i f \, dV \right|^2 = \bar{c}_i c_i - \bar{c}_i \iiint_V \overline{\psi}_i f \, dV - c_i \iiint_V \psi_i \overline{f} \, dV$$

$$+ \iiint_V \overline{\psi}_i f \, dV \iiint_V \psi_i \overline{f} \, dV$$

Hence, we see that only the second term depends on the c_i's and to make MSE least we should set the c_i's to

$$c_i = \iiint_V \overline{\psi}_i \, f \, dV = \langle \, \psi_i, f \, \rangle$$

Then our best n term approximation is

$$\sum_{i=1}^{n} \langle \, \psi_i, f \, \rangle \, \psi_i$$

and its mean square error is

$$\iiint_V |f| \, dV - \sum_{i=1}^{n} |c_i|^2 > 0$$

Upon increasing n, no c_i's already known need to be reevaluated, we have

$$\iiint_V |f|^2 \, dV - \sum_{i=1}^{\infty} |c_i|^2 \geq 0$$

and the infinite series converges, whereupon $c_i \to 0$ as $i \to \infty$.

The coefficients c_i are called the Fourier coefficients of f and we assume

$$\sum_{i=1}^{\infty} |c_i|^2 = \iiint_V |f|^2 \, dV$$

for all functions f of interest. Hence our series for f converges to f in the mean, i.e., the MSE vanishes as $n \to \infty$, and we ordinarily expect to have pointwise convergence almost everywhere in V.

The set of functions ψ_i is then said to be complete.

16.2   The Facts about the Solutions to the Eigenvalue Problem

We can go on and learn something about the eigenfunctions and the eigenvalues of ∇^2 by using Green's two theorems. If ψ and λ^2 satisfy the eigenvalue problem then so also do $\overline{\psi}$ and $\overline{\lambda^2}$ and hence on putting $\phi = \overline{\psi}$ in Green's second theorem we conclude that λ^2 must be real. Then on putting $\phi = \overline{\psi}$ in Green's first theorem we get

$$-\lambda^2 \iiint_V |\psi|^2 \, dV = \iint_S \overline{\psi} \vec{n} \cdot \nabla \psi \, dA - \iiint_V |\nabla \psi|^2 \, dV$$

$$= \iint_{S_3} -\beta |\psi|^2 \, dA - \iiint_V |\nabla \psi|^2 \, dV$$

This is an equation telling us the sign of λ^2. First, if $\beta > 0$, then $\lambda^2 = 0$ cannot be a solution and we have $\lambda^2 > 0$. But, if $\beta = 0$ and S_2 includes S_3, then $\lambda = 0$ and $\psi = $ constant might be a solution. Indeed if $S_2 = S$, and we have a Neumann problem, $\lambda^2 = 0$, $\psi = $ constant $\neq 0$ is one solution to our eigenvalue problem. Otherwise, $\beta > 0$ or $S_1 = S$, we must have $\lambda^2 > 0$. We go on and put $\phi = \overline{\psi}_i$, $\psi = \psi_j$ in Green's second theorem, where ψ_i and ψ_j are solutions to the eigenvalue problem corresponding to distinct eigenvalues, and learn that

$$\langle \psi_i, \psi_j \rangle = 0.$$

This, and the observation that any independent set of eigenfunctions corresponding to the same eigenvalue can be replaced by an orthogonal set of eigenfunctions, shows that the eigenvalue problem determines an orthogonal set of eigenfunctions in the inner product $\langle \phi, \psi \rangle = \iiint_V \overline{\phi} \psi dV$. In fact restricting ∇^2 to the class of functions on V satisfying homogeneous boundary conditions on $S = S_1 + S_2 + S_3$ we have $\langle \phi, \nabla^2 \psi \rangle = \langle \nabla^2 \phi, \psi \rangle$ and we say that ∇^2 is self-adjoint on that class of functions. In a way we have been very lucky. The boundary conditions of physical interest, the plain vanilla inner product and the differential operator ∇^2 have gotten together and given us simple answers to the important questions. In another inner product or for other boundary conditions or for another differential operator we would be required to determine an adjoint operator and adjoint boundary conditions to work out our theory. This comes up again in Lecture 19.

In the next lecture we turn to the question: how do we solve the eigenvalue problem? And we explain the method of separation of variables for doing this. The readers may satisfy themselves that there are places in the foregoing where it is important that the coefficient β in the Robin boundary condition be real and places where it is important that β be positive but nowhere is it required that β be constant. While this is so, in solving the eigenvalue problem by the method of separation of variables it will also be important that β be a constant, or at least be piecewise constant and constant on each coordinate surface.

16.3   The Critical Size of a Region Confining an Autothermal Heat Source

But before we go on to separation of variables we introduce the use of the eigenvalues of ∇^2 to estimate critical conditions in nonlinear problems.

To be definite, let heat be generated in a bounded region V of three dimensional space. To reach a steady condition where the heat lost to the surroundings balances the heat generated in V, the heat must be carried to the boundary of the region by conduction. The boundary is assumed to be in good contact with a heat bath held at a fixed temperature T_0.

The more distant the heat source is from the boundary, the higher the temperature must rise there to conduct it away. If the source is assigned in advance the heat generation can always be balanced by heat conduction. But if the source depends on the temperature, and increases in strength as the temperature increases, then there is a positive feed back and this may create a critical condition beyond which a steady solution cannot be found.

To see why this might be so we can study the problem as the region grows larger in size. Then heat is generated at greater and greater distances from the boundary and the temperature required to dissipate it must increase. The greatest temperature in the region then increases as the size of the region increases and, as this goes on, the source grows stronger. Depending on how fast the strength of the source increases as the temperature increases, there may be a critical size of the region beyond which the heat generated therein cannot be conducted steadily to the surroundings. This critical condition is called a runaway condition and it leads to a thermal explosion.

We suppose that the temperature is the only variable of interest and that the heat source is an exponential function of the temperature. Then in scaled variables we have

$$\nabla^2 u + \mu^2 e^u = 0 \ , \ \ \vec{r} \in V$$

and

$$u = 0 \ , \ \ \vec{r} \in S$$

where the size of the region appears in the constant μ^2 which indicates the strength of the source. This model is introduced by D.A. Frank-Kamenetskii in his book "*Diffusion and Heat Conduction in Chemical Kinetics.*"

In certain simple geometries this equation can be solved and the critical value of μ^2 can be determined. But that is not our aim here. What we do instead is use the eigenvalue problem for ∇^2 in V to estimate the critical value of μ^2. To do this we write our problem:

$$\nabla^2 u + \mu^2 f(u) = 0, \ \ \vec{r} \in V$$

and

$$u = 0 \ , \ \ \vec{r} \in S$$

where $f(u)$ denotes the nonlinear source of heat and where u and $f(u)$ have been scaled so that $f(u=0)=1$ and $\dfrac{df}{du}(u=0)=1$. Our interest is in how large μ^2 can be, consistent with a bounded solution u.

We assume that $f(u)$ can be written

$$f(u) = u + g(u)$$

where $g(u=0)=1$ and $\dfrac{dg}{du}(u=0)=0$ and where we must have $g(u) \geq 0 \, \forall u \geq 0$. Indeed if $\dfrac{d^2 f}{du^2} \geq 0 \, \forall u \geq 0$ then $g(u) \geq 1 \, \forall u \geq 0$.

The solutions to our problem must be non negative and our job is to estimate the range of values of the control variable μ^2 where this obtains. To do this we assume that we have a non negative solution to our problem

$$\nabla^2 u + \mu^2 u + \mu^2 g(u) = 0, \quad \vec{r} \in V$$

and

$$u = 0, \quad \vec{r} \in S$$

for some value of μ^2 and then introduce for comparison the eigenvalue problem for ∇^2 in V, viz.,

$$\nabla^2 \psi + \lambda^2 \psi = 0, \quad \vec{r} \in V$$

and

$$\psi = 0, \quad \vec{r} \in S$$

This problem determines a set of eigenfunctions $\psi_1, \psi_2, \cdots$ and a corresponding set of eigenvalues $\lambda_1^2, \lambda_2^2, \cdots$ where the eigenvalues are greater than zero. If we have $\lambda_1^2 < \lambda_2^2$ then ψ_1 must be singly signed and we take it to be non-negative. Weinberger has a simple argument for this but it can be seen to be true on physical grounds as the long time solution to an initial value problem in diffusion is

$$\langle \psi_1, c\,(t=0)\rangle\, e^{-\lambda_1^2 t}\psi_1$$

To determine a bound on μ^2 we observe that our second integration by parts formula tells us

$$\iiint\limits_V \left\{\psi_1 \nabla^2 u - u\nabla^2\psi_1\right\} dV = \iint\limits_S \overrightarrow{n}\cdot\left\{\psi_1\nabla u - u\nabla\psi_1\right\} dA$$

where the right hand side is zero by the boundary conditions on u and ψ_1. The left hand side is

$$\iiint\limits_V \left\{-\mu^2\psi_1 u - \mu^2\psi_1 g\,(u) + \lambda_1^2 u\psi_1\right\} dV$$

and as this must be zero the value of μ^2 corresponding to u must satisfy

$$\frac{\mu^2}{\lambda_1^2} = \frac{\iiint\limits_V u\psi_1 dV}{\iiint\limits_V u\psi_1 dV + \iiint\limits_V g\,(u)\,\psi_1 dV}$$

and hence we have

$$\mu^2 < \lambda_1^2.$$

This tells us that the critical value of μ^2 lies to the left of λ_1^2.

This is interesting. It tells us that the critical value of μ^2, a control variable in a nonlinear problem, cannot exceed λ_1^2, the slowest diffusion eigenvalue, where λ_1^2 can be determined by solving a linear eigenvalue problem. If we replace e^u by $1+u$, viz., linear heating, we expect to find

$$\mu_{crit}^2 = \lambda_1^2.$$

The heating problem now looks a lot like the eigenvalue problem.

16.4   Solvability

We may ask whether or not a problem presented to us is solvable. The question ordinarily comes up in a steady state problem such as

$$\nabla^2 \phi + \mu^2 \phi = Q \text{ in } V$$

where, for example,

$$\phi = 0 \text{ on } S$$

and Q is assigned throughout V.

We are not asking whether or not the expansion of Q in the eigenfunctions of ∇^2 makes sense. In fact Q ordinarily does not satisfy the same conditions on S as do the eigenfunctions and hence its series expansion most likely converges in norm, not pointwise.

Our problem is to find the coefficients in an expansion of ϕ in the eigenfunctions of ∇^2, viz.,

$$\phi = \sum c_i \psi_i$$

where $c_i = \langle \psi_i, \phi \rangle$. To do this we use our second integration by parts formula to obtain

$$\langle \nabla^2 \psi_i, \phi \rangle + \iint_S \left(\overline{\psi_i} \vec{n} \cdot \nabla \phi - \phi \vec{n} \cdot \nabla \overline{\psi_i} \right) dA + \mu^2 \langle \psi_i, \phi \rangle = \langle \psi_i, Q \rangle$$

Thus we have

$$\left(-\lambda_i^2 + \mu^2 \right) \langle \psi_i, \phi \rangle = \langle \psi_i, Q \rangle$$

and we conclude that so long as μ^2 is not one of the eigenvalues of ∇^2 we can find a solution to our problem. If, however, $\mu^2 = \lambda_i^2$ then Q must be perpendicular to every independent eigenfunction corresponding to λ_i^2. This is the solvability condition.

16.5 ∇^4

The linear differential operator $\nabla^4 = \nabla^2\nabla^2$ appears in problems in the slow flow of viscous fluids and in the deformation of elastic solids. Its eigenvalue problem is

$$\nabla^4\psi = \lambda^4\psi, \quad \vec{r} \in V$$

and any two of the three homogeneous conditions

$$\psi = 0$$

$$\vec{n} \cdot \nabla\psi = 0$$

$$\nabla^2\psi = 0$$

on each part of S. (Here, as earlier, the boundary conditions on the problem to be solved will determine the boundary conditions on ψ. The conditions listed are not all that are physically interesting and to these can be added their linear combinations.)

The integral formulas we need here can be obtained from Green's second theorem. First we put $\nabla^2\psi$ in place of ψ to get

$$\iiint_V \phi\nabla^4\psi\, dV = \iint_S \left\{ \phi\vec{n} \cdot \nabla\nabla^2\psi - \nabla^2\psi\vec{n} \cdot \nabla\phi \right\} dA + \iiint_V \nabla^2\psi\nabla^2\phi\, dV.$$

And then we put $\nabla^2\phi$ in place of ϕ, and use the result to rewrite the second term on the right hand side, to get

$$\iiint_V \phi\nabla^4\psi\, dV = \iint_S \left\{ \phi\vec{n} \cdot \nabla\nabla^2\psi - \nabla^2\psi\vec{n} \cdot \nabla\phi \right\} dA$$

$$+ \iint_S \left\{ \nabla^2\phi\vec{n} \cdot \nabla\psi - \psi\vec{n} \cdot \nabla\nabla^2\phi \right\} dA + \iiint_V \psi\nabla^4\phi\, dV$$

These two formulas can be used in solving equations in ∇^4 in just the same way that Green's first and second theorems can be used in solving the diffusion equation or any other equation in ∇^2. They are especially useful in exhibiting the way in which the sources of the field make their contribution to the field itself.

To get information about the eigenvalues and eigenfunctions of ∇^4 we first observe that if λ^4 and ψ satisfy the eigenvalue problem then so also do $\overline{\lambda}^4$ and $\overline{\psi}$. Hence putting $\overline{\psi}$ in place of ϕ in the second formula we discover that λ^4 must be real. Likewise putting $\overline{\psi}$ in place of ϕ in the first formula we get

$$\lambda^4 \iiint\limits_{V} |\psi|^2 \, dV = \iiint\limits_{V} |\nabla^2 \psi|^2 \, dV$$

and conclude that λ^4 is not negative. In both calculations the integrals over S vanish due to the conditions that we assume the eigenfunctions satisfy on S. To establish orthogonality we can go on and put $\phi = \overline{\psi}_i$, $\psi = \psi_j$ in the second formula, where ψ_i and ψ_j are solutions to the eigenvalue problem corresponding to different eigenvalues, and obtain

$$\langle \psi_i, \psi_j \rangle = \iiint\limits_{V} \overline{\psi}_i \psi_j \, dV = 0.$$

As real eigenvalues and orthogonal eigenfunctions correspond to self-adjoint operators we observe that ∇^4, restricted to the class of functions satisfying the homogeneous boundary conditions of the eigenvalue problem, is self-adjoint, viz.,

$$\langle \phi, \nabla^4 \psi \rangle = \langle \nabla^4 \phi, \psi \rangle.$$

The set of eigenfunctions determined by the eigenvalue problem for ∇^4 can be used in writing the solution to equations such as (This equation is not entirely made up, at least not by me. It is in the book "*Fractal Concepts in Surface Growth*" by Barabasi and Stanley.)

$$\frac{\partial c}{\partial t} = -\nabla^4 c + Q \;, \quad \vec{r} \in V$$

where the values of c and $\vec{n} \cdot \nabla c$ are specified on S $\forall \, t > 0$ and c is specified at $t = 0$ $\forall \, \vec{r} \in V$.

Indeed $\langle \psi_i, c \rangle$ is determined by solving

$$\frac{d}{dt} \langle \psi_i, c \rangle = -\iint\limits_S \left\{ \underline{\overline{\psi_i} \vec{n} \cdot \nabla \nabla^2 c} - \underline{\nabla^2 c \vec{n} \cdot \nabla \overline{\psi_i}} \right\} dA$$

$$-\iint\limits_S \left\{ \underline{\underline{\nabla^2 \overline{\psi_i} \vec{n} \cdot \nabla c}} - \underline{\underline{c \vec{n} \cdot \nabla \nabla^2 \overline{\psi_i}}} \right\} dA$$

$$- \lambda_i^4 \langle \psi_i, c \rangle + \langle \psi_i, Q \rangle$$

where the singly underlined terms vanish as we require $\psi_i = 0$ and $\vec{n} \cdot \nabla \psi_i = 0$ on S and the doubly underlined terms can be determined from the assigned values of c and $\vec{n} \cdot \nabla c$ on S.

A warning is appropriate: the eigenvalue problem

$$\nabla^4 \psi = \lambda^4 \psi$$

is not easy to solve. There are home problems in Lecture 17 which illustrate the difficulty.

16.6   Vector Eigenvalue Problems

To determine a vector field $\vec{v}$ defined in a region of space V and satisfying there an equation in ∇^2, we can expand our solution in the eigenfunctions of ∇^2.

The corresponding eigenvalue problem is then

$$\nabla^2 \vec{\psi} = -\lambda^2 \vec{\psi} \ , \quad \vec{r} \in V$$

where $\vec{\psi}$ satisfies homogeneous conditions on the boundary of V.

To derive some facts about the solutions to this problem, we first use

$$\nabla \cdot \left(\vec{\vec{T}} \cdot \vec{v} \right) = \left(\nabla \cdot \vec{\vec{T}} \right) \cdot \vec{v} + \vec{\vec{T}} : (\nabla \vec{v})^T$$

to get

$$\nabla \cdot \left(\nabla \vec{\psi} \cdot \vec{\phi} \right) = \nabla^2 \vec{\psi} \cdot \vec{\phi} + \underline{\nabla \vec{\psi} : \left(\nabla \vec{\phi} \right)^T}$$

and

$$\nabla \cdot \left(\nabla \vec{\phi} \cdot \vec{\psi} \right) = \nabla^2 \vec{\phi} \cdot \vec{\psi} + \underline{\nabla \vec{\phi} : \left(\nabla \vec{\psi} \right)^T}$$

where the underlined terms are equal as $\operatorname{tr}(AB) = \operatorname{tr}\left(B^T A^T\right)$; we then use

$$\iiint\limits_V \nabla \cdot \vec{v} \, dV = \iint\limits_S dA \, \vec{n} \cdot \vec{v}$$

to get our two integration by parts formulas

$$\iiint\limits_V \nabla^2 \vec{\psi} \cdot \vec{\phi} \, dV = \iint\limits_S dA \, \vec{n} \cdot \left(\nabla \vec{\psi} \cdot \vec{\phi} \right) - \iiint\limits_V \nabla \vec{\psi} : \left(\nabla \vec{\phi} \right)^T dV$$

and

$$\iiint\limits_V \nabla^2 \vec{\psi} \cdot \vec{\phi} \, dV = \iint\limits_S dA \, \vec{n} \cdot \left\{ \nabla \vec{\psi} \cdot \vec{\phi} - \nabla \vec{\phi} \cdot \vec{\psi} \right\} + \iiint\limits_V \nabla^2 \vec{\phi} \cdot \vec{\psi} \, dV$$

To go on we require that

$$\iint\limits_S dA \, \vec{n} \cdot \left\{ \nabla \vec{\psi} \cdot \vec{\phi} - \nabla \vec{\phi} \cdot \vec{\psi} \right\} = 0$$

whenever $\vec{\psi}$ and $\vec{\phi}$ satisfy the homogeneous form of the boundary conditions assigned to $\vec{v}$, then the second formula reduces to

$$\iiint\limits_V \nabla^2 \vec{\psi} \cdot \vec{\phi} \, dV = \iiint\limits_V \nabla^2 \vec{\phi} \cdot \vec{\psi} \, dV$$

If $\vec{\psi}$ is a solution of the eigenvalue problem corresponding to λ^2 so also is $\vec{\psi}$ corresponding to

λ^2. Then putting $\vec{\phi} = \overrightarrow{\psi}$ in the second formula shows that λ^2 must be real and hence that the real and imaginary parts of $\overrightarrow{\psi}$ must also be eigenfunctions corresponding to λ^2. And putting $\vec{\phi} = \overrightarrow{\psi}$ in the first formula shows that

$$-\lambda^2 \iiint_V \overrightarrow{\psi} \cdot \overrightarrow{\psi} \, dV = \iint_S dA \, \vec{n} \cdot \left(\nabla \overrightarrow{\psi} \cdot \overrightarrow{\psi} \right) - \iiint_V \nabla \overrightarrow{\psi} : \left(\nabla \overrightarrow{\psi} \right)^T dV$$

where the second integral on the right hand side is not negative and both integrals are zero if there is an eigenfunction such that $\nabla \overrightarrow{\psi} = \overrightarrow{0}$. If the boundary conditions are such that the first term vanishes, this formula shows that

$$\lambda^2 \geq 0;$$

otherwise the boundary conditions must be such that the sign of the right hand side is the sign of the second term if we must have $\lambda^2 \geq 0$.

The second formula establishes orthogonality. It shows that

$$\left\langle \overrightarrow{\psi}_i, \overrightarrow{\psi}_j \right\rangle = 0$$

whenever $\overrightarrow{\psi}_i$ and $\overrightarrow{\psi}_j$ are eigenfunctions corresponding to different eigenvalues where

$$\left\langle \overrightarrow{\psi}_i, \overrightarrow{\psi}_j \right\rangle = \iiint_V \overrightarrow{\psi}_i \cdot \overrightarrow{\psi}_j dV.$$

16.7   Home Problems

In each of these three problems there is an eigenvalue, denoted σ. You are to see if you can prove it is real.

1. You have an incompressible fluid at rest whose density varies upward: $\rho = \rho_0(z)$. The fluid is inviscid and you are to find out if the rest state is stable to small perturbations.

Upon perturbation, the base density is carried by the perturbation flow and you have

$$\rho_0 \frac{\partial \vec{v_1}}{\partial t} = -\nabla p_1 - \rho_1 g \vec{k}$$

$$\nabla \cdot \vec{v_1} = 0$$

and

$$\frac{\partial \rho_1}{\partial t} + \vec{v_1} \cdot \nabla \rho_0 = 0$$

Assuming solutions of the form

$$v_{x1} = \widehat{v}_{x1}(z) e^{\sigma t} e^{ik_x x} e^{ik_y y}$$

etc.

eliminate $\widehat{v}_{x1}$ and $\widehat{v}_{y1}$ in favor of $\widehat{v}_{z1}$ and $\widehat{p}_1$. Then eliminate $\widehat{p}_1$ obtaining

$$\frac{d}{dz}\left(\rho_0 \frac{d\widehat{v}_{z1}}{dz}\right) - k^2 \rho_0 \widehat{v}_{z1} = -\frac{k^2}{\sigma^2} g \frac{d\rho_0}{dz} \widehat{v}_{z1}$$

where $k^2 = k_x^2 + k_y^2$, where $\rho_0(z) > 0$ and where $\widehat{v}_{z1} = 0$ at $z = 0, H$.

Your job is to prove that $\sigma^2(k^2)$ is real no matter $\dfrac{d\rho_0}{dz}$ and that $\dfrac{d\rho_0}{dz} < 0$ implies $\sigma^2(k^2) < 0$.

2. To take viscosity into account, you again have

$$\rho = \rho_0(z), \quad \vec{v_0} = \vec{0}$$

and

$$\frac{dp_0}{dz} = -\rho_0 g$$

but now your perturbation problem is

$$\rho_0 \frac{\partial v_{x1}}{\partial t} = -\frac{\partial p_1}{\partial x} + \mu \nabla^2 v_{x1}$$

$$\rho_0 \frac{\partial v_{y1}}{\partial t} = -\frac{\partial p_1}{\partial y} + \mu \nabla^2 v_{y1}$$

$$\rho_0 \frac{\partial v_{z1}}{\partial t} = -\frac{\partial p_1}{\partial z} + \mu \nabla^2 v_{z1} - \rho_1 g$$

$$\frac{\partial v_{x1}}{\partial x} + \frac{\partial v_{y1}}{\partial y} + \frac{\partial v_{z1}}{\partial z} = 0$$

and

$$\frac{\partial \rho_1}{\partial t} + v_{z1} \frac{d\rho_0}{dz} = 0$$

Assume a solution

$$v_{x1} = \widehat{v}_{x1}(z) \, e^{ik_x x} \, e^{ik_y y} \, e^{\sigma t}$$

etc.

and eliminating $\widehat{v}_{x1}$ and $\widehat{v}_{y1}$, obtain

$$\rho_0 \, \sigma \left(-\frac{d\widehat{v}_{z1}}{dz} \right) = k^2 \widehat{p}_1 + \mu \left(\frac{d^2}{dz^2} - k^2 \right) \left(-\frac{d\widehat{v}_{z1}}{dz} \right)$$

which must be solved together with

$$\rho_0 \, \sigma \widehat{v}_{z1} = -\frac{d\widehat{p}_1}{dz} + \mu \left(\frac{d^2}{dz^2} - k^2 \right) \widehat{v}_{z1} - \widehat{\rho}_1 g$$

and

$$\sigma \widehat{\rho}_1 + \widehat{v}_{z1} \frac{d\rho_0}{dz} = 0$$

Eliminate $\widehat{\rho}_1$ and $\widehat{p}_1$ to derive an equation for $\widehat{v}_{z1}$ where $\widehat{v}_{z1} = 0 = \dfrac{d\widehat{v}_{z1}}{dz}$ at $z = 0, H$.

You now have an eigenvalue problem, where σ is the eigenvalue. Can you say anything about σ without a calculation?

3. You can account for viscosity more easily by assuming your fluid saturates a porous solid. Then you can use Darcy's law and you have

$$\mu \, \vec{v} = K \left(-\nabla p - \rho g \, \vec{k} \right)$$

$$\nabla \cdot \vec{v} = 0$$

and

$$\frac{\partial \rho}{\partial t} + \vec{v} \cdot \nabla \rho = 0$$

Assume you have a two dimensional problem whereupon your perturbation equations are

$$\mu \, v_{x1} = -K \frac{\partial p_1}{\partial x}$$

$$\mu \, v_{z1} = -K \frac{\partial p_1}{\partial z} - \rho_1 g$$

$$\frac{\partial v_{x1}}{\partial x} + \frac{\partial v_{z1}}{\partial z} = 0$$

and

$$\frac{\partial \rho_1}{\partial t} + v_{z1} \frac{d\rho_0}{dz} = 0$$

Writing

$$v_{x1} = \widehat{v}_{x1} (z) \sin kx \, e^{\sigma t}$$

$$v_{z1} = \widehat{v}_{z1} (z) \cos kx \, e^{\sigma t}$$

$$p_1 = \widehat{p}_1(z) \cos kx \, e^{\sigma t}$$

and

$$\rho_1 = \widehat{\rho}_1(z) \cos kx \, e^{\sigma t}$$

you have

$$\mu \widehat{v}_{x1} = Kk\widehat{p}_1$$

$$\mu \widehat{v}_{z1} = -K \frac{d\widehat{p}_1}{dz} - \rho_1 g$$

$$k\widehat{v}_{x1} + \frac{d\widehat{v}_{z1}}{dz} = 0$$

$$\sigma \widehat{\rho}_1 + \widehat{v}_{z1} \frac{d\rho_0}{dz} = 0$$

Hence you obtain

$$\frac{d^2\widehat{v}_{z1}}{dz^2} - k^2 \widehat{v}_{z1} = -\frac{k^2}{\sigma} \frac{g}{\mu} \frac{d\rho_0}{dz} \widehat{v}_{z1}$$

where $\widehat{v}_{z1} = 0$ at $z = 0, H$. This is an eigenvalue problem where σ is the eigenvalue. Assuming $\dfrac{d\rho_0}{dz}$ is a constant derive a formula for $\sigma(k^2)$.

Lecture 17

Separation of Variables

17.1   Separating Variables in Cartesian, Cylindrical and Spherical Coordinate Systems

In Lecture 16 we found that the eigenvalue problem that must be solved in order to solve the diffusion equation is:

$$\nabla^2 \psi + \lambda^2 \psi = 0, \qquad \vec{r} \in V$$

$$\psi = 0, \qquad \vec{r} \in S_1$$

$$\vec{n} \cdot \nabla \psi = 0, \qquad \vec{r} \in S_2$$

$$\vec{n} \cdot \nabla \psi + \beta \psi = 0, \qquad \vec{r} \in S_3$$

and we learned that the eigenvalues are real and that the eigenfunctions corresponding to different eigenvalues are orthogonal. We can add the term $V(\vec{r})\psi$ to the left hand side of $\nabla^2 \psi + \lambda^2 \psi = 0$, where $V(\vec{r})$ is real valued, and not change the conclusion that the eigenvalues are real and that the eigenfunctions are orthogonal in the plain vanilla inner product. We now turn to the method of solving this eigenvalue problem and present the details in three coordinates systems. The job begins here and is finished in Lecture 20.

The method we use to do this is called separation of variables. To see how it goes we suppose that we have an orthogonal coordinate system which is such that the bounding surface of the region V coincides piecewise with a finite number of coordinate surfaces. Then the first question to ask is this: in what form can we express the solutions to our eigenvalue problem?

If it works out, the method of separation of variables answers this question. The idea is to reduce a three dimensional problem to three one dimensional problems. In certain orthogonal coordinate systems this can be done. It is done by assuming that ψ can be written as the product of three functions, each depending on only one of the three coordinates, substituting this into $\nabla^2 \psi + \lambda^2 \psi = 0$, dividing by ψ and then determining which parts of the result must be constants. We begin by showing how this works out in the Cartesian, cylindrical and spherical coordinate systems.

Cartesian coordinate systems

To separate variables in Cartesian coordinates we substitute

$$\psi\left(x, y, z\right) = X\left(x\right) Y\left(y\right) Z\left(z\right)$$

into

$$\left(\nabla^2 + \lambda^2\right) \psi = \left(\frac{\partial}{\partial x^2} + \frac{\partial}{\partial y^2} + \frac{\partial}{\partial z^2} + \lambda^2\right) \psi = 0$$

and then divide by XYZ to get

$$\frac{1}{X}\frac{d^2 X}{dx^2} + \frac{d^2 Y}{dy^2} + \frac{d^2 Z}{dz^2} + \lambda^2 = 0$$

The first term depends only on x, the second only on y and the third only on z. Because these terms are independent of one another we conclude that each term must be equal to a constant. Denoting these undetermined constants by $-\alpha^2, -\beta^2$ and $-\gamma^2$, we have replaced $\left(\nabla^2 + \lambda^2\right) \psi = 0$

in Cartesian coordinates by the three equations

$$\frac{d^2 X}{dx^2} + \alpha^2 X = 0 \tag{1}$$

$$\frac{d^2 Y}{dy^2} + \beta^2 Y = 0 \tag{2}$$

and

$$\frac{d^2 Z}{dz^2} + \gamma^2 Z = 0 \tag{3}$$

Cylindrical coordinate systems

To separate variables in cylindrical coordinates we substitute

$$\psi(r, \theta, z) = R(r)\,\Theta(\theta)\,Z(z)$$

into

$$\left(\nabla^2 + \lambda^2\right)\psi = \left\{\frac{1}{r}\frac{\partial}{\partial r}\left(r\frac{\partial}{\partial r}\right) + \frac{1}{r^2}\frac{\partial^2}{\partial\theta^2} + \frac{\partial^2}{\partial z^2} + \lambda^2\right\}\psi = 0$$

and then divide by $R\Theta Z$ to get

$$\frac{1}{R}\frac{1}{r}\frac{d}{dr}\left(r\frac{dR}{dr}\right) + \frac{1}{r^2\Theta}\frac{d^2\Theta}{d\theta^2} + \frac{1}{Z}\frac{d^2 Z}{dz^2} + \lambda^2 = 0$$

We conclude first that $\dfrac{1}{Z}\dfrac{d^2 Z}{dz^2}$ must be a constant and then that $\dfrac{1}{\Theta}\dfrac{d^2\Theta}{d\theta^2}$ must be a constant. Denoting these constants by $-\gamma^2$ and $-m^2$ we have replaced $\left(\nabla^2 + \lambda^2\right)\psi = 0$ in cylindrical coordinates by the three equations

$$\frac{d^2 Z}{dz^2} + \gamma^2 Z = 0 \tag{4}$$

$$\frac{d^2\Theta}{d\theta^2} + m^2\Theta = 0 \tag{5}$$

and

$$\frac{1}{r}\frac{d}{dr}\left(r\frac{dR}{dr}\right) + \left(\lambda^2 - \gamma^2 - \frac{m^2}{r^2}\right)R = 0 \tag{6}$$

Spherical coordinate systems

To separate variables in spherical coordinates we substitute

$$\psi\left(r,\theta,\phi\right) = R\left(r\right)\Theta\left(\theta\right)\Phi\left(\phi\right)$$

into

$$\left(\nabla^2 + \lambda^2\right)\psi = \left\{\frac{1}{r^2}\frac{\partial}{\partial r}\left(r^2\frac{\partial}{\partial r}\right) + \frac{1}{r^2\sin\theta}\frac{\partial}{\partial\theta}\left(\sin\theta\frac{\partial}{\partial\theta}\right) + \frac{1}{r^2\sin^2\theta}\frac{\partial^2}{\partial\phi^2} + \lambda^2\right\}\psi$$

and divide by $R\Theta\Phi$ to get

$$\frac{1}{R}\frac{1}{r^2}\frac{d}{dr}\left(r^2\frac{dR}{dr}\right) + \frac{1}{r^2}\frac{1}{\Theta}\frac{1}{\sin\theta}\frac{d}{d\theta}\left(\sin\theta\frac{\partial\Theta}{\partial\theta}\right) + \frac{1}{r^2\sin^2\theta}\frac{1}{\Phi}\frac{d^2\Phi}{d\phi^2} + \lambda^2 = 0$$

Now we see that $\dfrac{1}{\Phi}\dfrac{d^2\Phi}{d\phi^2}$ must be a constant. Calling this $-m^2$, we then see that

$$\frac{1}{\Theta}\frac{1}{\sin\theta}\frac{d}{d\theta}\left(\sin\theta\frac{d\Theta}{d\theta}\right) - \frac{m^2}{\sin^2\theta}$$ must be a constant. Calling this constant $-\ell\left(\ell+1\right)$ we have

replaced $\left(\nabla^2 + \lambda^2\right)\psi = 0$ in spherical coordinates by

$$\frac{d^2\Phi}{d\phi^2} + m^2\Phi = 0 \tag{7}$$

$$\frac{1}{\sin\theta}\frac{d}{d\theta}\left(\sin\theta\frac{d\Theta}{d\theta}\right) + \left(\ell\left(\ell+1\right) - \frac{m^2}{\sin^2\theta}\right)\Theta = 0 \tag{8}$$

and

$$\frac{1}{r^2}\frac{d}{dr}\left(r^2\frac{dR}{dr}\right) + \left(\lambda^2 - \frac{\ell\left(\ell+1\right)}{r^2}\right)R = 0 \tag{9}$$

For Cartesian, cylindrical and spherical coordinate systems we have now reduced the problem of solving $\nabla^2\psi + \lambda^2\psi = 0$ to the problem of solving nine second order, linear, homogeneous ordinary differential equations, three for each coordinate system.

The homogeneous boundary conditions satisfied by ψ lead to homogeneous boundary conditions that must be satisfied by each of the factors making up ψ. And each of these factors satisfies a homogeneous ordinary differential equation depending on an undetermined constant. For arbitrary values of these constants the solutions must be zero. Our job will be to determine the special values of these constants for which the solutions are other than zero.

Indeed each of these problems is a one dimensional eigenvalue problem in its own right and some of them are one dimensional forms of eigenvalue problem for ∇^2, while all of them are one dimensional forms of the eigenvalue problem for $\nabla^2 + V(\vec{r})$.

In Lecture 19 we will describe in a little more detail the elementary facts about linear ordinary differential equations but for now we assume only that each of these equations has two independent solutions, observing that these two solutions can be written in many ways. Taking equation (1) as an example we can write its general solution as

$$A\cos\alpha x + B\sin\alpha x$$

or

$$Ae^{i\alpha x} + Be^{-i\alpha x}$$

or

$$A\cosh i\alpha x + B\sinh i\alpha x$$

These familiar functions satisfy our needs in Eqs. (1), (2), (3), (4), (5) and (7). Equations (6), (8) and (9) have solutions that are less familiar. For instance we denote by J_m and Y_m two independent solutions of Eq. (6) which is called Bessel's equation. And while these functions may not be as familiar as cosine and sine, Watson's book "*Theory of Bessel Functions*" has nearly 1,000 pages of information on these and related functions and so J_m and Y_m are very familiar to some people. The same is true of the solutions of Eqs. (8) and (9).

While the solutions of each of the nine equations are denoted by special symbols, in every instance the symbols stand for power series, either infinite or finite, or power series multiplied by familiar functions. The power series solutions are determined by what is called the method of Frobenius and we will show how this works by using Eq. (8) as an example in Lecture 20.

For now we observe only that $J_0(z)$ is the name assigned to the series $\displaystyle\sum_{m=0}^{\infty} \frac{(-1)^m \left(\frac{1}{2}z\right)^{2m}}{(m!)^2}$

which satisfies

$$\left(\frac{1}{z}\frac{d}{dz}z\frac{d}{dz} + 1\right)\psi = 0$$

$$\psi(z = 0) = 1$$

and

$$\psi'(z = 0) = 0$$

Likewise $\cos z$ is the name assigned to the series

$$\sum_{m=0}^{\infty} (-1)^m \frac{z^{2m}}{(m!)^2} = 1 - \frac{1}{2!}z^2 + \frac{1}{4!}z^4 - \cdots$$

which satisfies

$$\left(\frac{d^2}{dz^2} + 1\right)\psi = 0$$

$$\psi(z = 0) = 1$$

and

$$\psi'(z = 0) = 0$$

It is worth observing that when a function is defined by a power series it is, in fact, defined by the sequence of coefficients in the series which then must encode all of the properties of the function. To illustrate this idea we show in §17.4 how the zeros of $\cos z$ and $J_0(z)$ can be determined using the coefficients in their power series.

It is also worth observing that new technical difficulties come up as we move away from Cartesian coordinates. Eqs. (1), (2) and (3) are independent of one another. But in Eqs. (4), (5) and (6) m^2 must be determined in Eq. (5) before Eq. (6) can be solved while in Eqs. (7), (8) and (9) m^2 must be determined in Eq. (7) before Eq. (8) can be solved and $\ell(\ell+1)$ must be determined in Eq. (8) before Eq. (9) can be solved. Equations (5) and (6) correspond to spherical coordinates in two dimensions, Eqs. (7), (8) and (9) correspond to spherical coordinates in three dimensions and in §17.5 we observe that the pattern we see here obtains in spherical coordinates in four dimensions. Also in §17.5 we carry out separation of variables in elliptic cylinder coordinates where again something new happens that we have not seen heretofore. There are a dozen or so orthogonal coordinate systems where we can separate $(\nabla^2 + \lambda^2)\psi = 0$ and information on these can be found in Moon and Spencer's book "*Field Theory Handbook*," and in Morse and Feshbach's book "*Methods of Theoretical Physics*," as well as in many other books. Indeed a lot of information on orthogonal coordinate systems and on ∇^2 can be found in Pauling and Wilson's book "*Quantum Mechanics*" and Happel and Brenner's book "*Low Reynolds Number Hydrodynamics*." The titles of these books suggest the wide range of application of the method of separation of variables.

17.2   How the Boundary Conditions Fix the Eigenvalues

We turn now to an explanation of how we use Eqs. (1), ..., (9). The boundary conditions in a specific problem may be any of a large number of possibilities, yet we can illustrate the essential ideas by taking up a small number of concrete examples. We begin by observing that Eqs. (1), (2), (3), (4), (5) and (7) are identical and that we can write the general solution to each in terms of a linear combination of the functions cosine and sine. The boundary conditions for Eqs. (1), (2), (3) and (4) are ordinarily imposed on two coordinate surfaces and this may also be true for Eqs. (5) and (7), though often periodic conditions are imposed. To see how the boundary conditions select the solutions we use we take Eq. (1) as an example and work out the Dirichlet problem where c

is specified on the boundary of the region. Many more examples appear in Lecture 14. Then the solutions to Eq. (1) must satisfy

$$X\left(x=a\right)=0$$

and

$$X\left(x=b\right)=0$$

where the values of a and b are determined by the problem at hand.

Writing the general solution to Eq. (1) as

$$X = A\cos\alpha x + B\sin\alpha x$$

we must determine the constants A, B and α so that the conditions

$$0 = A\cos\alpha a + B\sin\alpha a$$

and

$$0 = A\cos\alpha b + B\sin\alpha b$$

or

$$\begin{pmatrix} \cos\alpha a & \sin\alpha a \\ \cos\alpha b & \sin\alpha b \end{pmatrix}\begin{pmatrix} A \\ B \end{pmatrix} = \begin{pmatrix} 0 \\ 0 \end{pmatrix}$$

are satisfied. Now for arbitrary values of α the only solution to this homogeneous equation is $A = 0 = B$. To get solutions other than $A = 0 = B$ we must find the special values of α that make the determinant of the matrix on the left hand side vanish. To each such value of α there are solutions such that $A \neq 0$ or $B \neq 0$ or $A \neq 0$ and $B \neq 0$ but there is only one independent solution as the rank of $\begin{pmatrix} \cos \alpha a & \sin \alpha a \\ \cos \alpha b & \sin \alpha b \end{pmatrix}$ is one.

To each value of α that satisfies

$$\cos \alpha a \sin \alpha b - \cos \alpha b \sin \alpha a = 0$$

there corresponds one independent solution of Eq. (1) and $X(x = a) = 0 = X(x = b)$. If $\cos \alpha a \neq 0$ it can be written

$$X = B \left\{ -\frac{\sin \alpha a}{\cos \alpha a} \cos \alpha x + \sin \alpha x \right\}$$

But not all values of α that make the determinant vanish produce new solutions. If α makes the determinant vanish so also does $-\alpha$ but α and $-\alpha$ determine the same eigenvalue α^2 and dependent eigenfunctions.

The simplest result obtains if $a = 0$ and $b = 1$ then

$$\left. \begin{array}{c} \alpha^2 = n^2\pi^2 \\ X = B \sin n\pi x \end{array} \right\} n = 1, 2, \ldots$$

Assuming we solve Eqs. (2) and (3) in a similar way we can determine the eigenvalues and eigenfunctions of a problem in Cartesian coordinates as

$$\lambda^2 = \alpha^2 + \beta^2 + \gamma^2$$

and

$$\psi = XYZ$$

where the arbitrary multiples in X, Y and Z will be determined so that $\langle \psi, \psi \rangle = 1$. And it is worth observing that only three sets of orthogonal functions are needed to solve problems in Cartesian coordinates. In this way Cartesian coordinates are special.

While the foregoing shows how we deal with Eqs. (5) and (7) when surfaces $\theta = constant$ or $\phi = constant$ separate the system from its surroundings, it often happens that the boundary of a region can be completely specified in terms that are independent of θ in Eq. (5) or ϕ in Eq. (7). Then we do not have surfaces on which we can specify physical boundary conditions and in place of this we must require the solution to our problem to be periodic in θ or ϕ of period 2π. This requirement is passed on to the eigenfunctions and then to their θ or ϕ dependent parts and so the boundary conditions for Eqs. (5) and (7) can be taken to be

$$\Theta(\theta + 2\pi) = \Theta(\theta)$$

or

$$\Phi(\phi + 2\pi) = \Phi(\phi)$$

17.3 Solving a Two Dimensional Diffusion Problem in Plane Polar Coordinates (Spherical Coordinates in Two Dimensions)

These ideas carry over to Eq. (6), (8) and (9), but we have not yet explained how to write the solutions to these equations. Before we do this we take up a simple concrete example which shows how the various parts of the solution fit together .

Suppose we wish to solve a 2-dimensional diffusion problem in a region bounded by concentric circles, i.e., the concentration is uniform in the z-direction in cylindrical coordinates and the only source is a specified nonzero initial solute distribution, denoted $c(t = 0)$. The inner circle is impermeable to solute, the outer circle is in perfect contact with a large solute free reservoir.

Thus we have to solve

$$\frac{\partial c}{\partial t} = \nabla^2 c = \frac{1}{r}\frac{\partial}{\partial r}\left(r\frac{\partial c}{\partial r}\right) + \frac{1}{r^2}\frac{\partial c}{\partial \theta^2} \qquad a < r < b, \quad 0 \le \theta < 2\pi$$

and

$$\frac{\partial c}{\partial r} \left(r = a\right) = 0 = c\left(r = b\right)$$

where $c\left(t = 0\right)$ is specified. The solution is

$$c\left(r, \theta, t\right) = \sum \left\langle\, \psi_i, c\left(t = 0\right)\,\right\rangle e^{-\lambda_i^2 t}\, \psi_i\left(r, \theta\right)$$

where the eigenvalue problem is

$$\nabla^2\psi + \lambda^2\psi = 0 \qquad a < r < b, \quad 0 \leq \theta < 2\pi$$

and

$$\frac{\partial\psi}{\partial r} \left(r = b\right) = 0 = \psi\left(r = a\right), \quad 0 \leq \theta < 2\pi$$

The expectation might be that we will have to piece together two sets of orthogonal functions in order build up the eigenfunctions we use to solve our problem. That would be true in Cartesian coordinates where the separated eigenvalue problems are not coupled, *but it is not ordinarily true, and it is not true here*.

To solve the eigenvalue problem we put $\psi = R\left(r\right)\Theta\left(\theta\right)$ and conclude that R and Θ satisfy

$$\frac{d^2\Theta}{d\theta^2} + m^2\Theta = 0$$

and

$$\frac{d^2 R}{dr^2} + \frac{1}{r}\frac{dR}{dr} - \frac{m^2}{r^2}R + \lambda^2 R = 0$$

where

$$\frac{dR}{dr}\left(r = a\right) = 0 = R\left(r = b\right)$$

We assume c, and therefore ψ, and therefore Θ to be periodic in θ and we assign to ψ and therefore to R homogeneous conditions corresponding to the boundary conditions assigned to c by the physics of the problem.

First we look at the θ part of the problem and we replace

$$\Theta\left(\theta\right) = \Theta\left(\theta + 2\pi\right)$$

by

$$\Theta\left(0\right) = \Theta\left(2\pi\right)$$

and

$$\Theta'\left(0\right) = \Theta'\left(2\pi\right)$$

as then we see that $\Theta''\left(0\right) = \Theta''\left(2\pi\right)$, etc.

Now writing Θ as

$$\Theta = A\cos m\theta + B\sin m\theta$$

we find A, B and m must satisfy

$$A = A\cos 2\pi m + B\sin 2\pi m$$

and

$$mB = -mA\sin 2\pi m + mB\cos 2\pi m$$

whereupon we have

$$\begin{pmatrix} \cos 2\pi m - 1 & \sin 2\pi m \\ -m\sin 2\pi m & m\left(\cos 2\pi m - 1\right) \end{pmatrix}\begin{pmatrix} A \\ B \end{pmatrix} = \begin{pmatrix} 0 \\ 0 \end{pmatrix}$$

And we see that only for special values of m does this system of homogeneous, linear, algebraic equations have solutions other that $A = 0 = B$. These special values of m are those that make the determinant of the matrix on the left had side vanish and as this determinant is

$$m\left(2 - 2\cos 2\pi m\right)$$

the solutions of interest correspond to

$$m = 0, \pm 1, \pm 2, \cdots$$

When m is $\pm 1, \pm 2, \cdots$ the rank of this matrix is zero and the system of equations has two independent solutions, the simplest choice being $\begin{pmatrix} 1 \\ 0 \end{pmatrix}$ and $\begin{pmatrix} 0 \\ 1 \end{pmatrix}$. So, to each integer value of m other than zero there corresponds the eigenvalue m^2 and the two eigenfunctions $\cos m\theta$ and $\sin m\theta$. As this obtains both for m and its negative we need admit only the values $m = 1, 2, \ldots$. When $m = 0$ we should write $\Theta = A + Bx$ whereupon we find only one periodic solution which we take to be $A = 1$, $B = 0$. So, to $m = 0$ there corresponds the eigenvalue 0 and the eigenfunction 1.

Now the simplest way to denote this set of eigenfunctions is to write the independent solutions corresponding to $m^2 = 0, 1, 4, \ldots$ as $e^{im\theta}$ and $e^{-im\theta}$, $m = 0, 1, 2, \ldots$ Doing this we exhibit the eigenfunctions satisfying periodic conditions as

$$\Theta_m = \frac{1}{\sqrt{2\pi}} e^{im\theta} \qquad m = \ldots, -2, -1, 0, 1, 2, \ldots$$

and we observe that

$$\int_0^{2\pi} \overline{\Theta}_m\left(\theta\right) \Theta_n\left(\theta\right) d\theta = \begin{cases} 0 & m \neq n \\ 1 & m = n \end{cases}$$

This can be seen by carrying out the integration but it can also be seen by observing that periodic boundary conditions make the term $\left[\overline{\phi} \dfrac{d\psi}{dx} - \psi \dfrac{d\overline{\phi}}{dx}\right]$ in our second integration by parts formula vanish.

In this way we deal with Eq. (5) in cylindrical coordinates and Eq. (7) in spherical coordinates. Now having established the values of m^2, viz., $0, 1, 4, \ldots$ we turn to the R equation and notice that we get a different R equation for each value of m^2 and that λ^2 appears only in the R equations. For each value of m^2 we denote the two independent solutions of the R equation by $J_m(\lambda r)$ and $Y_m(\lambda r)$, $m = 0, 1, 2, \ldots$ where J_m and Y_m denote independent solutions of Bessel's equation for nonnegative integer values of m.

Now there are many R equations, one corresponding to each fixed value of m^2, hence we write

$$R = A J_m(\lambda r) + B Y_m(\lambda r)$$

and seek to determine A, B and λ via the conditions at $r = a$ and $r = b$, viz.,

$$\lambda A\, J'_m(\lambda a) + \lambda B\, Y'_m(\lambda a) = 0$$

and

$$A\, J_m(\lambda b) + B\, Y_m(\lambda b) = 0$$

where J'_m denotes $\dfrac{d J_m(x)}{dx}$. Thus we have

$$\begin{pmatrix} \lambda\, J'_m(\lambda a) & \lambda\, Y'_m(\lambda a) \\ J_m(\lambda b) & Y_m(\lambda b) \end{pmatrix} \begin{pmatrix} A \\ B \end{pmatrix} = \begin{pmatrix} 0 \\ 0 \end{pmatrix}$$

and to each fixed value of m^2 the values of λ^2 can be determined by finding the values of λ that make the determinant of the matrix on the left hand side vanish. Indeed only for values of λ such that

$$\lambda\, J'_m(\lambda a)\, Y_m(\lambda b) - \lambda\, Y'_m(\lambda a)\, J_m(\lambda b) = 0$$

can constants other than $A = 0 = B$, and, therefore, solutions other than R $=0$, be determined. And we see that this equation must be solved for $m = 0, 1, 2, \ldots$.

To go on, we put a =0. By doing this we turn up a technical difficulty. The boundary of the region is now the circle $r = b$. And as this is the only surface on which we can assign a physical boundary condition we no longer have the two boundary conditions required to evaluate the two constants in the solution of the R equation. What gets us out of this is the discovery that $Y_m(r)$ is not bounded as $r \to 0$ and hence, upon requiring c to be bounded, and therefore ψ to be bounded, we must require R to be bounded and to achieve this we put $B = 0$. So if $a = 0$ and $b = 1$ we write

$$R = A\, J_m(\lambda r)$$

and determine λ via

$$J_m(\lambda) = 0$$

If λ is a root of this equation then so also is $-\lambda$ but λ and $-\lambda$ lead to the same eigenvalue and to dependent eigenfunctions as $J_m(-z) = \pm J_m(z)$. If zero is a root of this equation the corresponding eigenfunction is zero everywhere.

We let $\lambda_{|m|1}, \lambda_{|m|2}, \dots$ denote the positive roots of $J_{|m|}(\lambda) = 0$ and then organize the solutions to the eigenvalue problem in terms of m by assigning to each value of m, i.e., to $\dots, -2, -1, 0, 1, 2, \dots$ the eigenvalues

$$\lambda^2_{|m|1}, \lambda^2_{|m|2}, \dots$$

and the corresponding unnormalized eigenfunctions,

$$J_{|m|}\left(\lambda_{|m|1} r\right) \frac{e^{im\theta}}{\sqrt{2\pi}}, \qquad J_{|m|}\left(\lambda_{|m|2} r\right) \frac{e^{im\theta}}{\sqrt{2\pi}}, \qquad \dots$$

In this way, of the two eigenfunctions corresponding to the eigenvalue $\lambda^2_{|m|i}$, one is assigned to $|m|$ the other to $-|m|$. In problem 1 you will derive the factor normalizing the Bessel's functions.

Letting $\psi_{mi}(r, \theta)$ denote the normalized eigenfunction $R_{|m|i}(r)\,\Theta_m(\theta)$, where

$R_{|m|i}(r) \propto J_{|m|}\left(\lambda_{|m|i}\,r\right)$ we can write the solution to our problem as

$$c\left(r,\theta,t\right) = \sum_{m=-\infty}^{+\infty} \sum_{i=1}^{\infty} \left\langle\,\psi_{mi},\ c\left(t=0\right)\,\right\rangle e^{-\lambda_{|m|i}^2 t}\,\psi_{mi}\left(r,\theta\right)$$

where

$$\left\langle\,\psi_{mi},\ c\left(t=0\right)\,\right\rangle = \int_0^1 \int_0^{2\pi} \overline{R}_{|m|i}\left(r\right)\ \overline{\Theta}_m\left(\theta\right)\ c\left(t=0\right)\ dr\,r\,d\theta$$

and where this is

$$\left\langle\,\psi_{mi},\ c\left(t=0\right)\,\right\rangle = \int_0^1 \overline{R}_{|m|i}\left(r\right)\ r\,dr \int_0^{2\pi} \overline{\Theta}_m\left(\theta\right)\ c\left(t=0\right)\ d\theta$$

$$= \left\langle\,R_{|m|i},\ \left\langle\,\Theta_m,\ c\left(t=0\right)\,\right\rangle_\theta\,\right\rangle_r$$

The orthogonality works out as follows

$$\left\langle\,R_{|m|i}\,\Theta_m,\ R_{|n|j}\,\Theta_n\,\right\rangle = \left\langle\,R_{|m|i},\ R_{|n|j}\,\right\rangle_r \left\langle\,\Theta_m,\ \Theta_n\,\right\rangle_\theta$$

and this is zero if $m \neq n$ whereas it is $\left\langle\,R_{|m|i},\ R_{|m|j}\,\right\rangle_r$ if $m = n$ where

$$\left\langle\,R_{|m|i},\ R_{|m|j}\,\right\rangle_r = \begin{cases} 0, & i \neq j \\ 1, & i = j \end{cases}$$

So, corresponding to each value of m^2 we have a complete orthogonal set of functions of r and *indeed a different set for each value of m^2*. The completeness we assume; the orthogonality we infer from Lecture 15 or establish directly using the integration by parts formulas:

$$\int_0^1 \phi\frac{1}{r}\frac{d}{dr}\left(r\frac{d\psi}{dr}\right)r\,dr = \left[\phi r\frac{d\psi}{dr}\right]_0^1 - \int_0^1 r\frac{d\phi}{dr}\frac{d\psi}{dr}\,dr$$

and

$$\int_0^1 \phi\frac{1}{r}\frac{d}{dr}\left(r\frac{d\psi}{dr}\right)r\,dr = \left[\phi r\frac{d\psi}{dr} - r\frac{d\phi}{dr}\psi\right]_0^1 + \int_0^1 \frac{1}{r}\frac{d}{dr}\left(r\frac{d\phi}{dr}\right)\psi r\,dr$$

Thus if R and λ^2 satisfy

$$\frac{1}{r}\frac{d}{dr}\left(r\frac{dR}{dr}\right) + \left\{\lambda^2 - \frac{m^2}{r^2}\right\}R = 0$$

and

$$R(r=1) = 0$$

where R is required to be bounded as $r \to 0$ and where m^2 is fixed, then so also do $\overline{R}$ and $\overline{\lambda}^2$. On setting $\phi = \overline{R}$ and $\psi = R$ in the first and second formulas we find that λ^2 is real and positive. On setting $\phi = \overline{R}_{|m|i}$ and $\psi = R_{|m|j}$ in the second formula where $R_{|m|i}$ and $R_{|m|j}$ are two solutions corresponding to different values of λ^2 we find that

$$\int_0^1 \overline{R}_{|m|i}\, R_{|m|j}\, r\, dr = \left\langle R_{|m|i}, R_{|m|j} \right\rangle_{\mathrm{r}} = 0$$

and this is the orthogonality condition that we require.

What is going on here is this: the index m sorts out the θ variation of $c(t=0)$ and then the index i sorts out the corresponding r variation. Indeed we first expand an assigned initial solute distribution $c(t=0)$ in the set of functions

$$\left\{ \Theta_m(\theta) \right\}_{m=-\infty}^{+\infty}$$

as

$$c(t=0) = \sum_{m=-\infty}^{+\infty} \left\langle \Theta_m,\, c(t=0) \right\rangle_\theta \Theta_m(\theta)$$

{ The complex and real Fourier series are two forms of the same expansion. If $\sum_{m=-\infty}^{+\infty} c_m\, e^{im\phi} = \sum_{m=0}^{\infty} a_m \cos m\phi + b_m \sin m\phi$ then $a_m = c_m + c_{-m}$ and $b_m = ic_m - ic_{-m}$. }

This resolves $c(t=0)$ into its various angular pieces where each of the resulting coefficients $\left\langle \Theta_m,\, c(t=0) \right\rangle_\theta$ is a function of r. This function, defining the part of the r dependence of

$c\,(t=0)$ that corresponds to $\Theta_m\,(\theta)$, is then expanded in the set of functions $\left\{R_{|\,m|i}\,(r)\right\}_{i=1}^{\infty}$ and *this set is special to each value of* $|m|$.

The orthogonality of two functions $R_{|\,m|i}$ and $R_{|\,n|j}$, $m\neq n$ is never an issue.

As $c\,(t=0)$ is real we observe that $\left\langle\,\Theta_m,\quad c\,(t=0)\,\right\rangle_\theta\Theta_m\,(\theta)$ and $\left\langle\,\Theta_{-m},\ c\,(t=0)\,\right\rangle_\theta\Theta_{-m}\,(\theta)$ are complex conjugates and hence that the terms corresponding to $+m$ and $-m$ in our solution are complex conjugates. Therefore we can write $c\,(r,\theta,t)$ more simply as

$$c\,(r,\theta,t)=\sum_{m=0}^{\infty}\sum_{i=1}^{\infty}2\,\mathrm{Re}\left\{\left\langle\,R_{mi},\left\langle\,\Theta_m,\ c\,(t=0)\,\right\rangle_\theta\,\right\rangle_r R_{mi}\,(r)\,\Theta_m\,(\theta)\,e^{-\lambda_{|\,m|i}^2 t}\right\}$$

and we see that if $\left\langle\,\Theta_m,\ c\,(t=0)\,\right\rangle=0$ then $\left\langle\,\Theta_m,\ c\,\right\rangle=0\ \ \forall\,t>0$. Indeed if $\left\langle\,\Theta_m,\ c\,(t=0)\,\right\rangle=0,\quad\forall\,m$ other than $m=0$ then we have

$$c\,(r,t)=\sum_{i=1}^{\infty}\left\langle\,R_{0i},\ c\,(t=0)\,\right\rangle_r R_{0i}\,(r)\,e^{-\lambda_{0i}^2 t}$$

and so if $c\,(t=0)$ is uniform in θ then c itself is uniform in $\theta\ \forall\,t>0$.

Sketches of $J_0\,(z)\,,J_1\,(z)\,,J_2\,(z)\,,\ldots$

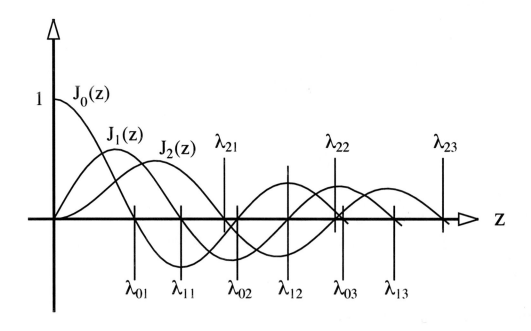

indicate that the positive zeros of $J_0, J_1, J_2, \ldots$ are ordered in the following way: The lowest is the first zero of J_0, the next lowest is the first positive zero of J_1, then the first positive zero of J_2, before the second zero of J_0.

Hence as t grows large the last remaining term in our solution corresponds to $m = 0, i = 1$ and this term is uniform in θ. The next to the last term corresponds to $m = 1, i = 1$, not to $m = 0, i = 2$. Indeed estimates of how large t must be before the series can be replaced by its first term are too short if we look at λ_{01} and λ_{02}. We need to look at λ_{01} and λ_{11}.

The terms corresponding to larger values of $|m|$ and i die out faster than the terms corresponding to smaller values of $|m|$ and i. This is the smoothing we associate with diffusion as the eigenfunctions corresponding to larger values of $|m|$ and i exhibit more oscillations and their contribution to the solution dies out faster.

The sketch below of $J_0\left(\lambda_{01}r\right)$, $J_0\left(\lambda_{02}r\right)$, $J_0\left(\lambda_{03}r\right)$, $\ldots$ shows that this set of orthogonal functions is constructed from $J_0\left(z\right)$ by scaling, in turn, its positive zeros, $z_1, z_2, z_3, \ldots$ to 1. Likewise $J_1\left(\lambda_{11}r\right)$, $J_1\left(\lambda_{12}r\right)$, $J_1\left(\lambda_{13}r\right)$, $\ldots$ is constructed from $J_1\left(z\right)$ in just this same way. Etc. This illustrates the rule that in a set of orthogonal functions each function can be identified by the number of its interior zeros. It also shows that the zeros of any two functions in such a set are nested.

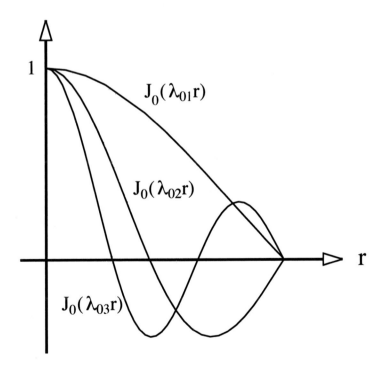

As the last remaining term in our solution as t grows large is a multiple of $J_0\left(\lambda_{01}r\right)$, it is important that $J_0\left(\lambda_{01}r\right)$ be singly signed. It is also important that $J_1\left(z=0\right)=0$ otherwise the eigenfunctions $J_1\left(\lambda_{1i}r\right)\cos\theta$ would be poorly behaved as $r\to 0$. Likewise $J_2\left(z=0\right)=0=J_2'\left(z=0\right)$ and this is important as the eigenfunctions exhibiting J_2 as a factor are multiplied by $\cos 2\theta$. Etc.

We have been a little lucky. Ordinarily, in orthogonal coordinate systems we have

$$dV = h_\xi\, h_\eta\, h_\zeta\, d\xi\, d\eta\, d\zeta$$

hence the inner product over V factors into one dimensional inner products only under special conditions. And these conditions are not satisfied in elliptic cylinder coordinates introduced in §17.5

Further, we have

$$\nabla^2 = \frac{1}{h_\xi\, h_\eta\, h_\zeta}\, \frac{\partial}{\partial\xi}\left(\frac{h_\eta\, h_\zeta}{h_\xi}\, \frac{\partial}{\partial\xi}\right) + \cdots$$

and for separation of variables to work out as simply as above places strong requirements on h_ξ, h_η and h_ζ, conditions not satisfied in all orthogonal coordinate systems.

In the next lecture we solve some problems where what we have done in this lecture is sufficient.

17.4   The Zeros of cos z and $J_0(z)$

To determine the zeros of $\cos z$ or $J_0\left(z\right)$ in terms of the coefficients in their power series expansions we observe that if $q\left(z\right)$ has a simple zero at z_0 then $q'(z_0)/q(z_0)$ has a simple pole there and its residue is 1. Then because the contour integral

$$\frac{1}{2\pi i}\oint_C \frac{\cos' w}{(w-z)\cos w}\, dw$$

is equal to the sum of the residues of its integrand at its poles inside C and because the integral vanishes as the diameter of C grows large, we get

$$0 = \frac{\cos' z}{\cos z} + \frac{1}{z_1 - z} + \frac{1}{-z_1 - z} + \cdots$$

where the zeros of $\cos z$ are real and where $0 < z_1 < z_2 < \cdots$ denote the positive zeros. Writing this as

$$\frac{\cos' z}{\cos z} = \frac{2z}{z^2 - z_1^2} + \frac{2z}{z^2 - z_2^2} + \cdots$$

or

$$-\frac{1}{2z} \frac{\cos' z}{\cos z} = \sum \frac{1}{z_i^2} \frac{1}{1 - \dfrac{z^2}{z_i^2}}$$

and expanding $\dfrac{1}{1 - \dfrac{z^2}{z_i^2}}$ as $\sum \left(\dfrac{z^2}{z_i^2}\right)^j$ when $|z^2| < |z_i^2|$ we get

$$-\frac{1}{2z} \frac{\cos' z}{\cos z} = \sum_{i=1}^{\infty} \frac{1}{z_i^2} \sum_{j=0}^{\infty} \left(\frac{z^2}{z_i^2}\right)^j$$

$$= \sum_{j=0}^{\infty} z^{2j} \sum_{i=1}^{\infty} \left(\frac{1}{z_i^2}\right)^{j+1}$$

where $\cos z$ is a function of z^2 and $\dfrac{d}{dz^2} \cos z = \cos' z \dfrac{1}{2z}$. Using this to write

$$-\frac{d}{dz^2} \cos z = \cos z \sum_{j=0}^{\infty} z^{2j} \sum_{i=1}^{\infty} \left(\frac{1}{z_i^2}\right)^{j+1}$$

and expanding both sides in powers of z^2, we can evaluate the sums $\sum_{i=1}^{\infty} \left(\dfrac{1}{z_i^2}\right)^{j+1}$. Thus, using the series defining $\cos z$, viz.,

$$\cos z = 1 - \frac{1}{2}z^2 + \frac{1}{24}z^4 - \frac{1}{720}z^6 + \cdots$$

and

$$\frac{d}{dz^2}\cos z = -\frac{1}{2} + \frac{1}{12}z^2 - \frac{1}{240}z^4 + \cdots$$

we get

$$\sum \frac{1}{z_i^2} = \frac{1}{2}$$

$$\sum \frac{1}{z_i^4} = \frac{1}{6}$$

etc.

Indeed $\sqrt{6} = 2.449$ is already a fair approximation to $z_1^2 = \dfrac{\pi^2}{4} = 2.467$.

The corresponding equations for J_0 are

$$\sum \frac{1}{z_i^2} = \frac{1}{4}$$

$$\sum \frac{1}{z_i^4} = \frac{1}{32}$$

$$\sum \frac{1}{z_i^6} = \frac{1}{192}$$

etc.

where $\sqrt[3]{192} = 5.769$ is a very good approximation to the square of the smallest positive zero of J_0.

This method was used by Rayleigh and goes back to Euler.

17.5 Separation of Variables in Two More Coordinate Systems

(I) Spherical coordinates in four dimensions

We let w, x, y and z denote rectangular Cartesian coordinates in four dimensions and define spherical coordinates via

$$w = r \cos \omega$$

$$x = r \sin \omega \, \cos \theta$$

$$y = r \sin \omega \, \sin \theta \, \cos \phi$$

$$z = r \sin \omega \, \sin \theta \, \sin \phi$$

Then we have $h_r = 1$, $h_\omega = r$, $h_\theta = r \sin \omega$ and $h_\phi = r \sin \omega \, r \sin \theta$ and hence

$$\nabla^2 = \frac{1}{r^3} \frac{\partial}{\partial r} \left(r^3 \frac{\partial}{\partial r} \right) + \frac{1}{r^2 \sin^2 \omega} \frac{\partial}{\partial \omega} \left(\sin^2 \omega \frac{\partial}{\partial \omega} \right)$$

$$+ \frac{1}{r^2 \sin^2 \omega \, \sin \theta} \frac{\partial}{\partial \theta} \left(\sin \theta \frac{\partial}{\partial \theta} \right)$$

$$+ \frac{1}{r^2 \sin^2 \omega \, \sin^2 \theta} \frac{\partial^2}{\partial \phi^2}$$

The result of substituting $\psi = R(r) \Omega(\omega) \Theta(\theta) \Phi(\phi)$ into $(\nabla^2 + \lambda^2) \psi$, dividing by $R\Omega\Theta\Phi$ and identifying terms which must be constant is

$$\frac{d^2\Phi}{d\phi^2} + m^2\Phi = 0$$

$$\frac{1}{\sin \theta} \frac{d}{d\theta} \left(\sin \theta \frac{d\Theta}{d\theta} \right) + \left\{ \ell(\ell+1) - \frac{m^2}{\sin^2 \theta} \right\} \Theta = 0$$

$$\frac{1}{\sin^2 \omega} \frac{d}{d\omega} \left(\sin^2 \omega \frac{d\Omega}{d\omega} \right) + \left\{ k(k+2) - \frac{\ell(\ell+1)}{\sin^2 \omega} \right\} \Omega = 0$$

and

$$\frac{1}{r^3}\frac{d}{dr}\left(r^3\frac{dR}{dr}\right) + \left\{\lambda^2 - \frac{k(k+2)}{r^2}\right\}R = 0$$

(II) Confocal elliptical cylinder coordinates

We define elliptical cylinder coordinates via

$$x = c\cosh\xi\,\cos\eta$$

$$y = c\sinh\xi\,\sin\eta$$

$$z = z$$

Then we find $h_\xi = h_\eta = c\sqrt{\sinh^2\xi + \sin^2\eta}$ and $h_z = 1$ and hence

$$\nabla^2 = \frac{1}{c^2\left(\sinh^2\xi + \sin^2\eta\right)}\left(\frac{\partial^2}{\partial\xi^2} + \frac{\partial^2}{\partial\eta^2}\right) + \frac{\partial^2}{\partial z^2}$$

Substituting $\psi = X(\xi)Y(\eta)Z(z)$ into $\left(\nabla^2 + \lambda^2\right)\psi = 0$ we get

$$\frac{d^2Z}{dz^2} + \gamma^2 Z = 0$$

$$\frac{d^2X}{d\xi^2} + \left\{c^2\sin^2\xi\left(\lambda^2 - \gamma^2\right) - m^2\right\}X = 0$$

$$\frac{d^2Y}{d\eta^2} + \left\{c^2\sin^2\eta\left(\lambda^2 - \gamma^2\right) + m^2\right\}Y = 0$$

We notice that λ^2 appears in two equations. This is new. And we see that the orthogonality does not factor, viz.,

$$\left\langle\,\psi_{ij},\,\psi_{k\ell}\,\right\rangle = \int\int \overline{X}_i(\xi)X_k(\xi)\overline{Y}_j(\eta)Y_\ell(\eta)\left(c^2\sinh^2\xi + c^2\sin^2\eta\right)d\xi d\eta$$

This is new.

17.6   Home Problems

1. Normalization of Bessel Functions

   To normalize Bessel's functions you need to evaluate integrals such as

   $$\int_0^1 J_m^2\left(\lambda r\right) r\, dr$$

   To do this put $\psi = J_m\left(\lambda r\right)$ and multiply

   $$\frac{1}{r}\frac{d}{dr}\left(r\frac{d\psi}{dr}\right) = \left\{-\lambda^2 + \frac{m^2}{r^2}\right\}\psi$$

   by $2r^2\dfrac{d\psi}{dr}$ to obtain

   $$2\left(r\frac{d\psi}{dr}\right)\frac{d}{dr}\left(r\frac{d\psi}{dr}\right) = 2\left\{-\lambda^2 r + m^2\right\}\psi\frac{d\psi}{dr}$$

   Now this is

   $$\frac{d}{dr}\left(r\frac{d\psi}{dr}\right)^2 = \left\{-\lambda^2 r^2 + m^2\right\}\frac{d}{dr}\left(\psi^2\right)$$

   hence you have

   $$\left[\left(r\frac{d\psi}{dr}\right)^2\right]_0^1 = -\lambda^2 \int_0^1 r^2\frac{d}{dr}\left(\psi^2\right)dr + \left[m^2\psi^2\right]_0^1$$

   and integrating by parts gets you a formula for

   $$\int_0^1 \psi^2 r\, dr$$

2. Heat is generated in a circle of radius R. The temperature at the edge is held fixed at $T = 0$.

   The rate of heat generation is a linear function of temperature, increasing as temperature increases. Hence we have

$$\nabla^2 T + \lambda^2 \left(1 + T\right) = 0$$

$$T = 0 \quad \text{at} \quad r = R$$

What is the greatest value of R at which there is a solution to our problem, at a fixed value of λ^2 ? { We could ask: what is the greatest value of λ^2 at which there is a solution to our problem at a fixed value of R ? }.

A cooling pipe of radius R_0 is introduced at the center of the circle. Its temperature is $T = 0$. By how much can R be increased?

It is not possible to center the cooing pipe precisely. What is its effect if it is off center by a small amount ε?

In the expansion

$$R = R_0 + \varepsilon R_1 \left(\theta\right) + \frac{1}{2} \varepsilon^2 R_2 \left(\theta\right) + \cdots$$

we have

$$R_1 = \cos \theta, \qquad R_2 = -\frac{\sin^2 \theta}{R_0}$$

3. Assume the temperature, $T > 0$, is specified at the cross section $z = 0$ of an infinite circular cylinder of radius R. The walls are held at $T = 0$. Find the temperature in the cylinder, $z > 0$, by solving

$$-\frac{\partial^2 T}{\partial z^2} = \frac{1}{r} \frac{\partial}{\partial r} r \frac{\partial T}{\partial r}$$

Do this by using the solutions to the eigenvalue problem

$$\frac{1}{r} \frac{d}{dr} r \frac{d\psi}{dr} + \lambda^2 \psi = 0, \qquad 0 < r < R, \quad \text{and} \quad \psi = 0 \text{ at } r = R$$

This should bring to mind the problem

$$\frac{\partial T}{\partial t} = \frac{1}{r}\frac{\partial}{\partial r}\, r \frac{\partial T}{\partial r}$$

where T is specified at $t = 0$.

4. The eigenvalue problem

$$\left(\frac{d^2}{dr^2} + \frac{1}{r}\frac{d}{dr}\right)^2 \psi = \lambda^4 \psi$$

where ψ is bounded at $r = 0$ and

$$\psi\left(r = 1\right) = 0$$

$$\psi'\left(r = 1\right) = 0$$

has the solution

$$\psi = A J_0\left(\lambda r\right) + B I_0\left(\lambda r\right)$$

due to

$$\left(\frac{d^2}{dr^2} + \frac{1}{r}\frac{d}{dr}\right) J_0\left(\lambda r\right) = -\lambda^2 J_0\left(\lambda r\right)$$

and

$$\left(\frac{d^2}{dr^2} + \frac{1}{r}\frac{d}{dr}\right) I_0\left(\lambda r\right) = \lambda^2 I_0\left(\lambda r\right)$$

Hence A, B and λ satisfy

$$A J_0\left(\lambda\right) + B I_0\left(\lambda\right) = 0$$

and

$$A\lambda J_0'(\lambda) + B\lambda I_0'(\lambda) = 0$$

whereupon, to have a solution such that A and B are not both zero, the λ's must satisfy

$$\lambda\left(J_0(\lambda)I_0'(\lambda) - J_0'(\lambda)I_0(\lambda)\right) = 0$$

and then we have

$$B = -\frac{AJ_0(\lambda)}{I_o(\lambda)}$$

or

$$\psi = A\left(J_0(\lambda r)I_0(\lambda) - I_0(\lambda r)J_0(\lambda)\right)$$

Your job is to prove

$$\int_0^1 r\,\psi_\lambda\,\psi_\mu\,dr = 0, \quad \lambda \neq \mu$$

and estimate the first few values of λ.

Denote by $W(\lambda)$ the Wronskian of $J_0(\lambda)$ and $I_0(\lambda)$ and show that

$$\frac{dW}{d\lambda} = -\frac{1}{\lambda}W + 2J_0(\lambda)I_0(\lambda)$$

where

$$W(\lambda) = J_0(\lambda)\frac{dI_0(\lambda)}{d\lambda} - \frac{dJ_0(\lambda)}{d\lambda}I_0(\lambda)$$

The λ's satisfy $W(\lambda) = 0$ and you will find that $W(\lambda)$ is not a very nice function.

5. Show that

$$\left(\frac{d^2}{dr^2} + \frac{1}{r}\frac{d}{dr}\right)^2 \psi = \lambda^4 \psi$$

has solutions

$$AJ_0(\lambda) + BY_0(\lambda) + CI_0(\lambda) + DK_0(\lambda)$$

and that

$$\frac{d^4\psi}{dz^4} = \lambda^4 \psi$$

has solutions

$$A \sin \lambda z + B \cos \lambda z + C \sinh \lambda z + D \cosh \lambda z$$

Each of these solutions has enough flexibility to satisfy four boundary conditions.

To solve

$$\left(\frac{\partial^2}{\partial r^2} + \frac{1}{r}\frac{\partial}{\partial r} + \frac{\partial}{\partial z^2}\right)^2 \psi = 4\lambda^4 \psi$$

observe that

$$\left(\frac{\partial^2}{\partial r^2} + \frac{1}{r}\frac{\partial}{\partial r} + \frac{\partial^2}{\partial z^2}\right) I_0(\lambda r) \sin \lambda z = 0$$

$$\left(\frac{\partial^2}{\partial r^2} + \frac{1}{r}\frac{\partial}{\partial r} + \frac{\partial^2}{\partial z^2}\right) I_0(\lambda r) \sinh \lambda z = 2\lambda^2 I_0(\lambda r) \sinh \lambda z$$

etc.

Hence you have solutions

$$\left(AJ_0(\lambda r) + BY_0(\lambda r)\right)\left(C \sin \lambda z + D \cos \lambda z\right)$$

and

$$\left(A I_0 \left(\lambda r \right) + B K_0 \left(\lambda r \right) \right) \left(C \sinh \lambda z + D \cosh \lambda z \right)$$

and also

$$\left(A J_0 \left(\sqrt{3}\, \lambda r \right) + B Y_0 \left(\sqrt{3}\, \lambda r \right) \right) \left(C \sinh \lambda z + D \cosh \lambda z \right)$$

etc.

But it is not easy to find ψ's with enough flexibility to satisfy the boundary conditions you are likely to meet. Show that the same problem arises if you are trying to solve

$$\nabla^4 \psi = \left(\frac{\partial^2}{\partial x^2} + \frac{\partial}{\partial y^2} \right)^2 = 0$$

viz., show that about the best you can do is

$$\psi = \{A \sin \lambda y + B \cos \lambda y\} \left\{ \widehat{A} \sinh \lambda x + \widehat{B} \cosh \lambda x + \widehat{C} \frac{1}{2\lambda} x \sinh \lambda x + \widehat{D} \frac{1}{2\lambda} x \cosh \lambda x \right\}$$

6. You are to solve

$$\nabla^4 \phi = f\left(x, y \right)$$

where $\phi = 0 = \phi_x$ at $x = -1$ and $x = 1$ and where ϕ and ϕ_y are specified as functions of x at $y = 0$ and $y = 1$.

You begin by solving the eigenvalue problem

$$\frac{d^4 \psi}{dx^4} = \lambda^4 \psi$$

where $\psi = 0 = \psi'$ at $x = -1$ and $x = 1$ and find even solutions,viz.,

$$\psi_\lambda = A \cos \lambda x + B \cosh \lambda x$$

where $\tan \lambda = -\tanh \lambda$ and odd solutions,viz.,

$$\psi_\lambda = C \sin \lambda x + D \sinh \lambda x$$

where $\tan \lambda = \tanh \lambda$. You then prove

$$\int_{-1}^{+1} \psi_{\lambda'} \psi_\lambda \, dx = 0 \qquad \lambda \neq \lambda'$$

Thus you write

$$\phi(x,y) = \sum c_\lambda(y) \psi_\lambda(x)$$

where

$$c_\lambda(y) = \int_{-1}^{+1} \psi_\lambda(x) \phi(x,y) \, dx$$

and then to find the equation satisfied by the c_λ's you multiply $\nabla^4 \phi = f$ by ψ_λ and integrate over $-1 \leq x \leq 1$ obtaining

$$\int_{-1}^{+1} \psi_\lambda \frac{\partial^4 \phi}{\partial x^4} \, dx + 2 \frac{d^2}{dy^2} \int_{-1}^{+1} \psi_\lambda \frac{\partial^2 \phi}{\partial x^2} \, dx + \frac{d^4}{dy^4} \int_{-1}^{+1} \psi_\lambda \phi \, dx = \int_{-1}^{+1} \psi_\lambda f \, dx$$

On carrying out integration by parts as many times as you need to, you discover one, and only one, technical difficulty: the term

$$2 \frac{d^2}{dy^2} \int_{-1}^{+1} \frac{d^2 \psi_\lambda}{dx^2} \phi \, dx$$

appears and it is not easy to write $\dfrac{d^2 \psi_\lambda}{dx^2}$ in terms of ψ_λ,

It turns out, in two dimensional problems, the theory of complex variables comes to your rescue: if you write $z = x + \imath y$ then

$$\nabla^2 (y f(z)) = 2 \imath f'(z)$$

in any region where $f(z)$ has a Taylor series.

7. You are solving

$$\frac{\partial c}{\partial t} = \nabla^2 c$$

on a bounded plane domain, $c(t = 0)$ assigned and $c = 0$ at the edge of the domain.

You introduce the eigenvalue problem

$$\nabla^2 \psi + \lambda^2 \psi = 0$$

on the domain and $\psi = 0$ at the edge and write your solution

$$c = \sum \langle \psi, c(t = 0) \rangle e^{-\lambda^2 t} \psi$$

Hence all you need are the solutions to the eigenvalue problem in order to estimate c.

First, suppose the domain is a circle of radius R_0. Find the two eigenvalues in control of the final stages of solute loss to the surroundings.

Then suppose the domain is a small displacement of the circle defined by

$$R(\theta) = R_0 + \varepsilon R_1 + \frac{1}{2} \varepsilon^2 R_2 + \cdots$$

where $R_1 = \cos\theta$, $R_2 = -\dfrac{\sin^2\theta}{R_0}$ and find the corrections to the above two eigenvalues in order to learn whether the diffusion of solute is faster or slower, i.e., find λ_1^2 and λ_2^2 in the series

$$\lambda^2 = \lambda_0^2 + \varepsilon \lambda_1^2 + \frac{1}{2} \varepsilon^2 \lambda_2^2 + \cdots$$

The hope is you find $\lambda_1^2 = 0 = \lambda_2^2$ for both eigenvalues and you are curious to know why this is so.

To see, take a circle of radius R_0, displace its center to $x = \varepsilon, y = 0$ and determine R_1 and R_2 in the expansion

$$R(\theta) = R_0 + \varepsilon R_1 + \frac{1}{2} \varepsilon^2 R_2 + \cdots$$

8. Elliptical cylinder coordinates are defined by

$$x = c \cosh \xi \, \cos \eta$$

$$y = c \sinh \xi \, \sin \eta$$

$$z = z$$

In the x, y plane the family of curves defined by holding ξ fixed, $0 \leq \xi < \infty$, is a family of confocal ellipses. The family defined by holding η fixed, $0 \leq \eta < 2\pi$, is a family of confocal hyperbolas. The two branches of each hyperbola correspond to four values of η. As $\xi \to \infty$, we have

$$\frac{y}{x} = \tan \eta$$

and hence the four values of η are the angles that the branches make with the positive x axis. The ellipses and hyperbolas are centered at $x = 0 = y$ and their foci lie at $x = \pm c, y = 0$. The curves $\eta = \frac{1}{3}\pi, \frac{2}{3}\pi, \frac{4}{3}\pi$ and $\frac{5}{3}\pi$ corresponding to two branchs of one hyperbola are shown in the sketch.

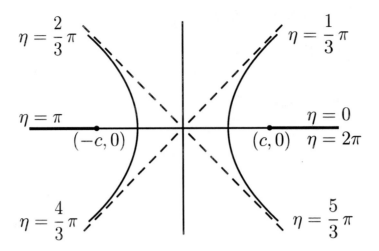

Write ∇^2 in this coordinate system and reduce the eigenvalue problem $\left(\nabla^2 + \lambda^2\right)\psi = 0$ to three one dimensional eigenvalue problems by separation of variables.

Suppose a solute distribution is assigned at $t = 0$ in the domain bounded by

$$\frac{x^2}{a^2} + \frac{y^2}{b^2} = 1, \qquad a^2 - b^2 = c^2$$

$$z = 0, \quad z = d$$

The planes $z = 0$ and $z = d$ are not permeable to solute.

Assuming that the solute concentration on the lateral surface of the elliptical cylinder does not depend on z, show that $c\,(t > 0)$ is independent of z iff $c\,(t = 0)$ is independent of z.

Suppose $\xi = \xi_1$ defines the elliptical cylinder, viz.,

$$a^2 = c^2 \cosh^2 \xi_1, \quad b^2 = c^2 \sinh^2 \xi_1$$

and suppose that our domain is in contact with a solute free reservoir so that $c\,(\xi = \xi_1) = 0$ for all $t > 0$. Then determine whether or not the solute concentration is always independent of η if it is initially independent of η. Moon and Spencer's book "*Field Theory Handbook:*

Including Coordinate Systems, Differential Equations, and their Solutions" is a useful reference.

9. A solute is initially distributed throughout an infinitely long pipe of radius R according to $c(t = 0) = c_0(r, z)$. Determine the solute concentration for all $t > 0$ by expanding c in the radial eigenfunctions of ∇^2 and then using the point source solution to the longitudinal diffusion equation to determine the coefficients of the eigenfunctions in the expansion. The wall of the pipe is impermeable to the solute.

10. Occasionally it is possible to use eigenfunctions in a region of simple shape to derive eigenfunctions in a not so simple region of interest. To do this we make linear combinations of the eigenfunctions we have in a way that satisfies the conditions on the boundary of the region of interest.

For example $\sin m\pi x \sin n\pi y$ and $\sin n\pi x \sin m\pi y$ are eigenfunctions of ∇^2 in the square, $0 \leq x \leq 1, 0 \leq y \leq 1$, satisfying $\psi = 0$ along $x = 0$, $x = 1$, $y = 0$ and $y = 1$. The eigenvalue is

$$\lambda^2 = \pi^2 (m^2 + n^2)$$

You are to derive eigenfunctions and eigenvalues of ∇^2 on the region $0 \leq x \leq 1$, $0 \leq y \leq 1-x$ satisfying $\psi = 0$ on $x = 0, y = 0$ and $x+y = 1$. Do this by making $x+y = 1$ a line where a linear combination of the eigenfunctions on the square, corresponding to a fixed eigenvalue, vanishes.

11. You are to solve the eigenvalue problem

$$\nabla^2 \psi + \lambda^2 \psi = 0, \qquad r \leq R(\theta), \quad 0 \leq \theta < 2\pi$$

where

$$\psi = 0 \quad \text{at} \quad r = R\left(\theta\right)$$

The region is an ellipse obtained by slightly displacing a circle of radius R_0, holding the area fixed.

Thus, you write

$$a = R_0 + \varepsilon$$

$$b = R_0 - \varepsilon + \frac{\varepsilon^2}{R_0} - \cdots$$

and

$$R\left(\theta\right) = R_0 + \varepsilon\, R_1\left(\theta\right) + \frac{1}{2}\varepsilon^2 R_2\left(\theta\right) + \cdots$$

and substitute these expansions into

$$b^2 x^2 + a^2 y^2 = a^2 b^2$$

where $x = R\cos\theta$, $y = R\sin\theta$, to find $R_1\left(\theta\right)$, $R_2\left(\theta\right)$, ... By doing this you should have $R_1\left(\theta\right) = \cos 2\theta$ and $R_2\left(\theta\right) =$ your job.

Then any eigenfunction and eigenvalue on the circle, viz., ψ_0, λ_0^2, can be corrected for a slight displacement by writing

$$\psi = \psi_0 + \varepsilon\,\psi_1 + \frac{1}{2}\varepsilon^2\,\psi_2 + \cdots$$

$$\lambda^2 = \lambda_0^2 + \varepsilon\,\lambda_1^2 + \frac{1}{2}\varepsilon^2\,\lambda_2^2 + \cdots$$

deriving the equations for ψ_1, λ_1^2 and ψ_2, λ_2^2 on the circle in the usual way and then deriving the conditions satisfied by ψ_1 and ψ_2 at $r = R_0$ by writing

$$\psi\big(r = R(\theta)\,\big) = \psi_0 + \varepsilon\left(\psi_1 + R_1 \frac{\partial \psi_0}{\partial r}\right)$$

$$+ \frac{1}{2}\varepsilon^2\left(\psi_2 + 2R_1 \frac{\partial \psi_1}{\partial r} + R_1^2 \frac{\partial^2 \psi_0}{\partial r^2} + R_2 \frac{\partial \psi_0}{\partial r}\right) + \cdots$$

where the RHS is evaluated at $r = R_0$. To see how this goes first take $\psi_0 = J_0(\lambda_0 r)$ where $J_0(\lambda_0 R_0) = 0$ and then $\psi_0 = J_1(\lambda_0 r)\cos\theta$ where $J_1(\lambda_0 R_0) = 0$.

You should notice that at each order 1st, 2nd, etc., the homogeneous problem is the zeroth order problem and it has a solution, viz., ψ_0, not zero. Hence a solvability condition must be satisfied. This determines λ_1^2 at first order, before ψ_1, λ_2^2 at second order, before ψ_2, etc. In solving for ψ_1, ψ_2, etc. you ought to expand the inhomogeneous terms in $1, \cos\theta, \cos 2\theta, \ldots$, eg., $\cos^2\theta = \frac{1}{2} + \frac{1}{2}\cos 2\theta$.

12. The setting for the free radical problem, see Lecture 14, is now a very long circular cylinder of radius R. What is the critical value of R in terms of k and D?

By how much can the critical value of R be increased if the cylinder is of length L and $c = 0$ at $z = 0$ and L.

13. Normalization of Bessel Functions

Define ψ by

$$\psi = A J_m(\lambda r) + B Y_m(\lambda r)$$

and notice that ψ satisfies

$$\frac{1}{r}\frac{d}{dr}(r\psi) + \left\{\lambda^2 - \frac{m^2}{r^2}\right\}\psi = 0$$

Multiply this equation by $r^2 \dfrac{d\psi}{dr}$ and integrate the product over $R_1 < r < R_2$. Derive a formula for

$$\int_{R_1}^{R_2} r\,\psi^2\,dr$$

in terms of ψ and ψ' at R_1 and R_2.

14. The frequencies of oscillation of a spinning column of inviscid fluid

   An inviscid fluid lies in a cylinder of radius R, spinning at angular velocity Ω.

   The equations are

   $$\frac{\partial \vec{v}}{\partial t} + \vec{v} \cdot \nabla \vec{v} = -\nabla p$$

   $$\nabla \cdot \vec{v} = 0$$

   where p denotes $\dfrac{p}{\rho}$.

   Show that

   $$v_r = 0, \qquad v_\theta = r\Omega, \qquad v_z = 0$$

   satisfies these equations.

   Introduce a small perturbation, write the perturbation equations and assume

   $$\left.\begin{aligned} v_{r1} &= \widehat{v}_r\,(r) \\ v_{\theta 1} &= \widehat{v}_\theta\,(r) \\ v_{z1} &= \widehat{v}_z\,(r) \\ p_1 &= \widehat{p}\,(r) \end{aligned}\right\} e^{i\,\sigma t}\,e^{i\,m\,\theta}\,e^{i\,k\,z}$$

to obtain

$$\imath\left(\sigma + m\,\Omega\right)\widehat{v}_r - 2\Omega\widehat{v}_\theta = -\frac{d\widehat{p}}{dr}$$

$$\imath\left(\sigma + m\,\Omega\right)\widehat{v}_\theta + 2\Omega\widehat{v}_r = -\frac{\imath\,m}{r}\,\widehat{p}$$

$$\imath\left(\sigma + m\,\Omega\right)\widehat{v}_z = -\imath\,k\,\widehat{p}$$

and

$$\frac{d\widehat{v}_r}{dr} + \frac{\widehat{v}_r}{r} + \frac{\imath\,m\widehat{v}_\theta}{r} + \imath\,k\widehat{v}_z = 0$$

This is an eigenvalue problem where σ is the eigenvalue, m and k are inputs.

Eliminate $\dfrac{\imath\,m\widehat{v}_\theta}{r} + \imath\,k\,\widehat{v}_z$ from the last equation and $\widehat{v}_\theta$ from the first two thereby obtaining two equations in $\widehat{v}_r$ and $\widehat{p}$. Then eliminate $\widehat{v}_r$, whereupon you have

$$\frac{d^2\widehat{p}}{dr^2} + \frac{1}{r}\frac{d\widehat{p}}{dr} + \left\{k^2\left(\frac{4\Omega^2}{(\sigma + m\Omega)^2} - 1\right) - \frac{m^2}{r^2}\right\}\widehat{p} = 0$$

and at $r = R$, $v_r = 0$ implies

$$\frac{d\widehat{p}}{dr} + \frac{2\,\Omega\,m}{\sigma + m\,\Omega}\frac{\widehat{p}}{r} = 0$$

Find the frequencies of oscillation, given k^2, in the simple case $m = 0$. At $m = 0$ we have a problem in σ^2 but at $m \neq 0$ it is a problem in $(\sigma + m\,\Omega)^2$.

15. Solve the Petri dish problem in a circular domain , assuming homogeneous Dirichlet conditions along the circumference.

16. A cold rod of radius κR_0 lies inside a hot pipe of radius R_0. The temperatures T_{cold} and

T_{hot} are held fixed and the temperature of the fluid in the annular region is

$$A + B \ln \frac{r}{R_0}$$

where

$$T_{\text{cold}} = A + B \ln \kappa$$

and

$$T_{\text{hot}} = A$$

The fluid and the cylinders are spinning at constant angular velocity, Ω, such that the fluid velocity is $\vec{\Omega} \times \vec{r}$ where $\vec{\Omega} = \Omega \vec{k}$ and $\vec{k}$ lies along the axis of the rod.

The density of the fluid depends on temperature via

$$\rho = \rho_{\text{ref}} \left(1 - \alpha \left(T - T_{\text{ref}} \right) \right)$$

and our base state exhibits an unstable density stratification.

Accounting for the variation of ρ only in the $\vec{v} \cdot \nabla \vec{v}$ terms derive the equations satisfied by a small perturbation of the base state, assume a solution

$$\left. \begin{array}{l} v_{r1} = \widehat{v}_r \left(r \right) \\ v_{\theta 1} = \widehat{v}_\theta \left(r \right) \\ p_1 = \widehat{p} \left(r \right) \\ T_1 = \widehat{T} \left(r \right) \end{array} \right\} e^{\sigma t} e^{\imath m \theta}$$

and derive the equations for $\widehat{v}_r$, $\widehat{v}_\theta$, $\widehat{p}$, and $\widehat{T}$. There is no gravity, no z variation, no v_z and for κ near 1 the base temperature is more or less linear.

Making this approximation find the critical value of $T_{\text{hot}} - T_{\text{cold}}$, i.e., the smallest value of $T_{\text{hot}} - T_{\text{cold}}$ such that $\sigma = 0$.

17. A hot rod of radius R_0 loses heat by conduction to a cold pipe of radius κR_0, $\kappa > 1$. Their temperatures are $T_h > T_c$. Derive a formula for the rate of heat loss. Move the rod off center by a small amount ε so that its surface is now

$$R(\theta) = R_0 + \varepsilon\, R_1(\theta) + \frac{1}{2}\, \varepsilon^2\, R_2(\theta) + \cdots$$

$$R_1(\theta) = \cos\theta, \qquad R_2(\theta) = -\frac{\sin^2\theta}{R_0}$$

and find out by how much the heat loss is changed.

18. The solutions to

$$\nabla^2\psi + \lambda^2\psi = 0 \qquad 0 \leq r \leq R_0, \qquad 0 \leq \theta < 2\pi$$

where

$$\psi = 0 \quad \text{at} \quad r = R_0$$

are

$$m = 0 \qquad J_0(\lambda r),\ J_0(\lambda R_0) = 0$$

$$m = 1 \qquad J_1(\lambda r)\cos\theta,\ J_1(\lambda R_0) = 0$$

etc.

If the circle is displaced into an ellipse of the same area, viz.,

$$r = R(\theta) = R_0 + \varepsilon R_1 + \cdots$$

where

$$ab = R_0^2$$

$$a = R_0 + \varepsilon$$

$$b = \frac{R_0^2}{a} = R_0 \left(1 - \frac{\varepsilon}{R_0} + \cdots \right)$$

we have

$$R_1 = \cos 2\theta, \qquad R_2 = \frac{1}{R_0} \left(-\frac{1}{2} - \cos 2\theta + \frac{3}{2} \cos 4\theta \right), \qquad \cdots$$

and we wish to solve

$$\nabla^2 \psi + \lambda^2 \psi = 0, \qquad \frac{x^2}{a^2} + \frac{y^2}{b^2} \le 1$$

where

$$\psi = 0, \quad \text{at} \quad \frac{x^2}{a^2} + \frac{y^2}{b^2} = 1$$

Writing the eigenvalues and eigenvectors on the ellipse

$$\lambda^2 = \lambda_0^2 + \varepsilon \lambda_1^2 + \cdots$$

$$\psi = \psi_0 + \varepsilon \psi_1 + \cdots$$

where λ_0^2 and ψ_0 are the corresponding eigenvalues and eigenvectors on the circle, derive the result

$$\forall \, \lambda_0^2 : \lambda_1^2 = 0 \quad \text{at} \quad m = 0$$

$$\forall \, \lambda_0^2 : \lambda_1^2 \ne 0 \quad \text{at} \quad m = 1$$

This second result ought to surprise you.

19. Your job is to estimate the solution to

$$\nabla^2 \psi + \lambda^2 \psi = 0, \quad \psi = 0 \quad \text{on all sides of the domain (see below)}$$

The curved side is specified by

$$y = Y(x) = Y_0 + \varepsilon Y_1(x) + \frac{1}{2} \varepsilon^2 Y_2(x) + \cdots$$

where $Y_1(x)$, $Y_2(x)$, $\cdots$ are inputs. You can make your job easy by assuming $Y_1(x) = \sin 2\pi x$, $Y_2(x) = 0$, ...

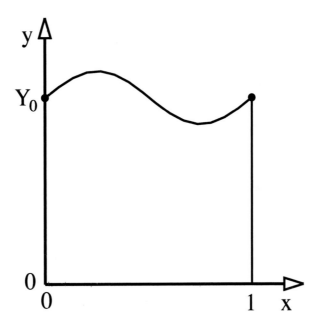

The reference domain is

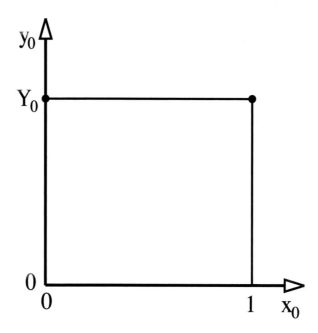

The solutions to

$$\nabla_0^2 \psi_0 + \lambda_0^2 \psi_0 = 0, \qquad \psi_0 = 0 \qquad \text{on all sides}$$

are

$$\psi_0 = \sin m\pi x_0 \, \sin n\pi \frac{y_0}{Y_0} \qquad m, n = 1, 2, \ldots$$

and

$$\lambda_0^2 = m^2\pi^2 + \frac{n^2\pi^2}{Y_0^2}$$

First select an eigenvalue to be corrected and write

$$\lambda^2 = \lambda_0^2 + \varepsilon\,\lambda_1^2 + \frac{1}{2}\,\varepsilon^2\,\lambda_2^2 + \cdots$$

where λ_0^2, ψ_0 correspond to definite values of m and n, say m_0, n_0, held fixed henceforth.

Then the $\lambda_1^2, \lambda_2^2, \ldots$ problems are, first,

$$\nabla_0^2 \psi_1 + \lambda_0^2 \psi_1 = -\lambda_1^2 \psi_0, \quad \psi_1 = 0 \quad \text{on three sides}$$

$$\psi_1 = -Y_1(x_0) \frac{\partial \psi_0}{\partial y_0}(x_0, y_0 = Y_0) \quad \text{on} \quad y_0 = Y_0$$

and, second,

$$\nabla_0^2 \psi_2 + \lambda_0^2 \psi_2 = -2\lambda_1^2 \psi_1 - \lambda_2^2 \psi_0, \quad \psi_2 = 0 \quad \text{on three sides}$$

$$\psi_2 = -2Y_1(x_0) \frac{\partial \psi_1}{\partial y_0}(x_0, y_0 = Y_0) - Y_1^2(x_0) \frac{\partial^2 \psi_0}{\partial y_0^2}(x_0, y_0 = Y_0)$$

$$- Y_2(x_0) \frac{\partial \psi_0}{\partial y_0}(x_0, y_0 = Y_0) \quad \text{on} \quad y_0 = Y_0$$

etc.

At each order the homogeneous problem has a solution, not zero. Hence a solvability condition must be satisfied and these conditions lead you to λ_1^2, λ_2^2, etc.

First, derive a formula for λ_1^2, viz.,

$$\int_0^1 dx_0 \left(-Y_1(x_0) \frac{\partial \psi_0}{\partial y_0}(x_0, y_0 = Y_0) \right) = \lambda_1^2 \int_0^{Y_0} \int_0^1 \psi_0^2 \, dx_0 \, dy_0$$

Second, solve for ψ_1 by deriving formulas for the coefficients A_{mn}, where

$$\psi_1 = \sum \sum A_{mn} \psi_{0mn}$$

and where

$$A_{mn} = \int \int \psi_{0mn} \psi_1 \, dx_0 \, dy_0$$

assuming $\int \int \psi_{0mn}^2 \, dx_0 \, dy_0 = 1$.

20. You wish to find the eigenvalues, λ^2, where

$$\nabla^2 \psi + \lambda^2 \psi = C$$

$$\psi = 0, \quad \text{at the sides}$$

and

$$\iint_A \psi \, dx dy = 0$$

The domain is the square:

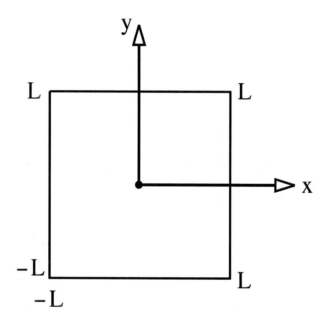

and ψ, C and λ^2 are the outputs. Your interest is the solutions where $C \neq 0$.

The problem

$$\nabla^2 \phi + \mu^2 \phi = 0$$

and

$$\phi = 0, \quad \text{at the sides}$$

has solutions

$$\sin m\pi \frac{x}{L} \sin n\pi \frac{y}{L}, \quad m, n = 1, 2, \ldots$$

$$\sin m\pi \frac{x}{L} \cos\left(n + \frac{1}{2}\right)\pi \frac{y}{L}, \quad m = 1, 2, \ldots \quad n = 0, 1, \ldots$$

$$\cos\left(m + \frac{1}{2}\right)\pi \frac{x}{L} \sin n\pi \frac{y}{L}, \quad m = 0, 1, \ldots \quad n = 1, 2, \ldots$$

all of which integrate to zero hence all of which are ψ's and λ^2's corresponding to $C = 0$.

It also has solutions

$$\cos\left(m + \frac{1}{2}\right)\pi \frac{x}{L} \cos\left(n + \frac{1}{2}\right)\pi \frac{y}{L}, \quad m, n = 0, 1, \ldots$$

none of which integrates to zero.

Expanding ψ in these solutions, at $C \neq 0$, derive

$$\sum_{m,n} \frac{\left(\iint \phi_{mn}\, dx dy\right)^2}{\lambda^2 - \mu_{mn}^2} = 0$$

and estimate the smallest λ^2.

Lecture 18

Two Stability Problems

Using what we did in Lecture 17 we can work out two stability problems: the Saffman-Taylor problem (P. G. Saffman, G. I. Taylor, Proc. Roy. Soc., Vol. 245, 312, 1958) and the Rayleigh-Taylor problem. (S. Chandrasekhar, *Hydrodynamic and Hydromagnetic Stability.*)

The setting for each is a cylindrical column of circular cross section bounding a porous solid. Fluid fills the pores and its velocity is given by Darcy's law.

We will need the balances, expressing the conservation laws, across a surface separating two phases, denoted (1) and (2):

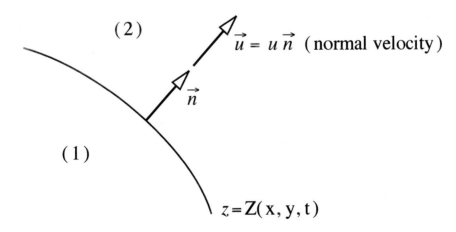

where

$$\vec{n} = \frac{\vec{k} - Z_x\vec{i} - Z_y\vec{j}}{\sqrt{1 + Z_x^2 + Z_y^2}}$$

$$u = \frac{Z_t}{\sqrt{1 + Z_x^2 + Z_y^2}}$$

and

$$2H = \frac{\left(1 + Z_y^2\right) Z_{xx} - 2Z_x Z_y Z_{xy} + \left(1 + Z_x^2\right) Z_{yy}}{\left(1 + Z_x^2 + Z_y^2\right)^{3/2}}$$

and where H denotes the mean curvature of the surface.

Assuming the phases are immiscible, neither crossing the surface, we have at $z = Z(x, y, t)$:

$$\vec{n} \cdot \vec{v}^{(1)} = u = \vec{n} \cdot \vec{v}^{(2)}$$

which can be written as

$$\left\{v_z - Z_x v_x - Z_y v_y\right\}^{(1)} = Z_t = \left\{v_z - Z_x v_x - Z_y v_y\right\}^{(2)}$$

and

$$-\vec{n}\vec{n} : \vec{\vec{T}}^{(1)} + \gamma 2H = -\vec{n}\vec{n} : \vec{\vec{T}}^{(2)}$$

which can be written, in the case of a Darcy fluid, as

$$p^{(1)} + \gamma 2H = p^{(2)}$$

where γ denotes the surface tension.

We introduce the notation

$$\nabla = \vec{k}\frac{\partial}{\partial z} + \nabla_H$$

and

$$\vec{v} = v_z \vec{k} + \vec{v}_H$$

and we plan to work in terms of $p^{(1)}, p^{(2)}$ and Z.

The present domain, bounded by the surface $z = Z(x, y, t)$ is obtained by a displacement of the reference domain, bounded by the surface $z = Z_0$.

Hence for a domain variable, say p, at the surface $z = Z$ we will write

$$p(Z) = p_0(Z_0) + \varepsilon \left(p_1(Z_0) + Z_1 \frac{dp_0}{dz}(Z_0) \right) + \cdots$$

assuming

$$Z = Z_0 + \varepsilon Z_1 + \cdots$$

where $p_0, p_1, \ldots$ are defined on the reference domain.

18.1   The Saffman-Taylor Problem

In this problem the stability of the surface separating two immiscible fluids is of interest, one fluid displacing the other in a porous rock. The flow is in the z direction at a speed U and gravity is not important. What is important is that the viscosity of the two fluids differs.

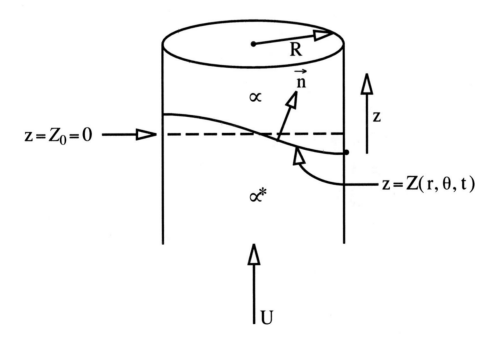

We introduce an observer moving at the velocity $U\vec{k}$. Then, in the moving frame, we write the nonlinear equations making up our model for the dynamics of the surface, $z = Z(r, \theta, t)$, separating the two phases. First, we have Darcy's law, which is not Galilean invariant, above and below the surface, viz.,

$$\frac{\mu}{K}\left(\vec{v} + \vec{U}\right) = -\nabla p, \qquad z > Z$$

and

$$\frac{\mu^*}{K}\left(\vec{v}^* + \vec{U}\right) = -\nabla p^*, \qquad z < Z$$

where $\nabla \cdot \vec{v} = 0 = \nabla \cdot \vec{v}^*$, where $\vec{U} = U\vec{k}$ and where K denotes the permeability of the porous solid, whereupon

$$\nabla^2 p = 0 = \nabla^2 p^*$$

At the side walls, $r = R$, no flow implies

$$\vec{n} \cdot \vec{v} = 0 = \vec{n} \cdot \nabla_H p, \qquad z > Z$$

and

$$\vec{n} \cdot \vec{v}^* = 0 = \vec{n} \cdot \nabla_H p^*, \qquad z < Z$$

and contact at right angles to the wall implies $\vec{n} \cdot \nabla_H Z = 0$. Far from the surface $z = Z$ the pressures p and p^* must be bounded.

At the surface, $z = Z$, we have

$$v_z - \vec{v}_H \cdot \nabla_H Z = Z_t = v_z^* - \vec{v}_H^* \cdot \nabla_H Z$$

and

$$p - p^* = \gamma 2H$$

All of the nonlinearities in our model appear at the surface.

Our base solution, denoted by the subscript zero, is

$$\vec{v}_0 = \vec{0} = \vec{v}_0^*$$

$$\frac{dp_0}{dz} = -\frac{\mu}{K}U$$

and

$$\frac{dp_0^*}{dz} = -\frac{\mu^*}{K}U$$

where the surface separating the two fluids lies at $z = Z_0 = 0$, defining the base domain.

Imposing a small displacement on our base solution and denoting the perturbation variables by the subscript 1, viz., $Z = Z_0 + \varepsilon Z_1$, we obtain the perturbation problem. It is defined on the base domain and we have

$$\nabla^2 p_1 = 0. \quad z > 0$$

and

$$\nabla^2 p_1^* = 0. \quad z < 0$$

At the side walls we have

$$v_{r1} = 0 \quad \therefore \quad \frac{\partial p_1}{\partial r} = 0 \quad \text{at} \quad r = R$$

$$v_{r1}^* = 0 \quad \therefore \quad \frac{\partial p_1^*}{\partial r} = 0 \quad \text{at} \quad r = R$$

and

$$\frac{\partial Z_1}{\partial r} = 0 \quad \text{at} \quad r = R$$

And at $z = 0$ we have

$$p_1 + Z_1 \frac{dp_0}{dz} - \left(p_1^* + Z_1 \frac{dp_0^*}{dz} \right) = \gamma \nabla_H^2 Z_1$$

$$-\frac{K}{\mu} \frac{\partial p_1}{\partial z} = v_{z1} = \frac{\partial Z_1}{\partial t} = v_{z1}^* = -\frac{K}{\mu^*} \frac{\partial p_1^*}{\partial z}$$

and

$$\int_0^{2\pi} \int_0^R Z_1 r \, dr d\theta = 0$$

where we assume no volume change on perturbation of the base surface.

This is a linear problem in p_1, p_1^* and Z_1, where each of these variables satisfies homogeneous Neumann conditions at $r = R$.

Hence, to solve it, we introduce the eigenvalue problem

$$\nabla_H^2 \psi + \lambda^2 \psi = 0$$

$$\frac{\partial \psi}{\partial r} = 0 \quad \text{at} \quad r = R$$

and ψ bounded at $r = 0$ and we write its solution

$$\psi = J_m \left(\lambda r \right) \cos m\theta$$

where λ is a root of

$$J_m' \left(\lambda R \right) = 0$$

and $J_m' \left(x \right)$ denotes $\frac{d}{dx} J_m \left(x \right)$.

Our plan is to determine the growth rate of surface displacements in the shape of any of the allowable eigenfunctions.

To do this we separate variables and write

$$p_1 = \widehat{p}_1(z)\, \psi(r,\theta)\, e^{\sigma t}$$

$$p_1{}^* = \widehat{p}_1{}^*(z)\, \psi(r,\theta)\, e^{\sigma t}$$

and

$$Z_1 = \widehat{Z}_1 \psi(r,\theta)\, e^{\sigma t}$$

whereupon we find

$$\frac{d\widehat{p}_1}{dz} - \lambda^2 \widehat{p}_1 = 0, \qquad z > 0$$

$$\frac{d\widehat{p}_1{}^*}{dz} - \lambda^2 \widehat{p}_1{}^* = 0, \qquad z < 0$$

and at $z = 0$ we have

$$-\frac{K}{\mu}\frac{d\widehat{p}_1}{dz} = \sigma \widehat{Z}_1 = -\frac{K}{\mu^*}\frac{d\widehat{p}_1{}^*}{dz}$$

$$\widehat{p}_1 + \widehat{Z}_1\left(-\frac{\mu}{K}U\right) - \left(\widehat{p}_1{}^* + \widehat{Z}_1\left(-\frac{\mu^*}{K}U\right)\right) = -\gamma\lambda^2 \widehat{Z}_1$$

and

$$\int_0^{2\pi}\int_0^R \widehat{Z}_1 \psi e^{\sigma t} r\, dr\, d\theta$$

Assuming $\widehat{p}_1$ is bounded as $z \to \infty$ and $\widehat{p}_1{}^*$ is bounded as $z \to -\infty$ we have

$$\widehat{p}_1 = A e^{-\lambda z}$$

and

$$\widehat{p}_1^* = A^* e^{\lambda z}$$

hence at $z = 0$ we find

$$\frac{K}{\mu} \lambda A = \sigma \widehat{Z}_1 = -\frac{K}{\mu^*} \lambda A^*$$

and

$$A + \widehat{Z}_1 \left(-\frac{\mu}{K} U\right) - \left(A^* + \widehat{Z}_1 \left(-\frac{\mu^*}{K} U\right)\right) = -\gamma \lambda^2 \widehat{Z}_1$$

The inputs to our problem are U and R, the output is σ and we have three linear homogeneous equations in A, A^* and $\widehat{Z}_1$ which have a non vanishing solution iff the determinant of the matrix of coefficients vanishes. This determines σ, the growth rate of a surface displacement in the shape ψ. The readers can work this out.

To determine the critical value of U, we set $\sigma = 0$ whereupon $A = 0 = A^*$ and we find

$$\frac{U}{K} (\mu - \mu^*) = \gamma \lambda^2$$

which tells us this: if $\mu^* > \mu$ there is no critical condition, i.e., the surface separating a more viscous fluid displacing a less viscous is stable to any small displacement. A critical value of U is possible iff $\mu^* < \mu$, i.e., a less viscous fluid displacing a more viscous fluid. A plot of the critical value of U vs λ^2, then, looks as follows

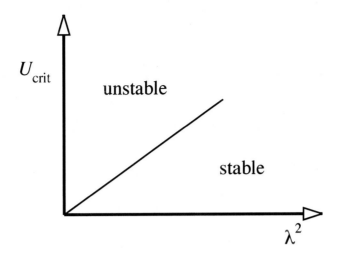

and if we mark the allowable values of λ^2 on the abscissa we see that the lowest allowable λ^2 sets the pattern of the instability.

The allowable λ^2's come from the roots of

$$J_0'(\lambda R) = 0, \qquad J_1'(\lambda R) = 0, \qquad J_2'(\lambda R) = 0 \qquad \text{etc.}$$

where we have

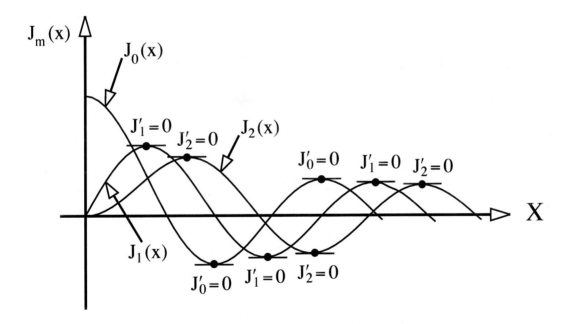

We look first at the eigenfunctions $\psi = J_0(\lambda r)$, where $J_0'(\lambda R) = 0$, and observe that $J_0'(x) = 0$ has a root at $x = 0$, hence we have a solution $\lambda = 0$ and $\psi = 1$. This solution does not satisfy

$$\int_0^{2\pi} \int_0^R \psi \, r \, dr d\theta = 0$$

and hence it is not marked on the diagram. Every other root of $J'_0(x) = 0$ is allowable because

$$\int_0^R J_0(\lambda r) \, r \, dr = \frac{1}{\lambda} r J_1(\lambda r) \Big|_0^R$$

and $J_1(x) = -J'_0(x)$.

All the eigenfunctions $J_m(\lambda r) \cos m\theta$, where $J'_m(\lambda R) = 0$, are allowable due to

$$\int_0^{2\pi} \cos m\theta \, d\theta = 0$$

and we observe that $J'_2(x)$, $J'_3(x)$, etc., all vanish at $x = 0$ but in each case the corresponding eigenfunction is zero.

Hence the lowest allowable value of λ^2 corresponds to $J'_1(\lambda R) = 0$ whence

$$U_{\text{crit}} = \frac{K\gamma}{(\mu - \mu*)} \lambda^2$$

where λR is the first root of $J'_1(x) = 0$.

Thus the pattern we should expect to see as we increase U in an experiment to just beyond U_{crit} should have a $\cos \theta$ dependence, viz.,

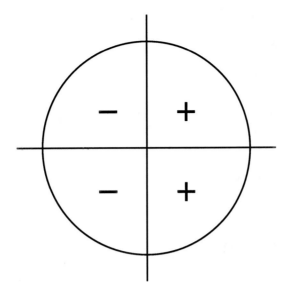

along with the radial dependence

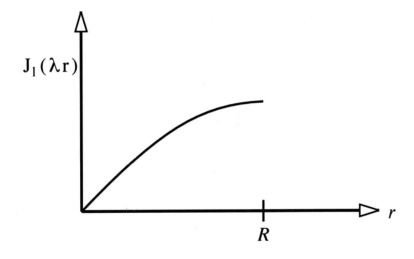

Now we can ask another question: at what value of R does the surface become unstable, given that U is fixed at a positive value?

From the above we have

$$\frac{U}{K\gamma}(\mu - \mu^*) = \lambda^2, \qquad \lambda^2 = \frac{x^2}{R^2}$$

where the x's are roots of $J'_1(x) = 0$.

If R is very small the right hand side is very large even for the smallest root of $J'_1(x) = 0$. Hence small diameter columns are stable unless U is very large. Upon increasing R we arrive at its critical value where

$$\frac{U}{K\gamma}(\mu - \mu*) = \frac{x_1^2}{R_{\text{crit}}^2}$$

and where x_1 is the smallest positive root of $J'_1(x) = 0$.

18.2   The Rayleigh-Taylor Problem

This problem does not differ from the Saffman-Taylor problem by much. Here the instability is caused by gravity and $\vec{g}$ takes the place of $\vec{U}$. Again we set the problem in a porous rock and use Darcy's law.

We have two fluids of different density lying in a gravitational field, the heavy fluid above the light fluid.

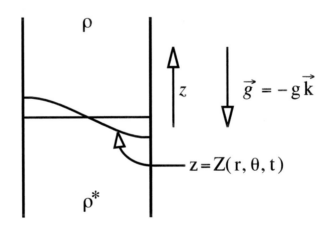

We write our equations in the laboratory frame as

$$\frac{\mu}{K}\vec{v} = -\nabla p + \rho\vec{g}, \quad \nabla \cdot \vec{v} = 0$$

for both fluids and hence we have

$$\nabla^2 p = 0, \quad z > Z$$

and

$$\nabla^2 p* = 0, \quad z < Z$$

The boundary conditions are as before:

$$\frac{\partial p}{\partial r} = 0, \quad \text{at} \quad r = R, \quad z > Z$$

and

$$\frac{\partial p*}{\partial r} = 0, \quad \text{at} \quad r = R, \quad z < Z$$

due to no flow across the side walls, p must be bounded as $z \to \infty$ and $p*$ must be bounded as $z \to -\infty$ and, at the surface $z = Z$,

$$\vec{n} \cdot \vec{v} = \frac{Z_t}{\sqrt{1 + Z_r^2 + \frac{1}{r^2}Z_\theta^2}} = \vec{n} \cdot \vec{v}*$$

due to no flow across the surface separating the fluids,

$$p - p* = \gamma 2H$$

and

$$\int_0^{2\pi} \int_0^R Z(r, \theta, t) \, r \, dr \, d\theta$$

At this point we do something a little different than before. We are going to change the boundary conditions satisfied by Z and require pinned edges in place of free edges, i.e., $Z = 0$ at $r = R$. Hence the boundary conditions satisfied by p_1, p_1* and Z_1 at the wall in the perturbed problem differ and will not allow us to separate variables as we did above, viz., p_1 and p_1* will be asking for one set of ψ's, corresponding to Neumann conditions, Z_1 will be asking for another set corresponding to Dirichlet conditions. Therefore we are limited in what we can do easily.

We do not ask for σ, instead we set σ to zero and look for the neutral condition. By setting σ to zero in the perturbation problem we have

$$\nabla^2 p_1 = 0 = \nabla^2 p_1{}^*$$

$$\frac{\partial p_1}{\partial r} = 0 = \frac{\partial p_1{}^*}{\partial r} \quad \text{at} \quad r = R$$

and

$$\frac{\partial p_1}{\partial z} = 0 = \frac{\partial p_1{}^*}{\partial z} \quad \text{at} \quad z = 0$$

where p_1 and $p_1{}^*$ must be bounded. Hence, at the critical value of R, p_1 and $p_1{}^*$ must be constants, but not necessarily zero as would be the case if the edges of the surface were free instead of pinned.

Then the equation for Z_1 is

$$p_1 + \frac{dp_0}{dz} Z_1 - \left(p_1{}^* + \frac{dp_0{}^*}{dz} Z_1 \right) = \gamma 2H_1$$

where, setting $\gamma C = p_1 - p_1{}^*$ and using

$$\frac{dp_0}{dz} = -\rho g, \qquad \frac{dp_0{}^*}{dz} = -\rho^* g \quad \text{and} \quad 2H_1 = \nabla_H^2 Z_1$$

we have

$$\gamma C - g\left(\rho - \rho^* \right) Z_1 = \gamma \nabla_H^2 Z_1$$

where

$$Z_1 = 0 \quad \text{at} \quad r = R, \quad Z_1 \text{ bounded at } r = 0$$

and

$$\int_0^{2\pi} \int_0^R Z_1 r \, dr d\theta = 0$$

This is a homogeneous problem in Z_1 and C and we are looking for the value of $\dfrac{g\,(\rho - \rho^*)}{\gamma}$ such that Z_1 is not zero.

Thus we write

$$\nabla_H^2 Z_1 + \lambda^2 Z_1 = C, \qquad Z_1 = 0 \quad \text{at} \quad r = R$$

and

$$\int_0^{2\pi} \int_0^R Z_1 r \, dr d\theta = 0$$

and we have an eigenvalue problem, the eigenvalue λ^2 corresponding to the eigenfunction Z_1, C. The question then is: for what value of R can $\dfrac{g}{\gamma}\,(\rho - \rho^*)$ be one of the eigenvalues of this problem?

Writing ψ in place of Z_1 we scale the problem and obtain

$$\nabla_H^2 \psi + \lambda^2 \psi = C$$

$$\psi = 0 \quad \text{at} \quad r = 1$$

and

$$\int_0^{2\pi} \int_0^1 \psi r \, dr d\theta = 0$$

where C denotes $C R^2$ and where λ^2 is now independent of R. Then

$$R^2 \frac{g}{\gamma}\,(\rho - \rho^*)$$

must be one of the λ^2's and we can look for critical values of R given $\dfrac{g}{\gamma}\,(\rho - \rho^*)$.

First we see that if $\rho^* > \rho$ the surface is stable to small perturbations for all values of R. This is the case of a light fluid lying above a heavy fluid. Then for $\rho > \rho^*$ and R very small, $R^2 \frac{g}{\gamma}\,(\rho - \rho^*)$ will be less than all λ^2's. And a heavy fluid lying above a light fluid will be stable to small perturbations.

As we increase R, the critical value of R will be reached when $R^2 \frac{g}{\gamma}(\rho - \rho^*)$ becomes equal to λ_1^2 the smallest eigenvalue among the set of λ^2's satisfying our eigenvalue problem.

For $m = 1, 2, \ldots$ we have solutions

$$\psi = J_m(\lambda r) \cos m\theta, \quad C = 0$$

where the λ's are the positive roots of $J_m(x) = 0$, viz.,

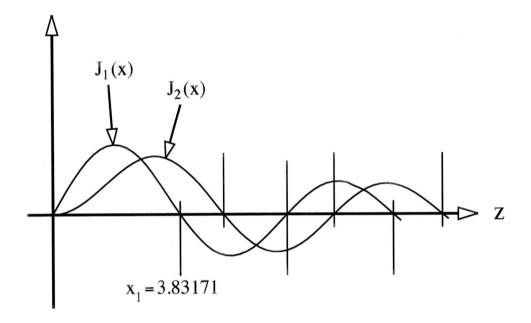

$$x_1 = 3.83171$$

The least of these is the first positive root of $J_1(x) = 0$.

This leaves only the case of axisymmetric disturbances, viz., $m = 0$, where we can not use $\int_0^{2\pi} \cos m\theta \, d\theta = 0$, $m = 1, 2, \ldots$ to easily conclude that $C = 0$. In fact at $m = 0$, C is not zero. Indeed at $m = 0$ we have

$$\psi = A J_0(\lambda r) + \frac{C}{\lambda^2}$$

which is the general solution to

$$\frac{d^2\psi}{dr^2} + \frac{1}{r}\frac{d\psi}{dr} + \lambda^2\psi = C$$

assuming ψ is bounded.

Now λ cannot be zero for then C must be zero and we have $\psi = A$, whereupon A must be zero. Hence to find the positive λ's we observe that

$$\psi\left(r = 1\right) = 0$$

and

$$\int_0^1 \psi r \, dr = 0$$

imply that

$$A J_0\left(\lambda\right) + \frac{C}{\lambda^2} = 0$$

and

$$A \int_0^1 J_0\left(\lambda r\right) r \, dr + \frac{1}{2}\frac{C}{\lambda^2} = 0$$

and we have two homogeneous equations in A and C.

Then using

$$\frac{d}{dr}\left(r J_1\left(\lambda r\right)\right) = \lambda r J_0\left(\lambda r\right)$$

and requiring a solution other than $A = 0 = C$ we obtain

$$\lambda J_0\left(\lambda\right) = 2 J_1\left(\lambda\right)$$

whose solutions are the values of λ at $m = 0$. The lowest solution lies to the right of x_1, the smallest positive root of $J_1\left(x\right) = 0$. Hence we conclude that the critical value of R is given by

$$R^2 \frac{g}{\gamma}\left(\rho - \rho*\right) = x_1^2$$

whence the surface will break in a non axisymmetric, viz., $m = 1$, mode.

We present a graph of $x\,J_0(x) - 2J_1(x)$ vs x and indicate the first few positive roots.

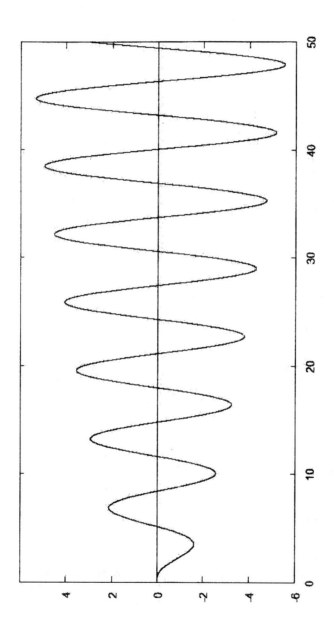

Roots of $xJ_0(x) - 2J_1(x) = 0$

0

5.1355

8.4172

11.6195

14.7960

17.9598

21.117

24.2701

27.4206

30.5692

33.7165

36.8628

40.0084

We also record the first positive zeros of $J_1(x)$ and $J_1'(x)$. They are

$$J_1(x) = 0: \qquad x_1 = 3.8317$$

$$J_1'(x) = 0: \qquad x_1 = 1.8412$$

Hence the case of pinned edges is most unstable to a $\cos\theta$ perturbation and, although both free and pinned edges are most unstable to a $\cos\theta$ perturbation, it takes a much larger value of R to destabilize the case where the edges are pinned.

18.3   Home Problems

1. Convection caused by an adverse temperature gradient.

You have two isothermal horizontal planes bounding a fluid. The lower one, at $z = 0$, is hot, the upper one, at $z = H$, is cold. The density of the fluid depends on its temperature via

$$\rho = \rho_{\text{ref}} \left\{ 1 - \alpha \left(T - T_{\text{ref}} \right) \right\}$$

Hence the fluid layer is unstably stratified, heavy over light.

The base solution is

$$\vec{v}_0 = \vec{0}, \qquad \frac{dT_0}{dz} = -\frac{T_H - T_C}{H}$$

Assume the problem is two dimensional, i.e., one horizontal dimension.

Your model is

$$\rho \frac{\partial \vec{v}}{\partial t} + \rho \vec{v} \cdot \nabla \vec{v} = -\nabla p + \mu \nabla^2 \vec{v} + \rho \vec{g}, \qquad \nabla \cdot \vec{v} = 0$$

and

$$\frac{\partial T}{\partial t} + \vec{v} \cdot \nabla T = \kappa \nabla^2 T$$

where we have no side walls and at $z = 0, H$ we have

$$v_z = 0, \qquad \frac{\partial v_z}{\partial x} + \frac{\partial v_x}{\partial z} = 0$$

corresponding to no flow and no shear.

Introduce a small perturbation of the base solution, eliminate p_1 by differentiation, eliminate v_{x1} by $\nabla \cdot \vec{v}_1 = 0$ and obtain

$$\rho \frac{\partial}{\partial t} \nabla^2 v_{z1} = \mu \nabla^2 \nabla^2 v_{z1} + \rho_{\text{ref}} \, \alpha g \, \frac{\partial^2 T_1}{\partial x^2}$$

and

$$\frac{\partial T_1}{\partial t} + v_{z1} \frac{dT_0}{dz} = \kappa \nabla^2 T_1$$

Your job is to find $\dfrac{dT_0}{dz}$ at neutral conditions where steady values of v_{z1} and T_1, not both zero, prevail. Dropping $\dfrac{\partial}{\partial t}$ and scaling, you can obtain

$$\nabla^2\nabla^2 v_{z1} - \Delta T\,\frac{\partial^2 T_1}{\partial x^2} = 0$$

and

$$\nabla^2 T_1 + v_{z1} = 0$$

where $T_1 = 0 = v_{z1}$ at $z = 0, 1$

$$\frac{\partial^2 v_{z1}}{\partial x^2} - \frac{\partial^2 v_{z1}}{\partial z^2} = 0 \qquad \text{at} \quad z = 0, 1$$

and where ΔT is a scaled temperature difference.

Eliminating T_1 you get

$$\nabla^2\nabla^2\nabla^2 v_{z1} + \Delta T\,\frac{\partial^2 v_{z1}}{\partial x^2} = 0$$

and you can solve this by separation of variables, viz.,

$$v_{z1} = A \sin n\pi z\,\cos kx$$

to find the critical value of ΔT as a function of n and k.

Your result should look like this, where $\sigma = 0$ curves are plotted as ΔT vs. k^2.

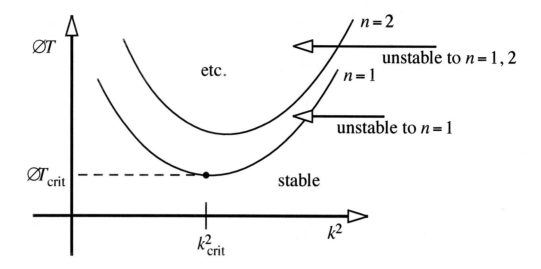

where k^2 is an input, telling you the horizontal wave length of the perturbation, $\dfrac{2\pi}{k}$. You will notice that very long and very short wave length disturbances are very stable.

Observe that

$$T_1 = B \sin n\pi z \, \cos kx$$

$$v_{x1} = C \cos n\pi z \, \sin kx$$

and

$$p_1 = D \cos n\pi z \, \cos kx$$

and find B, C and D in terms of A, n^2 and k^2.

Using integration by parts, workout

$$\int_0^1 \left\{ a \left(\frac{d^2}{dz^2} - k^2 \right)^3 b - b \left(\frac{d^2}{dz^2} - k^2 \right)^3 a \right\} dz$$

What is going on in this problem is this: An element of fluid in equilibrium, buoyancy balancing gravity, upon being given an upward displacement, finds itself surrounded by colder fluid of a higher density. Its increased buoyancy reinforces the displacement and the density stratification is unstable. This is offset by the fact that our element of fluid is now hotter than its new surroundings and it cools, its density increasing.

2. You have a porous rock bounded by a cylinder of constant cross section. At the top and bottom are planes held at constant temperature, viz., T_{hot} at $z = 0$, T_{cold} at $z = H$.

The side wall is an insulated, no-flow surface. The top and bottom planes are isothermal no-flow surfaces.

Your model is

$$\frac{\mu}{K} \vec{v} = -\nabla p - \rho g \vec{k}, \qquad \text{Darcy's law}$$

$$\nabla \cdot \vec{v} = 0$$

and

$$\frac{\partial T}{\partial t} + \vec{v} \cdot \nabla T = \kappa \nabla^2 T$$

where

$$\vec{n} \cdot \vec{v} = 0 \qquad \text{at all surfaces}$$

$$\vec{n} \cdot \nabla T = 0 \qquad \text{at the side walls}$$

$$T = T_{hot} \quad \text{at} \quad z = 0$$

and

$$T = T_{cold} \quad \text{at} \quad z = H$$

and where $\nabla \rho = -\rho_{ref} \, \alpha \, \nabla T$.

The base solution is

$$T_0 = T_0(z),$$

$$\vec{v}_0 = \vec{0},$$

$$\frac{dp_0}{dz} = -\rho(T_0)g,$$

$$\frac{dT_0}{dz} = -\frac{T_{\text{hot}} - T_{\text{cold}}}{H} < 0$$

You may proceed without declaring the shape of the cross section by writing

$$\vec{v} = v_z \vec{k} + \vec{v}_H$$

and

$$\nabla = \vec{k}\frac{\partial}{\partial z} + \nabla_H$$

whereupon

$$\nabla^2 = \frac{\partial^2}{\partial z^2} + \nabla_H^2$$

$$\nabla \cdot \vec{v} = \frac{\partial v_z}{\partial z} + \nabla_H \cdot \vec{v}_H$$

$$\frac{\mu}{K}v_z = -\frac{\partial p}{\partial z} - \rho g$$

and

$$\frac{\mu}{K}\vec{v}_H = -\nabla_H p$$

Then you can derive

$$\nabla^2 v_z = \frac{K}{\mu}\rho_{\text{ref}}\,\alpha\,g\nabla_H^2 T$$

and

$$\frac{\partial T}{\partial t} + v_z \frac{\partial T}{\partial z} + \vec{v}_H \cdot \nabla_H T = \kappa \nabla^2 T$$

And using $\vec{n} \cdot \vec{v}_H = 0 = \vec{n} \cdot \nabla_H T$ at the walls you have

$$\vec{n} \cdot \nabla_H v_z = 0 = \vec{n} \cdot \nabla_H T$$

You introduce a small perturbation of the base solution and derive the perturbation problem for v_{z1} and T_1. And, assuming a steady solution at the critical value of $\frac{dT_0}{dz}$, you solve your problem by separation of variables, viz.,

$$v_{z1} = \widehat{v}_{z1}(z)\, \psi(x,y)$$

$$T_1 = \widehat{T}_1(z)\, \psi(x,y)$$

where ψ is any solution to the eigenvalue problem on the cross section:

$$\nabla_H^2 \psi + \lambda^2 \psi = 0 \qquad \text{on the domain}$$

$$\vec{n} \cdot \nabla_H \psi = 0 \qquad \text{at the edge}$$

Derive the result:

$$\frac{K}{\mu}\, \rho_{\text{ref}}\, \alpha\, g\, \frac{1}{\kappa}\left(-\frac{dT_0}{dz}\right) = \frac{\left(\dfrac{n^2 \pi^2}{H^2} + \lambda^2\right)^2}{\lambda^2} \qquad n = 1, 2, \ldots$$

and draw a sketch, LHS vs. λ^2.

So far the cross section has not come into the problem. But at this point it determines the allowable values of λ^2 and these depend on the shape and diameter of the cross section.

Assume the cross section is a circle of radius R_0 and deduce the convection pattern seen at critical as R_0 increases from a small value to its critical value, at fixed $T_{\text{hot}} - T_{\text{cold}}$.

The dip in the plot of LHS vs. λ^2 allows you to see many patterns at the critical value of $T_{hot} - T_{cold}$. Set $n = 1$ and assume the cross section to be one dimensional having side walls at $x = 0$ and $x = L$. Then $\psi = \cos kx$ ($k = \lambda$) where the allowable values of k are $\dfrac{m\pi}{L}$, $m = 0, 1, 2, \ldots$.

For small values of L the most dangerous value of k corresponds to $m = 1$. As L increases show that the most dangerous value of k corresponds to increasing values of m and, therefore, that many patterns can be seen at the critical value of $T_{hot} - T_{cold}$ depending on the width of the cell.

3. Assume the cross section in Problem 2 to be a thin rectangle of length a and width b, $a \gg b$. Is it the value of a or b that controls the critical temperature difference?

4. Your job is to look again at the Rayleigh-Taylor problem, assuming Darcy's law tells you the velocity. Do this on an arbitrary cross section, writing

$$\nabla = \vec{k}\frac{\partial}{\partial z} + \nabla_H$$

and

$$\vec{v} = v_z \vec{k} + \vec{v}_H$$

and suppose that the surface is not pinned at the edge but contacts the side wall at right angles, viz.,

$$\vec{n} \cdot \nabla_H Z = 0$$

The question is: is there an effect of the fluid depths on the critical diameter of the cross section?

At infinite depths you have $p_1^{(1)} = 0 = p_1^{(2)}$ whereas at finite depths you have, instead, $p_1^{(1)} = c_1$, $p^{(2)} = c_2$. And your equation for Z_1 is then

$$c_2 - c_1 + g\left\{-\rho^{(2)} + \rho^{(1)}\right\} Z_1 = \gamma \nabla_H^2 Z_1$$

on the cross section.

Show that assuming

$$\iint\limits_{\text{cross section}} Z_1 \, dx dy = 0$$

implies $c_2 - c_1 = 0$ and therefore the depths are immaterial.

Do you think this would also be true if the edges were pinned, viz., $Z_1 = 0$ at the edges?

5. You are going to try to predict what you might see in a Rayleigh-Taylor experiment, assuming you see the pattern having the greatest growth rate.

You have a cylinder of circular cross section. The radius is denoted R. A heavy fluid, density ρ, lies above a light fluid, density $\rho\star$. The surface separating the fluids is denoted $z = Z(r, \theta, t)$ and at first the two fluids are at rest, being separated by the horizontal surface $z = Z_0 = 0$. You have $\vec{v_0} = 0 = \vec{v_0}\star$, $\dfrac{dp_0}{dz} = -\rho g$ and $\dfrac{dp_0^\star}{dz} = -\rho\star g$.

The domain equations are

$$\mu \vec{v} = K\left(-\nabla p - \rho g \, \vec{k}\right)$$

and

$$\nabla \cdot \vec{v} = 0$$

and therefore

$$\nabla^2 p = 0$$

for both fluids and at the side walls you have $\vec{n} \cdot \vec{v} = 0$ and therefore $\vec{n} \cdot \nabla p = 0$.

At the surface $z = Z(r, \theta, t)$ you have

$$v_z - Z_r v_r - \frac{Z_\theta}{r} v_\theta = Z_t = v_z^\star - Z_r v_r^\star - \frac{Z_\theta}{r} v_\theta^\star$$

and

$$p - p^\star = \gamma 2H$$

The surface is given a small perturbation, viz., $Z = Z_0 + \varepsilon\, Z_1$ and your perturbation equations are then

$$\nabla^2 p_1 = 0 = \nabla^2 p_1^\star$$

$$\frac{\partial p_1}{\partial r} = 0 = \frac{\partial p_1^\star}{\partial r} \quad \text{at} \quad r = R$$

and

$$\frac{\partial Z_1}{\partial r} = 0 \quad \text{at} \quad r = R$$

assuming the surface contacts the side walls at right angles.

At $z = 0$ you have

$$-\frac{K}{\mu} \frac{\partial p_1}{\partial z} = v_{z1} = Z_{1t} = v_{z1}^\star = -\frac{K}{\mu^\star} \frac{\partial p_1^\star}{\partial z}$$

and

$$p_1 + Z_1 \frac{dp_0}{dz} - \left(p_1^\star + Z_1 \frac{dp_0^\star}{dz} \right) = \gamma 2H_1 = \gamma \nabla_H^2 Z_1$$

Writing

$$p_1 = \widehat{p}_1(z)\, J_m(\lambda r)\, e^{im\theta} e^{\sigma t}$$

$$p_1^\star = \widehat{p}_1^\star (z) \, J_m (\lambda r) \, e^{im\theta} e^{\sigma t}$$

and

$$Z_1 = \widehat{Z}_1 J_m (\lambda r) \, e^{im\theta} e^{\sigma t}$$

where

$$J_m' (\lambda R) = 0$$

you have

$$\widehat{p}_1 = A e^{-\lambda z}$$

and

$$\widehat{p}_1^\star = A^\star e^{\lambda z}$$

where $\widehat{p}_1$ is assumed to be bounded as $z \to \infty$, likewise $\widehat{p}_1^\star$ is assumed to be bounded as $z \to -\infty$, leaving A, $A^\star$ and $\widehat{Z}_1$ to satisfy the conditions at $z = 0$, viz.

$$\frac{K}{\mu} A\lambda = \sigma \widehat{Z}_1 = -\frac{K}{\mu^\star} A^\star \lambda$$

and

$$A - A^\star = \left(-\gamma\lambda^2 + (\rho - \rho^\star)\, g \right) \widehat{Z}_1$$

whereupon to have a solution A, $A^\star$ and $\widehat{Z}_1$ not all zero you find

$$\frac{\sigma (\mu + \mu^\star)}{K} = \lambda \left(-\gamma\lambda^2 + (\rho - \rho^\star)\, g \right)$$

Your first job is to make certain all of the above is correct.

Your second job is to notice that σ vs λ is one curve, you can sketch it and you can observe that there is a greatest value of σ. The curve rises due to the kinematic condition then falls due to surface tension crossing zero at $\lambda^2 = \dfrac{(\rho - \rho^\star) g}{\gamma}$

Now the allowable λ's depend on R via

$$J'_m (\lambda R) = 0$$

Denoting by x_m the solution to $J'_m (x) = 0$ your λ's are: $\lambda = \dfrac{x_m}{R}$ and you have

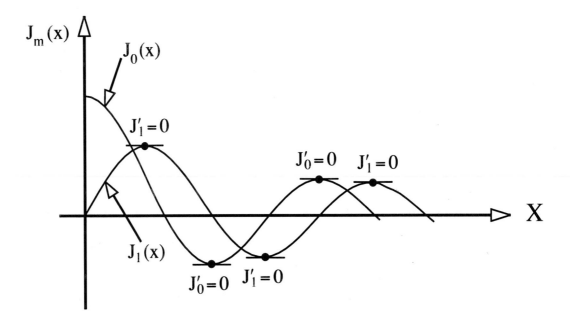

where $x = 0$ is ruled out by holding the volume constant on perturbation, viz.,

$$\int_0^R Z_1 r \, dr = 0$$

For a small value of R all the λ^2's lie to the right of $\dfrac{\Delta \rho g}{\gamma}$ and all perturbations are stable. As R increases they all move leftward but maintain their order. Soon the lowest moves to the left of $\dfrac{\Delta \rho g}{\gamma}$ and the problem is unstable to the corresponding perturbation. Upon increasing R we can make any of the λ^2's the fastest growing and that λ^2 determines the pattern you see.

Your third job is to satisfy yourself that all of the above is true and that first you will see an $m = 1$ pattern followed by an $m = 0$ pattern.

6. A heavy fluid lies above a light fluid. The two fluids are in hydrostatic equilibrium. The problem is two dimensional. The interface is horizontal, its ends are pinned and the width of the cell is such that the equilibrium is stable. The volume of the heavy fluid is $2LH$.

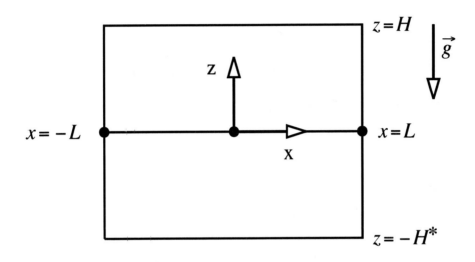

Now you add heavy fluid and remove light fluid resulting in

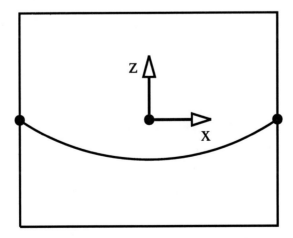

where the surface is denoted $z = Z_0(x)$ and where at first $Z_0(x) = 0$.

The eigenvalue problem, viz.,

$$P_1 - \lambda^2 \psi = \frac{d}{dx}\left(\frac{1}{(1 + Z_0^2(x))^{3/2}}\frac{d\psi}{dx}\right)$$

$$\psi = 0 \quad \text{at} \quad x = \pm L$$

and

$$\int_{-L}^{L} \psi\, dx = 0$$

appears upon investigating the stability of these solutions.

Multiplying by ψ and integrating over $-L < x < L$ you obtain

$$\lambda^2 = \frac{\displaystyle\int_{-L}^{L}\left(\frac{1}{(1 + Z_0^2(x))^{3/2}}\right)\psi_x^2\, dx}{\displaystyle\int_{-L}^{L}\psi^2\, dx}$$

This is called a Rayleigh quotient and the facts about Rayleigh quotients are explained in Weinberger's book.

Every trial function you put into the RHS, satisfying $\psi = 0$ at $x = \pm L$, $\int_{-L}^{L}\psi\, dx = 0$ gives an estimate of λ_1^2 lying above the true value of λ_1^2.

Your job is to satisfy yourself that λ_1^2 in the first picture lies above λ_1^2 in the second.

7. You have a cylinder of arbitrary cross section bounded above and below by parallel horizontal planes, one at $z = 0$, the other at $z = H$. The cylinder is filled with a porous solid whose free space is filled with a liquid.

The density of the liquid depends on its temperature, viz.,

$$\rho = \rho_{\text{ref}}\left(1 - \alpha(T - T_{\text{ref}})\right)$$

The lower plane is at temperature T_{hot}, the upper plane is at temperature T_{cold}. And you need to take into account only the temperature dependence of the density.

All walls are no flow, the vertical side walls are adiabatic and the upper and lower walls are isothermal.

The density is unstably stratified, heavy over light; and you are to, first, derive the critical value of T_{hot} at which flow sets in and then you are to increase T_{hot} slightly beyond its critical value and begin the process of estimating the steady flow above T_{hot} critical.

The model is

$$\vec{v} = -\frac{K}{\mu} \nabla p - \rho g \vec{k}, \qquad \nabla \cdot \vec{v} = 0$$

and

$$\frac{\partial T}{\partial t} + \vec{v} \cdot \nabla T = \kappa \nabla^2 T$$

where

$$T = T_{cold} \quad \text{at} \quad z = H$$

$$T = T_{hot} \quad \text{at} \quad z = 0$$

$$v_z = 0 \quad \text{at} \quad z = 0, H$$

and

$$\vec{n} \cdot \vec{v} = 0 = \vec{n} \cdot \nabla T \quad \text{at the side walls.}$$

Denoting the base solution by the subscript zero you have

$$\vec{v}_0 = \vec{0}$$

and

$$-\frac{dT_0}{dz} = \frac{T_{\text{hot }0} - T_{\text{cold }0}}{H}$$

Now set

$$\nabla = \vec{k}\,\frac{\partial}{\partial z} + \nabla_H$$

and

$$\vec{v} = v_z\,\vec{k} + \vec{v}_H$$

assume steady conditions and write the model

$$v_z = -\frac{K}{\mu}\frac{\partial p}{\partial z} - \rho g$$

$$\vec{v}_H = -\frac{K}{\mu}\,\nabla_H\,p$$

$$\frac{\partial v_z}{\partial z} + \nabla_H \cdot \vec{v}_H = 0$$

and

$$v_z\frac{\partial T}{\partial z} + \vec{v}_H \cdot \nabla_H T = \kappa\left(\frac{\partial^2}{\partial z^2} + \nabla_H^2\right)T$$

Then eliminating p derive

$$\left(\frac{\partial^2}{\partial z^2} + \nabla_H^2\right)v_z = -g\nabla_H^2\,\rho = \rho_{\text{ref}}\,\alpha\,g\nabla_H^2\,T$$

At the side walls $\vec{n} \cdot \vec{v} = 0$ implies $\vec{n} \cdot \nabla p = \vec{n} \cdot \nabla_H p = 0$ and hence $\vec{n} \cdot \nabla_H v_z = 0$. And our boundary conditions are

$$T = T_{\text{cold }0}, \quad v_z = 0 \quad \text{at} \quad z = H$$

$$T = T_{\text{hot }0}, \quad v_z = 0$$

$$\vec{n} \cdot \nabla_H v_z = 0 = \vec{n} \cdot \nabla_H T \quad \text{at the side walls}$$

To find the critical value of $T_{\text{hot }0}$, impose a small perturbation on the base solution, denote the perturbation variables by the subscript one and derive the perturbation problem at zero growth rate, viz.,

$$\left(\frac{\partial^2}{\partial z^2} + \nabla_H^2 \right) v_{z1} - \rho_{\text{ref}} \, g \alpha \nabla_H^2 T_1 = 0$$

$$\left(\frac{\partial^2}{\partial z^2} + \nabla_H^2 \right) T_1 + v_{z1} \frac{1}{\kappa} \left(-\frac{dT_0}{dz} \right) = 0$$

where

$$T_1 = 0 = v_{z1} \quad \text{at} \quad z = 0 = 0$$

and

$$\vec{n} \cdot \nabla_H v_z = 0 = \vec{n} \cdot \nabla_H T \quad \text{at the side walls.}$$

Your job is to find the values of $-\dfrac{dT_0}{dz}$ at which this problem has solutions other than $v_{z1} = 0 = T_1$

To do this introduce the eigenvalue problem on the cross section

$$\nabla_H^2 \psi + \lambda^2 \psi = 0$$

and

$$\vec{n} \cdot \nabla_H \psi = 0 \quad \text{at the edge}$$

and denote its solutions

$$\lambda_1^2, \lambda_2^2, \ldots$$

$$\psi_1, \psi_2, \ldots$$

These solutions depend on what the cross section is and if we denote its diameter by d then the λ^2's are multiples of d^{-2}. Henceforth by not specifying d you can view λ^2 as a continuous variable.

Assuming we have a perturbation in the shape of the eigenfunction ψ we can separate variables and write

$$v_{z1} = \widehat{v}_{z1}(z)\,\psi$$

and

$$T_1 = \widehat{T}_1(z)\,\psi$$

whereupon $\widehat{v}_{z1}$ and $\widehat{T}_1$ satisfy

$$\left(\frac{d^2}{dz^2} - \lambda^2\right)\widehat{v}_{z1} + \rho_{\text{ref}}\, g\, \alpha\, \lambda^2\, \widehat{T}_1 = 0$$

$$\left(\frac{d^2}{dz^2} - \lambda^2\right)\widehat{T}_1 + \widehat{v}_{z1}\frac{1}{\kappa}\left(-\frac{dT_0}{dz}\right) = 0$$

and

$$\widehat{v}_{z1} = 0 = \widehat{T}_1 \quad \text{at} \quad z = 0, H$$

Thus you have

$$\widehat{v}_{z1} = \beta \sin\frac{n\pi}{H}z$$

and

$$\widehat{T}_1 = \sin \frac{n\pi}{H} z$$

$$n = 1, 2, \ldots$$

Hence for each value of n you have the critical value of $-\dfrac{dT_0}{dz}$ as a function of λ^2, viz.,

$$\frac{\rho_{\text{ref}}\, \alpha\, g}{\kappa} \left(-\frac{dT_0}{dz} \right) = \frac{\left(\dfrac{n^2 \pi^2}{H^2} + \lambda^2 \right)^2}{\lambda^2}$$

The least critical value of $\left(-\dfrac{dT_0}{dz} \right)$ occurs at $n = 1$, $\lambda^2 = \dfrac{\pi^2}{H^2}$ where

$$\frac{\rho_{\text{ref}}\, \alpha\, g}{\kappa} \left(-\frac{dT_0}{dz} \right)_{\text{crit}} = 4 \frac{\pi^2}{H^2}$$

and

$$\beta = \frac{\left(\dfrac{\pi^2}{H^2} + \lambda^2 \right)}{\dfrac{1}{\kappa} \left(-\dfrac{dT_0}{dz} \right)}$$

Now you may advance T_{hot} from $T_{\text{hot }0}$ at critical by writing

$$T_{\text{hot}} = T_{\text{hot }0} + \frac{1}{2} \varepsilon^2 \, T_{\text{hot }2}$$

$$v_z = v_{z0} + \varepsilon \, v_{z1} + \frac{1}{2} \varepsilon^2 \, v_{z2}$$

and

$$T = T_0 + \varepsilon T_1 + \frac{1}{2} \varepsilon^2 \, T_2$$

In the Petrie dish problem in Lecture 15 you learned that the successful expansion of the control variable depends on the nonlinearity at hand.

Your job is to find out that the expansion above is the correct expansion by proving that solvability is satisfied at second order. Thus you have

$$v_{z0} = 0$$

$$T_0 = T_{\text{hot }0} + \left(T_{\text{cold }0} - T_{\text{hot }0} \right) \left(\frac{z}{H} \right)$$

$$v_{z1} = A \widehat{v}_{z1} \, \psi$$

and

$$T_1 = A \, \widehat{T_1} \, \psi$$

where $T_{\text{hot }0}$ is the critical value of T_{hot}, ψ corresponds to the critical value of λ^2 and $\widehat{v}_{z1}$ and $\widehat{T_1}$ are known from above.

What you are looking for is the value of A as a function of $T_{\text{hot }2}$

This can be found at third order if you can get through the second order problem without finding that $A = 0$ which would tell you that your expansion of T_{hot} is not correct.

At second order you should have

$$\left(\frac{\partial^2}{\partial z^2} + \nabla_H^2 \right) v_{z2} - \rho_{\text{ref}} \, \alpha \, g \, \nabla_H^2 \, T_2 = 0$$

$$\left(\frac{\partial^2}{\partial z^2} + \nabla_H^2 \right) T_2 + v_{z2} \frac{1}{\kappa} \left(-\frac{dT_0}{dz} \right) = \frac{2 \vec{v_1} \cdot \nabla T_1}{\kappa}$$

$$v_{z2} = 0 \quad \text{at} \quad z = 0, H$$

$$T_2 = 0 \quad \text{at} \quad z = H$$

and

$$T_2 = T_{\text{hot }2} \quad \text{at} \quad z = 0$$

And

$$v_{z2} = 0, \quad T_2 = T_{\text{hot 2}}\left(1 - \frac{z}{H}\right)$$

is a solution if $\dfrac{2\vec{v_1} \cdot \nabla T_1}{\kappa}$ is zero, so henceforth we set

$$T_2 = 0 \quad \text{at} \quad z = 0$$

Your problem then is this

$$\mathcal{L}\begin{pmatrix} v_{z2} \\ T_2 \end{pmatrix} = \begin{pmatrix} 0 \\ \dfrac{2\vec{v_1} \cdot \nabla T_1}{\kappa} \end{pmatrix}$$

and

$$v_{z2} = 0 = T_2 \quad \text{at} \quad z = 0, H$$

The corresponding homogeneous problem has a nonzero solution, hence a solvability condition must be satisfied, viz.,

$$\int_0^H dz \int_A dx\, dy \begin{pmatrix} v_{z1}^\star \\ T_1^\star \end{pmatrix}^T \begin{pmatrix} 0 \\ \dfrac{2\vec{v_1} \cdot \nabla T_1}{\kappa} \end{pmatrix} = 0$$

where

$$\mathcal{L}^\star \begin{pmatrix} v_{z1}^\star \\ T_1^\star \end{pmatrix} = \begin{pmatrix} 0 \\ 0 \end{pmatrix}$$

and

$$v_{z1}^\star = 0 = T_1^\star \quad \text{at} \quad z = 0, H$$

whereupon you have

$$v_{z1}^{\star} = \beta^{\star} \sin \frac{\pi z}{H} \psi$$

and

$$T_1^{\star} = \sin \frac{\pi z}{H} \psi$$

Now you need to work out $\vec{v_1} \cdot \nabla T_1$ to obtain

$$\vec{v_1} \cdot \nabla T_1 = A^2 \beta \frac{\pi}{H} \sin \frac{\pi z}{H} \cos \frac{\pi z}{H} \left\{ \nabla_H \cdot \frac{\psi \nabla_H \psi}{\lambda^2} \right\}$$

And then you can conclude that solvability is satisfied at second order due entirely to the z integration, viz.,

$$\int_0^H \sin \frac{\pi z}{H} \sin \frac{\pi z}{H} \cos \frac{\pi z}{H} \, dz = 0$$

Thus, if we had to, we could go on to third order in the expectation of finding A as a function of $T_{\text{hot } 2}$. This is the way the Petrie dish problem worked out for the cubic nonlinearity

Lecture 19

Ordinary Differential Equations

19.1 Boundary Value Problems in Ordinary Differential Equations

As we now know, using the method of separation of variables to solve the eigenvalue problem $\left(\nabla^2 + \lambda^2\right)\psi = 0$ reduces this problem to the solution of ordinary differential equations. In this lecture we present the elementary facts about second order, linear, ordinary differential equations.

We denote by L a second order, linear, differential operator acting on a class of functions defined on a finite interval and smooth enough so that we can use integration by parts in its ordinary form. We let x denote the independent variable, scale the interval of interest so that $0 \leq x \leq 1$ and write

$$Lu = a\left(x\right)\frac{d^2u}{dx^2} + b\left(x\right)\frac{du}{dx} + c\left(x\right)$$

We introduce the plain vanilla inner product

$$\langle\, u, v\,\rangle = \int_0^1 uv\, dx$$

and assume for now that u and v are real valued.

Integrating $ua\dfrac{d^2v}{dx^2}$ twice by parts and $ub\dfrac{dv}{dx}$ once, we can write $\langle\, u, Lv\,\rangle$ as

$$\langle\, u, Lv\,\rangle = \Big[\, a\left\{uv' - u'v\right\} - \left\{a' - b\right\}uv\,\Big]_0^1 + \langle\, L^*u, v\,\rangle$$

where a' denotes $\dfrac{da}{dx}$, etc. By doing this we introduce the operator L^*, associated to L and called its adjoint, where

$$L^*u = \frac{d^2}{dx^2}\left(au\right) - \frac{d}{dx}\left(bu\right) + cu$$

The adjoint, L^*, depends on the inner product we use.

A problem in second order, linear differential equations is specified in part by

$$Lu = f, \qquad 0 < x < 1$$

where $f\left(x\right)$ is an assigned function on the interval $(0, 1)$. To complete the specification of the problem two boundary conditions must be assigned. Most of the boundary conditions of physical interest can be taken into account by assigning values to two linear combinations of u and u' at the boundary, i. e., to two linear combinations of $u\left(x = 0\right)$, $u'\left(x = 0\right)$, $u\left(x = 1\right)$, $u'\left(x = 1\right)$. This includes both initial value and boundary value problems. We limit ourselves to unmixed boundary value problems and write the boundary conditions

$$\text{at} \quad x = 0: \qquad B_0 u = a_0 u + b_0 u' = g_0$$

and

$$\text{at} \quad x = 1: \qquad B_1 u = a_1 u + b_1 u' = g_1$$

where g_0 and g_1 are assigned real numbers.

Occasionally periodic conditions are imposed and these are of mixed type, viz.,

$$u\left(x = 0\right) - u\left(x = 1\right) = 0$$

and

$$u'(x = 0) - u'(x = 1) = 0$$

We will deal with these as exceptional cases.

A problem then is defined by the operators L, B_0 and B_1 and assigned sources $f(x)$, g_0 and g_1. Associated with this is the adjoint problem and to formulate the adjoint problem we need to identify the adjoint boundary operators B_0^* and B_1^* that go with the adjoint differential operator L^*. To do this write

$$\langle u, Lv \rangle - \langle L^*u, v \rangle = \left[a\{uv' - u'v\} - \{a' - b\}uv \right]_0^1$$

and define B_0^* and B_1^* such that $B_0^*u = 0 = B_1^*u$ and $B_0v = 0 = B_1v$ imply

$$\left[a\{uv' - u'v\} - \{a' - b\}uv \right]_0^1 = 0$$

To illustrate this: if $B_0v = v$ and $B_1v = v$ then $B_0^*u = u$ and $B_1^*u = u$. If $B_0v = v'$ and $B_1v = v'$ then $B_0^*u = \{-(au)' + bu\}$ and $B_1^*u = \{-(au)' + bu\}$.

Our job is to decide whether or not the problem

$$Lu = f, \quad 0 < x < 1$$

and

$$B_0u = g_0, \quad B_1u = g_1$$

has a solution and, if it does, to decide what it is. Naimark's book *"Linear Differential Operators"* deals with this, and more, but we do not need very many of Naimark's results as we deal only with second order differential operators and in this case a simplification obtains by which we need deal only with self adjoint differential operators.

To see why this is so, observe that Lv, L^*u and $uLv - L^*uv$ are

$$Lv = av'' + bv' + cv$$

$$L^*u = (au)'' - (bu)' + cu = au'' + (2a' - b)u' + (a'' - b' + c)u$$

and

$$uLv - L^*uv = \left\{ a(uv' - u'v) - (a' - b)uv \right\}'$$

Then if $b = a'$ we get

$$L^* = L$$

and

$$uLv - Luv = \left\{ a(uv' - u'v) \right\}'$$

whereupon we have $B_0^* = B_0$ and $B_1^* = B_1$. Thus, if $B_0u = 0 = B_0v$ we have

$$\begin{pmatrix} 0 \\ 0 \end{pmatrix} = \begin{pmatrix} B_0u \\ B_0v \end{pmatrix} = \begin{pmatrix} u(x=0) & u'(x=0) \\ v(x=0) & v'(x=0) \end{pmatrix} \begin{pmatrix} a_0 \\ b_0 \end{pmatrix}$$

and because a_0 and b_0 are not both zero, we conclude that

$$u(x=0)v'(x=0) - u'(x=0)v(x=0) = 0$$

Likewise if $B_1u = 0 = B_1v$, then

$$u(x=1)v'(x=1) - u'(x=1)v(x=1) = 0$$

This tells us that if $B_0^* = B_0$ and $B_1^* = B_1$ then $\left[a(uv' - u'v) \right]_0^1 = 0$.

Hence if $b = a'$, we get all of the following:

$$Lu = \frac{d}{dx}\left\{a\frac{du}{dx}\right\} + cu$$

and

$$L^* = L$$

$$B_0^* = B_0$$

and

$$B_1^* = B_1$$

When this is so, a problem defined by L, B_0 and B_1 is called self adjoint in the plain vanilla inner product (L is called self adjoint if $L^* = L$). We get all this by requiring only $b = a'$, but it must be observed that we have assumed special forms for B_0 and B_1, yet none of this depends on the values assigned to a_0, b_0, a_1 and b_1.

The condition $b = a'$ is important for two reasons. The first is that all of the ordinary differential operators coming from ∇^2 on separation of variables can be written as $\dfrac{1}{w(x)}$ times a self adjoint operator and hence are themselves self adjoint in the inner product

$$\langle\, u, v \,\rangle = \int_0^1 uvw\, dx$$

The second is that any second order linear differential operator, viz.,

$$Lu = au'' + bu' + cu$$

can be written

$$Lu = \frac{a}{d}\left\{\frac{d}{dx}\left(d\frac{du}{dx}\right) + \frac{c}{a}du\right\}$$

where

$$d(x) = e^{\int_{x_0}^{x} \frac{b}{a} d\xi}$$

and hence is self adjoint in a weighted inner product with $w = \dfrac{d}{a}$. Henceforth then we will assume that L is self adjoint in the plain vanilla inner product and write

$$Lu = \frac{d}{dx}\left(p\frac{du}{dx}\right) - qu$$

and

$$uLv - Luv = \frac{d}{dx}\left\{p\left(uv' - u'v\right)\right\}$$

The results we get will serve all of our purposes.

19.2   The Wronskian of Two Solutions to $Lu = 0$

The Wronskian of two functions u and v is denoted by W and is defined by

$$W = uv' - u'v = \det\begin{pmatrix} u & v \\ u' & v' \end{pmatrix}$$

Two functions u and v are linearly dependent if and only if their Wronskian vanishes.

Now we can write the problem $Lu = 0$ as

$$\frac{d}{dx}\begin{pmatrix} u \\ u' \end{pmatrix} = \begin{pmatrix} 0 & 1 \\ \dfrac{p}{q} & -\dfrac{p'}{p} \end{pmatrix}\begin{pmatrix} u \\ u' \end{pmatrix}$$

and then observe, as we discovered in Lecture 2, that if u and v are any two solutions of this equation their Wronskian satisfies

$$\frac{dW}{dx} = \text{tr} \begin{pmatrix} 0 & 1 \\ \dfrac{p}{q} & -\dfrac{p'}{p} \end{pmatrix} W = -\frac{p'}{p} W$$

This tells us that if u and v satisfy $Lu = 0$ then their Wronskian, multiplied by p, remains constant, i.e., $pW = const$. This is also a simple consequence of the formula

$uLv - Luv = \left\{ p \left(uv' - u'v \right) \right\}' = (pW)'$. So if p is not zero, then W is either always zero or never zero. And if pW is not zero and $p \to 0$ as $x \to x_0$ then $W \to \infty$ as $x \to x_0$ and at least one of u and v does not remain bounded as $x \to x_0$.

19.3 The General Solution to $Lu = f$

We can now write the general solution, i.e., no end conditions, to $Lu = f$ in terms of the solutions to $Lu = 0$. As Coddington and Levinson explain, there are always two independent solutions to $Lu = 0$ and every other solution can be expresses as a linear combination of any two independent solutions. We let u_1 and u_2 denote two independent solutions of $Lu = 0$, then $Lu_1 = 0 = Lu_2$, $W = u_1 u_2' - u_1' u_2$ does not vanish, pW is a nonzero constant and

$$u_0 = \frac{1}{pW} \left\{ -u_1 \left(x \right) \int_0^x u_2 \left(y \right) f \left(y \right) dy + u_2 \left(x \right) \int_0^x u_1 \left(y \right) f \left(y \right) dy \right\}$$

satisfies $Lu = f$. This can be verified by direct calculation. The general solution of $Lu = f$ is then

$$u = u_0 + c_1 u_1 + c_2 u_2$$

where c_1 and c_2 are two constants to be determined. We observe that $u_0 \left(x = 0 \right) = 0 = u_0' \left(x = 0 \right)$ and therefore $B_0 u_0 = 0$.

19.4 Solving the Homogeneous Problem $f = 0, g_0 = 0, g_1 = 0$

To solve the homogeneous problem

$$Lu = 0$$

and

$$B_0 u = 0, \qquad B_1 u = 0$$

we must determine c_1 and c_2 so that

$$u = c_1 u_1 + c_2 u_2$$

satisfies the boundary conditions. This requires

$$\begin{pmatrix} B_0 u_1 & B_0 u_2 \\ B_1 u_1 & B_1 u_2 \end{pmatrix} \begin{pmatrix} c_1 \\ c_2 \end{pmatrix} = \begin{pmatrix} 0 \\ 0 \end{pmatrix}$$

and we denote by D the determinant of the matrix of coefficients, viz.,

$$D = \det \begin{pmatrix} B_0 u_1 & B_0 u_2 \\ B_1 u_1 & B_1 u_2 \end{pmatrix}$$

If $D \neq 0$ then $c_1 = 0 = c_2$ is the unique solution to this system of equations and $u = 0$ is the unique solution to the homogeneous problem. If $D = 0$ then this system of equations has exactly one independent solution because not all $B_0 u_1$, $B_0 u_2$, $B_1 u_1$ and $B_1 u_2$ can vanish. So too then the homogeneous problem. The result is this: if $D \neq 0$, the homogeneous problem $Lu = 0$, $B_0 u = 0$, $B_1 u = 0$ has only the solution $u = 0$; if $D = 0$ the homogeneous problem has one independent solution.

Because u_1 and u_2 are independent solutions of $Lu = 0$, we can see that $B_0 u_1$, $B_0 u_2$, $B_1 u_1$ and $B_1 u_2$ cannot all be zero. Observe first that u_1 and u_2 do not depend on B_0 and B_1, being determined solely by L and the condition $W = u_1 u_2' - u_1' u_2 \neq 0$. Then to see that $B_0 u_1$ and $B_0 u_2$

cannot both be zero, write

$$
\begin{pmatrix} B_0 u_1 \\ B_0 u_2 \end{pmatrix} = \begin{pmatrix} u_1 \left(x = 0 \right) & u_1' \left(x = 0 \right) \\ u_2 \left(x = 0 \right) & u_2' \left(x = 0 \right) \end{pmatrix} \begin{pmatrix} a_0 \\ b_0 \end{pmatrix}
$$

and observe that $W \left(x = 0 \right)$ is not zero and that a_0 and b_0 cannot be both zero.

Now there is always a non zero solution to $Lu = 0$, $B_0 u = 0$ for, if neither $B_0 u_1$ nor $B_0 u_2$ is zero, a linear combination of u_1 and u_2 can be found which satisfies $B_0 u = 0$. And there is not another solution independent of this because, if there were, the Wronskian of the two solutions would vanish at $x = 0$,

Hence we always have one non zero solution, and it is the only independent solution, to $Lu = 0$, $B_0 u_0 = 0$. Likewise, there is one independent solution to $Lu = 0$, $B_1 u = 0$. And all this is true no matter the value of D. Then if D is zero, we have one independent solution of $Lu = 0$, $B_0 u = 0$, $B_1 u = 0$. If u_1 is this solution then neither $B_0 u_2$ nor $B_1 u_2$ can be zero.

These results depend on the boundary conditions being unmixed as both $\cos 2\pi x$ and $\sin 2\pi x$ satisfy

$$
\left(\frac{d^2}{dx^2} + 4\pi^2 \right) u = 0
$$

$$
u \left(0 \right) = u \left(1 \right)
$$

and

$$
u' \left(0 \right) = u' \left(1 \right)
$$

19.5 Solving the Inhomogeneous Problem

To solve the problem $Lu = f$ where $B_0 u = g_0$, $B_1 u = g_1$ we need to determine the values of the constants c_1 and c_2 in the general solution to $Lu = f$ so that the boundary conditions are satisfied. Substituting $u = u_0 + c_1 u_1 + c_2 u_2$ into the boundary conditions results in two equations in the two unknowns c_1 and c_2. Each solution to these equations produces a solution to our problem and by solving these equations we get every solution to our problem.

Thus we have

$$B_0 u_0 + c_1 B_0 u_1 + c_2 B_0 u_2 = g_0$$

and

$$B_1 u_0 + c_1 B_1 u_1 + c_2 B_1 u_2 = g_1$$

and hence

$$\begin{pmatrix} B_0 u_1 & B_0 u_2 \\ B_1 u_1 & B_1 u_2 \end{pmatrix} \begin{pmatrix} c_1 \\ c_2 \end{pmatrix} = \begin{pmatrix} g_0 - B_0 u_0 \\ g_1 - B_1 u_0 \end{pmatrix}$$

where if Bu is any linear combination of u and u', then

$$B\left(u_0 + c_1 u_1 + c_2 u_2\right) = Bu_0 + c_1 Bu_1 + c_2 Bu_2$$

and, as

$$u_0 = \frac{1}{pW}\left\{ -u_1 \int_0^x u_2 f \, dy + u_2 \int_0^x u_1 f \, dy \right\}$$

and

$$u_0' = \frac{1}{pW}\left\{ -u_1' \int_0^x u_2 f \, dy + u_2' \int_0^x u_1 f \, dy \right\}$$

we have

$$Bu_0 = \frac{1}{pW}\left\{ -Bu_1 \int_0^x u_2 f \, dy + Bu_2 \int_0^x u_1 f \, dy \right\}$$

Now, if D is not zero, the constants c_1 and c_2 can be determined uniquely and our problem has a unique solution; otherwise, to have a solution, a solvability condition must be satisfied and if the solvability condition is satisfied, the solution is not unique. The solvability condition for u is the solvability condition for c_1 and c_2.

Whether D is zero or not depends only on L, B_0 and B_1. It does not depend on how we select the two independent solutions of $Lu = 0$ denoted u_1 and u_2.

19.6   The Case $D \neq 0$

Our first result is this. The problem

$$Lu = f$$

and

$$B_0 u = g_0, \qquad B_1 u = g_1$$

has a solution and it is unique iff the problem

$$Lu = 0$$

and

$$B_0 u = 0, \qquad B_1 u = 0$$

has only the solution $u = 0$. This is the case $D \neq 0$,

To determine this unique solution we first must find two independent solutions of $Lu = 0$. Denoting these u_1 and u_2 where $W = u_1 u_2' - u_1' u_2$, $W(u_1, u_2) \neq 0$ and $pW = constant \neq 0$, we evaluate D. Then we write the solution

$$u = u_0 + c_1 u_1 + c_2 u_2$$

where

$$u_0 = \frac{1}{pW} \left\{ -u_1 \int_0^x u_2 f \, dy + u_2 \int_0^x u_1 f \, dy \right\}$$

and where

$$\begin{pmatrix} c_1 \\ c_2 \end{pmatrix} = \frac{1}{D} \begin{pmatrix} B_1 u_2 & -B_0 u_2 \\ -B_1 u_1 & B_0 u_1 \end{pmatrix} \begin{pmatrix} g_0 - B_0 u_0 \\ g_1 - B_1 u_0 \end{pmatrix}$$

and we observe that

$$B_0 u_0 = 0$$

and

$$B_1 u_0 = \frac{B_1 u_2}{pW} \langle u_1, f \rangle - \frac{B_1 u_1}{pW} \langle u_2, f \rangle$$

The simplest result obtains if u_1 and u_2 are chosen so that

$$B_0 u_1 = 0 = B_1 u_2$$

for then

$$c_1 = \frac{g_1}{B_1 u_1} + \frac{1}{pW} \langle u_2, f \rangle$$

and

$$c_2 = \frac{g_0}{B_0 u_2}$$

19.7   The Case $D = 0$

Our second result corresponds to the case $D = 0$ whereupon the problem $Lu = f$, $B_0u = g_0$, $B_1u = g_1$, may or may not have a solution. To decide whether or not it does we must find out if a solvability condition is satisfied. It is to this that we now turn.

The solvability condition for u is the solvability condition for c_1 and c_2. Now when $D = 0$, the rank of

$$\begin{pmatrix} B_0u_1 & B_0u_2 \\ B_1u_1 & B_1u_2 \end{pmatrix}$$

is one and the solvability condition for c_1 and c_2 is simply the requirement that the rank of

$$\begin{pmatrix} B_0u_1 & B_0u_2 & g_0 - B_0u_0 \\ B_1u_1 & B_1u_2 & g_1 - B_1u_1 \end{pmatrix}$$

be one also. This then is the requirement that

$$\det \begin{pmatrix} B_0u_1 & g_0 - B_0u_0 \\ B_1u_1 & g_1 - B_1u_0 \end{pmatrix} = 0 = \begin{pmatrix} B_0u_2 & g_0 - B_0u_0 \\ B_1u_2 & g_1 - B_1u_0 \end{pmatrix}$$

If one of these determinants is zero then so is the other as their first columns are dependent due to $D = 0$.

Now when $D = 0$ the homogeneous problem $Lu = 0$, $B_0u = 0$, $B_1u = 0$ has one independent solution and the solvability condition takes its simplest form if we take u_1 to be a nonvanishing solution to this problem. Then $B_0u_1 = 0 = B_1u_1$, $B_0u_2 \neq 0$, $B_1u_2 \neq 0$ and the solvability condition is

$$\det \begin{pmatrix} B_0u_2 & g_0 - B_0u_0 \\ B_1u_2 & g_1 - B_1u_0 \end{pmatrix} = 0$$

where $B_0 u_0 = 0$ and $B_1 u_0 = \dfrac{B_1 u_2}{pW} \langle u_1, f \rangle$. Hence we have

$$B_0 u_2 g_1 - B_1 u_2 g_0 = \frac{B_0 u_2 B_1 u_2}{pW} \langle u_1, f \rangle$$

or

$$\frac{(pW)_1}{B_1 u_2} g_1 - \frac{(pW)_0}{B_0 u_2} g_0 = \langle u_1, f \rangle$$

This, the simplest expression of the solvability condition, does not depend on how u_2 is selected once u_1 is set, so long as u_1 and u_2 are independent. Indeed, because W is the Wronskian of u_1 and u_2 and $B_1 u_1 = 0$, $\dfrac{W_1}{B_1 u_2}$ is unchanged if u_2 is replaced by a linear combination of u_1 and u_2. The same is true of $\dfrac{W_0}{B_0 u_2}$.

For homogeneous boundary conditions the solvability condition is

$$\langle u_1, f \rangle = 0$$

We can state this as: The problem

$$Lu = f$$

and

$$B_0 u = 0, \qquad B_1 u = 0$$

is solvable if and only if f is orthogonal to all solutions to

$$Lu = 0$$

and

$$B_0 u = 0, \qquad B_1 u = 0$$

If the solvability condition is satisfied, we have a solution, otherwise we do not. Assuming solvability is satisfied and u_1 is a solution of the homogeneous problem we have

$$\begin{pmatrix} 0 & B_0 u_2 \\ 0 & B_1 u_2 \end{pmatrix} \begin{pmatrix} c_1 \\ c_2 \end{pmatrix} = \begin{pmatrix} g_0 \\ g_1 - B_1 u_0 \end{pmatrix}$$

whereupon c_1 is arbitrary and

$$c_2 = \frac{g_0}{B_0 u_2} = \frac{g_1 - B_1 u_0}{B_1 u_2}$$

and our solution is

$$u_0 + c_2 u_2 + c u_1$$

where c is arbitrary. Likewise the solution to the homogeneous problem is

$$c u_1$$

As an example we may wish to determine whether or not the problem

$$\left(\frac{d^2}{dx^2} + a^2 \right) u = f, \qquad 0 < x < 1$$

and

$$u\left(x = 0\right) = 0, \qquad u\left(x = 1\right) = 0$$

has a solution. The differential operator $L = \dfrac{d^2}{dx^2} + a^2$ is self adjoint in the plain vanilla inner product and the functions $u_1 = \dfrac{1}{a} \sin ax$ and $u_2 = \cos ax$ are two independent solutions of $Lu = 0$. As $B_0 u = u$ and $B_1 u = u$ we find that D is

$$D = \frac{-1}{a} \sin a$$

and hence that D is not zero unless $a^2 = \pi^2, 2^2\pi^2, \ldots$. So for each value of a^2 other than $\pi^2, 2^2\pi^2, \ldots$ this problem has a solution for all functions $f(x)$. But if $a^2 = n^2\pi^2$, where n is a positive integer, the problem has a solution if and only if the function $f(x)$ satisfies

$$\int_0^1 \sin n\pi x f(x) \, dx = 0$$

19.8 The Green's Function

We suppose D is not zero so that our problem has a unique solution. We denote the two independent solutions of $Lu = 0$ by u_1 and u_2 where $W(u_1, u_2) \neq 0$. The simplest result obtains if

$$Lu_1 = 0, \qquad B_0 u_1 = 0$$

and

$$Lu_2 = 0, \qquad B_1 u_2 = 0$$

for then

$$\begin{pmatrix} c_1 \\ c_2 \end{pmatrix} = \frac{1}{D} \begin{pmatrix} B_1 u_2 & -B_0 u_2 \\ -B_1 u_1 & B_0 u_1 \end{pmatrix} \begin{pmatrix} g_0 - B_0 u_0 \\ g_1 - B_1 u_1 \end{pmatrix}$$

where

$$B_0 u_0 = 0$$

and

$$B_1 u_0 = \frac{1}{pW} \left\{ -B_1 u_1 \int_0^1 u_2 f \, dy + B_1 u_2 \int_0^1 u_1 f \, dy \right\}$$

and $\begin{pmatrix} c_1 \\ c_2 \end{pmatrix}$ is simply

$$\begin{pmatrix} c_1 \\ c_2 \end{pmatrix} = \frac{1}{D} \begin{pmatrix} 0 & -B_0 u_2 \\ -B_1 u_1 & 0 \end{pmatrix} \begin{pmatrix} g_0 \\ g_1 + \dfrac{1}{pW} B_1 u_1 \displaystyle\int_0^1 u_2 f \, dy \end{pmatrix}$$

where $D = -B_1 u_1 B_0 u_2$.

In the case $g_0 = 0 = g_1$ we have

$$c_1 = \frac{1}{pW} \int_0^1 u_2 f \, dy$$

and

$$c_2 = 0$$

and our solution is then

$$u = u_0 + c_1 u_1 = \frac{1}{pW} u_1 \int_x^1 u_2 f \, dy + \frac{1}{pW} u_2 \int_0^x u_1 f \, dy$$

which we can write

$$u = \int_0^1 g(x, y) f(y) \, dy$$

where g is called the Green's function for our problem and we have

$$g(x, y) = \begin{cases} \dfrac{1}{pW} u_1(y) u_2(x), & y < x \\[2ex] \dfrac{1}{pW} u_1(x) u_2(y), & x < y \end{cases}$$

Now at any y we see that

$$Lg = 0, \quad B_0 g = 0, \qquad x < y$$

and

$$Lg = 0, \quad B_1 g = 0, \qquad y < x$$

where

$$g(x, y) - g(x, y) = 0$$
$$x \to y^+ \quad x \to y^-$$

and

$$\frac{\partial g}{\partial x}(x, y) - \frac{\partial g}{\partial x}(x, y) = \frac{1}{p(y)}$$
$$x \to y^+ \qquad x \to y^-$$

and where we notice that

$$u_1(y) \, u_2'(y) - u_1'(y) \, u_2(y) = W(y)$$

Hence we can find g by solving the foregoing equations but we see that g is not an ordinary solution to $Lg = 0$ (as are u_1 and u_2) because $\dfrac{\partial^2 g}{\partial x^2}$ does not exist at $x = y$.

As an example, to solve

$$\frac{d^2 u}{dx^2} = f, \qquad 0 < x < 1$$

and

$$u(x = 0) = 0 = u(x = 1)$$

we can write

$$u(x) = \int_0^1 g(x, y) \, f(y) \, dy$$

where $g(x, y)$ satisfies

$$\frac{d^2 g}{dx^2} = 0, \qquad 0 < x < y$$

$$g(x = 0, y) = 0$$

$$\frac{d^2 g}{dx^2} = 0, \qquad y < x < 1$$

$$g(x = 1, y) = 0$$

$$g(y^+, y) - g(y^-, y) = 0$$

and

$$\frac{\partial g}{\partial x}(y^+, y) - \frac{\partial g}{\partial x}(y^-, y) = 1$$

whence

$$g(x, y) = -x(1 - y), \qquad 0 < x < y$$

$$= -y(1 - x), \qquad y < x < 1$$

For fixed y the function $g(x, y)$ is not an ordinary (smooth) solution of $Lu = 0$, $B_0 u = 0$, $B_1 u = 0$. The first derivative of g has a jump discontinuity at $x = y$ and so Lg is not defined there in an ordinary sense. The function g is called a generalized solution of $Lu = 0$, $B_0 u = 0$, $B_1 u = 0$. The only ordinary solution to this homogeneous equation when $D \neq 0$ is $u = 0$. There is more on this in §19.9

The reader can continue and determine how to use $g(x, y)$ to write the solution if g_0 and g_1 are not zero. This will complete the introduction to the Green's function when D is not zero. When D is zero it is possible to introduce a generalized Green's function and it is possible to do this in a way that produces a best approximation when a solution cannot be determined. We do not go into this but the main ideas can be found back in Lecture 4.

As a second example suppose we must find a solution to

$$\frac{1}{x}\frac{d}{dx}\left(x\frac{du}{dx}\right) = f, \qquad 0 < x < 1$$

where B_0 and B_1 are not as yet specified. Then we denote $\dfrac{d}{dx}\left(x\dfrac{d}{dx}\right)$ by L and work in the inner product: $\langle\, u, v\,\rangle = \displaystyle\int_0^1 uv\, dx$.

Doing this we find $L^* = L$ and $uLv - Luv = \{xW\}'$. As long as the boundary conditions remain unspecified we cannot determine $g\,(x, y)$ but we can write a general solution to our problem. Indeed $u_1 = 1$ and $u_2 = \ln x$ satisfy $Lu = 0$. And, as $W\,(u_1, u_2) = \dfrac{1}{x}$, u_1 and u_2 are independent and $pW = 1$. A particular solution to $\dfrac{1}{x}Lu = f$ is then

$$u_0 = -\int_0^x (\ln y)\, yf\,(y)\, dy + \ln x \int_0^x yf\,(y)\, dy$$

and, so long as f is a bounded function, u_0 and u_0' remain finite as $x \to 0$.

The general solution is

$$u = u_0 + c_1 + c_2\,(\ln x)$$

and the requirement that u remain bounded as $x \to 0$ replaces the boundary condition $B_0 u = g_0$. It is satisfied if and only if $c_2 = 0$.

Using this the reader can determine the Green's function for this problem and show that it satisfies

$$Lg = 0, \qquad 0 < x < y$$

$$g\,(x = 0, y) \qquad \text{finite}$$

$$Lg = 0, \qquad y < x < 1$$

$$B_1 g = 0$$

$$g\left(y^{+}, y\right) - g\left(y^{-}, y\right) = 0$$

and

$$\frac{\partial}{\partial x} g\left(y^{+}, y\right) - \frac{\partial}{\partial x} g\left(y^{-}, y\right) = \frac{1}{y}$$

The solution to

$$Lu = xf$$

u finite at $x = 0$

and

$$B_1 u = 0$$

is then

$$u\left(y\right) = \int_0^1 xf\left(x\right) g\left(x, y\right) \, dx$$

Note:

In §19.9 we introduce the delta function. We can use it to define the Green's function via

$$Lg = \delta\left(x - y\right)$$

$g\left(x, y\right)$ finite at $x = 0$

and

$$B_1 g = 0$$

Then we can use

$$\int_0^1 \{uLg - Lug\}\, dx = [xW]_0^1$$

to write the solution.

19.9   What is $\delta(x)$ Doing?

The short answer for us is nothing. This is, more or less, the first place where the symbol $\delta(x)$ has been used. But we have made great use of the integration by parts formula

$$\int_0^1 f\, \frac{dg}{dx}\, dx = \left[\, fg\, \right]_0^1 - \int_0^1 \frac{df}{dx}\, g\, dx$$

and we have not inquired whether this use is justified. To see why there might be a question let f be smooth and suppose first that g is continuous but that its derivative is not. For instance let g and $\frac{dg}{dx}$ be

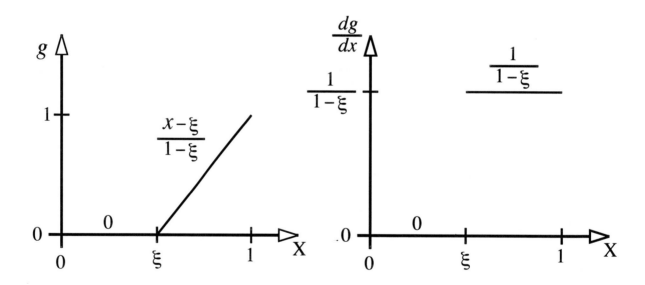

Then the left hand side of our integration by parts formula is

$$\text{LHS} = \int_0^\xi f\, \frac{dg}{dx}\, dx + \int_\xi^1 f\, \frac{dg}{dx}\, dx = \frac{1}{1-\xi} \int_\xi^1 f\, dx$$

whereas the right hand side is

$$\text{RHS} = f(1) - \int_{\xi}^{1} \frac{df}{dx} \frac{x-\xi}{1-\xi} \, dx = \frac{1}{1-\xi} \int_{\xi}^{1} f \, dx$$

As LHS = RHS we see that the discontinuity in $\frac{dg}{dx}$ does not require us to give up our integration by parts formula.

But suppose that g and $\frac{dg}{dx}$ are now

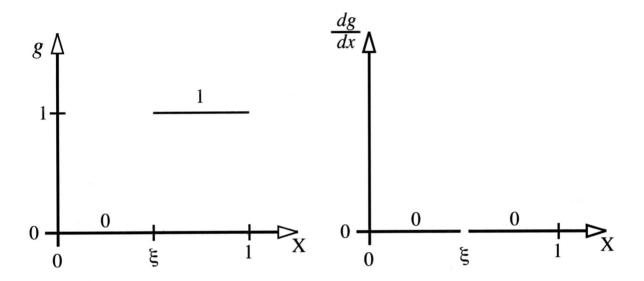

Then the right hand side is

$$\text{RHS} = f(1) - \int_{\xi}^{1} \frac{df}{dx} \, dx = f(\xi)$$

whereas the left hand side is ambiguous. If we suppose it to be zero then our integration by parts formula is incorrect. To account for the jump in g and to enlarge the class of functions for which our formula is correct we simply let the right hand side define the left hand side whenever the left hand side is ambiguous.

The easiest way to do this is to introduce the symbol $\delta(x)$ where

$$\int_a^b \delta(x)\,dx = 1, \quad a < 0 < b$$

$$\int_a^b \delta(x)\,dx = 0, \quad 0 < a < b \quad \text{or} \quad a < b < 0$$

Then

$$\int_a^b f(x)\,\delta(x - x_0)\,dx = f(x_0), \quad a < x_0 < b$$

$$= 0, \quad x_0 < a < b \quad \text{or} \quad a < b < x_0$$

and so if we write $\dfrac{dg}{dx} = \delta(x - \xi)$ in the second example we have

$$\text{LHS} = f(\xi)$$

and our integration by parts formula is restored.

What is really useful about this notation is that in terms of it the introduction of Green's functions can be formalized and simplified because we can now use our integration by parts formula to evaluate terms such as

$$\int uLg\,dx$$

where g is a Green's function. Its derivative takes a jump at a point where Lg is not defined. This integral then is like that in the second example just above.

To see how this goes let u satisfy

$$Lu = f(x), \quad 0 < x < 1$$

and

$$u(x = 0) = 0, \quad u(x = 1) = 0$$

and suppose that $g(x, \xi)$ can be determined so that

$$Lg = \delta(x - \xi)$$

and

$$g(x = 0, \xi) = 0, \qquad g(x = 1, \xi) = 0$$

Then as

$$\int_0^1 \left\{ gLu - uLg \right\} dx = \left[p\{gu' - ug'\} \right]_0^1 = 0$$

we have

$$u(\xi) = \int_0^1 g(x, \xi) f(x) \, dx$$

19.10   Turning a Differential Equation into an Integral Equation: Autothermal Heat Generation

To take a more interesting example, suppose that heat is released in a thin layer of fluid of width L bounded on either side by large reservoirs to which heat can be rejected. Where the fluid is in contact with the reservoirs, the fluid temperature is held at the reservoir temperature, T_0. Our problem is to determine whether or not the heat released can be balanced by heat conduction to the reservoirs. In the simplest case we get an interesting problem by assuming that the local heat source is autothermal with the temperature dependence of its rate given by the Arrhenius formula. Then scaling distance by the width of the fluid layer our problem is to determine u so that

$$\frac{d^2u}{dx^2} + \lambda\, e^{\gamma \frac{u}{1+u}} = 0$$

and

$$u\left(x = 0\right) = 0, \qquad u\left(x = 1\right) = 0$$

are satisfied where $u = \dfrac{T - T_0}{T_0}$, γ is a scaled activation energy and λ is a multiple of L^2.

This is not the kind of problem we have been talking about. The source is not specified in advance but instead is a function of the solution and a non-linear function at that. Nonetheless we can determine the Green's function to be

$$g\left(x, y\right) = -x\left(1 - y\right), \quad 0 < x < y$$

$$= -y\left(1 - x\right), \quad y < x < 1$$

as above and write

$$u\left(x\right) = \int_0^1 g\left(x, y\right) \left\{ -\lambda e^{\gamma \frac{u\left(y\right)}{1 + u\left(y\right)}} \right\} dy$$

Now we have not solved our problem, we have simply put it in another form. But this form is useful as $-g\left(x, y\right)$ is a non-negative function and using this integral equation it is easy to construct both bounds on and approximations to $u\left(x\right)$.

19.11   The Eigenvalue Problem

Writing L, B_0 and B_1 as

$$Lu = \frac{d}{dx}\left(p\frac{du}{dx}\right) - qu$$

$$B_0 u = a_0 u + b_0 u' \quad \text{at} \quad x = 0$$

and

$$B_1 u = a_1 u + b_1 u' \quad \text{at} \quad x = 1$$

we have

$$uLv = \frac{d}{dx}\left(up\frac{dv}{dx}\right) - p\frac{du}{dx}\frac{dv}{dx} - quv$$

and

$$uLv - Luv = \frac{d}{dx}\left\{p\left(uv' - u'v\right)\right\} = \frac{d}{dx}\{pW\}$$

Now, assuming a_0 and b_0 are not both zero and a_1 and b_1 are not both zero, $B_0u = 0 = B_0v$ and $B_1u = 0 = B_1v$ imply

$$W_0 = \{uv' - u'v\}\,(x = 0) = 0$$

and

$$W_1 = \{uv' - u'v\}\,(x = 1) = 0$$

Hence if u and v satisfy B_0u, B_1u, B_0v, B_1v all zero we have

$$\int_0^1 \{uLv - Luv\}\,dx = pW\Big|_0^1 = 0$$

If, say, $p\,(x = 0) = 0$ then instead of $B_0u = 0 = B_0v$, we only need u and v to remain bounded as $x \to 0$ to obtain the above formula.

Our integration by parts formulas are now

$$\int_0^1 uLv\,dx = puv'\Big|_0^1 - \int_0^1 \{pu'v' + quv\}\,dx$$

and

$$\int_0^1 \{uLv - Luv\}\,dx = \{p\left(uv' - u'v\right)\}\Big|_0^1$$

Our eigenvalue problem is

$$L\psi + \lambda^2\psi = 0, \quad 0 < x < 1$$

and

$$B_0\psi = 0 = B_1\psi$$

where the eigenfunctions are the non zero solutions, ψ, and the corresponding values of λ^2 are the eigenvalues.

For any fixed value of λ^2, the equation

$$Lu + \lambda^2 u = 0$$

has two independent solutions denoted $u_1\left(\lambda^2\right)$ and $u_2\left(\lambda^2\right)$, where $W\left(u_1\left(\lambda^2\right), u_2\left(\lambda^2\right)\right)$, does not vanish. The general solution of $\left(L + \lambda^2\right)\psi = 0$ is then

$$\psi = c_1 u_1\left(\lambda^2\right) + c_2 u_2\left(\lambda^2\right)$$

Each value of λ^2 such that $D\left(\lambda^2\right) = 0$ where

$$D\left(\lambda^2\right) = B_0 u_1\left(\lambda^2\right) B_1 u_2\left(\lambda^2\right) - B_1 u_1\left(\lambda^2\right) B_0 u_2\left(\lambda^2\right)$$

is an eigenvalue for then c_1 and c_2, not both zero, can be determined so that

$$\begin{pmatrix} B_0 u_1\left(\lambda^2\right) & B_0 u_2\left(\lambda^2\right) \\ B_1 u_1\left(\lambda^2\right) & B_1 u_2\left(\lambda^2\right) \end{pmatrix} \begin{pmatrix} c_1 \\ c_2 \end{pmatrix} = \begin{pmatrix} 0 \\ 0 \end{pmatrix}$$

and hence a solution to the eigenvalue problem other than $\psi = 0$ can be obtained. To each eigenvalue there will be one independent eigenfunction as the rank of $\begin{pmatrix} B_0 u_1\left(\lambda^2\right) & B_0 u_2\left(\lambda^2\right) \\ B_1 u_1\left(\lambda^2\right) & B_1 u_2\left(\lambda^2\right) \end{pmatrix}$ cannot be zero. The eigenvalues will be isolated as D will ordinarily be an analytic function of λ^2 and hence its zeros will be isolated. And D will ordinarily have infinitely many zeros.

Assuming p and q to be real valued functions and a_0, b_0, a_1 and b_1 to be real constants we see that if ψ is an eigenfunction corresponding to the eigenvalue λ^2 then so also $\overline{\psi}$ corresponding to $\overline{\lambda^2}$. Putting $u = \overline{\psi}$, $v = \psi$ in our second integration by parts formula and observing that $W_0\left(\overline{\psi}, \psi\right) = 0 = W_1\left(\overline{\psi}, \psi\right)$, we see that $\overline{\lambda^2} = \lambda^2$ or that λ^2 is real. Then putting $u = \overline{\psi}$, $v = \psi$ in our first formula we get

$$-\lambda^2 \int_0^1 |\psi|^2 \, dx = \left[p\,\overline{\psi}\,\psi' \right]_0^1 - \int_0^1 \left\{ p \left| \frac{d\psi}{dx} \right|^2 + q\,|\psi|^2 \right\} dx$$

Now as $B_0\psi = 0 = B_0\overline{\psi}$ we find that

$$p\,\overline{\psi}\,\psi'\,(x = 0) = \begin{cases} 0, & b_0 = 0 \\ -\dfrac{a_0}{b_0}p\,|\psi|^2\,(x = 0), & \text{otherwise} \end{cases}$$

and likewise

$$p\,\overline{\psi}\,\psi'\,(x = 1) = \begin{cases} 0, & b_1 = 0 \\ -\dfrac{a_1}{b_1}p\,|\psi|^2\,(x = 1), & \text{otherwise} \end{cases}$$

and hence if $p \geq 0$, $q \geq 0$, $a_1/b_1 \geq 0$ and $a_0/b_0 \leq 0$ it is certain that $\lambda^2 \geq 0$.

The second formula then shows that if the eigenfunctions ψ_1 and ψ_2 correspond to distinct eigenvalues, λ_1^2 and λ_2^2, we have

$$\left\langle \psi_1, \psi_2 \right\rangle = \int_0^1 \psi_1 \psi_2 \, dx = 0$$

This is the orthogonality condition.

Note: The eigenvalues are real and to each there is one independent eigenfunction. So we take the eigenfunction to be a real valued function. Under other boundary conditions it may be of some advantage to indroduce complex valued eigenfunctions. Then if ψ is an eigenfunction so also $\operatorname{Re}\psi$, $\operatorname{Im}\psi$ and $\overline{\psi}$.

The solution of $\nabla^2 \psi + \lambda^2 \psi = 0$ by separation of variables leads to eigenvalue problems of the form

$$\frac{1}{w} L\psi + \lambda^2 \psi = 0$$

where L is self-adjoint in the inner product

$$\langle\, u, v\,\rangle = \int_0^1 uv\, dx$$

In using the integration by parts formulas to again determine that the eigenvalues are real and not negative, we now put $L\psi = -\lambda^2 w\psi$ instead of $L\psi = -\lambda^2 \psi$ as above. Then in determining the sign of λ^2 we use

$$-\lambda^2 \int_0^1 w\,|\,\psi\,|^2\, dx = \left[\, p\,\overline{\psi}\,\psi'\,\right]_0^1 - \int_0^1 \left\{ p\left|\frac{d\psi}{dx}\right|^2 + q\,|\,\psi\,|^2 \right\} dx$$

and λ^2 remains non-negative if $w > 0$. The orthogonality condition is

$$\int_0^1 \psi_1 \psi_2 w\, dx = 0$$

which is not unexpected as $\dfrac{1}{w}L$ is self-adjoint in the inner product

$$\langle\, u, v\,\rangle = \int_0^1 uvw\, dx$$

19.12   Solvability Conditions

The simplest way to decide whether or not a problem has a solution is to determine if all of the steps required to write the solution can be carried out.

To see how this goes suppose the eigenvalue problem

$$L\psi + \lambda^2 \psi = 0, \quad 0 < x < 1$$

and

$$B_0\psi = 0, \qquad B_1\psi = 0$$

determines a complete orthogonal set of eigenfunctions and denote them $\psi_1, \psi_2, \ldots$ corresponding to the eigenvalues $0 \le \lambda_1^2 < \lambda_2^2 < \cdots$. Then to determine a function u satisfying

$$Lu = f, \quad 0 < x < 1$$

and

$$B_0u = g_0, \qquad B_1u = g_1$$

we write

$$u = \sum c_i\psi_i$$

and try to determine the coefficients $c_1, c_2, \cdots$ in this expansion. To find the equation satisfied by the coefficient $c_i = \langle \psi_i, u \rangle$, we multiply $Lu = f$ by ψ_i and integrate over $0 \le x \le 1$, getting

$$\int_0^1 \psi_i Lu \, dx = \int_0^1 \psi_i f \, dx$$

Then as

$$\int_0^1 \psi_i Lu \, dx = \left[pW \right]_0^1 + \int_0^1 L\psi_i u \, dx$$

where W is the Wronskian of ψ_i and u, we have

$$\langle \psi_i, f \rangle = \left[p\{\psi_i u' - \psi_i'u\} \right]_0^1 - \lambda_i^2 c_i$$

This is the result we need. It tells us this: in the expansion of u the coefficient of an eigenfunction corresponding to a non vanishing eigenvalue has one, and only one, value. But the coefficient of an eigenfunction corresponding to the eigenvalue zero either cannot be determined, whence a solution cannot be found, or is not unique, whence a solution can be found but it is not unique. The question of solvability then comes up only if zero is an eigenvalue. In the first instance

$$\langle \psi_1, f \rangle \neq \left[p \{\psi_1 u' - \psi_1' u\} \right]_0^1$$

whereas in the second

$$\langle \psi_1, f \rangle = \left[p \{\psi_1 u' - \psi_1' u\} \right]_0^1$$

This is the solvability condition,, but it is not in terms of B_0, B_1, g_0 and g_1.

To write it in terms of B_0 and B_1 we first use

$$\begin{pmatrix} 0 \\ g_0 \end{pmatrix} = \begin{pmatrix} B_0 \psi_1 \\ B_0 u \end{pmatrix} = \begin{pmatrix} \psi_1 \left(x = 0\right) & \psi_1' \left(x = 0\right) \\ u \left(x = 0\right) & u' \left(x = 0\right) \end{pmatrix} \begin{pmatrix} a_0 \\ b_0 \end{pmatrix}$$

to write a_0 and b_0 in terms of g_0 and $W_0 = W\left(x = 0\right)$. Doing this we get

$$a_0 = \frac{-g_0 \psi_1' \left(x = 0\right)}{W_0}$$

and

$$b_0 = \frac{g_0 \psi_1 \left(x = 0\right)}{W_0}$$

Likewise we get

$$a_1 = \frac{-g_1 \psi_1' \left(x = 1\right)}{W_1}$$

and

$$b_1 = \frac{g_1 \psi_1 \left(x = 1\right)}{W_1}$$

Then if a_0 and a_1 are not zero we can write the solvability condition as

$$\left\langle \psi_1, f \right\rangle = -p\,(x = 1)\,\psi_1'\,(x = 1)\,\frac{g_1}{a_1} + p\,(x = 0)\,\psi_1'\,(x = 0)\,\frac{g_0}{a_0}$$

The reader may wish to write our problem

$$\begin{pmatrix} L & 0 & 0 \\ 0 & B_0 & 0 \\ 0 & 0 & B_1 \end{pmatrix} \begin{pmatrix} u\,(x) \\ u\,(0) \\ u\,(1) \end{pmatrix} = \begin{pmatrix} f \\ g_0 \\ g_1 \end{pmatrix}$$

and look for an inner product in which

$$\begin{pmatrix} L & 0 & 0 \\ 0 & B_0 & 0 \\ 0 & 0 & B_1 \end{pmatrix}$$

is self adjoint. The solvability condition is then the requirement that

$$\begin{pmatrix} f \\ g_0 \\ g_1 \end{pmatrix}$$

be perpendicular to every vector in

$$\mathrm{Ker} \begin{pmatrix} L & 0 & 0 \\ 0 & B_0 & 0 \\ 0 & 0 & B_1 \end{pmatrix}$$

19.13   Home Problems

1. A reaction $A + B \rightarrow AB$ takes place in a layer of solvent separating reservoirs of A and B one on the left and the other on the right. Neither reservoir is permeable to the other reactant.

Our model is

$$\frac{d^2 c_A}{dx^2} - k\, c_A\, c_B = 0$$

$$\frac{d^2 c_B}{dx^2} - k\, c_A\, c_B = 0$$

where

$$c_A\,(x = 0) = c_A^\star, \quad \frac{dc_A}{dx}\,(x = 1) = 0, \quad \frac{dc_B}{dx}\,(x = 0) = 0, \quad c_B\,(x = 1) = c_B^\star$$

The constant k is assumed to be small.

To estimate the rate of production of AB write

$$c_A = c_{A0} + k\, c_{A1} + k^2 c_{A2} + \cdots$$

and

$$c_B = c_{B0} + k\, c_{B1} + k^2 c_{B2} + \cdots$$

where $c_{A0} = c_A^\star$, $c_{B0} = c_B^\star$ and $c_{A1}, c_{A2}, \ldots, c_{B1}, c_{B2}, \ldots$ satisfy

$$\frac{d^2 c_{A1}}{dx^2} - c_{A0}\, c_{B0} = 0$$

$$\frac{d^2 c_{B1}}{dx^2} - c_{A0}\, c_{B0} = 0$$

where

$$c_{A1}\,(x = 0) = 0, \quad \frac{dc_{A1}}{dx}\,(x = 1) = 0, \quad \frac{dc_{B1}}{dx}\,(x = 0) = 0, \quad c_{B1}\,(x = 1) = 0$$

and

$$\frac{d^2 c_{A2}}{dx^2} - c_{A0}\, c_{B1} - c_{A1}\, c_{B0} = 0$$

where

$$c_{A2}(x = 0) = 0, \qquad \frac{dc_{A2}}{dx}(x = 1) = 0$$

etc.

Derive the Green's function for the c_{A1}, c_{A2}, etc. problems and for the c_{B1}, c_{B2}, etc. problems. Use these Green's functions to find the first few terms in the two expansions.

2. Write the solution to the problem

$$0 = \frac{d^2u}{dx^2} + \lambda^2(1 + u)$$

where

$$u(x = -1) = 0 = u(x = +1)$$

by adding to the particular solution $u_0 = -1$ the general solution to the homogeneous equation

$$A\cos\lambda x + B\sin\lambda x$$

and then using the boundary conditions to find A and B.

Sketch the solution for a value of λ lying between 0 and $\frac{\pi}{2}$, then for λ lying between $\frac{\pi}{2}$ and $\frac{3\pi}{2}$, then $\frac{3\pi}{2}$ and $\frac{5\pi}{2}$, etc. The problem does not have a solution when $\lambda = \frac{\pi}{2}, \frac{3\pi}{2}, \ldots$.

To see why this is so look at the solvability condition for this problem. The corresponding homogeneous problem

$$0 = \frac{d^2u}{dx^2} + \lambda^2 u$$

where

$$u\left(x = -1\right) = 0 = u\left(x = +1\right)$$

has solutions that can be written

$$A \cos \lambda x + B \sin \lambda x$$

where A and B satisfy

$$\begin{pmatrix} \cos \lambda & -\sin \lambda \\ \cos \lambda & \sin \lambda \end{pmatrix} \begin{pmatrix} A \\ B \end{pmatrix} = \begin{pmatrix} 0 \\ 0 \end{pmatrix}$$

The determinant of the matrix on the LHS is

$$\cos \lambda \ \sin \lambda$$

To all values of λ such that $(\cos \lambda \ \sin \lambda) \neq 0$ the constants A and B are both zero and $u = 0$ is the only solution to the homogeneous problem. Our problem then has a unique solution.

To all values of λ such that $\cos \lambda = 0$, i.e., to $\frac{1}{2}\pi, \frac{3}{2}\pi, \ldots$, the constant B must be zero but A is indeterminate and the homogeneous problem has solutions $A \cos \lambda x$. Our problem then requires that a solvability condition be satisfied. Show that it is not satisfied.

To all values of λ such that $\sin \lambda = 0$, i.e., to $0, \pi, \ldots$, the constant A must be zero but B is indeterminate and the homogeneous problem has solutions $B \sin \lambda x$. When $\lambda = 0$ this is 0 but otherwise our problem again requires that a solvability condition be satisfied. Show that it is satisfied.

So when $\lambda = \frac{\pi}{2}, \frac{3\pi}{2}, \ldots$ our problem has no solution; when $\lambda = \pi, \ldots$ it has a solution but it is not unique.

To see what is going on solve the problem

$$\frac{\partial u}{\partial t} = \frac{\partial^2 u}{\partial x^2} + \lambda^2 (1 + u)$$

where

$$u(x = -1) = 0 = u(x = +1)$$

and where $u(t = 0) > 0$ is assigned. Do this when $\lambda = \frac{1}{2}\pi$ and π.

3. You have seen how the Green's function can be used to solve

$$Lu = f, \qquad B_0 u = 0, \qquad B_1 u = 0$$

use it to solve

$$Lu = 0, \qquad B_0 u = g_0, \qquad B_1 u = 0$$

4. We have

$$Lu = (pu')' - qu$$

and therefore

$$\frac{1}{p} Lu = u'' + \frac{p'}{p} u' - \frac{q}{p} u$$

Show that $\frac{1}{p} L$ is self adjoint in the inner product

$$\langle u, v \rangle = \int_0^1 uvp \, dx$$

5. We present three frictional heating problems for your enjoyment, all in cylindrical coordinates. In each case the temperature depends only on r as does the only non zero velocity component. The viscosity depends on temperature via

$$\frac{\mu\left(T_{\text{wall}}\right)}{\mu\left(T\right)} = 1 + \beta\left(T - T_{\text{wall}}\right)$$

And in each case a stress component is specified. The problems are written in terms of a scaled temperature

$$\beta\left(T - T_{\text{wall}}\right)$$

First: Sliding rod, $v_r = 0 = v_\theta$, $v_z = v_z\left(r\right)$

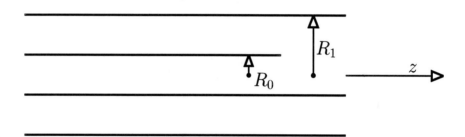

$$T\left(r = R_1\right) = T_{\text{wall}}$$

$$\frac{dT}{dr}\left(r = R_0\right) = 0$$

$$T_{rz}\left(r = R_0\right) \quad \text{input}$$

In terms of the scaled temperature our problem is

$$\frac{d^2T}{dr^2} + \frac{1}{r}\frac{dT}{dr} + \lambda^2 \frac{1}{r^2}\left(1 + T\right) = 0$$

$$\lambda^2 = \frac{\beta}{k\,\mu_{\text{wall}}}\, T_{rz}^2\,(r = R_0)\, R_0^2$$

Second: Poiseuille flow, $v_r = 0 = v_\theta$, $v_z = v_z\,(r)$

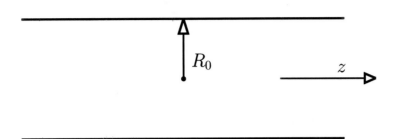

$$T\,(r = R_0) = T_{\text{wall}}$$

$$T\,(r = 0) \quad \text{bounded}$$

$$\frac{dp}{dz} \quad \text{input}$$

Our problem is

$$\frac{d^2 T}{dr^2} + \frac{1}{r}\frac{dT}{dr} + \lambda^2\, r^2\,(1 + T) = 0$$

$$\lambda^2 = \frac{\beta}{k\,T_{\text{wall}}}\,\frac{1}{4}\left(\frac{dp}{dz}\right)^2$$

Third: Spinning rod, $v_r = 0 = v_z$, $v_\theta = v_\theta\,(r)$

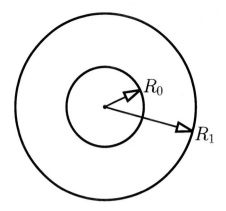

$$T\left(r = R_1\right) = T_{\text{wall}}$$

$$\frac{dT}{dr}\left(r = R_0\right) = 0$$

$$T_{r\,\theta}\left(r = R_0\right) \quad \text{input}$$

Our problem is

$$\frac{d^2T}{dr^2} + \frac{1}{r}\frac{dT}{dr} + \lambda^2\frac{1}{r^4}\left(1 + T\right) = 0$$

$$\lambda^2 = \frac{\beta}{k\,\mu_{\text{wall}}}\,T_{r\,\theta}^2\left(r = R_0\right) R_0^4$$

Your job is to solve these problems for increasing values of λ^2 and to find the value of λ^2 at which T becomes unbounded.

In the second and third problems, ψ, where

$$\frac{d^2\psi}{dr^2} + \frac{1}{r}\frac{d\psi}{dr} + \lambda^2\,r^2\,\psi = 0$$

or

$$\frac{d^2\psi}{dr^2} + \frac{1}{r}\frac{d\psi}{dr} + \lambda^2\frac{1}{r^4}\,\psi = 0$$

can be found by assuming

$$\psi = J_0\left(\mu r^p\right)$$

where in the second problem $p = 2$ and in the third $p = -1$ and where μ is to be found.

This is not true of the first problem which, however, is a Bernoulli equation.

6. This is a linear heating problem in Cartesian coordinates. Again we have

$$\frac{\mu_{\text{wall}}}{\mu\left(T\right)} = 1 + \beta\left(T - T_{\text{wall}}\right)$$

The fluid lying between two fixed plane walls at $x = L$ and $x = -L^\star$ is sheared by a plane wall at $x = 0$ moving to the right.

The stresses T_{xz} and $T_{xz}^\star$ are constants. But they are not independent. Our problem, in scaled temperature, is

$$\frac{d^2T}{dx^2} + \lambda^2\left(1 + T\right) = 0, \qquad T\left(x = L\right) = 0$$

$$\frac{d^2T^\star}{dx^2} + \left(\lambda^\star\right)^2\left(1 + T^\star\right) = 0, \qquad T\left(x = -L^\star\right) = 0$$

where

$$\lambda^2 = \frac{\beta}{k\,\mu_{\text{wall}}}\,T_{xz}^2, \qquad \left(\lambda^\star\right)^2 = \frac{\beta}{k\,\mu_{\text{wall}}}\left(T_{xz}^\star\right)^2$$

and where

$$T\left(x = 0\right) = T^\star\left(x = 0\right)$$

and

$$\frac{dT}{dx}\left(x = 0\right) = \frac{dT^\star}{dx}\left(x = 0\right)$$

Because the speed of the moving wall is common, T_{xz} and $T^{\star}_{xz}$ are not independent. Show that T_{xz} and $T^{\star}_{xz}$ have opposite signs and

$$-T_{xz} \int_0^L (1+T)\ dx = T^{\star}_{xz} \int_{-L^{\star}}^0 \left(1+T^{\star}\right)\ dx$$

Assume T_{xz} to be the input variable, determine the value of λ^2 at which the temperatures become infinite.

First do the T problem assuming $T(x=0) = 0$ and then $\dfrac{dT}{dx}(x=0) = 0$, before solving the $T, T^{\star}$ problem.

Lecture 20

Eigenvalues and Eigenfunctions of ∇^2 in Cartesian, Cylindrical and Spherical Coordinate Systems

20.1 Cartesian Coordinates: Sines, Cosines and Airy Functions

We have seen in some of our earlier examples that the information we can get out of the eigenvalues and the eigenfunctions is interesting in itself whether or not we plan to use them in an infinite series.

Of the nine one-dimensional problems arising upon separating variables in our three simple coordinate systems only problems (6), (8) and (9) might be unfamiliar, all the others are of the form

$$\frac{d^2 X}{dx^2} + \alpha^2 X = 0$$

and have a solution

$$X = A \sin \alpha x + B \cos \alpha x$$

where what we do next depends on the boundary conditions that must be satisfied.

The solutions of problems (6), (8) and (9) can be obtained by a method due to Fuchs and Frobenius which we will explain when we come to problem (8).

But first we can present another example of the importance of the eigenvalues themselves by looking at the simplest quantum mechanical problem.

The rules for writing Schrödinger's equation and the rules for interpreting its solutions are the postulates of quantum mechanics. They cannot be proved; they can only be shown to lead to conclusions that either agree or do not agree with experimental results. We obtain Schrödinger's equation in Cartesian coordinates, for a system of particles, if in the classical formula

$$H = T + V = E$$

where

$$T = \frac{1}{2m_1}\left\{p_{x_1}^2 + p_{y_1}^2 + p_{z_1}^2\right\} + \frac{1}{2m_2}\left\{p_{x_2}^2 + p_{y_2}^2 + p_{z_2}^2\right\} + \cdots$$

we replace $p_{x_1}, p_{y_1}, \cdots$ by $\frac{h}{2\pi i}\frac{\partial}{\partial x_1}, \frac{h}{2\pi i}\frac{\partial}{\partial y_1}, \cdots$ and E by $-\frac{h}{2\pi i}\frac{\partial}{\partial t}$, and then introduce a function Ψ for these differential operators to act on. It is easy to see how ∇^2 turns up. Indeed ∇^2 turns up for each particle and if we have N particles of the same mass we turn up ∇^2 in a $3N$ dimensional space.

We can restrict a particle having only kinetic energy to a box $0 < x < a$, $0 < y < a$, $0 < z < a$ by setting $V = 0$ inside the box and $V = \infty$ outside the box. Then Schrödinger's equation for this particle is

$$\frac{1}{2m}\left(\frac{h}{2\pi i}\right)^2 \nabla^2 \Psi = -\frac{h}{2\pi i}\frac{\partial \Psi}{\partial t}$$

where $\Psi = 0$ on the boundary of the box. The corresponding eigenvalue problem is

$$\frac{1}{2m}\left(\frac{h}{2\pi i}\right)^2 \nabla^2 \psi = E\psi$$

where $\psi = 0$ on the boundary of the box. It determines the allowable values of the energy of the particle in its stationary states. These values are important for they are the possible outcomes of a measurement of the energy of the particle. We introduce

$$\lambda^2 = \frac{8\pi^2 m}{h^2} E$$

and we write our problem

$$\nabla^2 \psi + \lambda^2 \psi = 0$$

where ψ vanishes at $x = 0, x = a, y = 0, y = a, z = 0$ and $z = a$ and we solve it by separation of variables. Writing $\psi = X(x) Y(y) Z(z)$ we find

$$\frac{d^2 X}{dx^2} + k_x^2 X = 0$$

$$X(x = 0) = 0 = X(x = a)$$

$$\frac{d^2 Y}{dy^2} + k_y^2 Y = 0$$

$$Y(y = 0) = 0 = Y(y = a)$$

and

$$\frac{d^2 Z}{dz^2} + k_z^2 Z = 0$$

$$Z(z = 0) = 0 = Z(z = a)$$

where $\lambda^2 = k_x^2 + k_y^2 + k_z^2$, where

$$k_x = \frac{\pi}{a} n_x, \quad k_y = \frac{\pi}{a} n_y, \quad k_z = \frac{\pi}{a} n_z$$

and where n_x, n_y and $n_z = 1, 2, \ldots$ Hence we have

$$\lambda^2 = \frac{\pi^2}{a^2} \left\{ n_x^2 + n_y^2 + n_z^2 \right\}$$

and

$$E = \frac{h^2}{8\pi^2 m} \lambda^2 = \frac{h^2}{8a^2 m} \left\{ n_x^2 + n_y^2 + n_z^2 \right\}$$

The points $\left(n_x,\ n_y,\ n_z \right)$, where $n_x,\ n_y$ and $n_z = 1, 2, \dots .$ lie at the nodes of a cubic lattice in the positive octant of quantum number space. To each point of the lattice there corresponds an eigenfunction, i.e., a quantum mechanical state. The energy of the state is $\frac{p^2}{2m}$, where $\vec{p} = \frac{h}{2a} \left\{ n_x \vec{i}_x + n_y \vec{j}_y + n_z \vec{k}_z \right\}$, and this is $\frac{h^2}{8ma^2}$ times the distance of the lattice point from the origin.

We can make this example a little more interesting by assuming that the particle inside the box is in a uniform gravitational field, $\vec{g} = -g\vec{k}$. Then we have $V = mgz$ and our problem is

$$\frac{1}{2m} \left(\frac{h}{2\pi i} \right)^2 \nabla^2 \psi + mgz\psi = E\psi$$

or

$$\nabla^2 \psi + \left(\lambda^2 - \frac{2m^2}{\hbar^2} z \right) \psi = 0$$

where $\lambda^2 = \frac{2mE}{\hbar^2}$, $\hbar = \frac{h}{2\pi}$ and $\left[\frac{2m^2 g}{\hbar^2} \right] = \frac{1}{L^3}$. Separating variables leads to the same X and Y problems as before but the Z problem is now

$$\frac{d^2 Z}{dz^2} + \left(k_z^2 - \frac{2m^2 g}{\hbar^2} z \right) Z = 0$$

where $Z(z = 0) = 0 = Z(z = a)$. Again $\lambda^2 = k_x^2 + k_y^2 + k_z^2$ but now the values of k_z^2 are new.

We can write the solution to our new Z problem in terms of Airy functions, cf., Bender and Orzag, *"Advanced Mathematical Methods for Engineers and Scientists."*

In terms of the Fuchs, Frobenius classification, the point $x = 0$ is an ordinary point of the homogeneous linear equations

$$\frac{d^2y}{dx^2} + y = 0$$

and

$$\frac{d^2y}{dx^2} - xy = 0$$

The second equation is called Airy's equation and two of its independent solutions are denoted $Ai\,(x)$ and $Bi\,(x)$ where these are the names of Taylor series about $x = 0$ having infinite radii of convergence.

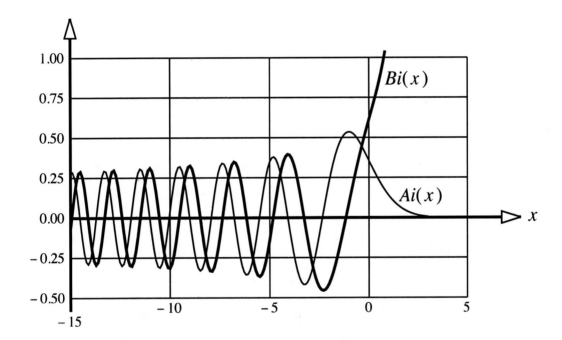

A sketch of $Ai\,(x)$ and $Bi\,(x)$ is presented above and we see that the Airy functions look like trigonometric functions to the left, exponential functions to the right.

Introducing a characteristic length, denoted β, where $\beta^3 \dfrac{2m^2g}{\hbar^2} = 1$, and replacing z by βz we have

$$\frac{d^2Z}{dz^2} + \left(\beta^2 k_z^2 - z\right) Z = 0$$

whence

$$Z = c_1 Ai\left(z - \beta^2 k_z^2\right) + c_2 Bi\left(z - \beta^2 k_z^2\right)$$

Thus, in order that the values of c_1 and c_2 are not both zero, k_z^2 must satisfy

$$Ai\left(-\beta^2 k_z^2\right) Bi\left(\frac{a}{\beta} - \beta^2 k_z^2\right) - Ai\left(\frac{a}{\beta} - \beta^2 k_z^2\right) Bi\left(-\beta^2 k_z^2\right) = 0$$

20.2   Cylindrical Coordinates: Bessel Functions

Solute Dispersion

Solute dispersion refers to the longitudinal spreading of a solute introduced into a flowing solvent stream. We have seen a simple example of this in our model of a chromatographic separation where we observed that the longitudinal dispersion of solute is due to a transverse variation of the solvent velocity.

For a solvent in straight line flow in a pipe our problem is at least two dimensional. One dimensional models are called dispersion equations and the coefficients appearing in these models are called dispersion coefficients.

Assuming the process takes place in a pipe of circular cross section, we use cylindrical coordinates and ∇^2 in cylindrical coordinates is what this lecture is about.

Suppose a solvent is in straight line flow in a long straight pipe of circular cross section. At time $t = 0$ a solute is injected into the solvent, its initial concentration being denoted $c\,(t = 0)$. Our job is to estimate its concentration some time later.

We align the z-axis with the axis of the pipe and denote the diameter of the pipe by $2a$. Then the solute concentration satisfies

$$\frac{\partial c}{\partial t} + v_z \frac{\partial c}{\partial z} = D\nabla^2 c, \qquad \theta \le r \le a, \quad 0 \le \theta < 2\pi, \quad -\infty < z < \infty$$

where

$$v_z = 2\,\overline{v}_z \left(1 - \frac{r^2}{a^2} \right)$$

where $\overline{v}_z$ denotes the average of v_z over the pipe cross section and where

$$\nabla^2 = \frac{1}{r}\frac{\partial}{\partial r}\left(r\frac{\partial}{\partial r} \right) + \frac{1}{r^2}\frac{\partial^2}{\partial \theta^2} + \frac{\partial^2}{\partial z^2}$$

Henceforth we write v in place of v_z.

Assuming that the wall of the pipe is impermeable to solute and that only a finite amount of solute is put into the solvent at $t = 0$, we require c to satisfy

$$\frac{\partial c}{\partial r}(r = a) = 0$$

and we assume c vanishes strongly as $z \to \pm\infty$.

Now solvent near the axis of the pipe is moving faster than solvent near the wall; the result is that solute is carried down stream by the solvent at different rates depending on where it initially resides on the cross section of the pipe. This convective distortion of the initial solute distribution creates transverse concentration gradients which tend to move solute from the axis to the wall at the leading edge of the distribution and in the opposite direction at the trailing edge.

The problem of determining c, or at least something about c, is called the problem of solute dispersion or Taylor dispersion. The first work leading to an understanding of this problem was reported by G.I. Taylor in his 1953 paper *"Dispersion of Solute Matter in Solvent Flowing Slowly Through a Tube."* We deal with this problem because it requires us to deal with the eigenfunctions of ∇^2 in cylindrical coordinates and because it requires us to solve the diffusion equation taking into account homogeneous sources. The latter two reasons are artifacts of our way of doing the problem and fit well into our sequence of lectures, Taylor did not require any of this.

The convective diffusion equation is a linear equation in c and as such its solution can be written in terms of the transverse eigenfunctions of ∇^2 and a special set of orthogonal functions in z. The result is not instructive. What invites the construction of models is that v is not constant but depends on r. As Taylor was able to measure the transverse average solute concentration, i.e., $\bar{c}$, where

$$\bar{c} = \frac{1}{\pi a^2} \int_0^a \int_0^{2\pi} cr \, dr d\theta$$

determining something about $\bar{c}$ as it depends on z and t became the goal of his and much subsequent work on this problem

The Action of the Flow by Itself on the Solute

To begin we inquire as to what the velocity field by itself is doing to the solute distribution in the pipe. If diffusion is set aside then c satisfies the purely convective equation

$$\frac{\partial c}{\partial t} = -v(r) \frac{\partial c}{\partial z}$$

and this is easy to solve for any assigned $c(t=0)$ as it is simply the statement that the value of c at the point r_0, θ_0, z_0 at the time $t = 0$ will be found also at the point $r_1 = r_0, \theta_1 = \theta_0,$ $z_1 = z_0 + v(r_0)t_1$ at the time t_1 and hence $c(t>0)$ can be calculated in terms of $c(t=0)$.

In the end the information we seek is independent of $c(t=0)$ and we can get it most easily by introducing the longitudinal power moments of c. Denoting these by c_m, where

$$c_m = \int_{-\infty}^{+\infty} z^m c \, dz, \quad m = 0, 1, 2, \ldots$$

we can derive the equation for c_m by multiplying the purely convective equation by z^m, integrating over z from $-\infty$ to $+\infty$, using integration by parts and discarding the terms evaluated at $z = \pm\infty$. Doing this for $m = 0, 1, 2, \ldots$ we get

$$\frac{\partial c_0}{\partial t} = 0$$

$$\frac{\partial c_1}{\partial t} = vc_0$$

$$\frac{\partial c_2}{\partial t} = 2vc_1$$

etc.

The moments of c can be determined recursively as

$$c_0 = c_0\,(t = 0)$$

$$c_1 = c_1\,(t = 0) + vc_0\,(t = 0)\,t$$

$$c_2 = c_2\,(t = 0) + 2vc_1\,(t = 0)\,t + v^2 c_0\,(t = 0)\,t^2$$

etc.

and their transverse averages calculated as

$$\bar{c}_0 = \bar{c}_0\,(t = 0)$$

$$\bar{c}_1 = \bar{c}_1\,(t = 0) + \left(\overline{vc_0\,(t = 0)}\right)t$$

$$\bar{c}_2 = \bar{c}_2\,(t = 0) + 2\left(\overline{vc_1\,(t = 0)}\right)t + \left(\overline{v^2 c_0\,(t = 0)}\right)t^2$$

etc.

To see how fast the solute is spreading in the axial direction due to nonuniform flow on the cross section we calculate the longitudinal variance of the transverse average solute concentration and determine how fast this is growing in time. Denoting the expected value of z^m by

$$\left\langle\, z^m\, \right\rangle = \frac{\int_{-\infty}^{+\infty} z^m \bar{c}\, dz}{\int_{-\infty}^{+\infty} \bar{c}\, dz} = \frac{\bar{c}_m}{\bar{c}_0}$$

we have

$$\left\langle (z - \langle z \rangle)^2 \right\rangle = \langle z^2 \rangle - \langle z \rangle^2 = \frac{\bar{c}_2}{\bar{c}_0} - \left(\frac{\bar{c}_1}{\bar{c}_0}\right)^2$$

which turns out to be

$$\left\{ \frac{\bar{c}_2}{\bar{c}_0} - \left(\frac{\bar{c}_1}{\bar{c}_0}\right)^2 \right\} (t = 0) + 2 \left\{ \frac{\overline{vc_1 (t = 0)}}{\bar{c}_0 (t = 0)} - \frac{\overline{vc_0 (t = 0)}}{\bar{c}_0 (t = 0)} \frac{\overline{c_1 (t = 0)}}{\bar{c}_0 (t = 0)} \right\} t +$$

$$\left\{ \frac{\overline{v^2 c_0 (t = 0)}}{\bar{c}_0 (t = 0)} - \left(\frac{\overline{vc_0 (t = 0)}}{\bar{c}_0 (t = 0)}\right)^2 \right\} t^2$$

If v were uniform on the cross section the variance would remain at its initial value. But a non-uniform v causes the variance to grow as t^2. To see how much of this might be due to the average motion we repeat the calculation in a fame moving at the average speed. To do this let

$$z' = z - \bar{v} t$$

and

$$t' = t$$

Then c satisfies

$$\frac{\partial c}{\partial t'} = -(v - \bar{v}) \frac{\partial c}{\partial z'}$$

and the new formula for the variance requires only that we put $v - \bar{v}$ in place of v in the formula we already have. As this leads to no change, we see that the variance is the same whether we examine solute spreading in a frame at rest or in a frame moving at the average speed of the solvent.

Now we know that diffusion and a linear growth of the variance go hand in hand, hence, if accounting for transverse diffusion can eliminate the quadratic term in the above formula, we can think about solute dispersion as a longitudinal diffusion process.

Assuming transverse diffusion can cut the growth of the longitudinal variance of the solute distribution from quadratic in time to linear in time, solute dispersion can be thought to be, on the average, a longitudinal diffusion process. But as the balance between longitudinal convection and transverse diffusion, on which elimination of the quadratic growth term depends, may take some time to be established and as this time may depend on how the solute is initially distributed, we need to view the representation of solute dispersion in terms of longitudinal diffusion as a long time representation.

We can arrive at Taylor's result viz.,

$$D_{\text{eff}} = \frac{1}{48}\frac{\bar{v}^2 a^2}{D}$$

where the reader can observe that D_{eff} is smaller as D is larger, by a route that takes us through familiar territory. To begin observe that if the model for $\bar{c}$ is

$$\frac{\partial \bar{c}}{\partial t} = D_{\text{eff}}\frac{\partial^2 \bar{c}}{\partial z^2} - V_{\text{eff}}\frac{\partial \bar{c}}{\partial z}$$

then the moments of $\bar{c}$, viz.,

$$\bar{c}_m = \int_{-\infty}^{+\infty} z^m \bar{c}\, dz$$

satisfy

$$\frac{d\bar{c}_0}{dt} = 0$$

$$\frac{d\bar{c}_1}{dt} = V_{\text{eff}}\bar{c}_0$$

$$\frac{d\bar{c}_2}{dt} = 2D_{\text{eff}}\bar{c}_0 + 2V_{\text{eff}}\bar{c}_1$$

$$\frac{d\bar{c}_3}{dt} = 6D_{\text{eff}}\bar{c}_1 + 3V_{\text{eff}}\bar{c}_2$$

etc.

and these equations tell us that we can determine V_{eff} and D_{eff} in terms of $\bar{c}_0, \bar{c}_1$ and $\bar{c}_2$ via

$$V_{\text{eff}} = \frac{1}{\bar{c}_0} \frac{d\bar{c}_1}{dt} = \frac{d}{dt} \left(\frac{\bar{c}_1}{\bar{c}_0} \right)$$

and

$$D_{\text{eff}} = \frac{1}{2\bar{c}_0} \left\{ \frac{d\bar{c}_2}{dt} - 2 \frac{\bar{c}_1}{\bar{c}_0} \frac{d\bar{c}_1}{dt} \right\} = \frac{1}{2} \frac{d}{dt} \left\{ \frac{\bar{c}_2}{\bar{c}_0} - \left(\frac{\bar{c}_1}{\bar{c}_0} \right)^2 \right\}$$

where $\dfrac{\bar{c}_1}{\bar{c}_0}$ and $\dfrac{\bar{c}_2}{\bar{c}_0} - \left(\dfrac{\bar{c}_1}{\bar{c}_0} \right)^2$ are the average and the variance of z distributed according to $\dfrac{\bar{c}(z,t)}{\bar{c}_0}$ at time t.

Now assuming axial symmetry, the solute distribution is in fact determined by

$$\frac{\partial c}{\partial t} = D \frac{1}{r} \frac{\partial}{\partial r} \left(r \frac{\partial c}{\partial r} \right) - v \frac{\partial c}{\partial z} + D \frac{\partial^2 c}{\partial z^2}$$

and

$$\frac{\partial c}{\partial r}(r = a) = 0$$

where $c(t = 0)$ is assigned. So to determine V_{eff} and D_{eff} we need only determine c_0, c_1 and c_2 and use their averages in the above formulas. This does not mean that the model is then correct, only that its first three power moments match those of the true solute distribution.

Before we go on we scale length, time and velocity by a, $\dfrac{a^2}{D}$ and $\dfrac{D}{a}$ then D_{eff} is scaled by D and we do not introduce new symbols. In terms of scaled variables the solute concentration satisfies

$$\frac{\partial c}{\partial t} = \frac{1}{r} \frac{\partial}{\partial r} \left(r \frac{\partial c}{\partial r} \right) - v \frac{\partial c}{\partial z} + \frac{\partial^2 c}{\partial z^2}$$

and

$$\frac{\partial c}{\partial r}(r = 1) = 0$$

where $c\,(t = 0)$ is assigned and where $v = 2\bar{v}\,(1 - r^2)$. The transverse average is now

$$\bar{u} = 2 \int_0^1 ur\,dr$$

and if we introduce an inner product via

$$\langle\, u, v \,\rangle = 2 \int_0^1 uvr\,dr$$

then

$$\bar{u} = \langle\, 1, u \,\rangle$$

The longitudinal moments of c, denoted by c_m and defined by

$$c_m = \int_{-\infty}^{+\infty} z^m c\,dz$$

satisfy

$$\frac{\partial c_0}{\partial t} = \frac{1}{r}\frac{\partial}{\partial r}\left(r\,\frac{\partial c_0}{\partial r} \right)$$

$$\frac{\partial c_1}{\partial t} = \frac{1}{r}\frac{\partial}{\partial r}\left(r\,\frac{\partial c_1}{\partial r} \right) + vc_0$$

$$\frac{\partial c_2}{\partial t} = \frac{1}{r}\frac{\partial}{\partial r}\left(r\,\frac{\partial c_2}{\partial r} \right) + 2vc_1 + 2c_0$$

etc.

which can be obtained by multiplying the equation satisfied by c by z^m, integrating this over z from $-\infty$ to $+\infty$, using integration by parts and then discarding terms evaluated at $z = \pm\infty$. The boundary conditions to be satisfied are

$$\frac{\partial c_0}{\partial r}\,(r = 1) = 0$$

$$\frac{\partial c_1}{\partial r}(r = 1) = 0$$

$$\frac{\partial c_2}{\partial r}(r = 1) = 0$$

etc.

and the initial conditions are determined by the moments of $c\,(t = 0)$.

The moment equations can be solved recursively. The equation for c_0 is simply a transverse diffusion equation in a region where the boundary is impermeable. So also the equation for c_1 but now there is a source depending on c_0; likewise for c_2, the source now depending on c_0 and c_1. Indeed the equation for c_m is a transverse diffusion equation having a source depending on c_{m-1} and c_{m-2}. The moment equations all take the same form. It is that of an unsteady, radial diffusion equation driven by an assigned source. The equations must be solved in sequence so that the sources can be determined before they are required. The power moments lead to this useful structure as the mth moment of $\dfrac{\partial c}{\partial z}$ is a multiple of the $m-1^{\text{st}}$ moment of c, etc. This lowering of the order of the moments moves the variable coefficient $v\,(r)$ into the source term in each equation.

Our eigenvalue problem is

$$\frac{1}{r}\frac{\partial}{\partial r}\left(r\frac{\partial \psi}{\partial r}\right) + \lambda^2 \psi = 0, \quad 0 \le r \le 1$$

and

$$\frac{\partial \psi}{\partial r}(r = 1) = 0$$

The eigenvalues, λ^2, are real and not negative and the eigenfunctions corresponding to different eigenvalues are orthogonal in the inner product

$$\langle\, u, v \,\rangle = 2\int_0^1 uvr\,dr$$

The eigenfunctions are thus solutions of Bessel's equation and because they must be bounded at $r = 0$, we have

$$\psi = A J_0 (\lambda r)$$

where the λ's must then be the roots of

$$J_0' (\lambda) = 0$$

Again, maybe for the third time, we say: $J_0 (z)$ is a power series in z^2, determined by Frobenius' method. The coefficients in the series tell us everything about J_0.

To every positive root, λ, there corresponds a negative root, $-\lambda$, but as $J_0 (z) = J_0 (-z)$ this root adds neither a new eigenvalue nor a new eigenfunction. Denoting the non-negative roots then as $\lambda_0, \lambda_1, \lambda_2, \ldots$ we have the corresponding normalized eigenfunctions $\psi_0, \psi_1, \psi_2, \ldots$, where

$$\psi_i = \frac{J_0 (\lambda_i r)}{\left\{ 2 \int_0^1 J_0^2 (\lambda_i r) \, r \, dr \right\}^{1/2}}$$

We observe that $\lambda_0 = 0$ and $\psi_0 = 1$ and hence transverse averages can be written

$$\bar{u} = \langle \, \psi_0, u \, \rangle$$

The first few eigenfunctions are sketched below

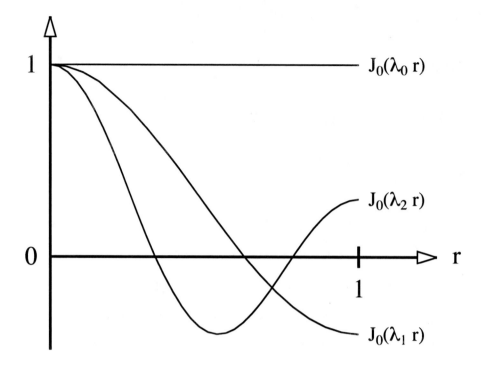

where each eigenfunction is a rescaling of $J_0(z)$, viz.,

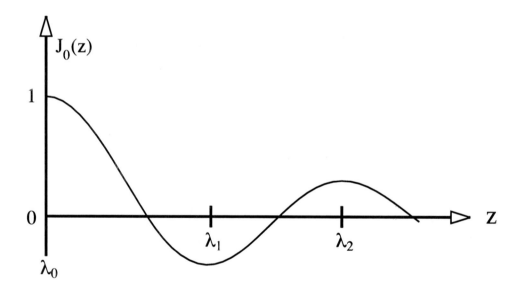

To find c_0 is straightforward. We expand c_0 as

$$c_0 = \sum_{j=0}^{\infty} \langle \psi_j, c_0 \rangle \psi_j$$

and obtain the Fourier coefficients of c_0 by multiplying the c_0 equation by $r\psi_j$ and integrating over $0 \leq r \leq 1$, viz.,

$$\left\langle \psi_j, \frac{\partial c_0}{\partial t} \right\rangle = \left\langle \psi_j, \frac{1}{r}\frac{\partial}{\partial r}\left(r\frac{\partial c_0}{\partial r} \right) \right\rangle$$

Then, because $\dfrac{\partial \psi_i}{\partial r}(r=1) = 0 = \dfrac{\partial c_0}{\partial r}(r=1)$, we have

$$\frac{d}{dt}\left\langle \psi_j, c_0 \right\rangle = -\lambda_j^2 \left\langle \psi_j, c_0 \right\rangle$$

and hence

$$c_0 = \sum_{j=0}^{\infty}\left\langle \psi_j, c_0(t=0) \right\rangle e^{-\lambda_j^2 t}\, \psi_j$$

whereupon

$$\bar{c}_0 = \left\langle \psi_0, c_0 \right\rangle = \left\langle \psi_0, c_0(t=0) \right\rangle = \bar{c}_0(t=0)$$

due to $\left\langle \psi_0, \psi_j \right\rangle = 0, \quad j = 1, 2, \ldots$ and $\lambda_0 = 0$. Thus $\bar{c}_0$ remains constant in time.

Turning to c_1, we must solve

$$\frac{\partial c_1}{\partial t} = \frac{1}{r}\frac{\partial}{\partial r}\left(r\frac{\partial c_1}{\partial r} \right) + vc_0$$

and

$$\frac{\partial c_1}{\partial r}(r=1) = 0$$

where $c_1(t=0)$ is assigned and where c_0 is now known. This is a diffusion equation driven by a homogeneous source. Again we expand the solution in the eigenfunctions $\psi_0, \psi_1, \psi_2, \ldots$ and determine the coefficients in the expansion. Here we write

$$c_1 = \sum_{j=0}^{\infty} \left\langle \psi_j, c_1 \right\rangle \psi_j$$

and determine $\left\langle \psi_j, c_1 \right\rangle$ by multiplying the c_1 equation by $r\psi_j$ and integrating over $0 \le r \le 1$, viz.,

$$\left\langle \psi_j, \frac{\partial c_1}{\partial t} \right\rangle = \left\langle \psi_j, \frac{1}{r}\frac{\partial}{\partial r}\left(r\frac{\partial c_1}{\partial r} \right) \right\rangle + \left\langle \psi_j, vc_0 \right\rangle$$

Then because $\dfrac{\partial \psi_j}{\partial r}(r = 1) = 0 = \dfrac{\partial c_1}{\partial r}(r = 1)$ we have

$$\frac{d}{dt}\left\langle \psi_j, c_1 \right\rangle = -\lambda_j^2 \left\langle \psi_j, c_1 \right\rangle + \left\langle \psi_j, vc_0 \right\rangle$$

and for each j, $j = 0, 1, 2, \ldots$, this is an equation determining the coefficients $\left\langle \psi_j, c_1 \right\rangle$. This equation contains a source on the right hand side but the source is completely known. Hence we find

$$\left\langle \psi_j, c_1 \right\rangle = \left\langle \psi_j, c_1(t=0) \right\rangle e^{-\lambda_j^2 t} + \int_0^t e^{-\lambda_j^2(t-\tau)} \left\langle \psi_j, vc_0(t=\tau) \right\rangle d\tau$$

and substituting for c_0 we get

$$\left\langle \psi_j, c_1 \right\rangle = \left\langle \psi_j, c_1(t=0) \right\rangle e^{-\lambda_j^2 t} + \sum_{k=0}^{\infty} \left\langle \psi_j, v\psi_k \right\rangle \left\langle \psi_k, c_0(t=0) \right\rangle \frac{e^{-\lambda_k^2 t} - e^{-\lambda_j^2 t}}{-\lambda_k^2 + \lambda_j^2}$$

where when $k = j$ we write $te^{-\lambda_j^2 t}$ in place of $\dfrac{e^{-\lambda_k^2 t} - e^{-\lambda_j^2 t}}{-\lambda_k^2 + \lambda_j^2}$. Thus we have determined c_1 as

$$c_1 = \sum_{j=0}^{\infty} \left\langle \psi_j, c_1(t=0) \right\rangle e^{-\lambda_j^2 t}\psi_j + \sum_{j=0}^{\infty}\sum_{k=0}^{\infty} \left\langle \psi_j, v\psi_k \right\rangle \left\langle \psi_k, c_0(t=0) \right\rangle \frac{e^{-\lambda_k^2 t} - e^{-\lambda_j^2 t}}{-\lambda_k^2 + \lambda_j^2}\psi_j$$

and we can go on and determine $\bar{c}_1$ as

$$\bar{c}_1 = \langle\, \psi_0,\, c_1 \,\rangle$$

$$= \langle\, \psi_0,\, c_1\,(t=0) \,\rangle + \sum_{k=0}^{\infty} \langle\, \psi_0,\, v\psi_k \,\rangle \langle\, \psi_k,\, c_0\,(t=0) \,\rangle \frac{e^{-\lambda_k^2 t} - 1}{-\lambda_k^2}$$

$$= \bar{c}_1\,(t=0) + \sum_{k=0}^{\infty} \overline{v\psi_k}\, \langle\, \psi_k,\, c_0\,(t=0) \,\rangle \frac{e^{-\lambda_k^2 t} - 1}{-\lambda_k^2}$$

where when $k = 0$ the third factor in the summand reduces to t.

We now have $c_0, \bar{c}_0, c_1$ and $\bar{c}_1$ and the reader can go on and obtain $c_2, \bar{c}_2, \ldots$ in a similar way; to do this simply requires a notational scheme to keep track of what is going on. Everything falls into place once such a scheme is invented.

It turns out that c_2 is not needed in order to derive a formula for $\bar{c}_2$ and hence for D_{eff}. To see this we average the equations satisfied by c_0, c_1 and c_2 obtaining

$$\frac{d\bar{c}_0}{dt} = \overline{\frac{1}{r}\frac{\partial}{\partial r}\left(r\,\frac{\partial c_0}{\partial r} \right)}$$

$$\frac{d\bar{c}_1}{dt} = \overline{\frac{1}{r}\frac{\partial}{\partial r}\left(r\,\frac{\partial c_1}{\partial r} \right)} + \overline{vc_0}$$

and

$$\frac{d\bar{c}_2}{dt} = \overline{\frac{1}{r}\frac{\partial}{\partial r}\left(r\,\frac{\partial c_2}{\partial r} \right)} + 2\overline{vc_1} + 2\bar{c}_0$$

Now ψ_0 and c_i satisfy homogeneous Neumann conditions at $r = 1$ and λ_0 is zero. This tells us that

$$\overline{\frac{1}{r}\frac{\partial}{\partial r}\left(r\,\frac{\partial c_i}{\partial r} \right)} = \left\langle\, \psi_0,\, \frac{1}{r}\frac{\partial}{\partial r}\left(r\,\frac{\partial c_i}{\partial r} \right) \,\right\rangle = \left\langle\, \frac{1}{r}\frac{d}{dr}\left(r\,\frac{d\psi_0}{dr} \right),\, c_i \,\right\rangle = 0$$

and hence that

$$\frac{d\bar{c}_0}{dt} = 0$$

$$\frac{d\bar{c}_1}{dt} = \overline{vc}_0$$

and

$$\frac{d\bar{c}_2}{dt} = 2\overline{vc}_1 + 2\bar{c}_0$$

These formulas simplify the determination of V_{eff} and D_{eff}. Putting them into

$$V_{\text{eff}} = \frac{1}{\bar{c}_0} \frac{d\bar{c}_1}{dt}$$

and

$$D_{\text{eff}} = \frac{1}{2\bar{c}_0} \left\{ \frac{d\bar{c}_2}{dt} - 2\frac{\bar{c}_1}{\bar{c}_0} \frac{d\bar{c}_1}{dt} \right\}$$

we get

$$V_{\text{eff}} = \frac{\overline{vc}_0}{\bar{c}_0}$$

and

$$D_{\text{eff}} = \frac{\overline{vc}_1}{\bar{c}_0} - \frac{\overline{vc}_0}{\bar{c}_0} \frac{\bar{c}_1}{\bar{c}_0} + 1$$

and this tells us that we need only $c_0, \bar{c}_0, c_1$ and $\bar{c}_1$ to determine V_{eff} and D_{eff}.

Now because

$$c_0 = \sum_{j=0}^{\infty} \left\langle \psi_j, c_0 (t = 0) \right\rangle e^{-\lambda_j^2 t} \psi_j$$

$$= \bar{c}_0 (t = 0) + \sum_{j=1}^{\infty} \left\langle \psi_j, c_0 (t = 0) \right\rangle e^{-\lambda_j^2 t} \psi_j$$

and

$$\bar{c}_0 = \bar{c}_0 (t = 0)$$

we find

$$\overline{vc_0} = \sum_{j=0}^{\infty} \left\langle \psi_j, \, c_0\,(t=0) \right\rangle e^{-\lambda_j^2 t} \, \overline{v\psi_j}$$

$$= \overline{c}_0\,(t=0)\,\overline{v} + O\left(e^{-\lambda_1^2 t}\right)$$

and we see that V_{eff} turns out to be time dependent. This results whenever $c_0\,(t=0)$ differs from its equilibrium value $\overline{c}_0\,(t=0)$. This difference weights the streamlines more or less heavily than their equilibrium weighting and as the streamlines differ in speed this leads to V_{eff} being other than $\overline{v}$. But for large enough values of t the equilibrium distribution of c_0 is attained and V_{eff} reaches a constant value. It is this that we now denote by V_{eff} and its value is

$$V_{\text{eff}} = \overline{v}$$

To go on and determine D_{eff} we need $\overline{vc_1}$ and $\overline{c}_1$. By using

$$c_1 = \sum_{j=0}^{\infty} \left\langle \psi_j, \, c_1\,(t=0) \right\rangle e^{-\lambda_j^2 t}\, \psi_j + \sum_{j=0}^{\infty}\sum_{k=0}^{\infty} \left\langle \psi_j, \, v\psi_k \right\rangle \left\langle \psi_k, \, c_0\,(t=0) \right\rangle \frac{e^{-\lambda_k^2 t} - e^{-\lambda_j^2 t}}{-\lambda_k^2 + \lambda_j^2}\, \psi_j$$

and

$$\overline{c}_1 = \overline{c}_1\,(t=0) + \sum_{k=0}^{\infty} \overline{v\psi_k} \left\langle \psi_k, \, c_0\,(t=0) \right\rangle \frac{e^{-\lambda_k^2 t} - 1}{-\lambda_k^2}$$

$$= \overline{c}_1\,(t=0) + \overline{vc_0}\,(t=0)\,t + \sum_{k=1}^{\infty} \overline{v\psi_k} \left\langle \psi_k, \, c_0\,(t=0) \right\rangle \frac{1}{\lambda_k^2} + O\left(e^{-\lambda_1^2 t}\right)$$

we find

$$\overline{vc_1} = \sum_{k=0}^{\infty} \left\langle \psi_j, c_1 \left(t = 0 \right) \right\rangle e^{-\lambda_j^2 t} \overline{v\psi_j}$$

$$+ \sum_{j=0}^{\infty} \sum_{k=0}^{\infty} \left\langle \psi_j, v\psi_k \right\rangle \left\langle \psi_k, c_0 \left(t = 0 \right) \right\rangle \frac{e^{-\lambda_k^2 t} - e^{-\lambda_j^2 t}}{-\lambda_k^2 + \lambda_j^2} \overline{v\psi_j}$$

$$= \overline{c_1 \left(t = 0 \right)} \, \overline{v} + O\left(e^{-\lambda_1^2 t} \right) + \overline{vc_0} \left(t = 0 \right) \overline{v} \, t$$

$$+ \sum_{k=1}^{\infty} \overline{v\psi_k} \left\langle \psi_k, c_0 \left(t = 0 \right) \right\rangle \frac{e^{-\lambda_k^2 t} - 1}{-\lambda_k^2} \overline{v}$$

$$+ \sum_{j=1}^{\infty} \left\langle \psi_j, v \right\rangle \overline{c}_0 \left(t = 0 \right) \frac{1 - e^{-\lambda_j^2 t}}{\lambda_j^2} \overline{v\psi_j} + O\left(te^{-\lambda_1^2 t} \right)$$

where the first of the last four terms corresponds to $j = 0$, $k = 0$, the second to $j = 0$, $k > 0$, the third to $j > 0$, $k = 0$ and the last to $j > 0$, $k > 0$. Using these formulas and writing $\left\langle \psi_j, v \right\rangle = \left\langle 1, \psi_j v \right\rangle = \overline{\psi_j v}$ we find

$$\overline{vc_1} \, \overline{c}_0 - \overline{vc_0} \, \overline{c}_1 = \sum_{j=1}^{\infty} \frac{\overline{v\psi_j} \, \overline{v\psi_j}}{\lambda_j^2} \overline{c}_0 \overline{c}_0 + O\left(te^{-\lambda_1^2 t} \right)$$

and hence, for large enough values of t, we we see that D_{eff} reaches a constant value. It is

$$D_{\text{eff}} = 1 + \sum_{j=1}^{\infty} \frac{\overline{v\psi_j} \, \overline{v\psi_j}}{\lambda_j^2}$$

This formula is correct for any v and for $v = 2\overline{v} \left(1 - r^2 \right)$ it reduces to

$$1 + \frac{1}{48} \overline{v}^2$$

which is Taylor's formula if 1 is added to take longitudinal diffusion into account.

To see this, we need to evaluated $\overline{v\psi_j} = \left\langle \psi_0, v\psi_j \right\rangle = \left\langle \psi_j, v \right\rangle$. Now because $v = 2\overline{v} \left(1 - r^2 \right)$

we find that

$$\frac{1}{r}\frac{d}{dr}\left(r\frac{dv}{dr}\right) = -8\overline{v}$$

where $v\left(r=1\right)=0$ and $\dfrac{dv}{dr}\left(r=1\right)=-4\overline{v}$. Then, integrating by parts, we have

$$\left\langle \psi_j, \frac{1}{r}\frac{d}{dr}\left(r\frac{dv}{dr}\right) \right\rangle = 2\left[\psi_j\, r\frac{dv}{dr} - r\frac{d\psi_j}{dr}\,v\right]_0^1 + \left\langle \frac{1}{r}\frac{d}{dr}\left(r\frac{d\psi_j}{dr}\right),\, v \right\rangle$$

and hence, for $j = 1, 2, \ldots$

$$\left\langle \psi_j,\, v \right\rangle = -\frac{8\overline{v}\psi_j\left(r=1\right)}{\lambda_j^2}$$

Now, using

$$\psi_j\left(r=1\right) = \frac{J_0\left(\lambda_j\right)}{\left\{2\displaystyle\int_0^1 J_0^2\left(\lambda_j r\right) r\, dr\right\}^{1/2}}$$

$$\int_0^1 J_0^2\left(\lambda_j r\right) r\, dr = \frac{1}{2}\left\{J_0'^2\left(\lambda\right) + J_0^2\left(\lambda\right)\right\}$$

and

$$J_0'\left(\lambda_j\right) = 0$$

we obtain

$$\psi_j\left(r=1\right) = 1$$

and our result is

$$\left\langle \psi_j,\, v \right\rangle = -\frac{8\overline{v}}{\lambda_j^2}$$

whereupon we have

$$\sum_{j=1}^{\infty} \frac{\overline{v\psi_j}\,\overline{v\psi_j}}{\lambda_j^2} = 64\overline{v}^2 \sum_{j=1}^{\infty} \frac{1}{\lambda_j^6}$$

The sum $\displaystyle\sum_{j=1}^{\infty} \frac{1}{\lambda_j^6}$ is found, cf., A. E. DeGance and L. E. Johns, "*Infinite Sums in the Theory of Dispersion of Chemically Reactive Solute*" SIAM J. Math. Anal. <u>18</u>, 473 (1987), to be $\dfrac{1}{3072}$, using the fact that $J_0'(z) = -J_1(z)$, and the formula for D_{eff}, when $v = 2\overline{v}(1 - r^2)$, is then

$$D_{\text{eff}} = 1 + \frac{1}{48}\overline{v}^2$$

Having now determined V_{eff} and D_{eff}, we must be careful not to claim too much for these results. Indeed we have not really matched the moments of the solution to the model equation to the moments of the true solute distribution, at least not yet. Of course we have not even looked at the moments for $i > 2$. But more than this, if we look at $\overline{c}_1$ where

$$\overline{c}_1 = \overline{c}_1(t = 0) + \sum_{k=0}^{\infty} \langle\, \psi_0,\, v\psi_k \,\rangle \,\langle\, \psi_k,\, c_0(t = 0) \,\rangle\, \frac{e^{-\lambda_k^2 t} - 1}{-\lambda_k^2}$$

we see that it can be written

$$\overline{c}_1 = \overline{c}_1(t = 0) + \overline{vc}_0(t = 0)\,t + \sum_{k=1}^{\infty} \langle\, \psi_0,\, v\psi_k \,\rangle \,\langle\, \psi_k,\, c_0(t = 0) \,\rangle\, \frac{e^{-\lambda_k^2 t} - 1}{-\lambda_k^2}$$

whereas from the model we get, assuming V_{eff} to be constant at its long time value,

$$\overline{c}_1(t = 0) + V_{\text{eff}}\,\overline{c}_0(t = 0)\,t$$

This cannot match the true $\overline{c}_1$ for all time but we can determine V_{eff}, $\overline{c}_0(t = 0)$ and $\overline{c}_1(t = 0)$ to match it to the true $\overline{c}_1$ as t grows large. To do this we let $V_{\text{eff}} = \overline{v}$ and retain $\overline{c}_0(t = 0)$ at its true value. But we cannot take $\bar{c}_1 (t = 0)$ to be its true value; we must instead take it to be

$$\bar{c}_1 (t = 0) + \sum_{k=1}^{\infty} \langle \psi_0, v\psi_k \rangle \langle \psi_k, c_0 (t = 0) \rangle \frac{1}{\lambda_k^2}$$

What we see then is that in using the model we must determine $\bar{c}_{\text{eff}} (t = 0)$ as well as V_{eff} and D_{eff}. If we do this we make the difference between the model prediction of $\bar{c}_1$ and the true $\bar{c}_1$, i.e.,

$$\sum_{k=1}^{\infty} \overline{v\psi_k} \langle \psi_k, c_0 (t = 0) \rangle \frac{e^{-\lambda_k^2 t}}{-\lambda_k^2},$$ to be $O\left(e^{-\lambda_1^2 t}\right)$; otherwise the difference will be $O(1)$. A similar observation pertains to the moment $\bar{c}_2$. To have the model get $\bar{c}_2$ right as t grows large, additional conditions are required to be satisfied by $\bar{c}_{\text{eff}} (t = 0)$.

The Zeros of $J_1(z)$

The Bessel function $J_1 (z)$ is defined to be the sum of the infinite series

$$\frac{1}{2} z \left\{ 1 - \frac{z^2}{2^2 (1! \, 2!)} + \frac{z^4}{2^2 (2! \, 3!)} - \cdots \right\} = \frac{1}{2} z \sum_{k=0}^{\infty} \frac{(-1)^k \left(\frac{z}{2}\right)^{2k}}{k! \, (k+1)!}$$

This series converges for all finite values of z as do the series obtained from it by term-by-term differentiation.

The zeros J_1 are real and simple and we denote them $0, \pm z_1, \pm z_2, \ldots$. The function $\frac{J_1'(z)}{J_1(z)}$ has simple poles at $z = 0, \pm z_1, \pm z_2, \ldots$ and its residue at each pole is $+1$. Thus on any closed contour C the value of the integral

$$\int_C \frac{1}{z - \zeta} \frac{J_1'(z)}{J_1(z)} \, dz$$

is the product of $2\pi i$ and the sum of the residues of the integrand at its poles inside C. In the limit as C grows large and encloses the entire complex plane, the integral vanishes and we get

$$0 = 2\pi i \left\{ \frac{J_1'(\zeta)}{J_1(\zeta)} + \frac{1}{0 - \zeta} + \frac{1}{z_1 - \zeta} + \frac{1}{-z_1 - \zeta} + \frac{1}{z_2 - \zeta} + \frac{1}{-z_2 - \zeta} + \cdots \right\}$$

and hence writing z in place of ζ, we have

$$\frac{1}{z} - \frac{J_1'(z)}{J_1(z)} = \frac{-2z}{-z_1^2 + z^2} + \frac{-2z}{-z_2^2 + z^2} + \cdots$$

whereupon

$$\frac{1}{2z}\left\{\frac{1}{z} - \frac{J_1'(z)}{J_1(z)}\right\} = \frac{1}{z_1^2\left\{1 - \dfrac{z^2}{z_1^2}\right\}} + \frac{1}{z_2^2\left\{1 - \dfrac{z^2}{z_2^2}\right\}} + \cdots$$

For any z such that $|z| < |z_1|$ where $|z_1| < |z_2| < \cdots$ we can expand the factors $\dfrac{1}{1 - \dfrac{z^2}{z_i^2}}$ on the right hand side to get

$$\frac{J_1(z) - zJ_1'(z)}{2z^2 J_1(z)} = \sum_{i=1}^{\infty}\frac{1}{z_i^2} + \sum_{i=1}^{\infty}\frac{1}{z_i^4}z^2 + \sum_{i=1}^{\infty}\frac{1}{z_i^6}z^4 + \cdots$$

and then, using the series for $J_1(z)$ to expand the left hand side in powers of z^2, we can match the coefficients of $1, z^2, z^4, \ldots$ on the two sides to get

$$\sum_{i=1}^{\infty}\frac{1}{z_i^2} = \frac{1}{8}$$

$$\sum_{i=1}^{\infty}\frac{1}{z_i^4} = \frac{1}{192}$$

$$\sum_{i=1}^{\infty}\frac{1}{z_i^6} = \frac{1}{3072}$$

etc.

These formulas can be used to estimate the zeros of $J_1(z)$; the third formula is of interest as it stands in evaluating D_{eff}.

Functions such as $\cos z$, $\sin z$, $J_0(z)$, $J_1(z)$, etc., defined by infinite sums converging for all finite values of z are generalizations of polynomials. As such they also have infinite product expansions which simplify this kind of work. The infinite product expansion of J_1 leads directly to the formula

$$\frac{J_1'(z)}{J_1(z)} = \frac{1}{z} + \frac{2z}{z^2 - z_1^2} + \frac{2z}{z^2 - z_2^2} + \cdots$$

which we got by the residue theorem.

20.3   Spherical Coordinates: Spherical Harmonics, the Method of Frobenius

The eigenvalue problem for ∇^2, viz., $\left(\nabla^2 + \lambda^2\right)\psi = 0$, when written out in spherical coordinates leads on separation of variables to three problems. Each of these determines one of the factors making up $\psi(r, \theta, \phi)$ as the product $R(r)\,\Theta(\theta)\,\Phi(\phi)$. Solving for $\Phi(\phi)$ is easy and will be reviewed here. But our main job in this lecture is the determination of $\Theta(\theta)$. This requires us to explain Frobenius' method and we use $\Theta(\theta)$ as an example of this. The determination of $R(r)$ is also easy and is not worked out in detail.

To get going we recall the multipole moment expansion of the electrical potential due to a set of point changes. This is of interest because the potential, denoted ϕ, satisfies $\nabla^2 \phi = 0$.

To write ϕ at a point P, whose position is denoted by $\vec{r}$, we use Coulomb's law (in rationalized MKS units), viz.,

$$\phi(\vec{r}) = \sum_{i=1}^{n} \frac{q_i}{4\pi\varepsilon_0 |\vec{r} - \vec{r}_i|}$$

Then, in terms of $r = |\vec{r}|$, $r_i = |\vec{r}_i|$ and θ_i, we have

$$|\vec{r} - \vec{r}_i|^2 = (\vec{r} - \vec{r}_i) \cdot (\vec{r} - \vec{r}_i)$$

$$= r^2 - 2r\,r_i \cos\theta_i + r_i^2$$

and hence we write

$$\frac{1}{|\vec{r} - \vec{r_i}|} = \frac{1}{r} \frac{1}{\sqrt{1 - 2\frac{r_i}{r} \cos\theta_i + \left(\frac{r_i}{r}\right)^2}}$$

so that if the field point is further from the origin then any source point we can use the expansion

$$\frac{1}{\sqrt{1+z}} = 1 + \frac{\left(-\frac{1}{2}\right)}{1!} z + \frac{\left(-\frac{1}{2}\right)\left(-\frac{3}{2}\right)}{2!} z^2 + \frac{\left(-\frac{1}{2}\right)\left(-\frac{3}{2}\right)\left(-\frac{5}{2}\right)}{3!} z^3 + \cdots$$

which holds whenever $|z| < 1$, and write

$$\frac{1}{\sqrt{1 - 2\frac{r_i}{r} \cos\theta_i + \left(\frac{r_i}{r}\right)^2}} = 1 + \frac{\left(-\frac{1}{2}\right)}{1!} \left\{ -2\frac{r_i}{r} \cos\theta_i + \left(\frac{r_i}{r}\right)^2 \right\}$$

$$+ \frac{\left(-\frac{1}{2}\right)\left(-\frac{3}{2}\right)}{2!} \left\{ -2\frac{r_i}{r} \cos\theta_i + \left(\frac{r_i}{r}\right)^2 \right\}^2 + \cdots$$

This requires $-1 < -2\frac{r_i}{r} \cos\theta_i + \left(\frac{r_i}{r}\right)^2 < 1$.

On rewriting this in ascending powers of $\frac{r_i}{r}$ we get

$$1 + \frac{r_i}{r} \cos\theta_i + \left(\frac{r_i}{r}\right)^2 \frac{1}{2} \left(3\cos^2\theta_i - 1 \right) + \cdots$$

Our interest in this lies in the functions 1, $\cos\theta_i$, $\frac{1}{2} \left(3\cos^2\theta_i - 1 \right)$, $\ldots$ obtained on expanding $\frac{1}{\{1+z\}^{1/2}}$ using the binomial theorem, writing $z = -2xy + y^2$ where $x = \cos\theta_i$ and $y = \frac{r_i}{r}$, and then arranging the result in a power series in y. We can write the expansion directly in powers of y as (see Linus Pauling and E. Bright Wilson, jr. *"Introduction to quantum mechanics, with applications to chemistry"*)

$$T(x,y) = \frac{1}{\{1 - 2xy + y^2\}^{1/2}} = \sum_{\ell=0}^{\infty} P_\ell(x) y^\ell$$

and then determine the sequence of functions $P_\ell, \ell = 0, 1, 2, \ldots$. To do this we must have $|x| \leq 1$ and $|y|$ small enough that the series converges uniformly in x. The function $T(x,y) = \dfrac{1}{\{1 - 2xy + y^2\}^{1/2}}$ is called the generating function for the functions $P_0(x), P_1(x), P_2(x), \ldots$ It turns up naturally in electrostatics and not only generates $P_0(x)$, $P_1(x), P_2(x), \ldots$ but also provides an easy way to uncover the important facts about these functions.

To begin we obtain $P_0(x)$ and $P_1(x)$ via

$$T(x, y = 0) = 1 = P_0(x)$$

and

$$\frac{\partial}{\partial y} T(x, y = 0) = x = P_1(x)$$

Then we can derive a recursion formula expressing any one function in terms of the previous two. Indeed using

$$T(x, y) = \frac{1}{\{1 - 2xy + y^2\}^{1/2}} = \sum P_\ell(x) y^\ell$$

and

$$\frac{\partial}{\partial y} T(x, y) = \frac{x - y}{\{1 - 2xy + y^2\}^{3/2}} = \sum \ell P_\ell(x) y^{\ell - 1}$$

we get

$$(x - y) \sum P_\ell(x) y^\ell = \{1 - 2xy + y^2\} \sum \ell P_\ell(x) y^{\ell - 1}$$

and requiring the coefficients of y^ℓ to agree on the two sides we get

$$(\ell + 1) P_{\ell+1}(x) - (2\ell + 1) x P_\ell(x) + \ell P_{\ell-1}(x) = 0$$

As $P_0(x) = 1$ and $P_1(x) = x$, this recursion formula can be used to obtain

$P_2(x) = \dfrac{3}{2}x^2 - \dfrac{1}{2}, \ldots$ And by doing this we can see that $P_\ell(x)$ is a polynomial of degree ℓ and that it is odd or even as ℓ is odd or even. The polynomials so defined are called Legendre polynomials and they are fundamental to work in spherical coordinate systems.

We can use the generating function to obtain the differential equation satisfied by $P_\ell(x)$. To do this we observe that

$$\frac{\partial T}{\partial x}(x, y) = \frac{y}{\{1 - 2xy + y^2\}^{3/2}} = \sum P'_\ell(x) y^\ell$$

and hence that

$$y \sum P_\ell(x) y^\ell = \{1 - 2xy + y^2\} \sum P'_\ell(x) y^\ell$$

and again requiring the coefficients of y^ℓ to agree on the two sides we get

$$P'_{\ell+1}(x) - 2x P'_\ell(x) + P'_{\ell-1}(x) - P_\ell(x) = 0$$

Then multiplying this by $\ell+1$ and subtracting the result from the derivative of the recursion formula leads to

$$x P'_\ell(x) - P'_{\ell-1}(x) - \ell P_\ell(x) = 0$$

Likewise we find

$$P'_{\ell+1}(x) - x P'_\ell(x) - (\ell + 1) P_\ell(x) = 0$$

and subtracting the first, multiplied by x, from the second, written with $\ell - 1$ in place of ℓ we obtain

$$\left(1 - x^2\right) P'_\ell(x) + \ell x P_\ell(x) - \ell P_{\ell-1}(x) = 0$$

which on differentiation results in

$$\frac{d}{dx}\left\{\left(1 - x^2\right)\frac{d}{dx}P_\ell(x)\right\} + \ell(\ell + 1) P_\ell(x) = 0$$

upon replacing $\ell x P'_\ell(x) - \ell P'_{\ell-1}(x)$ by $\ell^2 P_\ell(x)$. This is called Legendre's differential equation and it is satisfied by the Legendre polynomials. These polynomials satisfy the orthogonality condition

$$\int_{-1}^{+1} P_\ell P_{\ell'} \, dz = 0, \quad \ell \neq \ell', \quad \ell, \ell' = 0, 1, 2, \ldots$$

To see this we use

$$\int_{-1}^{+1} P \frac{d}{dz} \left\{ (1-z^2) \frac{dQ}{dz} \right\} dz - \int_{-1}^{+1} Q \frac{d}{dz} \left\{ (1-z^2) \frac{dP}{dz} \right\} dz =$$
$$\left[(1-z^2) \left\{ P \frac{dQ}{dz} - Q \frac{dP}{dz} \right\} \right]_{-1}^{+1} = 0$$

which holds for any smooth and bounded functions P and Q. Then setting $P = P_\ell$ and $Q = P_{\ell'}$ produces the required result.

To make use of the orthogonality condition:

$$\int_{-1}^{+1} P_\ell(z) P_{\ell'}(z) \, dz = \left\{ \int_{-1}^{+1} P_\ell^2(z) \, dz \right\} \delta_{\ell\ell'}$$

we need to evaluate the integrals

$$\int_{-1}^{+1} P_\ell^2(z) \, dz$$

To get these we write

$$P_\ell(z) = \frac{1}{\ell} \left\{ (2\ell - 1) z P_{\ell-1}(z) - (\ell - 1) P_{\ell-2}(z) \right\}$$

multiply by $P_\ell(z)$ and integrate from -1 to $+1$. Due to the orthogonality of P_ℓ and $P_{\ell-2}$ we get

$$\int_{-1}^{+1} P_\ell^2(z) \, dz = \frac{2\ell - 1}{\ell} \int_{-1}^{+1} P_\ell(z) z P_{\ell-1}(z) \, dz$$

Then as $zP_\ell(z) = \dfrac{1}{2\ell+1}\left\{(\ell+1)P_{\ell+1}(z) + \ell P_{\ell-1}(z)\right\}$ we see that

$$\int_{-1}^{+1} P_\ell^2(z)\,dz = \frac{2\ell-1}{2\ell+1}\int_{-1}^{+1} P_{\ell-1}^2(z)\,dz$$

and using this recursively we get

$$\int_{-1}^{+1} P_\ell^2(z)\,dz = \frac{2}{2\ell+1}$$

The Angular Eigenfunctions of ∇^2 in Spherical Coordinates: The Use of Frobenius' Method

We will rediscover the functions $P_\ell(z)$ in the course of solving for the eigenfunctions of ∇^2 in spherical coordinates; it is to this that we now turn. Writing ∇^2 in spherical coordinates and substituting $\psi(r,\theta,\phi) = R(r)\Theta(\theta)\Phi(\phi)$ into the eigenvalue problem $\left(\nabla^2 + \lambda^2\right)\psi = 0$ we find that $R(r), \Theta(\theta)$ and $\Phi(\phi)$ satisfy

$$\frac{d^2\Phi}{d\phi^2} + m^2\Phi = 0$$

$$\frac{1}{\sin\theta}\frac{d}{d\theta}\left(\sin\theta\,\frac{d\Theta}{d\theta}\right) + \left\{\beta^2 - \frac{m^2}{\sin^2\theta}\right\}\Theta = 0$$

and

$$\frac{1}{r^2}\frac{d}{dr}\left(r^2\frac{dR}{dr}\right) + \left\{\lambda^2 - \frac{\beta^2}{r^2}\right\}R = 0$$

These three ordinary differential equations turn up in the order listed and $-m^2$ and $-\beta^2$ are separation constants introduced in the course of carrying out the separation of variables algorithm. The values of m^2, β^2 and λ^2 must be determined as well as expressions for the functions Φ, Θ and R. The problems must be solved in order, because m^2 in the first carries over to the second and β^2 in the second carries over to the third. Each of these equations is a one-dimensional eigenvalue problem in its own right but is not a definite problem until boundary conditions are assigned. The boundary conditions derive from those satisfied by ψ which in turn come from the physical problem of interest.

The ϕ and θ equations are common to many problems and we work out their solutions in some detail.

We suppose that physical boundary conditions are assigned only on surfaces where r is constant. Hence we use symmetry and boundedness conditions to derive the dependence of ψ on θ and ϕ and therefore Θ on θ and Φ on ϕ.

Thus Φ must be periodic in ϕ of period 2π and we write the solution to

$$\frac{d^2\Phi}{d\phi^2} + m^2\Phi = 0$$

as

$$\Phi = A\cos m\phi + B\frac{1}{m}\sin m\phi$$

whereupon using $\Phi(0) = \Phi(2\pi)$ and $\Phi'(0) = \Phi'(2\pi)$ we have, for A, B and m,

$$\begin{pmatrix} \cos 2\pi m - 1 & \dfrac{1}{m}\sin 2\pi m \\ -m\sin 2\pi m & \cos 2\pi m - 1 \end{pmatrix} \begin{pmatrix} A \\ B \end{pmatrix} = \begin{pmatrix} 0 \\ 0 \end{pmatrix}$$

Only values of m such that

$$2 - 2\cos 2\pi m = 0$$

lead to A and B other than $A = 0 = B$.

Hence we must have $m = 0, \pm 1, \pm 2, \ldots$. If $m = 0$, the rank of the matrix on the LHS is 1, otherwise it is 0. At $m = 0$, we write $\begin{pmatrix} A \\ B \end{pmatrix} = \begin{pmatrix} 1 \\ 0 \end{pmatrix}$; at $m = +1, +2, \ldots$ we write $\begin{pmatrix} A \\ B \end{pmatrix} = \begin{pmatrix} 1 \\ 0 \end{pmatrix}$ and $\begin{pmatrix} 0 \\ 1 \end{pmatrix}$

Corresponding to $m = 0$, we have $m^2 = 0$, $\Phi = 1$; corresponding to $m = +1, +2, \ldots$, we have $m^2 = 1, 4, \ldots$ and $\Phi = \cos m\phi$ and $\sin m\phi$. Corresponding to $m = -1, -2, \ldots$ there is no new information.

However what we do is this: we replace $\cos m\phi$ and $\sin m\phi$ with two linear combinations, $e^{im\phi}$ and $e^{-im\phi}$. Then we assign $e^{im\phi}$ to m and let m run through $\cdots, -2, -1, 0, 1, 2, \cdots$. So for $m = 0, \pm 1, \pm 2, \ldots$ we have

$$\Phi_m(\phi) = \frac{1}{\sqrt{2\pi}}\, e^{im\phi}$$

where

$$\int_0^{2\pi} \overline{\Phi}_m \Phi_n \, d\phi = \begin{cases} 1, & m = n \\[2mm] 0, & m \neq n \end{cases}$$

and where the coefficients c_m in an expansion

$$f(\phi) = \sum_{m=-\infty}^{+\infty} c_m \Phi_m(\phi)$$

are given by

$$c_m = \int_0^{2\pi} \overline{\phi}_m(\phi)\, f(\phi)\, d\phi = \int_0^{2\pi} \phi_{-m}(\phi)\, f(\phi)\, d\phi$$

Now having obtained the values of m^2, we turn to the θ equation:

$$\frac{1}{\sin\theta}\frac{d}{d\theta}\left(\sin\theta\frac{d\Theta}{d\theta}\right) + \left\{\beta^2 - \frac{m^2}{\sin^2\theta}\right\}\Theta = 0$$

where $0 \leq \theta \leq \pi$. To solve this equation we introduce $z = \cos\theta$ so that z runs from $+1$ to -1 as θ runs from 0 to π. Then writing $P(z)$ in place of $\Theta(\theta) = P(\cos\theta)$ we find that $P(z)$ satisfies

$$\frac{d}{dz}\left\{\left(1 - z^2\right)\frac{dP}{dz}\right\} + \left\{\beta^2 - \frac{m^2}{1 - z^2}\right\}P = 0, \quad -1 \leq z \leq 1$$

or

$$\left(1 - z^2\right)\frac{d^2 P}{dz^2} - 2z\frac{dP}{dz} + \left\{\beta^2 - \frac{m^2}{1 - z^2}\right\}P = 0, \quad -1 \leq z \leq 1$$

The Method of Frobenius

Our aim is to derive the solution to this equation by obtaining the coefficients in a series expansion of $P(z)$ in powers of z.

But first we outline the method we use, due to Fuchs and Frobenius, for obtaining solutions to equations of the form

$$y^{(m)} + p_{m-1}(x)\, y^{(m-1)} + \cdots + p_1(x)\, y^{(1)} + p_0(x)\, y = 0$$

A point x_0 is called an ordinary point if all $p_i(x)$ have Taylor series expansions about x_0. For example $x = 0$ is an ordinary point of

$$\frac{d^2 y}{dx^2} + y = 0$$

and the reader may seek a solution to this equation in the form

$$y = \sum_{n=0}^{\infty} a_n x^n$$

producing the series for $\cos x$ if $a_0 = 1$, $a_1 = 0$ and for $\sin x$ if $a_0 = 0$, $a_1 = 1$. The coefficients in these series tell us everything we might want to know about the functions denoted $\sin$ and $\cos$.

A point x_0 is called a regular singular point if

$$(x - x_0)^n p_0(x),\ (x - x_0)^{n-1} p_1(x),\ \cdots,\ (x - x_0)\, p_{n-1}(x)$$

all possess Taylor series expansions about x_0. If x_0 is a regular singular point, we can find a solution

$$y = (x - x_0)^\alpha \sum a_n (x - x_0)^n, \quad a_0 \neq 0$$

In the case of second order equations, where x_0 is a regular singular point, we write

$$y'' + \frac{p(x)}{x - x_0} y' + \frac{q(x)}{(x - x_0)^2} y = 0$$

where $p(x)$ and $q(x)$ have Taylor series expansions about $x = x_0$, viz.,

$$p = \sum_{n=0} p_n (x - x_0)^n$$

and

$$q = \sum_{n=0} q_n (x - x_0)^n$$

Substituting

$$y = (x - x_0)^\alpha \sum a_n (x - x_0)^n = \sum a_n (x - x_0)^{n+\alpha}$$

$$y' = \alpha \sum a_n (x - x_0)^{n+\alpha-1} + \sum n a_n (x - x_0)^{n+\alpha-1}$$

$$y'' = \alpha (\alpha - 1) \sum a_n (x - x_0)^{n+\alpha-2} + 2\alpha \sum n a_n (x - x_0)^{n+\alpha-2}$$
$$+ \sum n(n-1) a_n (x - x_0)^{n+\alpha-2}$$

into our differential equation we obtain

$$\alpha(\alpha - 1) \sum a_n (x - x_0)^{n+\alpha-2} + 2\alpha \sum n a_n (x - x_0)^{n+\alpha-2}$$
$$+ \sum n(n-1) a_n (x - x_0)^{n+\alpha-2}$$
$$+ \sum p_n (x - x_0)^n \sum (\alpha + n) a_n (x - x_0)^{n+\alpha-2}$$
$$+ \sum q_n (x - x_0)^n \sum a_n (x - x_0)^{n+\alpha-2} = 0$$

and equating the coefficients of $(x - x_0)^{n + \alpha - 2}$, $\quad n = 0, 1, 2, \ldots$ to zero we have

$$n = 0 : \quad \left(\alpha^2 + (p_0 - 1)\,\alpha + q_0 \right) a_0 = 0$$

$$n = 1, 2, \ldots : \quad \left((\alpha + n)^2 + (p_0 - 1)\,(\alpha + n) + q_0 \right) a_n =$$

$$- \sum_{k=0}^{n-1} \left((\alpha + k)\,p_{n-k} + q_{n-k} \right) a_k$$

Hence, at $n = 0$, $a_0 \neq 0$ implies $P(\alpha) = \alpha^2 + (p_0 - 1)\,\alpha + q_0 = 0$, (called the indicial equation) whereas for $n = 1, 2, \ldots$ we have

$$P(\alpha + n)\,a_n = - \sum_{k=0}^{n-1} \left((\alpha + k)\,p_{n-k} + q_{n-k} \right) a_k$$

which leads to $a_1, a_2, \ldots, a_n$ terms of a_0 so long as $P(\alpha + n)$ is not zero.

We denote by α_1 and α_2 the two roots of $P(\alpha) = 0$ and assume $\operatorname{Re} \alpha_1 \geq \operatorname{Re} \alpha_2$. Then for $\alpha = \alpha_1$ we have $P(\alpha + n) \neq 0$ for $n = 1, 2, \ldots$ because α_2 is the only root, other than α_1, of $P(\alpha) = 0$ and $\alpha_2 \neq \alpha_1 + n$. Hence we always have one solution to our equation.

As an example, we solve

$$y'' + \frac{1}{x}\,y' - \left(1 + \frac{\nu^2}{x^2} \right) y = 0$$

which has a regular singular point at $x = 0$. We substitute $y = x^\alpha \sum a_n x^n$ obtaining

$$\sum_{n=0} (\alpha + n)\,(\alpha + n - 1)\,a_n\,x^{n + \alpha - 2} + \sum_{n=0} (\alpha + n)\,a_n\,x^{n + \alpha - 2}$$

$$- \sum_{n=2} a_{n-2}\,x^{n + \alpha - 2} - \nu^2 \sum_{n=0} x^{n + \alpha - 2} = 0$$

hence we have

$$\left(\alpha^2 - \nu^2 \right) a_0 = 0$$

$$\left((\alpha+1)^2 - \nu^2 \right) a_1 = 0$$

$$\left((\alpha+n)^2 - \nu^2 \right) a_n = a_{n-2}, \quad n = 2, \ldots$$

where $a_0 \neq 0$ implies $P(\alpha) = (\alpha^2 - \nu^2) = 0$, therefore, $\alpha = \pm\nu$ and $\alpha_1 = \nu$, $\alpha_2 = -\nu$ assuming $\mathrm{Re}\,\nu > 0$. Thus $P(\alpha_1 + n) \neq 0$, $n = 1, 2, \ldots$ and we have $a_1, a_3, a_5, \ldots$ all zero and

$$a_{2n} = \frac{a_{2n-2}}{2^2 n (\nu + n)} = \frac{a_{2n-4}}{2^4 n (n-1)(\nu+n)(\nu+n-1)}$$

etc.

where, chosing $a_0 = \dfrac{2^{-\nu}}{\Gamma(\nu+1)}$, we have the series

$$\sum_{n=0} \frac{\left(\frac{1}{2}x\right)^{2n+\nu}}{n!\,\Gamma(\nu+n+1)}, \quad \Gamma(z) = z\Gamma(z-1)$$

which has an infinite radius of convergence and defines the functions $I_\nu(x)$.

If $\alpha_1 - \alpha_2$ is not an integer, we can find a second solution by using α_2 in place of α_1. But often $\alpha_1 - \alpha_2$ is an integer and this presents a technical difficulty.

If $\alpha_1 = \alpha_2$ (case $\nu = 0$ above) we can solve the recursion formula for a_n as a function of a_0 and α. Then

$$y(x, \alpha) = (x - x_0)^\alpha \sum a_n(\alpha) (x - x_0)^n$$

is not a solution unless $\alpha = \alpha_1$ and we have

$$\left\{ \frac{d^2}{dx^2} + \frac{p(x)}{(x - x_0)}\frac{d}{dx} + \frac{q(x)}{(x - x_0)^2} \right\} y(x, \alpha) = q_0 (x - x_0)^{\alpha - 2} P(\alpha)$$

where at $\alpha = \alpha_1$, $P(\alpha) = 0$, the RHS $= 0$ and $y(x, \alpha_1)$ is a solution. Now differentiating with respect to α the RHS is a linear combination of $P(\alpha)$ and $P'(\alpha)$ and if $\alpha_1 = \alpha_2$, $P(\alpha) =$

$(\alpha - \alpha_1)^2$ and the new RHS is zero at $\alpha = \alpha_1$. Thus our second solution is

$$\frac{\partial}{\partial \alpha} y(x, \alpha) \bigg|_{\alpha = \alpha_1} = y(x, \alpha_1) \ln(x - x_0) + \sum \frac{d}{d\alpha} a_n(\alpha) \bigg|_{\alpha = \alpha_1} (x - x_0)^{n + \alpha_1}$$

Much more than this sketch can be found in "*Advanced Mathematical Methods for Scientists and Engineers*" by Carl M. Bender and Steven A. Orszag and in "*An Introduction to the Theory of Functions of a Complex Variable*" by E. T. Copson.

The reader can work out the closely related problem, again Bessel's equation,

$$\frac{d^2 y}{dx^2} + \frac{1}{x} \frac{dy}{dx} + \left\{ 1 - \frac{m^2}{x^2} \right\} y = 0$$

where $x = 0$ is a regular singular point. Assuming $m = \ldots, -2, -1, 0, 1, 2, \ldots$ and writing

$$y = x^\alpha \sum a_n x^n$$

you should have

$$y' = \alpha x^{\alpha - 1} \sum a_n x^n + x^\alpha \sum n a_n x^{n-1}$$

and

$$y'' = \alpha (\alpha - 1) x^{\alpha - 2} \sum a_n x^n + 2\alpha x^{\alpha - 1} \sum n a_n x^{n-1} + x^\alpha \sum n(n-1) a_n x^{n-2}$$

and upon substituting and multiplying by x^2 you should obtain

$$\left\{ \alpha(\alpha - 1) + \alpha + (x^2 - m^2) \right\} x^\alpha \sum a_n x^n + \{ 2\alpha + 1 \} x^{\alpha + 1} \sum n a_n x^{n-1}$$

$$+ x^{\alpha + 2} \sum n(n-1) a_n x^{n-2} = 0$$

The coefficient of x^α on the LHS is $(\alpha^2 - m^2) a_0$ and assuming $a_0 \neq 0$, you have $\alpha = \pm m$. The coefficient of $x^{\alpha+n}$ is

$$a_n \left(n^2 + 2\alpha n \right) - a_{n-2}$$

and hence you have the recursion formula

$$a_n = \frac{-1}{n^2 + 2\alpha n} \, a_{n-2}$$

and this determines the even numbered coefficients in the series in terms of a_0.

The odd numbered coefficients, multiples of a_1, are zero. For example, setting $m = 0, a_0 = 1$ you have a power series solution for all finite x. The series is called $J_0(x)$ where

$$J_0(x) = 1 - \frac{1}{2^2} \, x^2 + \frac{1}{2^2 \, 4^2} \, x^4 - \frac{1}{2^2 \, 4^2 \, 6^2} \, x^6 + \cdots$$

A second independent solution can be obtained. Because the Wronskian of two solutions is $\dfrac{1}{x}$, this second solution will diverge logarithmically as $x \to 0$.

Back to the Solution of the θ-Equation

Now we return to our problem

$$\frac{d}{dz} \left\{ \left(1 - z^2 \right) \frac{dP}{dz} \right\} + \left(\beta^2 - \frac{m^2}{1 - z^2} \right) P = 0, \quad -1 \le z \le 1$$

where $m = \ldots, -2, -1, 0, 1, 2, \ldots$, and observe that it has regular singular points at $z = \pm 1$. Because our aim is to expand P in a power series in z about $z = 0$ and to make use of the condition that P must be bounded, we need to determine what is going on near $z = \pm 1$ by investigating the indicial equation at each of these points. To see what happens at $z = +1$ we introduce $x = 1 - z$ and write $R(x) = P(z)$, translating the singular point to $x = 0$. Then our equation is

$$\frac{d}{dx} \left\{ x \left(2 - x \right) \frac{dR}{dx} \right\} + \left(\beta^2 - \frac{m^2}{x \left(2 - x \right)} \right) R = 0$$

and substituting

$$R = x^\alpha \sum a_n x^n, \quad a_0 \neq 0$$

into

$$x\left(2-x\right)\left\{ x\left(2-x\right)\frac{d^2 R}{dx^2} + \left(2-2x\right)\frac{dR}{dx} \right\} + \left\{ x\left(2-x\right)\beta^2 - m^2 \right\} R = 0$$

we find the coefficient of x^α on the left hand side to be

$$\left(4\alpha^2 - m^2\right) a_0$$

and for this to be zero when a_0 is not zero, we have

$$\alpha^2 = \frac{1}{4}\, m^2$$

or

$$\alpha = \pm \frac{|m|}{2}$$

Likewise at $z = -1$ we find

$$\alpha = \pm \frac{|m|}{2}$$

Because we are looking for bounded solutions on $-1 \leq z \leq 1$ we discard the factors $\left(1 - z\right)^{-\frac{|m|}{2}}$ and $\left(1 + z\right)^{-\frac{|m|}{2}}$ and assume that $P\left(z\right)$ can be written

$$P\left(z\right) = \left(1-z\right)^{\frac{|m|}{2}} \left(1+z\right)^{\frac{|m|}{2}} G\left(z\right)$$

$$= \left(1 - z^2\right)^{\frac{|m|}{2}} G\left(z\right)$$

where $G(z)$ is a power series in z about $z = 0$. The differential equation satisfied by G is then

$$\left(1 - z^2\right) G'' - 2\{\,|m| + 1\,\}zG' + \left(\beta^2 - |m|\{|m| + 1\}\,\right)G = 0$$

The points $z = \pm 1$ remain regular singular points but the corresponding indicial equations require $\alpha = 0$. And so we look for a solution in the form of an ordinary power series about $z = 0$ where the singular points $z = \pm 1$ bound the interval of interest. Writing

$$G = \sum a_n z^n$$

$$G' = \sum n a_n z^{n-1}$$

and

$$G'' = \sum n\,(n - 1)\,a_n z^{n-2}$$

and substituting into the equation satisfied by G we get

$$\sum n\,(n-1)\,a_n z^{n-2} - \sum n\,(n-1)\,a_n z^n - 2\{\,|m| + 1\,\}\sum n a_n z^n$$
$$+ \left(\beta^2 - |m|\{|m| + 1\}\,\right)\sum a^n z^n = 0$$

Observing that

$$\sum_{n=0}^{\infty} n\,(n-1)\,a_n\,z^{n-2} = \sum_{n=0}^{\infty} (n+2)\,(n+1)\,a_{n+2}\,z^n$$

and setting the coefficient of z^n on the left hand side to zero we discover the two term recursion formula

$$a_{n+2} = -\frac{\beta^2 - |m|\{|m| + 1\} - 2n\{\,|m| + 1\,\} - n\,(n-1)}{(n+2)\,(n+1)}\,a_n$$
$$= \frac{(n + |m|)\,(n + |m| + 1) - \beta^2}{(n+2)\,(n+1)}\,a_n$$

in which $a_2, a_4, \ldots$ can be determined sequentially once a_0 is assigned and $a_3, a_5, \ldots$ can be determined sequentially once a_1 is assigned. The simplest way to obtain two independent solutions is to take them to be the even series corresponding to $a_0 = 1, a_1 = 0$ and the odd series corresponding to $a_0 = 0, a_1 = 1$. So for any fixed values of $|m|$ and β^2 we have two series, one an even function of z, the other an odd function of z. Now as Copson explains in his book *"An Introduction to the Theory of Functions of a Complex Variable"* the radius of convergence of a power series $\sum a_n z^n$ is $\dfrac{1}{\varlimsup |a_n|^{1/n}}$. For both our series this is 1 and so for any fixed values of $|m|$ and β^2 both our even and our odd series converge for all z such that $|z| < 1$. But both series diverge when $z = \pm 1$.

So, for fixed values of $|m|$ and β^2 we ordinarily do not get a bounded solution on $-1 \leq z \leq 1$. But for fixed values of $|m|$ there is the possibility that special values of β^2 lead to a bounded solution. These values of β^2 are those that make one series or the other terminate in a finite number of terms. For each fixed value of $|m|$ we can select a sequence of values of β^2 that terminate the even series in $1, 2, \ldots$ terms and a sequence of values of β^2 that terminate the odd series in $1, 2, \ldots$ terms. To each such value of β^2 one series terminates in a polynomial, while the other series does not terminate and is discarded. (The boundedness condition works here just like it did in cylindrical coordinates where we discarded a solution to Bessel's equation on the same grounds.)

So for each fixed value of $|m|$ we get an infinite sequence of polynomial solutions. The polynomial whose highest power is z^ν, $\nu = 0, 1, 2, \ldots$, corresponds to

$$\beta^2 = (\nu + |m|)(\nu + |m| + 1)$$

For each even value of ν the even series terminates in a polynomial and we discard the divergent odd series; for each odd value of ν the odd series terminates in a polynomial and we discard the divergent even series.

Again, to each fixed value of $|m|$ there corresponds to each value of β^2 such that

$$\beta^2 = (\nu + |m|)(\nu + |m| + 1), \quad \nu = 0, 1, 2, \ldots$$

a polynomial solution to the equation for G whose highest power is z^ν, and it is odd or even as ν is odd or even. The polynomial is completely determined by the recursion formula up to a constant factor. If we write

$$\ell = \nu + |m|$$

then

$$\beta^2 = \ell(\ell + 1)$$

and, due to $\nu = 0, 1, \ldots$, we have

$$\ell = |m|, \; (|m| + 1), \; (|m| + 2), \; \ldots$$

The polynomial is of degree $\ell - |m|$.

Going back to the equation for $P(z)$, and hence to the equation for $\Theta(\theta)$, we see that to each fixed value of $|m|, |m| = 0, 1, 2, \ldots$, we have determined a sequence of eigenvalues β^2, where

$$\beta^2 = \ell(\ell + 1)$$

and where

$$\ell = |m| + \nu, \quad \nu = 0, 1, 2, \ldots$$

To each such eigenvalue we have one independent eigenfunction and it is of the form

$$P(z) = \left(1 - z^2\right)^{\frac{|m|}{2}} \times \left\{ \begin{array}{l} \text{an even or odd polynomial in } z \\ \text{of degree } \quad \nu = \ell - |m| \end{array} \right\}$$

To each value of $|m|$ and to each ℓ the polynomial is defined by the recursion formula

$$a_{\nu+2} = \frac{(\nu + |m|)(\nu + |m| + 1) - \ell(\ell + 1)}{(\nu + 1)(\nu + 2)} a_\nu$$

$\ell = |m|, (|m| + 1), \ldots$, whence

$$\ell = |m| \qquad \text{leads to} \quad a_0,$$

$$\ell = |m| + 1 \quad \text{leads to} \quad a_1 z,$$

$$\ell = |m| + 2 \quad \text{leads to} \quad a_0 + a_2 z^2, \text{etc.}$$

We can write out these eigenfunctions using the recursion formula for the coefficients in the polynomials, but there is a better way of presenting our results. Before we do this we establish the orthogonality of two eigenfunctions corresponding to different eigenvalues. Let $|m|$ be fixed and let $P(z; \ell, m)$ denote the bounded solution of

$$\frac{d}{dz}\left\{ (1 - z^2) \frac{d}{dz} P \right\} + \left\{ \ell(\ell + 1) - \frac{m^2}{1 - z^2} \right\} P = 0$$

corresponding to $\ell = |m|, (|m| + 1), \ldots$ Then as

$$\int_{-1}^{+1} P \frac{d}{dz}\left\{ (1 - z^2) \frac{dQ}{dz} \right\} dz - \int_{-1}^{+1} Q \frac{d}{dz}\left\{ (1 - z^2) \frac{dP}{dz} \right\} dz =$$

$$\left[P(1 - z^2) \frac{dQ}{dz} - Q(1 - z^2) \frac{dP}{dz} \right]_{-1}^{+1} = 0$$

for any smooth functions P and Q bounded on $-1 \le z \le 1$, we see that

$$\int_{-1}^{+1} P(z; \ell, m) \, P(z; \ell', m) \, dz = 0$$

where ℓ and ℓ' are distinct values taken from the sequence $|m|, (|m| + 1), (|m| + 2), \ldots$

We emphasize the fact that to each value of $|m|$ there corresponds an infinite set of polynomials. And the sets of polynomials differ as $|m|$ differs. The orthogonality is a condition satisfied by the polynomials in each one of these sets.

We return to our problem as originally written and put $|m| = 0$. Then it is

$$\frac{d}{dx}\left\{ (1 - z^2) \frac{dP}{dz} \right\} + \beta^2 P = 0$$

and its bounded solutions on $-1 \le z \le +1$ are even or odd polynomials of degree $\ell, \ell = 0, 1, 2, \ldots$ corresponding to $\beta^2 = \ell(\ell + 1)$. Now the Legendre polynomials $P_0, P_1, P_2, \ldots$, introduced earlier, are even or odd polynomials of degree ℓ, $\ell = 0, 1, 2, \ldots$, satisfying

$$\frac{d}{dx}\left\{\left(1 - z^2\right)\frac{dP}{dz}\right\} + \ell\left(\ell + 1\right)P = 0$$

and as polynomial solutions of a fixed degree of this equation are unique up to a constant factor we can take the solutions to our problem for $|m| = 0$ to be

$$P_\ell\left(z\right), \quad \ell = 0, 1, 2, \ldots$$

This we do henceforth

We can use the Legendre polynomials to define the associated Legendre functions of degree ℓ and order $|m|$ via

$$P_\ell^{|m|} = \left(1 - z^2\right)^{\frac{|m|}{2}}\frac{d^{|m|}}{dz^{|m|}}P_\ell\left(z\right)$$

where $|m| = 0, 1, 2, \ldots$, where, to each value of $|m|$, $\ell = |m|, \left(|m| + 1\right), \left(|m| + 2\right), \ldots$ and where, as $P_\ell\left(z\right)$ is an even or odd polynomial of degree ℓ, $\dfrac{d^{|m|}}{dz^{|m|}}P_\ell\left(z\right)$ is an even or odd polynomial of degree $\ell - |m|$. Then by differentiating

$$\frac{d}{dx}\left\{\left(1 - z^2\right)\frac{d}{dz}P_\ell\left(z\right)\right\} + \ell\left(\ell + 1\right)P_\ell\left(z\right) = 0$$

$|m|$ times we get

$$\left(1 - z^2\right)\frac{d^{|m| + 2}}{dz^{|m| + 2}}P_\ell\left(z\right) - 2\left\{|m| + 1\right\}z\frac{d^{|m| + 1}}{dz^{|m| + 1}}P_\ell\left(z\right)$$
$$+ \left\{\ell\left(\ell + 1\right) - |m|\left\{|m| + 1\right\}\right\}P_\ell\left(z\right) = 0$$

and using $P_\ell^{|m|} = \left(1 - z^2\right)^{\frac{|m|}{2}}\dfrac{d^{|m|}}{dz^{|m|}}P_\ell\left(z\right)$ we find

$$\left(1 - z^2\right)\frac{d^2 P_\ell^{|m|}}{dz^2} - 2z\frac{dP_\ell^{|m|}}{dz} + \left\{\ell\left(\ell + 1\right) - \frac{m^2}{1 - z^2}\right\}P_\ell^{|m|} = 0$$

and we see that the associated Legendre functions are bounded solutions of our differential equation, viz.,

$$\frac{1}{\sin\theta}\frac{d}{d\theta}\left(\sin\theta\frac{d\Theta}{d\theta}\right) + \left(\ell(\ell+1) - \frac{m^2}{\sin^2\theta}\right)\Theta = 0$$

where $z = \cos\theta$ and $P(z) = \Theta(\theta)$. And as $P_\ell^{|m|}(z)$ is $(1-z^2)^{\frac{|m|}{2}}$ times an even or odd polynomial of degree $\ell - |m|$ the associated Legendre functions must be constant multiples of the solutions we determined by Froebenius' method.

What we find then is this: bounded solutions of

$$\frac{d}{dz}\left\{(1-z^2)\frac{dP}{dz}\right\} + \left\{\beta^2 - \frac{m^2}{1-z^2}\right\}P = 0$$

on the interval $-1 \le z \le +1$ corresponding to $|m| = 0, 1, 2, \ldots$ can be found for $\beta^2 = \ell(\ell+1)$, $\ell = |m|, (|m|+1), \ldots$, and can be written

$$P_\ell^{|m|} = \left(1-z^2\right)^{\frac{|m|}{2}}\frac{d^{|m|}}{dz^{|m|}}P_\ell(z)$$

where $P_\ell(z), \ell = 0, 1, 2, \ldots$ are the Legendre polynomials. Indeed all that we may wish to know about the associated Legendre functions can be determined from the Legendre polynomials. For instance the integrals of their squares,

$$\int_{-1}^{+1}\left(P_\ell^{|m|}(z)\right)^2 dz$$

required for normalization, can be found, after some work, to be

$$\frac{2}{2\ell+1}\frac{\{\ell+|m|\}!}{\{\ell-|m|\}!}$$

This establishes the angular eigenfunctions of ∇^2. So, while the radial parts of the eigenfunctions of ∇^2 remain to be determined, we now have a complete picture of the angular part. Defining

$Y_{\ell m}(\theta, \phi)$ by

$$Y_{\ell m}(\theta, \phi) = \sqrt{\frac{2\ell + 1}{2} \frac{\{\ell - |m|\}!}{\{\ell + |m|\}!}} \, P_\ell^{|m|}(\cos \theta) \, \Phi_m(\phi)$$

$$m = \ldots, -2, -1, 0, 1, 2, \ldots$$

$$\ell = |m|, (|m| + 1), \ldots$$

we have a complete set of orthogonal functions defined on the surface of a sphere. The orthogonality works like this. Because

$$\int_0^{2\pi} \int_0^\pi \overline{Y}_{\ell m} Y_{\ell' m'} \sin \theta \, d\theta d\phi$$

$$= \sqrt{\ell, m} \sqrt{\ell', m'} \int_0^\pi P_\ell^{|m|}(\cos \theta) \, P_{\ell'}^{|m'|}(\cos \theta) \sin \theta \, d\theta \int_0^{2\pi} \overline{\Phi}_m(\phi) \Phi_{m'}(\phi) \, d\phi$$

$$= \sqrt{\ell, m} \sqrt{\ell', m'} \int_{-1}^{+1} P_\ell^{|m|}(z) \, P_{\ell'}^{|m'|}(z) \, dz \int_0^{2\pi} \overline{\Phi}_m(\phi) \Phi_{m'}(\phi) \, d\phi$$

where

$$\sqrt{\ell, m} = \sqrt{\frac{2\ell + 1}{2} \frac{\{\ell - |m|\}!}{\{\ell + |m|\}!}}$$

and

$$\sqrt{\ell', m'} = \sqrt{\frac{2\ell' + 1}{2} \frac{\{\ell' - |m|\}!}{\{\ell' + |m|\}!}}$$

we look at the second integral first. It is either 1 or 0 as m is or is not equal to m'. Only if m is equal to m' need we to look at the first integral. It is either 1 or 0 as ℓ is or is not equal to ℓ'.

The functions $Y_{\ell m}$ are called spherical harmonics and as they are eigenfunctions of ∇^2 on the surface of a sphere, viz.,

$$\nabla^2 Y_{\ell m} = -\frac{\ell(\ell + 1)}{r^2} Y_{\ell m}$$

we can use them to solve problems there.

For instance if a solute is distributed over the surface of a sphere at $t = 0$ according to $c\,(t = 0)$ and then comes to equilibrium by diffusion, to determine $c\,(t > 0)$ we must solve

$$\frac{\partial c}{\partial t} = \nabla^2 c, \qquad 0 \leq \phi \leq 2\pi, \quad 0 \leq \theta \leq \pi$$

where $c\,(t = 0)$ is assigned and c is required to be periodic in ϕ and bounded in θ. Then $c\,(t > 0)$ is given by

$$c\,(\theta, \phi, t) = \sum_{m=-\infty}^{+\infty} \sum_{\ell=|m|}^{\infty} c_{\ell m} Y_{\ell m}\,(\theta, \phi)\, e^{-\ell\,(\ell+1)\,t}$$

where

$$c_{\ell m} = \int_0^{2\pi} \int_0^{\pi} \overline{Y}_{\ell m}\,(\theta, \phi)\, c\,(t = 0) \sin\theta\, d\theta d\phi$$

and where distance and time are scaled so that the radius of the sphere is one unit of length and the diffusivity is one unit of length^2/time. The readers can determine that $c\,(t > 0)$ is independent of ϕ for all $t > 0$ if $c\,(t = 0)$ is independent of ϕ and then go on and investigate the long time dependence of c on θ for arbitrary $c\,(t = 0)$.

The double sum $\sum_{m=-\infty}^{+\infty} \sum_{\ell=|m|}^{\infty}$ is often written $\sum_{\ell=0}^{\infty} \sum_{m=-\ell}^{\ell}$ where a given value of ℓ corresponds to $2\ell + 1$ values of m, viz., the values $m = -\ell, \ -\ell+1, \ \ldots, \ \ell-1, \ \ell$. This is sometimes important. Indeed Schrödinger's wave equation for the stationary states of an electron moving relative to a nucleus of charge Ze to which it is bound is

$$\nabla^2 \psi + \frac{8\pi^2 \mu}{h^2}\,\{E - V\}\,\psi = 0$$

where E is the energy of the electron, assuming the nucleus is fixed, $\mu = \dfrac{m_1 m_2}{m_1 + m_2}$ is the reduced mass, m_1 and m_2 being the masses of the particles, and r, θ and ϕ are the spherical coordinates of the electron taking the nucleus to be at the origin of our coordinate system. Both particles are assumed to be point particles and the potential energy is that due to the Coulombic attraction of their electric charges, i.e.,

$$V = -\frac{Ze^2}{r}$$

This the Coulomb's law in unrationalized electrostatic units.

The solutions to this eigenvalue problem are then of the form

$$\psi(r, \theta, \phi) = R(r) Y_{\ell m}(\theta, \phi)$$

where $R(r)$ satisfies

$$\frac{1}{r^2}\frac{d}{dr}\left(r^2\frac{dR}{dr}\right) + \left\{\frac{8\pi^2\mu}{h^2}(E - V(r)) - \frac{\ell(\ell+1)}{r^2}\right\} R = 0$$

where $m = 0, \pm 1, \pm 2, \ldots$ and $\ell = |m|, (|m|+1), (|m|+2), \ldots$. Thus a fundamental problem in atomic and molecular physics is reduced to the solution of a one dimensional eigenvalue problem.

The eigenvalues are independent of m. So, corresponding to $\ell = 0$, we have $m = 0$ and one eigenfunction, corresponding to $\ell = 1$, we have $m = -1, 0, 1$ and three eigenfunctions, etc.

The Radial Part of the Eigenfunctions

The eigenfunctions of ∇^2 in spherical coordinates, periodic in ϕ and bounded in θ, take the form

$$\Psi = R(r) Y_{\ell m}(\theta, \phi)$$

where $R(r)$ satisfies

$$\frac{1}{r^2}\frac{d}{dr}\left(r^2\frac{dR}{dr}\right) + \left\{\lambda^2 - \frac{\ell(\ell+1)}{r^2}\right\} R = 0$$

and where $m = 0, \pm 1, \pm 2, \ldots$ and $\ell = |m|, (|m|+1), (|m|+2), \ldots$. This equation can be reduced to Bessel's equation and solved in terms of Bessel functions. We observe that if $\ell = 0$ two independent solutions are

$$\frac{\sin \lambda r}{\lambda r} \quad \text{and} \quad \frac{\cos \lambda r}{\lambda r}$$

where the first, but not the second, is bounded as $r \to 0$. Likewise for each value of $\ell, \ell = 0, 1, 2, \ldots$, two independent solutions can be written in terms of $\sin \lambda r$, $\cos \lambda r$ and powers of λr. These can be used to solve diffusion problems in a sphere when $c\,(t = 0)$ and all other sources of solute are assigned. The reader can determine that

$$\frac{\sin \lambda r}{(\lambda r)^2} - \frac{\cos \lambda r}{\lambda r}$$

is the bounded solution when $\ell = 1$.

Solutions to $\nabla^2 c = 0$, Laplace's Equation, and $\nabla^2 c + Q = 0$, Poisson's Equation, in Spherical Coordinates

It may be worth a sentence or two to explain that as

$$\nabla^2 r^\ell = \frac{1}{r^2} \ell\,(\ell + 1)\, r^\ell$$

and

$$\nabla^2 \frac{1}{r^{\ell+1}} = \frac{1}{r^2} \ell\,(\ell + 1)\, \frac{1}{r^{\ell+1}}$$

the function

$$\left\{ A_{\ell m} r^\ell + B_{\ell m} \frac{1}{r^{\ell+1}} \right\} Y_{\ell m}\,(\theta, \phi)$$

satisfies Laplace's equation.

To solve Laplace's equation in spherical coordinates we introduce the inner product:

$$\langle\, f, g\,\rangle = \int_0^{2\pi} \int_0^{\pi} \overline{f} g \sin \theta \, d\theta d\phi$$

Then for all f and g periodic in ϕ and bounded in θ we can derive

$$\langle\, f, \nabla^2 g \,\rangle = \langle\, \nabla^2 f, g \,\rangle$$

Hence to solve

$$\nabla^2 c = 0, \qquad R_1 \le r \le R_2, \quad 0 \le \theta \le \pi, \quad 0 \le \phi \le 2\pi$$

where physical conditions are assigned on the spheres $r = R_1$ and $r = R_2$ we write

$$c = \sum\sum c_{\ell m}\,(r)\,Y_{\ell m}\,(\theta, \phi)$$

where the radial part of the solution is denoted $c_{\ell m}\,(r)$ and where

$$c_{\ell m}\,(r) = \langle\, Y_{\ell m}, c \,\rangle$$

The equation satisfied by $c_{\ell m}\,(r)$ is then obtained by multiplying Laplace's equation by $\overline{Y}_{\ell m}$ and integrating over θ and ϕ, viz.,

$$0 = \langle\, Y_{\ell m}, \nabla^2 c \,\rangle$$

$$= \frac{1}{r^2}\frac{d}{dr}r^2\frac{d}{dr}\langle\, Y_{\ell m}, c \,\rangle + \langle\, Y_{\ell m}, \nabla^2_{\theta,\phi}\, c \,\rangle$$

$$= \frac{1}{r^2}\frac{d}{dr}r^2\frac{d}{dr}\langle\, Y_{\ell m}, c \,\rangle + \langle\, \nabla^2 Y_{\ell m}, c \,\rangle$$

This leads to the differential equation

$$\frac{1}{r^2}\frac{d}{dr}r^2\frac{d}{dr}\langle\, Y_{\ell m}, c \,\rangle - \frac{\ell\,(\ell+1)}{r^2}\langle\, Y_{\ell m}, c \,\rangle = 0$$

and, using

$$\langle\, Y_{\ell m}, c \,\rangle = A_{\ell m}r^\ell + \frac{B_{\ell m}}{r^{\ell+1}}$$

we get

$$c = \sum \sum \left\{ A_{\ell m} r^\ell + \frac{B_{\ell m}}{r^{\ell+1}} \right\} Y_{\ell m}(\theta, \phi)$$

where $A_{\ell m}$ and $B_{\ell m}$ can be determined using the conditions assigned on $r = R_1$ and $r = R_2$. This will work in solving Poisson's equation as well, viz.,

$$0 = \nabla^2 c + Q(r, \theta, \phi)$$

The differential equation determining the radial part of the solution is then

$$\frac{1}{r^2}\frac{d}{dr} r^2 \frac{d}{dr} \langle Y_{\ell m}, c \rangle - \frac{\ell(\ell+1)}{r^2} \langle Y_{\ell m}, c \rangle + \langle Y_{\ell m}, Q \rangle = 0$$

To do any other problem, where, for instance, c is driven by an initial condition or by an initial condition and a volume source, brings us back to

$$\nabla^2 \psi + \lambda^2 \psi = 0$$

where

$$\left\{ \nabla^2 + \lambda^2 \right\} R(r) Y_{\ell m} = \left\{ \frac{1}{r^2}\frac{d}{dr} r^2 \frac{d}{dr} + \lambda^2 - \frac{\ell(\ell+1)}{r^2} \right\} R Y_{\ell m}$$

and where a homogeneous condition must be satisfied by ψ and hence by R on the surface of a sphere (or on two spheres) corresponding to a physical condition imposed on c there. The angular parts of this, the $Y_{\ell m}$'s, are now known, only the radial part part remains to be determined, viz., the solutions to

$$\left\{ \frac{1}{r^2}\frac{d}{dr} r^2 \frac{d}{dr} + \lambda^2 - \frac{\ell(\ell+1)}{r^2} \right\} R = 0$$

and this can be done by extending the observations on page 589 to $\ell = 2, 3, \ldots$.

We notice that λ^2 depends on ℓ but not on m.

20.4   Small Amplitude Oscillations of an Inviscid Sphere

We wish to find the frequencies of small amplitude oscillations of a sphere of inviscid fluid.

The radius of the sphere is denoted R_0 and we introduce a displacement so that its surface is $r = R\left(\theta, \phi, t\right).$

We have

$$\rho \frac{\partial \vec{v}}{\partial t} = -\nabla p, \quad \nabla \cdot \vec{v} = 0$$

and hence

$$\nabla^2 p = 0$$

Ignoring the effect of the outside fluid we have at $r = R\left(\theta, \phi, t\right)$

$$p = -\gamma 2H$$

and

$$v_r - \frac{R_\theta}{R} v_\theta - \frac{R_\phi}{R \sin \theta} v_\phi = \frac{\partial R}{\partial t}$$

Now introducing a small displacement of the rest state, viz., the state $p_0 = \gamma \dfrac{2}{R_0}$, $\vec{v_0} = \vec{0}$, we write

$$R\left(\theta, \phi, t\right) = R_0 + \varepsilon \, R_1\left(\theta, \phi, t\right)$$

Then to order ε we have

$$v_r = \varepsilon v_{r1}$$

$$p = p_0 + \varepsilon \, p_1$$

and

$$2H = -\frac{2}{R_0} + \varepsilon \left\{ \frac{2}{R_0^2} + \frac{1}{R_0^2} \, \nabla_{\theta\phi}^2 \right\} R_1$$

where

$$\nabla_{\theta\phi}^2 = \frac{1}{\sin\theta} \frac{\partial}{\partial\theta} \left(\sin\theta \, \frac{\partial}{\partial\theta} \right) + \frac{1}{\sin^2\theta} \frac{\partial^2}{\partial\phi^2}$$

Thus our small amplitude equations are

$$\rho \frac{\partial v_{r1}}{\partial t} = -\frac{\partial p_1}{\partial r}$$

$$\nabla^2 p_1 = 0$$

$$v_{r1} = \frac{\partial R_1}{\partial t} \quad \text{at} \quad r = R_0$$

and

$$p_1 = -\frac{\gamma}{R_0^2} \left\{ 2 + \nabla_{\theta\phi}^2 \right\} R_1 \quad \text{at} \quad r = R_0$$

and we have solutions

$$v_{r1} = \widehat{v}_{r1}(r) \, Y_{\ell m}(\theta, \phi) \, e^{\sigma t}$$

$$p_1 = \widehat{p}_1(r) \, Y_{\ell m}(\theta, \phi) \, e^{\sigma t}$$

and

$$R_1 = \widehat{R}_1 \, Y_{\ell m}(\theta, \phi) \, e^{\sigma t}$$

where $\widehat{v}_{r1}$, $\widehat{p}_1$, and $\widehat{R}_1$ satisfy

$$\rho \, \sigma \widehat{v}_{r1} = -\frac{d\widehat{p}_1}{dr}$$

$$\frac{1}{r^2}\frac{d}{dr}\left(r^2\frac{d\widehat{p}_1}{dr}\right) - \frac{\ell(\ell+1)}{r^2}\widehat{p}_1 = 0$$

$$\widehat{v}_{r1} = \sigma\widehat{R}_1 \quad \text{at} \quad r = R_0$$

and

$$\widehat{p}_1 = -\frac{\gamma}{R_0^2}\{2 - \ell(\ell+1)\}\,\widehat{R}_1 \quad \text{at} \quad r = R_0$$

and we see that

$$\widehat{p}_1 = A\,r^\ell$$

where

$$\ell = |m|, |m| + 1, \ldots$$

Thus we obtain

$$\sigma^2 = -\frac{\gamma}{\rho R_0^3}\,\ell(\ell-1)(\ell+2)$$

Now to order ε the volume of the sphere is

$$V_0 + \varepsilon 3R_0^2\widehat{R}_1 e^{\sigma t}\int_0^{2\pi}\int_0^\pi Y_{\ell\mathbf{m}}(\theta,\phi)\sin\theta\,d\theta\,d\phi$$

and for $|m| = 1, 2, \ldots, \ell \geq |m|$, the integral is zero. But $m = 0, \ell = 0$ must be ruled out assuming the volume of the sphere remains fixed on perturbation.

But what about $\ell = 1$ which is possible at $m = 0$ and $|m| = 1$? Suppose we have a sphere of radius R_0 centered at $\{\varepsilon, 0, 0\}$ and we wish to write this

$$r = R(\theta,\phi) = R_0 + \varepsilon R_1(\theta,\phi)$$

Then, substituting

$$x = R \sin\theta \cos\phi$$

$$y = R \sin\theta \sin\phi$$

$$z = R \cos\theta$$

$$R = R_0 + \varepsilon\, R_1$$

into

$$(x - \varepsilon)^2 + y^2 + z^2 = R_0^2$$

we find

$$R_1 = \sin\theta \cos\phi$$

Likewise if a sphere of radius R_0 is centered at $(0, \varepsilon, 0)$ we have

$$R_1 = \sin\theta \sin\phi$$

and if our sphere is centered at $(0, 0, \varepsilon)$ we find $R_1 = \cos\theta$. Now $\sin\theta \cos\phi$, $\sin\theta \sin\phi$ and $\cos\theta$ are eigenfunctions of $\nabla^2_{\theta,\phi}$ corresponding to $|m| = 1, \ell = 1$ and $m = 0, \ell = 1$. Hence all of the small displacements at $\ell = 1$ correspond simply to moving the sphere off center and hence lead to $\sigma = 0$.

20.5   The Solution to Poisson's Equation

Coulomb's law tells us that the electrostatic potential at the field point $\vec{r}$ due to electric charges q_i concentrated at the source points $\vec{r}_i$ is

$$\phi\left(\vec{r}\right) = \frac{1}{4\pi\varepsilon_0}\sum\frac{q_i}{|\vec{r}-\vec{r}_i|}$$

If the charge is distributed continuously, instead of discretely, due to an assigned charge density ρ, then the electrostatic potential is

$$\phi\left(\vec{r}\right) = \frac{1}{4\pi\varepsilon_0}\iiint\frac{\rho\left(\vec{r}\,'\right)}{|\vec{r}-\vec{r}_i|}dV'$$

where the source at $\vec{r}\,'$ is $\rho\left(\vec{r}\,'\right)dV'$ and the sum over point charges is replaced by an integral over the charge density.

This formula is the solution to Poisson's equation

$$\nabla^2\phi = -\rho/\varepsilon_0$$

To see this we write our second integration by parts formula in the form

$$\iiint\limits_V\phi\nabla^2\psi\,dV = \iint\limits_{S_2}dA\vec{n}\cdot\{\phi\nabla\psi-\psi\nabla\phi\} + \iint\limits_{S_1}dA\vec{n}\cdot\{\phi\nabla\psi-\psi\nabla\phi\} + \iiint\limits_V\psi\nabla^2\phi\,dV$$

where S_1 and S_2 bound a region V. To use this we assume the origin lies inside S_1 which in turn lies inside S_2. Then setting $\psi = \frac{1}{r}$ so that $\nabla\psi = -\frac{1}{r^2}\vec{i}_r$ and $\nabla^2\psi = 0$, we get

$$0 = -\iint\limits_{S_2}dA\vec{n}\cdot\left\{\frac{\phi}{r^2}\vec{i}_r + \frac{1}{r}\nabla\phi\right\} - \iint\limits_{S_1}dA\vec{n}\cdot\left\{\frac{\phi}{r^2}\vec{i}_r + \frac{1}{r}\nabla\phi\right\} + \iiint\limits_V\frac{1}{r}\nabla^2\phi\,dV$$

We now let S_2 be a sphere of very large radius and we let S_1 be a sphere of radius ε where $\varepsilon\to 0$. Then by requiring that $\phi\to 0$ at least as fast as $\frac{1}{r}$ as $r\to\infty$, the first term on the right hand side vanishes and by requiring that ϕ and $\nabla\phi$ remain bounded as $r\to 0$, the second reduces to

$4\pi\phi\left(\vec{0}\right)$. Using this our integration by parts formula simplifies to

$$\phi\left(\vec{0}\right) = \frac{1}{4\pi}\iiint\limits_V \frac{-\nabla^2\phi}{r}\,dV$$

and so, if ϕ satisfies Poisson's equation, $\nabla^2\phi = -\dfrac{\rho}{\varepsilon_0}$, we get

$$\phi\left(\vec{0}\right) = \frac{1}{4\pi\varepsilon_0}\iiint\limits_V \frac{\rho\left(\vec{r}\right)}{r}\,dV$$

which establishes the formula written earlier.

By doing this we have discovered the Green's function for the operator ∇^2 when its domain is the set of functions defined throughout all space, required to vanish as $r \to \infty$ at least as fast as $\dfrac{1}{r}$. Indeed we observe that

$$\iint\limits_S dA\vec{n}\cdot\nabla\left\{\frac{1}{r}\right\} = \iiint\limits_V \nabla^2\left\{\frac{1}{r}\right\}\,dV = 0$$

whenever S is the complete surface bounding any region V that does not include the origin. Then when S is any surface enclosing the origin we conclude that

$$\iint\limits_S dA\vec{n}\cdot\nabla\left\{\frac{1}{r}\right\} = -\iint\limits_{S_\varepsilon} dA\vec{n}\cdot\nabla\left\{\frac{1}{r}\right\} = -4\pi$$

where S_ε is a sphere of radius ε and $\vec{n} = -\vec{i_r}$ thereon.

By using this we see that the function g, where

$$g = \frac{1}{D}\frac{1}{4\pi}\frac{1}{r}\cdot 1$$

and where the factor 1 is written because it has physical dimensions $\dfrac{M}{T}$, satisfies

$$\nabla^2 g = 0, \qquad \forall\vec{r}\neq\vec{0}$$

and

$$\iint_S dA\vec{n} \cdot \{-D\nabla g\} = 1$$

where S is any surface enclosing the origin. It is therefore the concentration field resulting when a steady point source of unit strength, i.e., of strength 1 in units $\dfrac{M}{T}$, is established at $\vec{r} = \vec{0}\ \forall t$.

This is the Green's function for ∇^2. If a mass source is distributed via a continuous source density so that $Q(\vec{r})\,dV$ units of mass per unit of time is introduced at $\vec{r}$ then by superposition

$$c(\vec{r}) = \iiint_V \frac{1}{D4\pi\,|\vec{r} - \vec{r}'|}\,Q(\vec{r}')\,dV'$$

satisfies

$$D\nabla^2 c + Q = 0$$

20.6   Home Problems

1. A quantum ball of mass m in a uniform gravitational field is illustrated below

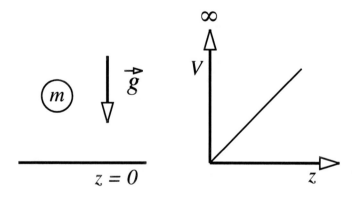

where $V = gz$, $z > 0$ and where $\psi = 0$ at $z = 0$ and $\psi \to 0$ as $z \to \infty$.

You are to find the energies of the ball by solving

$$\frac{d^2\psi}{dz^2} = \frac{2m}{\hbar^2}(gz - E)\psi$$

Replacing z by αz where $\dfrac{\alpha^3 2mg}{\hbar^2} = 1$ and setting $\dfrac{\alpha^2 2mg}{\hbar^2}E = \varepsilon$ you have

$$\frac{d^2\psi}{dz^2} = (z - \varepsilon)\psi$$

The solution is

$$\psi = CA_i(z - \varepsilon) + D\dot{B}_i(z - \varepsilon)$$

where $A_i(x)$ and $B_i(x)$ are sketched early in Lecture 20.

2. An incompressible fluid is in steady straight line flow in a long straight pipe of rectangular cross section. Determine $v_z(x, y)$ where

$$\nabla^2 v_z = \frac{1}{\mu}\frac{dp}{dz} < 0$$

where $\dfrac{dp}{dz}$ is an input and where v_z is zero at $x = \pm a$ and $y = \pm b$.

Determine Q, the volumetric flow rate, and then determine the values of a and b so that the volumetric flow rate is greatest at a fixed cross sectional area.

3. Instead of expanding v_z in the eigenfunctions of ∇^2 as you might have done in Problem 1 solve

$$\nabla^2 v_z = \frac{\partial^2 v_z}{\partial x^2} + \frac{\partial^2 v_z}{\partial y^2} = a^2 \frac{1}{\mu}\frac{\partial p}{\partial z}$$

where

$$v_z(x = 0, 1) = 0 = v_z(y = 0, 1)$$

by writing

$$v_z = \sum_{n=1}^{\infty} c_n(y) \sin n\pi x$$

and determining the function $c_n(y)$, $n = 1, 2, \ldots$ where

$$c_n(y) = \int_0^1 2(\sin n\pi x) \; v_z(x, y) \; dx$$

This series is the sum of products of functions of x times functions of y, unlike the solution in Problem 1.

4. An incompressible fluid is in steady straight line flow in a long straight pipe. The axial velocity depends on x and y and satisfies

$$\nabla^2 v_z = \frac{1}{\mu} \frac{\partial p}{\partial z}, \qquad (x, y) \in A$$

and

$$v_z = 0, \qquad (x, y) \in P$$

where $\dfrac{dp}{dz}$ is a negative constant, A denotes the fixed cross section of the pipe and P denotes its perimeter.

Polynomial solutions to this equation can be obtained for a variety of cross sections using

$$\nabla^2 \{1, x, y\} = \{0, 0, 0\}$$

$$\nabla^2 \{x^2, xy, y^2\} = \{2, 0, 2\}$$

$$\nabla^2 \{x^3, x^2 y, xy^2, y^3\} = \{6x, 2y, 2x, 6y\}$$

etc.

Determine the solution to the problem when the cross section of the pipe is bounded by:

(i) the circle: $x^2 + y^2 = a^2$

(ii) the ellipse: $\dfrac{x^2}{a^2} + \dfrac{y^2}{b^2} = 1$

(iii) the equilateral triangle:

$$y = -\frac{1}{2\sqrt{3}}\, a$$

$$y = \sqrt{3}\, x + \frac{1}{\sqrt{3}}\, a$$

$$y = -\sqrt{3}\, x + \frac{1}{\sqrt{3}}\, a$$

Show that for the same pressure gradient and the same area, the circular cross section carries the greatest volumetric flow.

Show that polynomial solutions cannot be found when the cross section is a square.

5. This has to do with the nonlinear heat generation problem presented at the end of Lecture 16.

Let the region V in which a heat generating material resides a be a rectangular parallelepipde of side lengths a, b and c. Its volume is then abc. Determine the eigenvalues of ∇^2 in this region when the eigenfunctions are required to vanish on its boundary. Observe that

$$\lambda_1^2 = \pi^2 \left\{ \frac{1}{a^2} + \frac{1}{b^2} + \frac{1}{c^2} \right\}$$

Holding the volume of the region fixed show that a cube might be thought to be the most dangerous shape. Is this result expected on physical grounds?

6. A solvent is in straight line flow in a long straight pipe of rectangular cross section. The pipe is aligned with the axis $0z$ of a Cartesian coordinate system and $-a < x < a$, $-b < y < b$, $-\infty < z < \infty$.

The longitudinal velocity of the solvent, denoted $v_z\,(x,y)$, is known from Problem 1.

A solute is introduced at $t = 0$. Denote its initial distribution by $c\,(t = 0)$ and determine V_{eff} and D_{eff} as t grows large in terms of $v_z\,(x,y)$. Assume $c\,(t = 0)$ is independent of x and y.

Define the inner product and the transverse average by

$$\langle\, u, v\,\rangle = \frac{1}{A} \iint_A uv\, dA$$

and

$$\bar{u} = \frac{1}{A} \iint_A u\, dA$$

then

$$\bar{u} = \langle\, 1, u\,\rangle$$

7. Suppose c satisfies $\vec{n} \cdot \nabla c = 0$ on the boundary of a region V. Show that

$$\overline{\nabla^2 c} = \frac{1}{V} \iiint_V \nabla^2 c\, dV = 0$$

8. Two planes, one at $x = 0$ and the other at $x = L$, are held at temperature T_0. A plane at $y = 0$ is held at temperature $T_1 > T_0$.

Derive a formula for the steady temperature field in the region $0 < x < L$, $0 < y < \infty$.

To estimate the heat that must be supplied to establish this temperature field find $\dfrac{\partial T}{\partial y}$ along the hot plane $y = 0$.

First differentiate the formula for $T\,(x,y)$ term by term, observe that the result is a series that diverges at $y = 0$ and conclude that, while the series produces T, it does not produce everything that we might want to know about the problem.

Devise a way of determining $\dfrac{\partial T}{\partial y}\,(y=0)$.

9. Solve the differential equation

$$\frac{d}{dz}\left\{\left(1-z^2\right)\frac{dP}{dz}\right\}+n\left(n+1\right)P=0, \qquad -1\le z\le 1$$

where n is a non-negative integer, by expanding P in powers of z:

$$P=a_0+a_1 z+a_2 z^2+\cdots$$

10. Define spherical coordinates in four dimensions by

$$z=r\cos\omega$$

$$w=r\sin\omega\,\cos\theta$$

$$x=r\sin\omega\,\sin\theta\,\cos\phi$$

$$y=r\sin\omega\,\sin\theta\,\sin\phi$$

and derive

$$\nabla^2=\frac{1}{r^3}\frac{\partial}{\partial r}\left(r^3\frac{\partial}{\partial r}\right)+\frac{1}{r^2\sin^2\omega}\frac{\partial}{\partial\omega}\left(\sin^2\omega\frac{\partial}{\partial\omega}\right)$$

$$+\frac{1}{r^2\sin^2\omega\,\sin\theta}\frac{\partial}{\partial\theta}\left(\sin\theta\frac{\partial}{\partial\theta}\right)+\frac{1}{r^2\sin^2\omega\,\sin^2\theta}\frac{\partial^2}{\partial\phi^2}$$

Separate variables in the four dimensional eigenvalue problem

$$\left(\nabla^2+\lambda^2\right)\psi=0$$

and obtain four one dimensional eigenvalue problems. The θ and ϕ equations are just the θ

and ϕ equations that come up in three dimensions. The ω equation is new, being

$$\frac{1}{\sin^2 \omega} \frac{d}{d\omega} \left(\sin^2 \omega \frac{d\Omega}{d\omega} \right) + \left\{ \beta^2 - \frac{\ell(\ell+1)}{\sin^2 \omega} \right\} \Omega = 0$$

where $\ell(\ell+1)$ comes from the θ equation.

Let $z = \cos\omega$, $P(z) = \Omega(\omega)$ and reduce this equation to

$$\left(1 - z^2\right) \frac{d^2 P}{dz^2} - 3z \frac{dP}{dz} + \left\{ \beta^2 - \frac{\ell(\ell+1)}{1 - z^2} \right\} P = 0$$

Let $x = 1 - z$ (and $x = 1 + z$) and solve the indicial equation to find

$$s = \frac{1}{2}\ell, \qquad -\frac{1}{2}(\ell+1)$$

Write $P(z) = (1 - z^2)^{\ell/2} G(z)$ and show that G satisfies

$$\left(1 - z^2\right) \frac{d^2 G}{dz^2} - (2\ell + 3) z \frac{dG}{dz} + \left\{ \beta^2 - \ell(\ell+2) \right\} G = 0$$

Then let $G = a_0 + a_1 z + a_2 z^2 + \cdots$, derive the recursion formula

$$a_{v+2} = \frac{(v+\ell)(v+\ell+2) - \beta^2}{(v+1)(v+2)} a_v$$

and conclude that

$$\beta^2 = k(k+2), \quad k = \ell, \ell+1$$

There is a pattern in the separation constants in spherical coordinates:

in two dimensions: $m(m+0)$

in three dimensions: $m(m+0)$

                         $\ell(\ell+1)$

in four dimensions: $m\,(m+0)$

$$\ell\,(\ell+1)$$

$$k\,(k+2)$$

Denote the spherical harmonics in four dimensions by $Y_{k\ell m}$. They satisfy

$$\nabla^2_{\omega\theta\phi}\,Y_{k\ell m} = -\frac{k\,(k+2)}{r^2}\,Y_{k\ell m}$$

Show that

$$r^k\,Y_{k\ell m} \quad \text{and} \quad \frac{1}{r^{k+2}}\,Y_{k\ell m}$$

satisfy Laplace's equation in four dimensions.

Almost everything that needs to be known, in any number of dimensions, can be inferred by pursuing the pattern that is emerging.

11. A dye drop in the shape of North and South America is absorbed on the surface of a sphere of water in which it is insoluble. The dye cannot escape into the surrounding air but being subjected to collisions by the water molecules it can diffuse over the surface of the water. Find a formula for the diffusive homogenization of the dye. Scale length and time so that $D = 1$ and $R = 1$ where R is the radius of the sphere of water. Then the surface concentration of dye satisfies

$$\frac{\partial c}{\partial t} = \nabla^2 c, \qquad 0 \le \theta \le \pi, \quad 0 \le \phi \le 2\pi$$

where $c\,(t = 0)$ is assigned.

Show that as t grows large the dye becomes uniformly distributed over the surface of the sphere. Find the long time limiting concentration of the dye. What is the θ and ϕ dependence of the non-uniform terms that die out most slowly? If $c\,(t = 0)$ depends only on θ, is this symmetry maintained for all $t > 0$?

12. In the Debye model for the reorientation of rigid rods by Brownian motion, the direction of a rod can be specified by a unit vector lying along its axis and hence by a point (θ, ϕ) lying on the surface of the unit sphere. If the initial orientation of the rods is specified by the probability density $p\,(t = 0)$, then the probability density $p\,(t > 0)$ satisfies

$$\frac{\partial p}{\partial t} = D\,\nabla_{\theta\phi}^2\, p$$

where $[D] = \dfrac{1}{T}$. Write a formula for $p\,(t > 0)$.

If in the initial orientation, the rods are concentrated at (θ_0, ϕ_0) then

$$p\,(t = 0) = \frac{\delta\,(\theta - \theta_0)\,\delta\,(\phi - \phi_0)}{\sin\theta_0}$$

where the delta function is introduced in Lecture 19, Appendix 3. Show that

$$p\,(t > 0) = \sum_{m=-\infty}^{+\infty} \sum_{\ell=|m|}^{\infty} \overline{Y}_{\ell m}\,(\theta_0, \phi_0)\,Y_{\ell m}\,(\theta, \phi)\,e^{-\ell\,(\ell + 1)\,Dt}$$

This is a transition probability. It is the probability that a rod acquires the orientation (θ, ϕ) at time t, given its initial orientation is (θ_0, ϕ_0).

13. The energy of a state, denoted E, is a solution to

$$\nabla^2\psi = -\frac{8\pi^2\mu}{h^2}\,E\psi$$

where

$$\iiint_V \overline{\psi}\psi\,dV < \infty$$

and ψ vanishes strongly as $|\vec{r}| \to \infty$. Prove that $E > 0$.

14. This is a 2-dimensional heat conduction problem, based on the 1-dimensional problem presented in Lecture 14

We have a region $0 < z < H$, $0 < x < L$ in which the temperature is specified at $t = 0$. The wall at $z = H$ is at a fixed temperature, the walls at $x = 0$ and $x = L$ are insulated. The wall at $z = 0$ is in perfect contact with a well stirred reservoir. The uniform temperature of the reservoir is denoted T_0. Hence we have $T(z = 0) = T_0$ for all $x \in (0, L)$ whereupon T is uniform in x at $z = 0$ but $\dfrac{\partial T}{\partial z}$ need not be uniform.

At $z = 0$ we also have

$$\int_0^L \frac{\partial T}{\partial z} \, dx = \text{constant} \times \frac{dT_0}{dt}$$

Our problem is to find $T(x, z, t)$ where

$$\frac{\partial T}{\partial t} = \kappa \nabla^2 T$$

on the domain.

First we scale our problem and then introduce an eigenvalue problem to aid us in solving the scaled problem.

Thus we have

$$\frac{\partial T}{\partial z} = \nabla^2 T \qquad 0 < x < 1, \quad 0 < z < H$$

$$T = 0 \quad \text{at} \quad z = H$$

$$\frac{\partial T}{\partial x} = 0 \quad \text{at} \quad x = 0, 1$$

and

$$\int_0^1 \frac{\partial T}{\partial z} \, dx = C \frac{\partial T}{\partial t} \quad \text{at} \quad z = 0$$

Our eigenvalue problem is then

$$\nabla^2 \psi + \lambda^2 \psi = 0$$

$$\psi = 0 \quad \text{at} \quad z = H$$

$$\frac{\partial \psi}{\partial x} = 0 \quad \text{at} \quad x = 0, 1$$

$$\int_0^1 \frac{\partial \psi}{\partial z} \, dx + C\lambda^2 \psi = 0 \quad \text{at} \quad z = 0$$

First, the eigenvalue problem must be solved, and we do this by separation of variables, viz., we write

$$\psi(x, z) = Z(z) \cos kx$$

$$k = 0, \pi, 2\pi, \ldots$$

and obtain the ψ's and λ's. Then the ψ's and λ's must be used to obtain $T(x, z, t)$ where $T(x, z, t = 0)$ is specified.

Our integration by parts formulas can be used to light our path, i.e., to tell us the inner product we ought to be using.

15. The Stokes' equation for the slow flow of a constant density fluid is

$$\nabla^2 \vec{v} = \frac{1}{\mu} \nabla p, \qquad \nabla \cdot \vec{v} = 0$$

Derive the equations

$$\nabla^2 p = 0$$

and

$$\nabla^2 \left(\frac{1}{2\mu} \vec{r}\, p \right) = \frac{1}{\mu} \nabla p$$

Denote by p_n, χ_n and ϕ_n solid spherical harmonics, viz.,

$$\left(A_n r^n + B_n r^{-(n+1)} \right) Y_{nm}$$

Then we have

$$p = \sum p_n$$

and you are to derive the result that

$$\vec{v} = \sum \left\{ \nabla \times (\vec{r} \chi_n) + \nabla \phi_n + \frac{n+3}{2\mu(n+1)(2n+3)} r^2 \nabla p_n \right.$$
$$\left. - \frac{n}{\mu(n+1)(2n+3)} \vec{r}\, p_n \right\}$$

satisfies

$$\nabla^2 \vec{v} = \frac{1}{\mu} \nabla p \quad \text{and} \quad \nabla \cdot \vec{v} = 0$$

This result appears in Lamb's book "*Hydrodynamics.*"

16. Stokes' equation for slow flow past a sphere has no solution in two dimensions. This is Stokes' paradox. It has a solution in three dimensions, but no first order corrections to account for non zero Reynolds number. This is Whitehead's paradox.

Your job is to see what you can find out about slow flow past a four dimensional sphere.

First you ought to derive the symmetry conditions and then ask if a streamfunction can be found.

17. We have a sphere of radius R centered at the origin lying in an unbounded region. We impose a constant temperature gradient at a great distance from the sphere. The thermal conductivities of the sphere and its surroundings are k_S and k.

Far away from the sphere we have

$$\frac{\partial T}{\partial z} = \frac{dT_0}{dz}, \qquad \frac{\partial T}{\partial x} = 0 = \frac{\partial T}{\partial y}$$

No temperature is specified at any point in the problem and we have $\nabla^2 T = 0$ inside and outside the sphere. The axisymmetric solutions to this equation are

$$A_0 + \frac{B_0}{r} + \left(A_1 r + \frac{B_1}{r^2} \right) \cos\theta + \left(A_2 r^2 + \frac{B_2}{r^3} \right) \left(\frac{3}{2} \cos^2\theta - \frac{1}{2} \right) + \text{etc.}$$

Dropping the constant A_0, derive the formulas

$$T = \frac{3k}{k_S + 2k} \frac{dT_0}{dz} r \, \cos\theta$$

and

$$T = \frac{dT_0}{dz} r \, \cos\theta + \frac{dT_0}{dz} \left(\frac{k - k_S}{k_S + 2k} \right) \frac{R^3}{r^3} r \, \cos\theta$$

for the temperatures inside and outside the sphere.

18. D. J. Jeffrey, "Conduction through a random suspension of spheres," *Proc. Roy. Soc. London*, A335 (1973) 355, proposes to determine the effective conductivity of a dilute suspension of spheres in the following way:

Write

$$\langle -\vec{q} \rangle = k_{\text{eff}} \langle \nabla T \rangle$$

where

$$\left\langle \, (\,) \, \right\rangle = \lim_{V \to \infty} \frac{1}{V} \iiint\limits_V (\,) \, dV$$

and then write

$$\left\langle \, -\vec{q} \, \right\rangle = \frac{1}{V} \left\{ \iiint\limits_{V_{\text{spheres}}} k_S \nabla T \, dV + \iiint\limits_{V - V_{\text{spheres}}} k \nabla T \, dV \right\}$$

$$= \frac{1}{V} \left\{ \iiint\limits_V k \nabla T \, dV + \iiint\limits_{V_{\text{spheres}}} (k_S - k) \, \nabla T \, dV \right\}$$

$$= k \left\langle \, \nabla T \, \right\rangle + \frac{1}{V} \iiint\limits_{V_{\text{spheres}}} (k_S - k) \, \nabla T \, dV$$

Set

$$\vec{\Delta}_i = \iiint\limits_{V_{\text{sphere } i}} (k_S - k) \, \nabla T \, dV$$

whereupon

$$\left\langle \, -\vec{q} \, \right\rangle = k \left\langle \, \nabla T \, \right\rangle + \frac{1}{V} \left\{ \vec{\Delta}_1 + \vec{\Delta}_2 + \cdots \right\}$$

and then assuming all $\vec{\Delta}$'s to be the same, we have for n spheres in a volume V

$$\left\langle \, -\vec{q} \, \right\rangle = k \left\langle \, \nabla T \, \right\rangle + \frac{n}{V} \, \vec{\Delta}_1$$

Assuming the spheres are dilute and do not interact with one another, we can use

$$\nabla T_S = \frac{3k}{k_S + 2k} \frac{dT_0}{dz} \, \vec{k}$$

where $z = r\cos\theta$, which we found in an earlier problem.

Setting $\langle\,\nabla T\,\rangle = \dfrac{dT_0}{dz}\,\vec{k}$ derive:

$$\frac{k_{\text{eff}}}{k} = 1 + 3\phi\,\frac{k_S - k}{k_S + 2k}$$

where $\phi = \dfrac{n\dfrac{4}{3}\pi R^3}{V}$ is the volume fraction spheres.

19. In cylindrical coordinates the Bessel's functions $I_m\,(\lambda r)$ and $K_m\,(\lambda r)$ satisfy

$$\left\{\frac{d^2}{dr^2} + \frac{1}{r}\frac{d}{dr} - \frac{m^2}{r^2} - \lambda^2\right\}\{I_m\,(\lambda r)\} = 0$$

and

$$\left\{\frac{d^2}{dr^2} + \frac{1}{r}\frac{d}{dr} - \frac{m^2}{r^2} - \lambda^2\right\}\{K_m\,(\lambda r)\} = 0$$

where $K_m\,(\lambda r)$ is not bounded at $\lambda r = 0$.

Assume you need to solve

$$\nabla^2\vec{v} = \nabla p, \qquad \nabla\cdot\vec{v} = 0$$

Show that

$$\nabla^2 p = 0 \quad\text{and}\quad \nabla^4\vec{v} = \vec{0}$$

However we would be wise to stay away from ∇^4.

Assume p is bounded at $r = 0$ and is periodic in θ and show that

$$p = I_m\,(\lambda r)\,e^{i\,m\theta}e^{i\,\lambda z}$$

satisfies $\nabla^2 p = 0$ for any λ and for $m = 0, \pm1, \pm2, \ldots.$

Now in cylindrical coordinates your problem is

$$\nabla^2 v_z = \frac{\partial p}{\partial z}$$

$$\nabla^2 v_r - \frac{v_r}{r^2} - \frac{2}{r^2}\frac{\partial v_\theta}{\partial \theta} = \frac{\partial p}{\partial r}$$

$$\nabla^2 v_\theta - \frac{v_\theta}{r^2} + \frac{2}{r^2}\frac{\partial v_r}{\partial \theta} = \frac{1}{r}\frac{\partial p}{\partial \theta}$$

and

$$\frac{\partial v_r}{\partial r} + \frac{v_r}{r} + \frac{1}{r}\frac{\partial v_\theta}{\partial \theta} + \frac{\partial v_z}{\partial z} = 0$$

where $\vec{v} = v_r\,\vec{\imath}_r + v_\theta\,\vec{\imath}_\theta + v_z\,\vec{\imath}_z$

Observe that

$$\psi = I_{\sqrt{m^2+1}}(\lambda r)\; e^{\imath\, m\theta}\; e^{\imath\, \lambda z}$$

satisfies

$$\left(\nabla^2 - \frac{1}{r^2}\right)\psi = 0$$

and, if there is no θ variation, observe that

$$v_z = I_0(\lambda r)\, e^{\imath\, \lambda z}$$

$$v_r = I_1(\lambda r)\, e^{\imath\, \lambda z}$$

and

$$v_\theta = I_1(\lambda r)\, e^{\imath\, \lambda z}$$

satisfy the homogeneous problem $\nabla^2 \vec{v} = \vec{0}$.

Assuming

$$p = I_m\left(\lambda r\right)\,\cos m\theta\,\sin\lambda z$$

you have

$$\nabla^2 v_z = \lambda\,I_m\left(\lambda r\right)\,\cos m\theta\,\cos\lambda z$$

Then write

$$v_z = f\left(r, m, \lambda\right)\,\cos m\theta\,\cos\lambda z$$

and show that

$$\left(\frac{d^2}{dr^2} + \frac{1}{r}\frac{d}{dr} - \frac{m^2}{r^2} - \lambda^2\right)f = \lambda\,I_m\left(\lambda r\right)$$

and that a particular solution is

$$f = \frac{1}{2\lambda}\,r\,\frac{d}{dr}\,I_m\left(\lambda r\right)$$

20. A cylindrical column of inviscid fluid is rigid body rotation about its axis of symmetry at angular velocity Ω. Thus you have $v_r = 0$, $v_\theta = r\Omega$ and $v_z = 0$. A small axisymmetric perturbation is introduced and your job is to find the frequencies of small amplitude oscillations.

Your equations are

$$\frac{\partial \vec{v}}{\partial t} + \vec{v} \cdot \nabla \vec{v} = -\nabla p, \quad \nabla \cdot \vec{v} = 0$$

where $\dfrac{p}{\rho}$ has been replaced by p.

And you need to derive the equations corresponding to $m = 0$, viz.,

$$\frac{\partial v_{r1}}{\partial t} - 2\Omega\, v_{\theta 1} = -\frac{\partial p_1}{\partial r}$$

$$\frac{\partial v_{\theta 1}}{\partial t} + 2\Omega\, v_{r1} = 0$$

$$\frac{\partial v_{z1}}{\partial t} = -\frac{\partial p_1}{\partial z}$$

$$\frac{\partial v_{r1}}{\partial r} + \frac{v_{r1}}{r} + \frac{\partial v_{z1}}{\partial z} = 0$$

Then seeking solutions $v_{r1} = \widehat{v}_{r1}(r)\, e^{i\omega t} e^{ikz}$, etc. you can eliminate $\widehat{v}_{\theta 1}$ in favor of $\widehat{v}_{r1}$ and $\widehat{v}_{z1}$ in favor of $\widehat{p}_1$ and arrive at an equation for $\widehat{v}_{r1}$:

$$\frac{d^2 \widehat{v}_{r1}}{dr^2} + \frac{1}{r}\frac{d\widehat{v}_{r1}}{dr} - \frac{1}{r^2}\widehat{v}_{r1} + k^2\left(\frac{4\Omega^2}{\omega^2} - 1\right)\widehat{v}_{r1} = 0$$

where $\widehat{v}_{r1} = 0$ at $r = R$ and $\widehat{v}_{r1}$ is bounded at $r = 0$.

This eigenvalue problem tells you the values of ω^2 as they depend on k^2. Photographs illustrating what you have found are presented by D. Fultz, *J. Meteorology*, <u>16</u> 199 (1959).

21. Your problem is to solve

$$\frac{\partial c}{\partial t} = \nabla^2 c, \qquad 0 < x < 1, \quad 0 < y < a$$

where $c\,(t=0)$ is assigned.

Two cases are of interest

    a) $c=0$ at the edge of the rectangle

       and

    b) $\vec{n}\cdot\nabla c=0$ at the edge of the rectangle.

In case a) you are to deduce the fact that if $a\gg1$, the x variation dies out quickly leaving the y variation in control of the loss of solute to the surroundings. If $a\ll1$, it is the reverse, i.e., the long direction is slow.

In case b) there is no solute loss, the initial solute distribution is simply working its way to uniformity. You have eigenfunctions

- independent of x and y,

- independent of x and dependent on y

- independent of y and dependent on x

   and

- dependent on both x and y

For $a\gg1$ and for $a\ll1$ does the x or y variation control the final approach to uniformity.

22. Suppose we inject a decomposing solute into a solvent in straight line flow in a long pipe of circular cross section.

The solute concentration decreases due to a first order reaction and we have, in scaled variables,

$$\frac{\partial c}{\partial t}=\frac{1}{r}\frac{\partial}{\partial r}\left(r\frac{\partial c}{\partial r}\right)-v\frac{\partial c}{\partial z}+\frac{\partial^2 c}{\partial z^2}-kc$$

and

$$\frac{\partial c}{\partial r}(r=1)=0$$

where $v = 2\bar{v}(1 - r^2)$.

Our model is

$$\frac{\partial \bar{c}}{\partial t} = D_{\text{eff}} \frac{\partial^2 \bar{c}}{\partial z^2} - V_{\text{eff}} \frac{\partial \bar{c}}{\partial z} - K_{\text{eff}} \bar{c}$$

First derive

$$K_{\text{eff}} = -\frac{1}{\bar{c}_0} \frac{d\bar{c}_0}{dt}$$

$$V_{\text{eff}} = \frac{d}{dt}\left(\frac{\bar{c}_1}{\bar{c}_0}\right)$$

etc.

Then show that

$$c_0(k > 0) = c_0(k = 0) \, e^{-kt}$$

$$c_1(k > 0) = c_1(k = 0) \, e^{-kt}$$

etc.

and conclude that $K_{\text{eff}} = k$ and that V_{eff} and D_{eff} are independent of k.

Estimate the fraction of solute remaining at the time V_{eff} and D_{eff} become nearly constant.

23. Your tennis ball, having a diameter $2R$ and a wall thickness L is filled with air at pressure $P_0 > P_{\text{atm}}$. The air diffuses across the wall and your tennis ball goes flat. Assume all the pressure drop is across the wall and diffusion through the wall is steady. Then write

$$\nabla \cdot (c\,\vec{v}) = 0$$

$$c = \frac{P}{RT}$$

and

$$\vec{v} = \frac{K}{\mu} \nabla P \qquad \text{(Darcy's law)}$$

where K denotes the permeability of the wall, assumed to be a porous solid.

At constant temperature and assuming one dimensional diffusion derive the equation

$$\frac{d}{dr}\left(Pr^2\frac{dP}{dr}\right) = 0$$

solve it and derive a formula for the time at which the pressure in your tennis ball falls to $\dfrac{P_0 + P_{\text{atm}}}{2}$.

24. A solid sphere of radius R_0 and density c_S dissolves sparingly in a solvent. It is in equilibrium with the solvent at solids concentration c^*, where

$$c^* = c^*_\infty - \mathcal{A}\gamma 2H$$

and where c^*_∞ is the concentration in equilibrium with c_S at a plane surface.

Writing

$$R = R_0 + \varepsilon R_1$$

and assuming no ϕ variation we have

$$2H_0 = -\frac{2}{R_0}$$

and

$$2H_1 = \frac{1}{R_0^2}\left\{2R_1 + \frac{\partial^2 R_1}{\partial\theta^2} + \frac{\cos\theta}{\sin\theta}\frac{\partial R_1}{\partial\theta}\right\}$$

Introduce a small perturbation to the system at rest at concentration c_0^* corresponding to $R = R_0$, denote the perturbation variables by the subscript 1 and write

$$\frac{\partial c_1}{\partial t} = D\nabla^2 c_1$$

$$c_1 = c_1^* = \mathcal{A}\gamma 2H_1 \quad \text{at} \quad r = R_0$$

and

$$(c_S - c_0^*)\frac{\partial R_1}{\partial t} = D\frac{\partial c_1}{\partial r} \quad \text{at} \quad r = R_0$$

where c_1^* does not appear in the third equation because $\dfrac{\partial R_0}{\partial t} = 0$

Your aim is to find out how fast the perturbation dies out.

Assume a solution

$$c_1 = \widehat{c}_1(r)\, P_\ell(\cos\theta)\, e^{\sigma t}$$

and

$$R_1 = \widehat{R}_1 P_\ell(\cos\theta)\, e^{\sigma t}$$

and derive the domain equation for $\widehat{c}_1(r)$. Its solutions are denoted

$$j_\ell\left(r\sqrt{\frac{-\sigma}{D}}\right) \quad \text{and} \quad y_\ell\left(r\sqrt{\frac{-\sigma}{D}}\right)$$

and are called spherical Bessel functions.

The case $\ell = 0$ does not maintain the volume of the sphere fixed and the case $\ell = 1$ is neutral, so set $\ell = 2$. Then a technical difficulty arises, viz., $j_2(r)$ and $y_2(r)$ do not vanish as fast as you would like as $r \to \infty$.

To get some idea of what is going on drop σ on the domain on the grounds that $\dfrac{\partial R}{\partial t}$ is controlling the equilibration.

Then

$$\widehat{c_1} = Ar^2 + \frac{B}{r^3}$$

and you can derive a formula for σ.

25. Your job is to find the shape of a sphere spinning at angular velocity $\vec{\Omega} = \Omega \vec{k}$. To do this you need the velocity of the fluid:

$$\vec{v} = r \sin \theta \Omega \, \vec{i_\phi}$$

and its pressure:

$$p = \gamma \frac{2}{R_0} + \left(\frac{1}{2} \rho r^2 \sin^2 \theta + C \right) \Omega^2$$

Now writing

$$R = R_0 + \Omega^2 R_1 + \frac{1}{2} \Omega^4 R_2 + \cdots$$

and observing that the volume of the sphere is

$$V = \frac{1}{3} \int_0^{2\pi} \int_0^\pi R^3 \sin \theta \, d\theta d\phi$$

and must remain constant, independent of Ω^2, you conclude, to first order in Ω^2:

$$\int_0^{2\pi} \int_0^\pi R_1 \sin \theta \, d\theta d\phi = 0$$

At $r = R(\theta, \phi)$ you have $p + \gamma 2H = 0$ and hence to first order in Ω^2 you have, at $r = R_0$,

$$p_1 + \gamma 2H_1 = 0$$

where

$$2H_1 = \frac{1}{R_0^2}\left(2 + \frac{1}{\sin\theta}\frac{d}{d\theta}\sin\theta\frac{d}{d\theta}\right)R_1$$

and where you must have $m = 0$.

Now you can find R_1 and hence you will have R to first order in Ω^2.

Answer:

$$R = R_0 + \Omega^2\frac{\rho}{\gamma}R_0^4\left(\frac{2}{3} - 2\cos^2\theta\right)$$

At $m = 0$, the $Y_{\ell m}$'s are the P_ℓ's: $P_0 = 1$, $P_1 = \cos\theta$, $P_2 = \frac{3}{2}\cos^2\theta - \frac{1}{2}$

To go to second order in Ω^2, you will need $2H_2$. At $r = R_0$ you will find

$$2R_1\frac{\partial p_1}{\partial r} + \gamma 2H_2 = 0$$

because p_2, $\dfrac{d^2 p_0}{dr^2}$ and $\dfrac{dp_0}{dr}$ are all zero.

26. An inviscid fluid confined to a circle of radius R_0 by the surface tension acting at its edge is spinning at a constant angular velocity, Ω.

   The velocity of the fluid and its pressure are given by

   $$\vec{v_0} = r\Omega\,\vec{i_\theta}$$

   and

   $$\frac{dp_0}{dr} = \rho\Omega^2 r$$

   where $p_0 = \dfrac{\gamma}{R_0}$ at $r = R_0$

Your job is to find the frequency of oscillation, σ, assuming the surface is given a small displacement, viz., $R = R_0 + \varepsilon R_1$ and assuming there is no z variation and $v_z = 0$.

You have

$$\frac{\partial \vec{v}}{\partial t} + \vec{v} \cdot \nabla \vec{v} = -\nabla \frac{p}{\rho}, \qquad \nabla \cdot \vec{v} = 0$$

and, at $r = R$,

$$v_r - \frac{R_\theta}{R} v_\theta = R_t$$

and

$$\frac{p}{\rho} + \frac{\gamma}{\rho} 2H = 0$$

where

$$2H = \frac{1}{\left(1 + \dfrac{R_\theta^2}{R^2}\right)^{3/2}} \left(\frac{R_{\theta\theta}}{R^2} - \frac{1}{R} - \frac{2R_\theta^2}{R^3} \right)$$

Write the perturbation problem and assume a solution

$$\left. \begin{array}{rcl} v_{r1} &=& \widehat{v}_{r1}(r) \\ v_{\theta 1} &=& \widehat{v}_{\theta 1}(r) \\ p_1 &=& \widehat{p}_1(r) \\ R_1 &=& \widehat{R}_1 \end{array} \right\} e^{im\theta} e^{i\sigma t}$$

Eliminate $\dfrac{\widehat{p}_1}{\rho}$ by differentiation and use $im\,\widehat{v}_{\theta 1} = -\dfrac{d}{dr}\left(r\widehat{v}_{r1} \right)$ to eliminate $\widehat{v}_{\theta 1}$ arriving at

$$\left(\frac{d^2}{dr^2} + \frac{3}{r}\frac{d}{dr} + \frac{1 - m^2}{r^2} \right) \widehat{v}_{r1} = 0$$

Retain the bounded solution, observing that $m = 0$ is ruled out if the area of the circle is held fixed on perturbation.

Obtain a formula for σ^2 and notice that $\sigma^2 = 0$ at $m = 1$. At $\Omega = 0$ your formula is

$$\sigma^2 = \frac{\gamma}{\rho R_0^3} m (m - 1) (m + 1)$$

Answer:

$$(\sigma + m\Omega)^2 - (\sigma + m\Omega) 2\Omega + m^2\Omega^2 + \frac{\gamma}{\rho R_0^3} m (m - 1)^2 = 0$$

27. The mechanical energy equation may tell you something important about a problem before you try to solve it, eg., the oscillating drop problem. As a simple example, we may have a fluid occupying a volume V bounded by a surface S where we omit the fluid outside S.

Then in V we have

$$\rho \frac{\partial \vec{v}}{\partial t} + \rho \vec{v} \cdot \nabla \vec{v} = \nabla \cdot \overset{\Rightarrow}{T}, \qquad \nabla \cdot \vec{v} = 0$$

and on S we have

$$\vec{n} \cdot \vec{v} = u$$

$$-\vec{n}\,\vec{n} : \overset{\Rightarrow}{T} + \gamma\, 2H = 0$$

$$-\vec{t}\,\vec{n} : \overset{\Rightarrow}{T} = 0$$

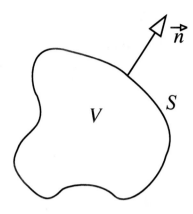

The plan is: dot the equation with $\vec{v}$, use

$$\left(\nabla \cdot \overset{\Rightarrow}{T}\right) \cdot \vec{v} = \nabla \cdot \left(\overset{\Rightarrow}{T} \cdot \vec{v}\right) - \left(\overset{\Rightarrow}{T}\right)^T : \nabla \vec{v},$$

integrate over V, use

$$\overset{\Rightarrow}{I} : \nabla \vec{v} = \nabla \cdot \vec{v} = 0,$$

use Leibnitz rule and write

$$\overset{\Rightarrow}{T} = -p\,\overset{\Rightarrow}{I} + 2\mu\,\overset{\Rightarrow}{D},$$

where

$$\nabla \vec{v} = \overset{\Rightarrow}{D} + \overset{\Rightarrow}{W}, \qquad \overset{\Rightarrow}{D} : \overset{\Rightarrow}{W} = 0$$

to derive

$$\frac{d}{dt}\int_V \frac{1}{2}\rho\vec{v}^2\,dV = \int_S dA\,\gamma\,2Hu - 2\mu\int_V dV\,\overset{\Rightarrow}{D} : \overset{\Rightarrow}{D}$$

Now, because

$$\frac{d}{dt} \int_S dA = - \int_S dA \, 2Hu$$

you have

$$\frac{d}{dt} \int_V \frac{1}{2} \rho \vec{v}^2 \, dV + \frac{d}{dt} \int_S \gamma \, dA = -2\mu \int_V \overset{\Rightarrow}{D} : \overset{\Rightarrow}{D} \, dV$$

And you have a constraint:

$$\frac{d}{dt} \int_V dV = \int_S dA \, u = 0$$

Now if you wish, you can easily include gravity in your formula by adding $+\rho \vec{g}$, $\vec{g} = -\nabla \phi$, to the RHS at the beginning.

Suppose you have a spherical drop at rest in free space, no gravity. Denote its radius by R_0 its volume by V_0. You give it a small displacement from rest. Prove that the drop returns to rest

Index

CPSIA information can be obtained
at www.ICGtesting.com
Printed in the USA
FFOW01n1519200718
47486566-50790FF